T0183927

Lecture Notes in Computer Science   11353

Commenced Publication in 1973
Founding and Former Series Editors:
Gerhard Goos, Juris Hartmanis, and Jan van Leeuwen

More information about this series at http://www.springer.com/series/7407

Roberto Battiti · Mauro Brunato
Ilias Kotsireas · Panos M. Pardalos (Eds.)

Learning and Intelligent Optimization

12th International Conference, LION 12
Kalamata, Greece, June 10–15, 2018
Revised Selected Papers

Springer

Editors
Roberto Battiti ⓘ
University of Trento
Trento, Italy

Mauro Brunato ⓘ
University of Trento
Trento, Italy

Ilias Kotsireas ⓘ
Wilfrid Laurier University
Waterloo, ON, Canada

Panos M. Pardalos ⓘ
University of Florida
Gainesville, FL, USA

ISSN 0302-9743 ISSN 1611-3349 (electronic)
Lecture Notes in Computer Science
ISBN 978-3-030-05347-5 ISBN 978-3-030-05348-2 (eBook)
https://doi.org/10.1007/978-3-030-05348-2

Library of Congress Control Number: 2018963070

LNCS Sublibrary: SL1 – Theoretical Computer Science and General Issues

This Springer imprint is published by the registered company Springer Nature Switzerland AG
The registered company address is: Gewerbestrasse 11, 6330 Cham, Switzerland

Preface

This volume edited by Roberto Battiti, Mauro Brunato, Ilias Kotsireas, and Panos Pardalos contains peer-reviewed papers from the 12th Learning and Intelligent International Optimization Conference (LION-12) held in Kalamata, Greece, during June 10–15, 2018.

The LION-12 conference has continued the successful series of the constantly expanding and worldwide recognized LION events (LION-1: Andalo, Italy, 2007; LION-2 and LION-3: Trento, Italy, 2008 and 2009; LION-4: Venice, Italy, 2010; LION-5: Rome, Italy, 2011; LION-6: Paris, France, 2012; LION-7: Catania, Italy, 2013; LION-8: Gainesville, USA, 2014; LION-9: Lille, France, 2015; LION-10: Ischia, Italy, 2016; LION-11: Nizhny Novgorod, Russia, 2017). This edition was organized by Panos Pardalos, Center for Applied Optimization, University of Florida (USA) and Ilias Kotsireas, CARGO Lab, Wilfrid Laurier University (Canada), who were conference general chairs.

Like its predecessors, the LION-12 international meeting explored advanced research developments in such interconnected fields as mathematical programming, global optimization, machine learning, and artificial intelligence. The location of LION-12 in Kalamata, Greece, was an excellent occasion to meet researchers and consolidate research and personal links.

More than 70 participants took part in the LION-12 conference. A total of 43 papers were accepted for oral presentation.

The following four plenary lecturers shared their current research directions with the LION-12 participants:

- Lefteris Kirousis, Department of Mathematics, National and Kapodistrian University of Athens, Greece. "Algorithmic Aspects of the Lovász Local Lemma — Can a Needle Be Efficiently Located in a Haystack? Sometimes!"
- George Michailidis, Professor and Director of the Informatics Institute, University of Florida, United States. "Fast Randomized Algorithms for Tensor Operations and Their Applications."
- Yaroslav D. Sergeyev, Dipartimento di Ingegneria Informatica, Modellistica, Elettronica e Sistemistica, Università della Calabria, Rende (CS), Italy — Department of Software and Supercomputing Technologies, Lobachevsky State University of Nizhni Novgorod, Russia. "Deterministic Lipschitz Global Optimization Algorithms and Their Comparison with Metaheuristics."
- Michael N. Vrahatis, Computational Intelligence Laboratory (CILab), Department of Mathematics, University of Patras, GR-26110 Patras, Greece. "Generalizations of the Bolzano Theorem for Tackling Problems with Imprecise Information."

Moreover, a tutorial was also presented during the conference:

- "Visual Analytics for High-Dimensional Data Exploration and Engineering Design Optimization" by Alfred Inselberg, Professor, School of Mathematical Sciences, Tel Aviv University and Senior Fellow San Diego Supercomputing Center, and Timoleon Kipouros, Senior Research Associate, Engineering Design Centre, Department of Engineering, University of Cambridge.

A total of 28 long papers and 12 short papers were accepted for publication in this LNCS volume after thorough peer reviewing (up to three review rounds for some manuscripts) by the members of the LION-12 Program Committee and independent reviewers.

These papers describe advanced ideas, technologies, methods, and applications in optimization and machine learning.

The editors thank all the participants for their dedication to the success of LION-12 and are grateful to the reviewers for their valuable work. The support of the Springer LNCS editorial staff is greatly appreciated.

The editors express their gratitude to the organizers and sponsors of the LION-12 international conference:

- Center for Applied Optimization at the University of Florida University of Florida, USA
- Wilfrid Laurier University, Canada
- CARGO Lab (Computer Algebra Research Group of Wilfrid Laurier University)
- APM Institute for the Advancement of Physics and Mathematics.

Their support was essential for the success of this event.

September 2018

Roberto Battiti
Mauro Brunato
Ilias Kotsireas
Panos Pardalos

Organization

Technical Program Committee

Chairs

Roberto Battiti	Università degli Studi di Trento, Italy
Mauro Brunato	Università degli Studi di Trento, Italy
Ilias Kotsireas	Wilfrid Laurier University, Canada
Panos Pardalos	University of Florida, USA

Members

Dachuan Xu	Beijing University of Technology, China
Yaroslav Sergeyev	Università della Calabria, Italy
Horst Samulowitz	IBM, USA
Meinolf Sellmann	GE Research, USA
Konstantinos E. Parsopoulos	University of Ioannina, Greece
Andre Augusto	Cire University of Toronto Scarborough, Canada
Helena Ramalhinho	Universitat Pompeu Fabra, Spain
Paola Festa	Università di Napoli Federico II, Italy
Tias Guns	Vrije Universiteit Brussel (VUB), Belgium
Thomas Stützle	Université Libre de Bruxelles (ULB), Belgium
Marc Schoenauer	Inria Saclay Île-de-France, France
Lars Kotthoff	University of British Columbia, Canada
Martin Golumbic	University of Haifa, Israel
Luca Di Gaspero	Università di Udine, Italy
Bernd Bischl	LMU Munich, Germany
Andre de Carvalho	University of São Paulo, Brazil
Dario Landa-Silva	The University of Nottingham, UK
Youssef Hamadi	LIX, France
Francesca Rossi	IBM, Italy
Frank Hutter	University of Freiburg, Germany
George Katsirelos	MIAT, INRA, France
John Chinneck	Carleton University, Ottawa, Canada
Barry O'Sullivan	University College Cork, Ireland
Marie-Eleonore Marmion	Université de Lille, France
Mikhail Posypkin	FRC CSC RAS, Russia
Michael Khachay	Krasovsky Institute of Mathematics and Mechanics, Russia
Tatiana Tchemisova	University of Aveiro, Portugal
Silvano Martello	Università di Bologna, Italy
Oleg Prokopyev	University of Pittsburgh, USA

Oleg Khamisov	Melentiev Institute of Energy Systems, Russia
Yury Kochetov	Sobolev Institute of Mathematics, Russia
Nenad Mladenovic	Mathematical Institute SANU, Serbia
Renato De Leone	Università di Camerino, Italy
Antanas Žilinskas	Vilnius University, Lithuania
Adil Erzin	Sobolev Institute of Mathematics, Russia
Valeria Ruggiero	Uiversità di Ferrara, Italy
Giovanni Fasano	Università Ca' Foscari di Venezia, Italy
Jun Pei	Hefei University of Technology, China
Hoai An Le Thi	Université de Lorraine, France
Michael Trick	Carnegie Mellon University, USA
Eric D. Taillard	University of Applied Science Western Switzerland, Switzerland
Bistra Dilkina	Georgia Institute of Technology, USA
Christian Blum	Spanish National Research Council, Spain
David Gao	Federation University, Australia
Alexander Strekalovskiy	Matrosov Institute for System Dynamics and Control Theory SB RAS, Russia
Luca Zanni	Università di Modena e Reggio Emilia, Italy
Anatoly Zhigljavsky	Cardiff University, UK
Petr Vilím	IBM Czech, Czech Republic
Thomas Pock	Graz University of Technology, Austria
Sonia Cafieri	École Nationale de l'Aviation Civile, France
Julius Žilinskas	Vilnius University, Lithuania
Daniele Vigo	Università di Bologna, Italy
Evgeni Nurminski	FEFU, Russia
Gerardo Toraldo	Università di Napoli Federico II, Italy
Vladimir Grishagin	Nizhni Novgorod State University, Russia
Massimo Roma	Università La Sapienza di Roma, Italy
Kaisa Miettinen	University of Jyväskylä, Finland
Daniela Lera	Università di Cagliari, Italy
Remigijus Paulavicius	Imperial College London, UK
Dmitri Kvasov	Università della Calabria, Italy
Antonio Fuduli	Università della Calabria, Italy
Annabella Astorino	ICAR-CNR, Italy
Martin J. Geiger	Helmut Schmidt University, Germany

Local Organizing Committee Chair

Dimitris Souravlias	Helmut Schmidt University, Germany

Web Chair

Andrea Mariello	Università degli Studi di Trento, Italy

Contents

Accelerated Randomized Coordinate Descent Algorithms for Stochastic
Optimization and Online Learning.............................. 1
   Akshita Bhandari and Chandramani Singh

Spiral Search Method to GPU Parallel Euclidean Minimum Spanning
Tree Problem 16
   Wen-Bao Qiao and Jean-Charles Créput

A Cooperative Learning Approach for the Quadratic Knapsack Problem 31
   Eduardo Lalla-Ruiz, Eduardo Segredo, and Stefan Voß

An Improved BTK Algorithm Based on Cell-Like P System
with Active Membranes.................................... 36
   Linlin Jia, Laisheng Xiang, and Xiyu Liu

A Simple Algorithmic Proof of the Symmetric Lopsided Lovász
Local Lemma 49
   Lefteris Kirousis and John Livieratos

Creating a Multi-iterative-Priority-Rule for the Job Shop Scheduling
Problem with Focus on Tardy Jobs via Genetic Programming............ 64
   Georg E. A. Froehlich, Guenter Kiechle, and Karl F. Doerner

A Global Optimization Algorithm for Non-Convex
Mixed-Integer Problems.................................... 78
   Victor Gergel, Konstantin Barkalov, and Ilya Lebedev

Massive 2-opt and 3-opt Moves with High Performance GPU Local Search
to Large-Scale Traveling Salesman Problem 82
   Wen-Bao Qiao and Jean-Charles Créput

Instance-Specific Selection of AOS Methods for Solving Combinatorial
Optimisation Problems via Neural Networks....................... 98
   Teck-Hou Teng, Hoong Chuin Lau, and Aldy Gunawan

CAVE: Configuration Assessment, Visualization and Evaluation.......... 115
   André Biedenkapp, Joshua Marben, Marius Lindauer, and Frank Hutter

The Accuracy of One Polynomial Algorithm for the Convergecast
Scheduling Problem on a Square Grid with Rectangular Obstacles 131
   Adil Erzin and Roman Plotnikov

An Effective Heuristic for a Single-Machine Scheduling Problem
with Family Setups and Resource Constraints.   141
  Júlio C. S. N. Pinheiro, José E. C. Arroyo, and Ricardo G. Tavares

Learning the Quality of Dispatch Heuristics Generated
by Automated Programming.   154
  Andrew J. Parkes, Neema Beglou, and Ender Özcan

Explaining Heuristic Performance Differences for Vehicle Routing
Problems with Time windows.   159
  Jeroen Corstjens, An Caris, and Benoît Depaire

Targeting Well-Balanced Solutions in Multi-Objective Bayesian
Optimization Under a Restricted Budget   175
  D. Gaudrie, R. Le Riche, V. Picheny, B. Enaux, and V. Herbert

How *Grossone* Can Be Helpful to Iteratively Compute Negative
Curvature Directions   180
  *Renato De Leone, Giovanni Fasano, Massimo Roma,
  and Yaroslav D. Sergeyev*

Solving Scalarized Subproblems within Evolutionary Algorithms
for Multi-criteria Shortest Path Problems   184
  Jakob Bossek and Christian Grimme

Exact and Heuristic Approaches for the Longest Common Palindromic
Subsequence Problem   199
  Marko Djukanovic, Günther R. Raidl, and Christian Blum

Multi-objective Performance Measurement: Alternatives to PAR10
and Expected Running Time.   215
  Jakob Bossek and Heike Trautmann

Algorithm Configuration: Learning Policies for the Quick Termination
of Poor Performers   220
  Daniel Karapetyan, Andrew J. Parkes, and Thomas Stützle

Probability Estimation by an Adapted Genetic Algorithm
in Web Insurance   225
  Anne-Lise Bedenel, Laetitia Jourdan, and Christophe Biernacki

Adaptive Multi-objective Local Search Algorithms for the Permutation
Flowshop Scheduling Problem   241
  *Aymeric Blot, Marie-Éléonore Kessaci, Laetitia Jourdan,
  and Patrick De Causmaecker*

Portfolio Optimization via a Surrogate Risk Measure: Conditional
Desirability Value at Risk (CDVaR) 257
  İ. İlkay Boduroğlu

Rover Descent: Learning to Optimize by Learning to Navigate
on Prototypical Loss Surfaces.................................... 271
  Louis Faury and Flavian Vasile

Analysis of Algorithm Components and Parameters: Some Case Studies 288
  Nguyen Dang and Patrick De Causmaecker

Optimality of Multiple Decision Statistical Procedure for Gaussian
Graphical Model Selection 304
  Valery A. Kalyagin, Alexander P. Koldanov, Petr A. Koldanov,
  and Panos M. Pardalos

Hyper-Reactive Tabu Search for MaxSAT 309
  Carlos Ansótegui, Britta Heymann, Josep Pon, Meinolf Sellmann,
  and Kevin Tierney

Exact Algorithms for Two Quadratic Euclidean Problems of Searching
for the Largest Subset and Longest Subsequence.................... 326
  Alexander Kel'manov, Sergey Khamidullin, Vladimir Khandeev,
  and Artem Pyatkin

A Restarting Rule Based on the Schnabel Census for Genetic Algorithms ... 337
  Anton V. Eremeev

Intelligent Pump Scheduling Optimization in Water Distribution Networks... 352
  Antonio Candelieri, Riccardo Perego, and Francesco Archetti

Detecting Patterns in Benchmark Instances of the Swap-Body Vehicle
Routing Problem.. 370
  Dimitris Souravlias and Sandra Huber

Evolutionary Deep Learning for Car Park Occupancy Prediction
in Smart Cities 386
  Andrés Camero, Jamal Toutouh, Daniel H. Stolfi, and Enrique Alba

Asymptotically Optimal Algorithm for the Maximum m-Peripatetic
Salesman Problem in a Normed Space............................. 402
  E. Kh. Gimadi and O. Yu. Tsidulko

Computational Intelligence for Locating Garbage Accumulation
Points in Urban Scenarios 411
  Jamal Toutouh, Diego Rossit, and Sergio Nesmachnow

Fully Convolutional Neural Networks for Mapping Oil Palm
Plantations in Kalimantan.   427
   Artem Baklanov, Michael Khachay, and Maxim Pasynkov

Calibration of a Water Distribution Network with Limited Field Measures:
The Case Study of Castellammare di Stabia (Naples, Italy).   433
   Armado Di Nardo, Michele Di Natale, Anna Di Mauro,
   Giovanni Francesco Santonastaso, Andrea Palomba,
   and Stefano Locoratolo

Combinatorial Methods for Testing Communication Protocols
in Smart Cities   437
   Dimitris E. Simos, Ludwig Kampel, and Murat Ozcan

Pseudo-pyramidal Tours and Efficient Solvability of the Euclidean
Generalized Traveling Salesman Problem in Grid Clusters   441
   Michael Khachay and Katherine Neznakhina

Constant Factor Approximation for Intersecting Line Segments with Disks. . .   447
   Konstantin Kobylkin

Scheduling Deteriorating Jobs and Module Changes with Incompatible Job
Families on Parallel Machines Using a Hybrid SADE-AFSA Algorithm.   455
   Yuwei Sun, Xiaofei Qian, and Siwen Liu

Author Index   473

Accelerated Randomized Coordinate Descent Algorithms for Stochastic Optimization and Online Learning

Akshita Bhandari$^{(\boxtimes)}$ and Chandramani Singh

Department of ESE, Indian Institute of Science Bangalore, Bangalore, India
{akshita,chandra}@iisc.ac.in

Abstract. We propose accelerated randomized coordinate descent algorithms for stochastic optimization and online learning. Our algorithms have significantly less per-iteration complexity than the known accelerated gradient algorithms. The proposed algorithms for online learning have better regret performance than the known randomized online coordinate descent algorithms. Furthermore, the proposed algorithms for stochastic optimization exhibit as good convergence rates as the best known randomized coordinate descent algorithms. We also show simulation results to demonstrate performance of the proposed algorithms.

1 Introduction

Convex optimization problems are at the heart of many machine learning algorithms. These problems are typically huge in scale due to either large number of samples or large number of features or both. Gradient descent type algorithms are standard approaches to solve these problems. However, these algorithms are computationally expensive, i.e., have huge per-iteration complexity, in large scale problems. We thus require alternative iterative algorithms where

1. Only one feature parameter (or, a block of parameters) is updated at each iteration. These are coordinate descent algorithms [1]; these check per-iteration computation complexity when the number of features are humongous.
2. Data samples are processed one (or, one block) at a time. This is preferred when data sets have enormous samples, and is necessitated in scenario where samples are availed in real time. Processing randomly chosen samples from the whole available data set leads to stochastic gradient type algorithms [2], whereas processing samples in real time is referred to as online convex optimization or online learning [3].

None of these alternatives is adequate if both, the number of samples as well as the number of features, are huge.

We propose iterative descent algorithms that first choose a sample and then randomly choose a feature, and update the corresponding parameter only based

© Springer Nature Switzerland AG 2019
R. Battiti et al. (Eds.): LION 12 2018, LNCS 11353, pp. 1–15, 2019.
https://doi.org/10.1007/978-3-030-05348-2_1

on the chosen sample. In other words, we propose algorithms that combine characteristics of coordinate descent and stochastic gradient descent (or, online gradient descent) algorithms. It is well known that stochastic coordinate descent algorithms suffer from slow convergence due to variance of stochastic gradients of random samples. On the other hand, randomized coordinate descent algorithms have same convergence rate as gradient descent but worse constants. We have proposed "accelerated" gradient algorithms in order to alleviate these deficiencies.

1.1   Related Work

Stochastic gradient descent (SGD) [2] was introduced by Nemirovski et al. where as Online convex optimization and the associated projected gradient (OGD) [3] were introduced by Zinkevich. Hu et al. derived accelerated versions of SGD and OGD; they refer to these algorithms as Stochastic Accelerated Gradi-Ent (SAGE) [4]. Recently, Roux et al. proposed stochastic average gradient (SAG) [5] and Johnson and Zhang proposed Stochastic variation reduced gradient (SVRG) [6], both aimed at improving the convergence rate of SGD. Langford et al. [7] and McMahan and Streeter studied delay tolerant OGD algorithms [8] where parameter updates are based on stale gradient information.

Cyclic block coordinate descent algorithms were introduced by Luo and Tseng [9,10]. Nesterov proposed randomized block coordinate descent (RBCD) [1] algorithms for large scale optimization problems. Fercoq and Richtarik [11] and Singh et al. [12] proposed accelerated randomized block coordinate algorithms. More recently, Allen-zhu et al. [13] proposed faster accelerated coordinate descent methods in which sampling frequencies depend on coordinate wise smoothness parameters (i.e., Lipschitz parameters of the corresponding partial derivatives).

Dang and Lan proposed stochastic block mirror descent (SBMD) [14] which combines SGD and RBCD. Similar algorithms were proposed by Wang and Banerjee [15], Hua et al. [16], Zhao et al. [17] and Zhang and Gu [18] also, who called their algorithms ORBCD, R-BCG, MRBCD and ASBCD, respectively. ORBCD and MRBCD were enhanced by variance reduction techniques to attain linear rate of convergence for strongly convex loss functions. On the other hand, ASBCD used optimal sampling of training data samples to achieve the same.

More recently, there has been interest in machine learning settings in which training data and features are distributed across nodes of a computing cluster, or more generally, of a network. Nathan and Klabjan [19] proposed an algorithm where nodes parallelly update (possibly overlapping) blocks of feature parameters based on locally available data samples; this was seen as a combination of SVRG and block coordinate descent. Konecny et al. [20] studied algorithms for nodes connected through a network; this scenario was referred to as federated learning.

1.2 Our Contribution

We have proposed two accelerated randomized coordinate descent algorithms for stochastic optimization and online learning, which we refer to as SARCD and OARCD, respectively. Expectedly, these algorithms have significantly less per-iteration computation complexity than accelerated gradient descent algorithms, e.g., SAGE. Moreover, the proposed algorithms have the following properties.

1. SARCD for general convex objective functions exhibits convergence rate $\mathcal{O}(\frac{n}{\sqrt{T}})$ which matches the best known rates.
2. SARCD for strongly convex objective functions exhibits convergence rate $\mathcal{O}(\frac{n}{T})$ which is strictly better than $\mathcal{O}(\frac{n\log(T)}{T})$ rate of ORBCD.
3. OARCD for general convex loss functions yields regret bound $\mathcal{O}(\sqrt{nT})$ which is strictly better than $\mathcal{O}(n\sqrt{T})$, the regret bound of ORBCD and R-BCG.
4. OARCD for strongly convex loss functions yields regret bound $\mathcal{O}(n\log(T))$ which is strictly better than $\mathcal{O}(n^2\log(T))$, the regret bound of ORBCD.

The proposed algorithms can easily be generalized to accelerated randomized block coordinate descent algorithms.

2 The Learning Problems

We consider machine learning models characterized by input-output pairs, model parameters (or, feature parameters) and a loss function. The model parameters are used to estimate or predict outputs (also called *labels*) from the corresponding inputs (also called *features*). The loss function is used to measure discrepancy between the predicted and the actual outputs. To illustrate, let $\xi = (\xi_i, \xi_o)$ be an input-output pair and $y \in \mathbb{R}^n$ be a vector of the model parameters. Then ξ_i and y yield an estimate $\hat{\xi}_o$ of ξ_o. Clearly, the loss function, which provides a measure of discrepancy between $\hat{\xi}_o$ and ξ_o, can be seen as a mapping from the couple (y, ξ) to real numbers; let us denote this function as $l(\cdot, \cdot)$. For example, considering l_2-losses,

$$l(y, \xi) = \|\hat{\xi}_o - \xi_o\|^2.$$

Machine learning aims at identifying the model parameters that minimize the losses for all input-output pairs. We make this notion precise in the following two subsections which focus on two different premises.

2.1 Stochastic Optimization

Let us assume that we have a collection of input-output pairs, also called training samples. An input-output pair may appear more than once in the collection, and different pairs can have different relative frequencies. We aim at determining the model parameters that minimize the average loss over all the input-output pairs. Towards this, we let ξ denote a random input-output pair with appropriate distribution and consider the optimization problem[1]

$$\min_y \{f(y) \equiv \mathbb{E}_\xi[l(y, \xi)]\}.$$

Let $g(y, \xi) = \nabla_y l(y, \xi)$. We assume that

1. $f : \mathbb{R}^n \to \mathbb{R}$ is convex and differentiable. Moreover, we assume that $\nabla f(y)$ is Lipschitz continuous with parameter L,
2. $g(y, \xi)$ is an unbiased estimator of $\nabla f(y)$, i.e., $\mathbb{E}_\xi[g(y, \xi)] = \nabla f(y)$,
3. $f(\cdot)$ is strongly convex with parameter $\mu \geq 0$; $\mu > 0$ yields better iteration complexity.[2]

In Sect. 3, we propose an algorithm, SARCD, to solve the above problem. We establish convergence rates of SARCD for general convex loss functions (or, cost functions) and strongly convex loss functions in Theorems 1 and 2, respectively.

2.2   Online Learning

Here the input-output pairs arrive sequentially in steps and the model parameters are updated after each step. More precisely, we start with an arbitrary modelling parameter vector y_1. Assuming that we have model parameters y_t on arrival of the tth input-output pair ξ_t, we incur a loss $f_t(y_t) \equiv l(y_t, \xi_t)$ and update y_t to y_{t+1} based on $f_t(\cdot)$. For a given $T > 0$, we aim at generating a sequence of model parameters $y_1, y_2, \ldots, y_T$ that minimize the T-step "regret" $R(T)$, defined as

$$R(T) = \sum_{t=1}^{T} f_t(y_t) - \min_y \sum_{t=1}^{T} f_t(y).$$

Here we assume that $f_t : \mathbb{R}^n \to \mathbb{R}$ are convex and differentiable for all $t \geq 1$. Moreover, we also assume that, for all $t \geq 1$, $\nabla f_t(\cdot)$ are Lipschitz continuous with parameter L and $f_t(\cdot)$ are strongly convex with parameter $\mu \geq 0$.

In Sect. 4, we propose an algorithm, OARCD, to solve the above learning problem. We provide regret bounds of OARCD for general convex loss functions and strongly convex loss functions in Theorems 3 and 4, respectively.

3   Stochastic Accelerated Randomized Coordinate Descent

SARCD is the "coordinate descent" version of SAGE and an "accelerated" version of ORBCD. In other words, it is an iterative algorithm in which, at each iteration, we randomly choose an input-output pair and then a coordinate, and

---

[1] The proposed algorithm does not need the distribution of ξ.
[2] We allow $\mu = 0$ to accommodate general convex loss functions. Strong convexity warrants $\mu > 0$.

update only this coordinate of the vector of model parameters. As is typical of accelerated gradient methods, we update two other sequences $\{x_t\}$ and $\{z_t\}$ apart from $\{y_t\}$, and we also maintain two parameter sequences $\{\alpha_t\}$ and $\{L_t\}$ (see Sect. 2.1, [1,4]). Further, we use two constants $a(n)$ and $b(n)$ which we later set to achieve best convergence results. We let ξ_t indicate the random input-output pair chosen at tth iteration; $\{\xi_t\}$ are i.i.d. We also use a random diagonal matrix $Q_t \in \{0,1\}^{n \times n}$ to indicate the coordinate chosen at tth iteration; each Q_t has only one nonzero entry and $\{Q_t\}$ are i.i.d. Formally, our algorithm is as follows.

Algorithm 1 Stochastic Accelerated Randomized Coordinate Descent

---

Input: Sequences $\{L_t\}$ and $\{\alpha_t\}$

Initialize: $y_{-1} = z_{-1} = 0$

for $t = 0$ to T **do**

$\quad x_t = (1 - \alpha_t)y_{t-1} + \alpha_t z_{t-1}$

$\quad y_t = \arg\min_x \{\langle a(n)Q_t g(x_t, \xi_t), x - x_t\rangle + \frac{L_t}{2}\|x - x_t\|^2\}$

$\quad z_t = z_{t-1} - \frac{a(n)b^2(n)}{nL_t\alpha_t + \mu a(n)b(n)}[\frac{nL_t}{a(n)b(n)}(x_t - y_t) + \frac{\mu}{b(n)}(z_{t-1} - x_t)]$

end for

Output: y_T

Clearly, ξ_t and Q_t are independent and are also independent of x_t. Let $\Delta_t = a(n)Q_t(g(x_t, \xi_t) - \nabla f(x_t))$. Let $\delta_t = L_t(x_t - y_t) = a(n)Q_t g(x_t, \xi_t)$ be the gradient mapping involved in updating y_t. Also, let $\mathcal{H}_t$ denote the history of the algorithm until time t. More explicitly,

$$\mathcal{H}_t = (\xi_0, Q_0, \xi_1, Q_1, \ldots, \xi_{t-1}, Q_{t-1}).$$

Notice that $(x_l, y_l, z_l, l = 0, \ldots, t-1)$ and x_t are functions of $\mathcal{H}_t$. We first establish the following lower bound on $f(x)$.

Lemma 1. *For $t \geq 0$,*

$$f(x) \geq \mathbb{E}[f(y_t)|\mathcal{H}_t] + \frac{n}{a(n)}\mathbb{E}[\langle \delta_t, x - x_t\rangle|\mathcal{H}_t] + \frac{n}{a(n)}\mathbb{E}[\langle \Delta_t, y_t - x\rangle|\mathcal{H}_t]$$

$$+ \frac{\frac{2}{a(n)}L_t - L}{2L_t^2}\mathbb{E}[\|\delta\|^2|\mathcal{H}_t] + \frac{(n-1)}{a(n)L_t}\mathbb{E}[\langle \Delta_t, \delta_t\rangle|\mathcal{H}_t] + \frac{\mu}{2}\|x - x_t\|^2.$$

Proof. Please refer to the technical report [21]. $\qquad\square$

Proposition 1. *Assume that $\|g(x_t, \xi_t) - \nabla f(x_t)\| \leq \sigma$ and $L_t > a(n)L$ for all $t \geq 0$. Then, for all $t \geq 0$,*

$$\mathbb{E}[f(y_t) - f(x)|\mathcal{H}_t]$$

$$\leq (1 - \alpha_t)(f(y_{t-1}) - f(x)) + \frac{\sigma^2}{2n(\frac{L_t}{a(n)} - L)} + \frac{n\alpha_t}{a(n)}\mathbb{E}[\langle \Delta_t, x - z_{t-1}\rangle|\mathcal{H}_t]$$

$$+ \frac{nL_t\alpha_t^2}{2a(n)}\mathbb{E}[\|x - z_{t-1}\|^2 - \|x - z_t\|^2|\mathcal{H}_t] - \frac{\mu\alpha_t}{2}\mathbb{E}[\|x - z_t\|^2|\mathcal{H}_t]$$

Proof. Let us first notice that (see Algorithm 1)

$$z_t = \arg\min_x \left(\langle b(n)\delta_t, x - x_t \rangle + \frac{L_t \alpha_t}{2} \|x - z_{t-1}\|^2 + \frac{\mu a(n)b(n)}{2n} \|x - x_t\|^2 \right),$$

and also that the objective function in this minimization problem is strongly convex with parameter $(L_t \alpha_t + \frac{\mu a(n)b(n)}{n})$. Hence,

$$\langle b(n)\delta_t, x - x_t \rangle + \frac{L_t \alpha_t}{2} \|x - z_{t-1}\|^2 + \frac{\mu a(n)b(n)}{2n} \|x - x_t\|^2$$
$$\geq \langle b(n)\delta_t, z_t - x_t \rangle + \frac{L_t \alpha_t}{2} \|z_t - z_{t-1}\|^2 + \frac{\mu a(n)b(n)}{2n} \|z_t - x_t\|^2$$
$$+ \frac{L_t \alpha_t}{2} \|x - z_t\|^2 + \frac{\mu a(n)b(n)}{2n} \|x - z_t\|^2.$$

Using this in Lemma 1,

$$f(x) \geq \mathbb{E}[f(y_t)|\mathcal{H}_t] + \frac{n}{a(n)}\mathbb{E}[\langle \delta_t, z_t - x_t \rangle|\mathcal{H}_t] - \frac{n}{a(n)}\mathbb{E}[\Delta_t, x - y_t|\mathcal{H}_t]$$
$$+ \frac{nL_t\alpha_t}{2a(n)b(n)}\mathbb{E}[\|z_t - z_{t-1}\|^2 + \|x - z_t\|^2 - \|x - z_{t-1}\|^2|\mathcal{H}_t]$$
$$+ \frac{\frac{2}{a(n)}L_t - L}{2L_t^2}\mathbb{E}[\|\delta_t\|^2|\mathcal{H}_t] + \frac{n-1}{a(n)L_t}\mathbb{E}[\langle \Delta_t, \delta_t \rangle|\mathcal{H}_t] + \frac{\mu}{2}\mathbb{E}[\|x - z_t\|^2|\mathcal{H}_t]$$

$$(1)$$

where we have dropped the term $\frac{\mu}{2}\|z_t - x_t\|^2$ from the right hand side without affecting the inequality. Also, substituting $x = y_{t-1}$ in Lemma 1,

$$f(y_{t-1}) \geq \mathbb{E}[f(y_t)|\mathcal{H}_t] - \frac{n}{a(n)}\mathbb{E}[\langle \Delta_t, y_{t-1} - y_t \rangle|\mathcal{H}_t] + +\frac{n}{a(n)}\mathbb{E}[\langle \delta_t, y_{t-1} - x_t \rangle|\mathcal{H}_t]$$
$$+ \frac{\frac{2}{a(n)}L_t - L}{2L_t^2}\mathbb{E}[\|\delta\|^2|\mathcal{H}_t] + \frac{(n-1)}{a(n)L_t}\mathbb{E}[\langle \Delta_t, \delta_t \rangle|\mathcal{H}_t],$$

$$(2)$$

where again we have dropped the term $\frac{\mu}{2}\|y_{t-1} - x_t\|^2$ from the right hand side. Now, multiplying (1) by α_t and (2) by $(1 - \alpha_t)$ and adding,

$$\alpha_t f(x) + (1 - \alpha_t)f(y_{t-1})$$
$$\geq \mathbb{E}[f(y_t)|\mathcal{H}_t] + \frac{n\alpha_t}{a(n)}\mathbb{E}[\langle \delta_t, z_t - x_t \rangle|\mathcal{H}_t] - \frac{n\alpha_t}{a(n)}\mathbb{E}[\Delta_t, x - y_t|\mathcal{H}_t]$$
$$+ \frac{nL_t\alpha_t^2}{2a(n)b(n)}\mathbb{E}[\|z_t - z_{t-1}\|^2 + \|x - z_t\|^2 - \|x - z_{t-1}\|^2|\mathcal{H}_t]$$
$$- \frac{n(1 - \alpha_t)}{a(n)}\mathbb{E}[\langle \Delta_t, y_{t-1} - y_t \rangle|\mathcal{H}_t] + \frac{n(1 - \alpha_t)}{a(n)}\mathbb{E}[\langle \delta_t, y_{t-1} - x_t \rangle|\mathcal{H}_t]$$
$$+ \frac{\frac{2}{a(n)}L_t - L}{2L_t^2}\mathbb{E}[\|\delta\|^2|\mathcal{H}_t] + \frac{(n-1)}{a(n)L_t}\mathbb{E}[\langle \Delta_t, \delta_t \rangle|\mathcal{H}_t] + \frac{\alpha_t\mu}{2}\mathbb{E}[\|x - z_t\|^2|\mathcal{H}_t].$$

Rearranging the terms,

$$\mathbb{E}[f(y_t) - f(x)|\mathcal{H}_t]$$

$$\leq (1 - \alpha_t)(f(y_{t-1}) - f(x)) - \frac{\frac{2}{a(n)}L_t - L}{2L_t^2}\mathbb{E}[\|\delta\|^2|\mathcal{H}_t] - \frac{(n-1)}{a(n)L_t}\mathbb{E}[\langle \Delta_t, \delta_t\rangle|\mathcal{H}_t]$$

$$- \frac{nL_t\alpha_t^2}{2a(n)b(n)}\mathbb{E}[\|z_t - z_{t-1}\|^2|\mathcal{H}_t] + A + B$$

$$+ \frac{nL_t\alpha_t^2}{2a(n)b(n)}\mathbb{E}[\|x - z_{t-1}\|^2 - \|x - z_t\|^2|\mathcal{H}_t] - \frac{\mu\alpha_t}{2}\mathbb{E}[\|x - z_t\|^2|\mathcal{H}_t],$$

where

$$A = -\frac{n}{a(n)}\mathbb{E}[\langle \delta_t, \alpha_t(z_t - x_t) + (1 - \alpha_t)(y_{t-1} - x_t)\rangle|\mathcal{H}_t]$$

$$\text{and } B = \frac{n}{a(n)}\mathbb{E}[\langle \Delta_t, \alpha_t(x - y_t) + (1 - \alpha_t)(y_{t-1} - y_t)\rangle|\mathcal{H}_t].$$

We can see that

$$A = -\frac{n}{a(n)}\mathbb{E}[\langle \delta_t, \alpha_t(z_t - z_{t-1})\rangle + \langle \delta_t, \alpha_t(z_{t-1} - x_t) + (1 - \alpha_t)(y_{t-1} - x_t)\rangle|\mathcal{H}_t]$$

$$= \frac{n}{a(n)}\mathbb{E}[\langle \delta_t, \alpha_t(z_{t-1} - z_t)\rangle|\mathcal{H}_t]$$

$$\leq \frac{n}{a(n)}\mathbb{E}\left[\frac{\|\delta_t\|^2}{2nL_t} + \frac{nL_t\alpha_t^2}{2}\|z_t - z_{t-1}\|^2\Big|\mathcal{H}_t\right],$$

where we use the update rule of x_t to get the second equality (see Algorithm 1) and then the Young's inequality.[3] Further,

$$B = \frac{n}{a(n)}\mathbb{E}[\langle \Delta_t, \alpha_t x + (1 - \alpha_t)y_{t-1} - x_t\rangle + \langle \Delta_t, x_t - y_t\rangle|\mathcal{H}_t]$$

$$= \frac{n}{a(n)}\mathbb{E}\left[\alpha_t\langle \Delta_t, x - z_{t-1}\rangle + \frac{\langle \Delta_t, \delta_t\rangle}{L_t}\Big|\mathcal{H}_t\right],$$

where we again use the update rule of x_t in the last equality. Using the above bound on A, the expression for B and Cauchy–Schwartz inequality (to infer $\langle \Delta, \delta\rangle \leq \|\Delta_t\|\|\delta_t\|$), and setting $b(n) = \frac{1}{n}$,

$$\mathbb{E}[f(y_t) - f(x)|\mathcal{H}_t] \leq (1 - \alpha_t)(f(y_{t-1}) - f(x)) - \frac{\frac{1}{a(n)}L_t - L}{2L_t^2}\mathbb{E}[\|\delta\|^2|\mathcal{H}_t]$$

$$+ \frac{1}{a(n)L_t}\mathbb{E}[\|\Delta_t\|\|\delta_t\||\mathcal{H}_t] + \frac{n\alpha_t}{a(n)}\mathbb{E}[\langle \Delta_t, x - z_{t-1}\rangle|\mathcal{H}_t]$$

$$+ \frac{n^2 L_t\alpha_t^2}{2a(n)}\mathbb{E}[\|x - z_{t-1}\|^2 - \|x - z_t\|^2|\mathcal{H}_t] - \frac{\mu\alpha_t}{2}\mathbb{E}[\|x - z_t\|^2|\mathcal{H}_t].$$

---

[3] The Young's inequality states that $\langle x, y\rangle \leq \frac{\|x\|^2}{2a} + \frac{a\|y\|^2}{2}$ for any $a > 0$.

We now set $a = \frac{\frac{L_t}{a(n)} - L}{2}, b = \frac{\|\Delta_t\|}{a(n)}, \theta = \frac{\|\delta_t\|}{L_t}$, and use the fact that $-a\theta^2 + b\theta \leq \frac{b^2}{4a}, a, b \geq 0$, to get

$$
\begin{aligned}
\mathbb{E}[f(y_t) - f(x)|\mathcal{H}_t] \leq &(1 - \alpha_t)(f(y_{t-1}) - f(x)) \\
&+ \frac{1}{2a^2(n)(\frac{L_t}{a(n)} - L)}\mathbb{E}[\|\Delta_t\|^2|\mathcal{H}_t] + \frac{n\alpha_t}{a(n)}\mathbb{E}[\langle\Delta_t, x - z_{t-1}\rangle|\mathcal{H}_t] \\
&+ \frac{n^2 L_t \alpha_t^2}{2a(n)}\mathbb{E}[\|x - z_{t-1}\|^2 - \|x - z_t\|^2|\mathcal{H}_t] - \frac{\mu\alpha_t}{2}\mathbb{E}[\|x - z_t\|^2|\mathcal{H}_t].
\end{aligned}
$$

Finally, we get the desired result by using the bound $\mathbb{E}[\|\Delta_t\|^2|\mathcal{H}_t] \leq \frac{a^2(n)\sigma^2}{n}$. $\square$

Let x^* be the optimal solution of the optimization problem (Sect. 2.1). Then

$$
\begin{aligned}
\mathbb{E}_{\mathcal{H}_t, \xi_t, Q_t}[\langle\Delta_t, x^* - z_{t-1}\rangle|\mathcal{H}_t] &= \mathbb{E}_{\mathcal{H}_t}[\mathbb{E}_{\xi_t, Q_t}\langle\Delta_t, x^* - z_{t-1}\rangle|\mathcal{H}_t] \\
&= \mathbb{E}_{\mathcal{H}_t}[\langle\mathbb{E}_{\xi_t, Q_t}[\Delta_t|\mathcal{H}_t], x^* - z_{t-1}\rangle] = 0
\end{aligned}
$$

because $\mathbb{E}_{\xi_t, Q_t}[\Delta_t|\mathcal{H}_t] = 0$ owing to unbiasedness of $g(y, \xi)$. We set $x = x^*$ in Proposition 1 and take expectation on both the sides to obtain the following corollary.

Corollary 1. *Assume that* $\|g(x_t, \xi_t) - \nabla f(x_t)\| \leq \sigma$ *and* $L_t > a(n)L$ *for all* $t \geq 0$. *Then, for all* $t \geq 0$,

$$
\begin{aligned}
\mathbb{E}[f(y_t) - f(x^*)] \leq &(1 - \alpha_t)\mathbb{E}[f(y_{t-1}) - f(x^*)] + \frac{\sigma^2}{2n(\frac{L_t}{a(n)} - L)} \\
&+ \frac{n^2 L_t \alpha_t^2}{2a(n)}\mathbb{E}[\|x^* - z_{t-1}\|^2 - \|x^* - z_t\|^2] - \frac{\mu\alpha_t}{2}\mathbb{E}[\|x^* - z_t\|^2].
\end{aligned}
$$

We now appropriately set α_t, L_t and $a(n)$ to obtain rapid convergence of $\mathbb{E}[f(y_T)]$ to $f(x^*)$. We first consider the case when $\mu = 0$, i.e., f is not strongly convex.

Theorem 1. *Assume that that* $\mu = 0$, $\|g(x_t, \xi_t) - \nabla f(x_t)\| \leq \sigma$ *and* $\mathbb{E}[\|x^* - z_t\|^2] \leq D^2$. *Set* $a(n) = n, \alpha_t = \frac{2}{t+2}$ *and* $L_t = b(t + 1)^\beta + a(n)L$ *for* $t \geq 0$, *where* $\beta = \frac{3}{2}$ *and* $b > 0$ *is a constant. Then*

$$
\mathbb{E}[f(y_T) - f(x^*)] \leq \frac{2n^2 D^2 L}{T^2} + \left(2nD^2 b + \frac{4\sigma^2}{3b}\right)\frac{1}{\sqrt{T}}.
$$

Proof. Please refer to the technical report [21].

We now consider the case when $\mu > 0$, i.e., f is strongly convex.

Theorem 2. *Assume that that* $\mu > 0, \|g(x_t, \xi_t) - \nabla f(x_t)\| \leq \sigma$ *and* $\mathbb{E}[\|x^* - z_t\|^2] \leq D^2$ *for some* $D > 0$. *Set* $\alpha_0 = 1, L_0 = a(n)L + \frac{a(n)\mu}{n^2}$ *and* $\alpha_t =$

$\sqrt{\lambda_{t-1} + \frac{\lambda_{t-1}^2}{4}} - \frac{\lambda_{t-1}}{2}, L_t = a(n)L + \frac{a(n)\mu}{n^2\lambda_{t-1}}$ *for* $t \geq 1$ *where* $\lambda_0 = 1$ *and* $\lambda_t = \prod_{k=1}^{t}(1 - \alpha_k)$ *for* $t \geq 1$. *Then*

$$\mathbb{E}[f(y_T) - f(x^*)] \leq \frac{2(n^2L + \mu)D^2}{(T+2)^2} + \frac{2n\sigma^2}{(T+2)\mu}\left(1 + \frac{2\ln(T+1)}{T+2}\right).$$

Proof. Please refer to the technical report [21]. □

Remark 1. 1. Settings of L_t and α_t in Theorems 1 and 2 do not require knowledge of σ and the number of iterations T.
2. The convergence rate bound in the case of strongly convex objective functions is independent of the choice of $a(n)$.

4 Online Accelerated Randomized Coordinate Descent

OARCD can also be seen as the "coordinate descent" version of the SAGE-based online learning algorithm and an "accelerated" version of ORBCD for online learning. Here, at each step t, we encounter an input-output pair and incur a loss $f_t(y_t)$. We then randomly choose a coordinate of y_t, and update only this coordinate based on $f_t(\cdot)$. As in the case of SARCD, we maintain two other sequences, $\{x_t\}$ and $\{z_t\}$, and two parameter sequences, $\{\alpha_t\}$ and $\{L_t\}$, and also use two constants $a(n)$ and $b(n)$ to achieve optimal regret bounds. We again use a random diagonal matrix $Q_t \in \{0, 1\}^{n \times n}$ to indicate the coordinate chosen at tth step. Formally, this algorithm is as follows.

Let $\delta_t = L_t(x_t - y_t) = a(n)Q_t\nabla f_{t-1}(x_t)$ and $\mathcal{H}_t$ denote the history of the algorithm until time t. We first establish the following lower bound on $f_{t-1}(x)$.

Lemma 2. *For* $t \geq 1$,

$$f_{t-1}(x) \geq \mathbb{E}[f_{t-1}(y_t)|\mathcal{H}_t] + \frac{n}{a(n)}\mathbb{E}[\langle\delta_t, x - x_t\rangle|\mathcal{H}_t]$$

$$+ \frac{\frac{2}{a(n)}L_t - L}{2L_t^2}\mathbb{E}[\|\delta\|^2|\mathcal{H}_t] + \frac{\mu}{2}\|x - x_t\|^2.$$

Algorithm 2 Online Accelerated Randomized Coordinate Descent

Input: Sequences $\{L_t\}$ and $\{\alpha_t\}$
Initialize: $y_0 = z_0 = 0$
for $t = 1, 2, \ldots,$ do
$\quad x_t = (1 - \alpha_t)y_{t-1} + \alpha_t z_{t-1}$
$\quad y_t = \arg\min_x\{\langle a(n)Q_t\nabla f_t(y_t), x - x_t\rangle + \frac{L_t}{2}\|x - x_t\|^2\}$
$\quad z_t = z_{t-1} - \frac{a(n)b^2(n)\alpha_t}{nL_t + a(n)b(n)\alpha_t\mu}[\frac{nL_t}{a(n)b(n)}(x_t - y_t) + \frac{\mu}{b(n)}(z_{t-1} - x_t)]$
end for

Proof. Please refer to the technical report [21].

Proposition 2. *Assume that* $\|\nabla f_t(x)\| \leq R$ *and* $L_t > L$ *for all* $t \geq 1$. *Then, for all* $t \geq 1$,

$$\mathbb{E}[f_{t-1}(y_{t-1})] - f_{t-1}(x)$$

$$\leq \frac{a(n)R^2}{2n(1-\alpha_t)(L_t - L)} + \frac{nL_t}{2a(n)\alpha_t}\mathbb{E}[\|x - z_{t-1}\|^2 - \|x - z_t\|^2]$$

$$- \frac{nL_t}{2a(n)}\mathbb{E}[\|z_t - y_t\|^2] + \frac{n((1-\alpha_t^2)L_t - \alpha_t(1-\alpha_t)L)}{2a(n)}\mathbb{E}[\|y_{t-1} - z_{t-1}\|^2]$$

$$+ \frac{a(n)(2(n-1)L_t + (a(n)-n)L)R^2}{2nL_t^2} - \frac{\mu}{2}\mathbb{E}[\|x - z_t\|^2].$$

Proof. From the update equation of z_t (see Algorithm 2),

$$z_t = \arg\min_x \left(\langle b(n)\delta_t, x - x_t \rangle + \frac{L_t}{2\alpha_t}\|x - z_{t-1}\|^2 + \frac{\mu a(n)b(n)}{2n}\|x - x_t\|^2 \right).$$

The objective function in this minimization problem is strongly convex with parameter $(\frac{L_t}{\alpha_t} + \frac{\mu a(n)b(n)}{n})$. Hence,

$$\langle b(n)\delta_t, x - x_t \rangle + \frac{L_t}{2\alpha_t}\|x - z_{t-1}\|^2 + \frac{\mu a(n)b(n)}{2n}\|x - x_t\|^2$$

$$\geq \langle b(n)\delta_t, z_t - x_t \rangle + \frac{L_t}{2\alpha_t}\|z_t - z_{t-1}\|^2 + \frac{\mu a(n)b(n)}{2n}\|z_t - x_t\|^2$$

$$+ \frac{L_t}{2\alpha_t}\|x - z_t\|^2 + \frac{\mu a(n)b(n)}{2n}\|x - z_t\|^2.$$

Using this in Lemma 2,

$$\mathbb{E}[f_{t-1}(y_t)|\mathcal{H}_t] - f_{t-1}(x) \leq \frac{n}{a(n)}\mathbb{E}[\langle \delta_t, x_t - z_t \rangle|\mathcal{H}_t]$$

$$- \frac{nL_t}{2a(n)b(n)\alpha_t}\mathbb{E}[\|z_t - z_{t-1}\|^2 + \|x - z_t\|^2 - \|x - z_{t-1}\|^2|\mathcal{H}_t]$$

$$- \frac{\frac{2}{a(n)}L_t - L}{2L_t^2}\mathbb{E}[\|\delta_t\|^2|\mathcal{H}_t] - \frac{\mu}{2}\mathbb{E}[\|x - z_t\|^2|\mathcal{H}_t], \tag{3}$$

where we have dropped the term $-\frac{\mu}{2}\|z_t - x_t\|^2$ from the right hand side without affecting the inequality.

On the other hand,

$$\frac{L_t}{2}(\|z_t - x_t\|^2 - \|z_t - y_t\|^2) = \frac{L_t}{2}(\|z_t - x_t\|^2 - \|z_t - x_t + x_t - y_t\|^2)$$

$$= \frac{L_t}{2}\left(2\langle x_t - z_t, x_t - y_t \rangle - \frac{\|\delta_t\|^2}{L_t^2} \right)$$

$$= \langle \delta_t, x_t - z_t \rangle - \frac{\|\delta_t\|^2}{2L_t}.$$

Hence, using the update equation of $x(t)$ (see Algorithm 2),

$$\langle \delta_t, x_t - z_t \rangle$$

$$= \frac{L_t}{2}(\|z_t - (1-\alpha_t)y_{t-1} - \alpha_t z_{t-1}\|^2 - \|z_t - y_t\|^2) + \frac{\|\delta_t\|^2}{2L_t}$$

$$= \frac{L_t}{2}(\|z_t - z_{t-1} + (1-\alpha_t)(z_{t-1} - y_{t-1})\|^2 - \|z_t - y_t\|^2) + \frac{\|\delta_t\|^2}{2L_t}$$

$$\leq \frac{L_t}{2}\left(\frac{\alpha_t\|z_t - z_{t-1}\|^2}{\alpha_t^2} + (1-\alpha_t)\|z_{t-1} - y_{t-1}\|^2 - \|z_t - y_t\|^2\right) + \frac{\|\delta_t\|^2}{2L_t},$$

where the inequality follows from convexity of $\|\cdot\|^2$. Using this inequality in (3) with $b(n) = 1$,

$$\mathbb{E}[f_{t-1}(y_t)|\mathcal{H}_t] - f_{t-1}(x) \leq \frac{nL_t}{2a(n)}\mathbb{E}[(1-\alpha_t)\|z_{t-1} - y_{t-1}\|^2 - \|z_t - y_t\|^2|\mathcal{H}_t]$$

$$+ \frac{nL_t}{2a(n)\alpha_t}\mathbb{E}[\|x - z_{t-1}\|^2 - \|x - z_t\|^2|\mathcal{H}_t]$$

$$+ \frac{\frac{(n-2)L_t}{a(n)} + L}{2L_t^2}\mathbb{E}[\|\delta_t\|^2|\mathcal{H}_t] - \frac{\mu}{2}\mathbb{E}[\|x - z_t\|^2|\mathcal{H}_t]. \quad (4)$$

Further, from convexity of $f_{t-1}(\cdot)$,

$$f_{t-1}(y_{t-1}) - f_{t-1}(y_t)$$

$$\leq \langle \nabla f_{t-1}(y_{t-1}), y_{t-1} - y_t \rangle$$

$$\leq \frac{a(n)\|\nabla f_{t-1}(y_{t-1})\|^2}{2n(1-\alpha_t)(L_t - L)} + \frac{n(1-\alpha_t)(L_t - L)\|y_{t-1} - y_t\|^2}{2a(n)}$$

$$\leq \frac{a(n)R^2}{2n(1-\alpha_t)(L_t - L)} + \frac{n(1-\alpha_t)(L_t - L)\|y_{t-1} - x_t + x_t - y_t\|^2}{2a(n)}$$

$$= \frac{a(n)R^2}{2n(1-\alpha_t)(L_t - L)} + \frac{n(1-\alpha_t)(L_t - L)\|\alpha_t(y_{t-1} - z_{t-1}) + x_t - y_t\|^2}{2a(n)}$$

$$= \frac{a(n)R^2}{2n(1-\alpha_t)(L_t - L)} + \frac{n\alpha_t(1-\alpha_t)(L_t - L)\|y_{t-1} - z_{t-1}\|^2}{2a(n)} + \frac{n(L_t - L)\|\delta_t\|^2}{2a(n)L_t^2},$$

where the second inequality follows from Young's inequality, the third inequality from the bound on $\|\nabla f_{t-1}(y_{t-1})\|$, the first equality from the update rule of x_t and the second equality from convexity of $\|\cdot\|^2$. Taking conditional expectation in the above inequality and adding with (4) we get

$$\mathbb{E}[f_{t-1}(y_{t-1})|\mathcal{H}_t] - f_{t-1}(x)$$

$$\leq \frac{a(n)R^2}{2n(1-\alpha_t)(L_t - L)} + \frac{nL_t}{2a(n)\alpha_t}\mathbb{E}[\|x - z_{t-1}\|^2 - \|x - z_t\|^2|\mathcal{H}_t]$$

$$- \frac{nL_t}{2a(n)}\mathbb{E}[\|z_t - y_t\|^2|\mathcal{H}_t] + \frac{n((1-\alpha_t^2)L_t - \alpha_t(1-\alpha_t)L)}{2a(n)}\mathbb{E}[\|y_{t-1} - z_{t-1}\|^2|\mathcal{H}_t]$$

$$+ \frac{\frac{2(n-1)L_t}{a(n)} + \frac{(a(n)-n)L}{a(n)}}{2L_t^2}\mathbb{E}[\|\delta\|^2|\mathcal{H}_t] - \frac{\mu}{2}\mathbb{E}[\|x - z_t\|^2|\mathcal{H}_t].$$

Finally we get the desired result by using the bound $\mathbb{E}[\|\delta_t\|^2|\mathcal{H}_t] \leq \frac{a^2(n)R^2}{n}$ and then taking expectation. $\qquad\square$

Let us assume that x^* minimizes $\sum_{t=1}^T f_t(y)$ (see (Sect. 2.2)). We now set α_t, L_t and $a(n)$ to obtain best regret bounds. As in case of SARCD, we first consider $\mu = 0$.

Theorem 3. *Assume that* $\mu = 0$, $\|\nabla f_t(x)\| \leq R$ *and* $\|x^* - z_t\| \leq D$ *for* $t \geq 1$. *Set* $a(n) = \sqrt{n}$, $\alpha_t = \alpha$ *and* $L_t = \alpha\sqrt{t-1}L + L$, *where* $\alpha \in (0,1)$ *is a constant, and* $G = ((1 - \alpha^2)L_2 - \alpha(1 - \alpha)L)\|y_1 - z_1\|^2$. *Then the regret of OARCD can be bounded as*

$$\sum_{t=1}^T (\mathbb{E}[f_t(y_t)] - f_t(x^*)) \leq \left(\frac{2R^2}{\alpha L} + \frac{LD^2}{2}\right)\sqrt{nT} + \frac{R^2}{(1-\alpha)\alpha L}\sqrt{\frac{T}{n}} + \left(\frac{(\alpha+1)LD^2}{2\alpha} + \frac{G}{2}\right)\sqrt{n}.$$

Proof. Please refer to the technical report [21].

Finally, we consider the case when $\mu > 0$.

Theorem 4. *Assume that* $\mu > 0$, $\|\nabla f_t(x)\| \leq R$ *and* $\|x^* - z_t\| \leq D$ *for* $t \geq 1$. *Set* $a(n) = n$, $\alpha_t = \alpha$ *and* $L_t = \alpha\mu t + L$, *where* $\alpha \in (0,1)$ *is a constant, and* $G = ((1 - \alpha^2)L_2 - \alpha(1 - \alpha)L)\|y_1 - z_1\|^2$. *Then the regret of OARCD can be bounded as*

$$\sum_{t=1}^T (\mathbb{E}[f_t(y_t)] - f_t(x^*)) \leq \frac{R^2\ln(T+1)}{2(1-\alpha)\alpha\mu} + \frac{nR^2\ln(T+1)}{\alpha\mu} + \frac{D^2}{2\alpha}(2\alpha\mu + L) + \frac{G}{2}.$$

Proof. Please refer to the technical report [21].

5   Numerical Evaluation

The algorithm proposed in this paper OARCD gives an improvement over ORBCD [15,16] as shown in red in the Figs. 1, 2 and 3. The regret is much lower for both classification as well as regression. Table 1 shows the dataset description taken from UCI Repository.

We have performed several experiments on RCV1 dataset also. We have observed that regularization is needed for the well posedness of the problem. Adaptive coordinated descent method is also used for setting the learning rates as a function of the sum of the gradients which reduced the regret significantly. We have observed that normalization and choosing L as sum of the squares of maximum features are related in the sense that if we do not normalize but set L to be the sum of the squares of maximum features then also it gives the same result as we obtained after normalizing and setting L to be the number of features. We don't require per coordinate Lipschitz continuity if normalization is done. Online gradient methods will not work in cases where number of features are more. In the dataset 3 and 4 as shown in Tables 1 and 2, OGD [3] and SAGE [4] took more than 1 min while OARCD proposed in this paper took only 6 seconds to compute the regret. Also, OARCD shows improvent in accuracy and performs less number of mistakes than ORBCD [15,16] as shown in Table 2. When we add more number of examples, we see that the regret becomes constant since the difference between the best algorithm upto time t and the online algorithm becomes less. Figure 4 shows the comparison between the loss of APPROX [11] and SARCD proposed in this paper.

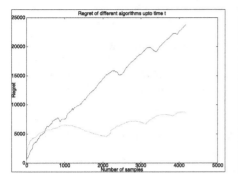 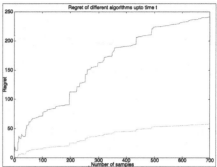

Fig. 1. Regret comparison on abalone dataset

Fig. 2. Regret comaprison on breast-cancer dataset

Table 1. Dataset description

Datasets	# features	# examples	Type
Abalone	7	4177	Regression
Breast cancer	9	699	Classification
Dorothea	100000	1950	Classification
RCV1	47236	20242/677, 399	Classification

Table 2. Accuracy and number of mistakes

Algorithm	Accuracy	# mistakes
OARCD on abalone	91.32	–
OARCD on breastcancer	94.423462%	22
OARCD on dorothea	90.25%	78
OARCD on RCV1	89%	115

6   Conclusion

We have proposed two accelerated randomized coordinate descent algorithms for stochastic optimization and online learning, respectively. Our algorithms exhibit performance as good as the best known randomized coordinate descent algorithms and yield strictly better regret bounds in case of online learning.

Our ongoing and future work entails extending these algorithms to regularized loss functions. We would like to investigate adaptation of feature selection probabilities to coordinate wise smoothness parameters. We would also like to consider online learning problems where update of model parameters take considerable time, and so, the updated parameters are available only after a certain, potentially random, number of data samples have passed.

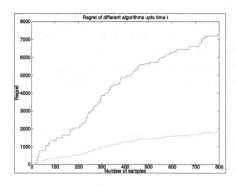
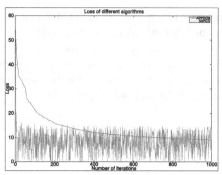

Fig. 3. Regret comaprison on dorothea dataset

Fig. 4. Loss of different algorithms on abalone dataset

Acknowledgments. The second author acknowledges support of INSPIRE Faculty Research Grant (DSTO-1363).

References

1. Nesterov, Y.: Efficiency of coordinate descent methods on huge-scale optimization problems. SIAM J. Optim. **22**, 341–362 (2012)
2. Nemirovski, A., Juditsky, A., Lan, G., Shapiro, A.: Robust stochastic approximation approach to stochastic programming. SIAM J. Optim. **19**, 1574–1609 (2009)
3. Zinkevich, M.: Online convex programming and generalized infinitesimal gradient ascent. In: Proceedings of the 20th International Conference on Machine Learning (ICML-03), pp. 928–936 (2003)
4. Hu, C., Pan, W., Kwok, J.: Accelerated gradient methods for stochastic optimization and online learning. Advances in Neural Information Processing Systems, pp. 781–789 (2009)
5. Le Roux, N., Schmidt, M., Bach, F.: A stochastic gradient method with an exponential convergence rate for finite training sets. In: Neural Information Processing Systems (2012)
6. Johnson, R., Zhang, T.: Accelerating stochastic gradient descent using predictive variance reduction. In: Advances in Neural Information Processing Systems, pp. 315–323 (2013)
7. Langford, J., Smola, A., Zinkevich, M.: Slow learners are fast. Adv. Neural Inf. Process. Syst. **22**, 2331–2339 (2009)
8. McMahan, B., Streeter, M.: Delay-tolerant algorithms for asynchronous distributed online learning. In: Advances in Neural Information Processing Systems, pp. 2915–2923 (2014)
9. Luo, Z.-Q., Tseng, P.: On the convergence of the coordinate descent method for convex differentiable minimization. J. Optim. Theory Appl. **72**, 735 (2002)
10. Tseng, P.: Convergence of a block coordinate descent method for non differentiable minimization. J. Optim. Theory Appl. **109**, 475–494 (2001)
11. Fercoq, O., Richtarik, P.: Accelerated, parallel, and proximal coordinate descent. SIAM J. Optim. **25**, 1997–2023 (2015)

12. Singh, C., Nedic, A., Srikant, R.: Random block-coordinate gradient projection algorithms. In: Decision and Control (CDC), pp. 185–190. IEEE (2014)
13. Allen-Zhu, Z., Qu, Z., Richtarik, P., Yuan, Y.: Even faster accelerated coordinate descent using non-uniform sampling. In: International Conference on Machine Learning, pp. 1110-1119 (2016)
14. Deng, Q., Lan, G., Rangarajan, A.: Randomized block subgradient methods for convex nonsmooth and stochastic optimization (2015)
15. Wang, H., Banerjee, A.: Randomized block coordinate descent for online and stochastic optimization (2014)
16. Hua, X., Kadomoto, S., Yamshita, N.: Regret analysis of block coordinate gradient methods for online convex programming (2015)
17. Zhao, T., Yu, M., Wang, Y., Arora, R., Liu, H.: Accelerated mini-batch randomized block coordinate descent method. In: Advances in neural information processing systems, pp. 3329–3337 (2014)
18. Zhang, A., Gu, Q.: Accelerated stochastic block coordinate descent with optimal sampling. In: KDD, pp. 2035–2044 (2016)
19. Nathan, A., Klabjan, D.: Optimization for large-scale machine learning with distributed features and observations. In: Perner, P. (ed.) MLDM 2017. LNCS (LNAI), vol. 10358, pp. 132–146. Springer, Cham (2017). https://doi.org/10.1007/978-3-319-62416-7_10
20. Konecny, J., McMahan, H., Ramage, D., Richtarik, P.: Federated optimization: distributed machine learning for on-device intelligence (2016)
21. Bhandari, A., Singh, C.: Accelerated randomized coordinate descent algorithms for stochastic optimization and online learning (2018). arXiv:1806.01600

Spiral Search Method to GPU Parallel Euclidean Minimum Spanning Tree Problem

Wen-Bao Qiao[✉] and Jean-Charles Créput

Le2i, CNRS, Arts et Métiers, University Bourgogne Franche-Comté, Besançon, France
rapidbao@outlook.com

Abstract. We present both sequential and data parallel approaches to build hierarchical minimum spanning forest (MSF) or trees (MST) in Euclidean space (EMSF/EMST) for applications whose input N points are uniformly or boundedly distributed in the Euclidean space. The sequential approach takes $O(N)$ time complexity through combining Borůvka's algorithm with an improved component-based neighborhood search algorithm, namely sliced spiral search, which is a newly proposed improvement of Bentley's spiral search for finding a component graph's closest outgoing point on 2D plane. We also propose a k-d search technique to extend this kind of search into 3D space. The data parallel approach includes a newly proposed two direction breadth-first search (BFS) implementation on graphics processing unit (GPU), which is aimed for selecting a spanning tree's shortest outgoing edge. The GPU parallel approaches assign N threads with one thread associated to one input point, one thread occupies $O(1)$ local memory and the whole algorithm occupies $O(N)$ global memory. Experiments are conducted on point set of both uniformly distributed data sets and TSPLIB database. We evaluate computation time of the proposed approaches on more than 40 benchmarks with size N growing up to 10^5 points.

Keywords: Parallel Euclidean minimum spanning tree · Spiral search · Sliced spiral search · Minimum spanning forest · GPU data clustering

1 Introduction

Solutions to build hierarchical minimum spanning forest or tree (MSF/MST) provide basis to many other algorithms, like nearest insertion and Chirstofid's heuristics for Traveling Salesman Problems (TSP). Given a complete weighted undirected graph $G = (V, E)$ with N vertexes and predefined $N \times N$ edges, the traditional sequential MST algorithms like Prim's (1957) [1], Kruskal's (1956) [2] or Borůvka's (1926) [3] algorithms take $O(ElogV)$ time complexity. The classical *Borůvka's Algorithm* [3] has a natural attribute to maintain hierarchical minimum spanning forest (MSF) from bottom-to-up in divide-and-conquer mode. It

© Springer Nature Switzerland AG 2019
R. Battiti et al. (Eds.): LION 12 2018, LNCS 11353, pp. 16–30, 2019.
https://doi.org/10.1007/978-3-030-05348-2_2

works by iteratively finding a shortest outgoing edge for each subtree (component) and merging these subtrees into a new larger one.

Many attempts exist to address the MST problem in parallel or distributed way when the input is a graph with explicit edge list [4–8]. They are all variations of the well-known Prim's, Kruskal's or Borůvka's algorithms.

However, considering MST in its Euclidean version (EMST) when the input data is a set of points in the Euclidean space $\Re^d$, with no edge list as input, and where weights between any two nodes is the Euclidean distance between them, very few parallel approaches directly address EMST except for some efficient sequential algorithms like these works using delaunay triangulation [9], using voronoi diagram [10], using k-d tree and dual tree [11]. More precisely, we do not find an efficient parallel/distributed algorithm that directly addresses the EMST without predefining the $N \times N$ edge list that takes $O(N^2)$ time complexity before starting the parallel algorithms.

Since no edge list is given as an input, and the beginning graph $G = (V, \emptyset)$, $(v_0, v_1, ...v_n \in V)$ is implicitly complete, two key problems need to be considered for EMST implementation. Firstly, find the shortest **outgoing edge** for one entire sub-tree component. Secondly, how to preserve $O(1)$ local memory for each GPU thread and linear global memory for large size input. These two key problems restrict the selection of basic tools for building parallel EMST. The first problem can be turned into following **two sub-steps**, while the second problem is influenced by concrete parallel algorithms to solve these two sub-steps:

- Find the graph's closest **outgoing point** to a vertex $v_i \in V$;
- Graph search to collect the shortest **outgoing edge** within one component.

The first sub-step can be solved by using classical nearest neighbor search algorithms (NNS) like k-d tree [12] or Elias method [13, 14] with searching filters. These algorithms are originally used to find the closest Euclidean point to one vertex of same color. K-d tree is often preferred because its average $O(logN)$ time complexity for searching the closest point. But during parallelism of the recursive k-d tree data structure, it is complex to both consider the requirement of memory occupation and decentralized control with which each point's nearest neighbor search works independently [15–17]. Rather than dealing with parallelism of the recursive k-d tree data structure, we turn to Elias' closest point finding approaches [13, 14], like Bentley spiral search [18], that we think to be a natural candidate approach for massively parallel closest point findings that follows data parallelism and decentralized control. ***Elias' approaches*** [13, 14] indicate a category of nearest neighbor searching algorithms that work on dividing the Euclidean space into congruent and non-overlapping sub-regions, cells, or bins. Each cell contains a list of the points that fall within its boundaries. When a query point q comes in, the algorithm firstly searches the cell where query point q is located, then passes to search these neighbor cells that are close to the starting cell. ***Bentley's*** **local spiral search** [18] belongs to a kind of Elias' approaches, which accesses neighbor cells in a spiral manner. Once one point is found, it is guaranteed that there is no need to search any other cells that do not intersect the circle of radius equals to the distance to the first point found and centered at the query point [18]. Balanced k-d tree has a sequential construction cost in $O(NlogN)$ time, Elias' approaches have sequential $O(N)$ construction cost. Here, we expect to exploit such geometric properties into the GPU CUDA platform [19] and their improvement adapted to EMST closest outgoing point finding problems.

The second sub-step can be achieved by graph search algorithms like breadth-first search (BFS). Classical sequential BFS operates on a *frontier* data structure (usually queues or vectors) that contains a set of vertexes to be visited next, and keeps this *frontier* data as local variables. Since CUDA programming on GPU allows arbitrarily coalesced and scattered memory access from multiple threads, each thread can run one independently sequential BFS to collect the shortest outgoing edge within one component. This kind of kernel launch works under the assumption that GPU can provide sufficient local memory for each independent BFS implementation, which can be found in the work done by Zhou et al. [15] for constructing parallel k-d tree.

Another kind of GPU parallel BFS implementation proposed by Harish et al. [5] is straightforward to process vertexes in the current frontier in parallel. Its merit lies in that their implementation assigns one thread to one vertex and eliminates the need for queues for each independent classical BFS process [5]. However, this kind of one direction GPU BFS implementation is not specially designed to select the shortest outgoing edge within one component, because each frontier node accesses independently and simultaneously to the memory so that communication can not happen in parallel. Here, we proposed a two direction GPU BFS implementation to solve the communication problem.

The following paper is organized as this: Sect. 2 presents the standard distributed graph representation for parallel EMST computing. Section 3 presents preliminaries related to the proposed methods. Section 4 presents our proposed approaches for building EMST both sequentially and in parallel, which includes the newly proposed spiral search methods to find graph's closest outgoing point, and the two direction GPU BFS implementation. Section 5 presents the experimental results, while Sect. 6 concludes the paper.

2   Component-Based Graph Representation

The hierarchical EMSF or EMSTs are represented by a variation of adjacency list, namely an adjacency list where each vertex only possesses a collection of all its adjacency neighboring vertexes. We call this kind of data structure as Doubly Linked Vertex List (DLVL) in this paper for easy reference since it follows that the graph is doubly linked as each node of a given edge has a link to its connected node. This doubly linked property provides key support to the proposed parallel EMSF/EMST algorithm, since the algorithm needs to access one whole independent sub-tree from any one of its vertex, and for easy connection of any two adjacent sub-trees.

In EMST graph representation, each node of the DLVL graph data structure contains a bounded size buffer for memorizing its adjacent neighbor nodes since maximum degree of a EMST is bounded [20] under the assumption that every point's position is unique in the Euclidean space.

3 Preliminaries

Bounded Cellular Partition for the Closest Point Finding. As shown in illustrated in Fig. 1, the Euclidean space containing the input N points in a rectangular region of Euclidean $length, width$ is partitioned into cellular adapted to this region and each cell contains up-bounded points. In case of a uniform input data distribution, each cell only theoretically contains one point. Considering GPU implementation, the partition of the space is modeled by an array of cells. Each input data point can compute its cell location in constant time [21].

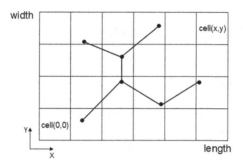

Fig. 1. Cellular partition of 2D Euclidean plane used by Elias' NNS approaches [13, 14].

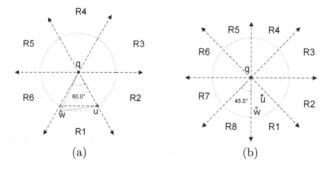

Fig. 2. Uniqueness property. (a) A partition of $\mathbb{R} - q$ into 6 congruently circulator sector partitions such that for any two points u and w that lie in the same partitioned region $\mathbb{R}_1$, either $\|wu\| < \|wq\|$ or else $\|uw\| < \|uq\|$, except for the equal case where $\|uq\| = \|wq\|$. (b) A partition of $\mathbb{R} - q$ into 8 congruently sector partitions.

Uniqueness Property. Given a query point q on Euclidean plane, a partitioned region $\mathbb{R}$ has the *uniqueness property* with respect to q if for every pair of points $u, w \in \mathbb{R}$, $\|wu\| < max(\|wq\|, \|uq\|)$ [20]. As shown in Fig. 2, if it exists the closest point u to q in a partitioned sub-region $R_{i,(6<i)}$ centering at the query point q, then there is not else point w who satisfy $\|wq\| < \|wu\|$. We call this kind of one circulator sector partition $R_{i,(6<i)}$ as one *slice* or *slab* in order to distinguish it from the preceding cellular partition.

Considering points on Euclidean plane, the maximum partitioned region centering at point q is the circulator sector with 60 degree as shown in Fig. 2(a), while the equal case where $\|uq\| = \|wq\|$ should be carefully treated. To exclude the equal distance cases, congruently divided circulator sectors with 45 degree is applied as shown in Fig. 2(b). Under this *uniqueness property*, if the query point q has found the closest point in a single sliced region $R_{i,(6<i)}$, for example point u at the R_1 in Fig. 2(b), then there will not be any more point lying on the same sliced region R_1 that satisfies $\|wq\| < \|wu\|$.

4   GPU Parallel EMSF Algorithm

One iteration of the proposed EMST algorithm consists in augmenting a EMSF by finding shortest outgoing edges, until the overall spanning tree is reached, which is achieved by four main steps: Find component's closest outgoing point from each vertex; Find shortest outgoing edge within each component; Connect sub-trees; Compact sub-trees. In its parallel version, each step can be executed as Kernels in GPU platform with CUDA programming [19]. Our contribution lies in that an improvement of Bentley's spiral search to find component graph's closest outgoing point and a GPU two-direction parallel breadth-first search to find components' shortest outgoing edge based on CUDA platform, and the first attempt to deal with parallel EMST totally on GPU.

At the very beginning of the proposed GPU parallel EMSF algorithm, each vertex is treated as one component of the initial spanning forest and has its empty bounded buffer for memorizing its MST neighbor links. Besides, each vertex contains a root identifier represented by itself in order to identify the component, which is managed according to disjoint set data structure.

4.1   Find Component's Closest Outgoing Point from Each Vertex

The first step consists in finding component's closest outgoing point to a vertex $v_i \in V$ in order to find the shortest outgoing edge for this component at next step, which is shorted as *Find minimum 1*. This could be done with Bentley's spiral search to find a point' closest outgoing neighbor by adding a filter.

However, as small EMSTs in the hierarchical EMSF grow with different iterations while the cellular partition of the input Euclidean space does not change, searching one point's closest outgoing point with pure Bentley's spiral search will lead to $O(N/2)$ complexity in the worst case. To alleviate this situation when finding the component's closest outgoing point, the uniqueness property can be applied during Bentley's spiral search.

Sliced Spiral Search on Cellular Partition. To reduce computation at this *Find Minimum 1* step when using pure Bentley's spiral search (shown in Fig. 3), we propose *sliced spiral search*, illustrated in Fig. 4, with consideration of the uniqueness property and the fact that each node of one same component tries to find the same component's shortest outgoing edge. Once the query point q finds the closest point p_i in a sliced region $R_i(i = 1, ..., 8)$ centering at q, shown in Fig. 2, no-matter this p_i lies in same component with the query point or not, it does not need to search later cells that totally locate in this region R_i any more, further more, it does not need to search closest outgoing point for this query point in this sliced region for later different executions of this *Find Minimum 1* step, and it does not need to repeat searching these neighbor cells being searched in previous iterations. Based on data parallelism on GPU, massive points' closest outgoing point findings can be executed in parallel on the same cellular partition.

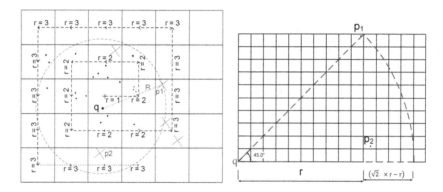

Fig. 3. Spiral search manner to access neighbor cells. r indicates the searching radius on cellular partition centering at the first cell where the query point q locates. R indicates Euclidean radius centering at current q on the Plane.

K-d Search on Cellular Partition. The above spiral search manner can hardly be extended into 3D or 4-dimension space. To solve this drawback, we propose a K-d search manner to access neighbor cells slab by slab centering at current cell where the query point locates, for Euclidean neighborhood search based on partitioning space into non-overlapping and congruent cells or bins. Taking 2D cellular partition shown in Fig. 1 as an example, each cell has four edges. The K-d search vertically accesses neighbor cells from each edge of a cell and extends to horizontal neighbor cells until it meets the boundaries of one slab centering at current cell where the query point locates. This procedure can be illustrated in Fig. 5 where we take the two of eight slabs centering at the current cell as examples.

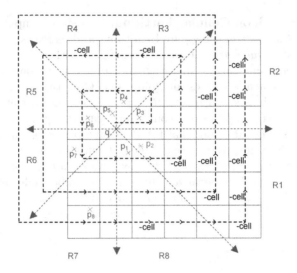

Fig. 4. Sliced spiral search on 2D cellular partition. The dark black dash line with arrows indicates the way in which one spiral search accesses neighbor cells. The cells marked as $-cell$ are not searched for current query point q since its closest point has already been found in corresponding slab centering at current query point q.

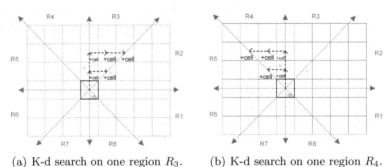

(a) K-d search on one region R_3.   (b) K-d search on one region R_4.

Fig. 5. The K-d search manner to access neighbor cells slab by slab on cellular partition. One search enters into every slab region $R_i, (6 < i)$ centering at current cell where the query point q locates. For example, the cells marked as $+cell$ are searched for current query point q in the specific corresponding slab region R_i.

4.2 Find Shortest Outgoing Edge Within Each Component with GPU Two Direction Parallel Breadth First Search (BFS)

The second step corresponds to traverse into a spanning subtree in order to collect its shortest outgoing edge. Once found, the node possessing the shortest outgoing edge is marked as winner node for connecting to its corresponding subtree at next step. We adopt a classical graph search algorithm called Breadth-first Search (BFS) to accomplish this task.

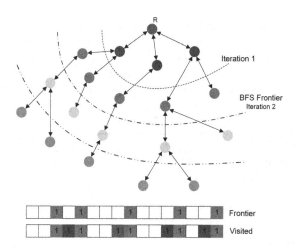

Fig. 6. GPU two direction parallel breadth-first search that follows the GPU one direction BFS proposed by Harish et al. [5] and works on DLVL graph representation to collect components' shortest outgoing edges in parallel (see DLVL in Sect. 2).

For applications that have to eliminate the need for queues used by the classical BFS algorithm, we propose a GPU two direction parallel BFS algorithm illustrated in Fig. 6 that works on the DLVL graph representation and follows the GPU one direction BFS implementation proposed by Harish et al. [5]. This decision ensures the EMST algorithm only takes $O(N)$ global memory, and each thread occupies constant local memory. It should be two-direction because it has to start from the root of a spanning tree to traverse all leaves and return back to the root in order to collect shortest outgoing edge within one component.

GPU Implementation. The overall algorithm flow of this two direction GPU parallel BFS is shown in Algorithm 1. This GPU two direction BFS implementation uses several arrays with same size of N to collect the shortest outgoing edge for each component and assigns one thread to treat one node. These arrays and their initial values are listed in Table 1.

The first two kernels (Algorithms 2, 3) broadcast each node v_i's parent P_{a_i} and its number of sons $nSon_{a_i}$ from the root to each leaves. At each iteration of Kernel 2 (Algorithm 3), the frontier node $F_a[i]$ changes its role to be visited node $V_a[i]$ and authorizes its unvisited sons as next frontiers. The third kernel 3 (Algorithm 4) collects the shortest outgoing edge from leaves to root. At each iteration, each active leaf whose $nSon_a[i] == 0$ subtracts its parent's $nSon_a[j]$ with CUDA Atomic operation, marks itself as visited $V_a[i] = 1$, selects shortest outgoing edge among itself, its sons and the sons' Son_a, then writes the selected outgoing edge to $Son_a[i]$ for next iteration to be collected by its parent.

Avoid Circles with Parallel Choosing Rules. To avoid circles during parallel computation, separate choosing rules should be respected at the two independent [*Find Minimum*] steps in case of equal outgoing distance. [*Find Minimum 1*] should respect that outgoing vertex v_i possessing the least identifier i has the first priority. [*Find Minimum 2*] should respect that the edge possessing the minimum lexicography order of one edge's two vertexes' identifiers $(i, j), i < j$ is selected. Here i, j are 1-Dimensional identifiers of the input points.

Algorithm 1: Overview of the GPU two direction parallel BFS.

  1: Create and initialize working arrays on CUDA device.
  2: Invoke Kernel 1 shown in Algorithm 2
  3: Synchronize
  4: **while** not termination **do**
  5:    Invoke Kernel 2 shown in Algorithm 3
  6:    Synchronize
  7:    Judge termination
  8:    Synchronize
  9: **end while**
10: **while** not termination **do**
11:    Invoke kernel 3 shown in Algorithm 4
12:    Synchronize
13:    Judge termination
14:    Synchronize
15: **end while**

Table 1. Variables and their usage in GPU two direction Parallel BFS implementation.

Variable	Purpose	Initial value
A_a	Array holds the roots (or source nodes)	0
F_a	Array holds vertexes to be visited next	0
V_a	Array holds nodes already visited	0
$nSon_a$	Array holds number of sons for each vertex	∞
Dis_a	Array holds each vertex' outgoing distance	∞
P_a	Array holds parent of each vertex.	$(-1, -1)$
Son_a	Array holds candidate son vertex for each parent	$(-1, -1)$

Algorithm 2: Kernel 1. GPU parallel BFS in first direction: start from source vertexes.

1: $tid \leftarrow$ getThreadID
2: **if** $A_a[tid]$ **then**
3:     $V_a[tid] \leftarrow$ true
4:     $nSon_a[tid] \leftarrow$ number of Links
5:     **for** all links nid of tid **do**
6:         $F_a[nid] \leftarrow$ true
7:         $P_a[nid] \leftarrow$ tid
8:     **end for**
9: **end if**

Algorithm 3: Kernel 2. GPU parallel BFS in first direction: start from frontier vertexes.

1: $tid \leftarrow$ getThreadID
2: **if** $F_a[tid]$ **then**
3:     $V_a[tid] \leftarrow$ true
4:     $F_a[tid] \leftarrow$ false
5:     $nSon_a[tid] \leftarrow$ (number of Links) - 1
6:     **for** all links nid of tid **do**
7:         **if** $nSon_a[nid] == \infty$ **then**
8:             $F_a[nid] \leftarrow$ true
9:             $P_a[nid] \leftarrow$ tid
10:        **end if**
11:    **end for**
12: **end if**

4.3 Connect Graph

After the preceding *Find Minimum 2* step, each component finds its closest outgoing vertex v_j in parallel and has its winner vertex w_{v_i} pointing to v_j. Here, this step builds graph edges by adding links to these two corresponding vertexes' adjacency list in parallel, while at the same time avoiding repeated links in each node's adjacency list. Three kinds of relationship exist between w_{v_i} and v_j, which are listed below, but only the last case leads to repeat links.

- First, v_j is not a winner vertex;
- Second, v_j is also a winner vertex but its corresponding node is not w_{v_i};
- Third, v_j is also a winner vertex and its corresponding node is w_{v_i}.

For both sequential and parallel implementation of this step, the winner node w_{v_i} of a component adds its corresponding vertex v_j to its own adjacency list, while the node v_j adds w_{v_i} under the above three conditions. Parallel implementation of this step uses atomic operations to update these nodes' adjacency list, since two or more components may find the same vertex as their closest outgoing nodes. After this *Connect Graph* step, small components have already been connected and become new bigger components.

Algorithm 4: Kernel 3. GPU parallel BFS in reverse direction: start from vertex whose number of sons has been counted to zero.

1: $tid \leftarrow$ getThreadID
2: **if** $nSon_a[tid] == 0$ **then**
3:    $nSon_a[tid] \leftarrow \infty$; $V_a[tid] \leftarrow$ true; Point $v_{parent} \leftarrow P_a[tid]$
4:    Atomic subtraction $nSon_a[v_{parent}]$ - 1
5:    Select shortest outgoing edge (tid)
6:    **if** $Son_a[tid]$ exist **then**
7:       Select shortest outgoing edge $(Son_a[tid])$
8:    **end if**
9:    **for** all links nid of tid **do**
10:       **if** $P\_a[nid] == tid$ **then**
11:          Select shortest outgoing edge (nid)
12:          **if** $Son_a[nid]$ exist **then**
13:             Select shortest outgoing edge $(Son_a[nid])$
14:          **end if**
15:       **end if**
16:    **end for**
17:    $Son_a[tid] \leftarrow$ selected vertex possessing the shortest outgoing edge
18:    **if** tid is root **then**
19:       Register the selected shortest outgoind edge
20:    **end if**
21: **end if**

4.4   Compact Graph

The fourth step called *compact graph* executes the components' union operation on disjoint set data structure [22]. The role of this step is to update these disjoint components' root identifiers, which will allow distinguishment between components during the *Find Minimum* steps.

This step follows the classical sequential and parallel union-find solutions. The sequential version with path compression operation only takes $O(\alpha(N))$ time, $\alpha(N)$ is the inverse Ackermann function. In parallel version of this paper, we construct a root merging graph[1] in order to find one uniform new root among the connected sub-trees and then do the union operation. The path compression step can be operated in parallel by associating one thread to one input point.

---

[1]   https://stanford.edu/~rezab/classes/cme323/S15/projects/parallel_union_find_presentation.pdf.

5 Experiments

We provide three test versions of the proposed EMST algorithm for comparing. We also implement sequential Prim's algorithm and Delaunay triangulation EMST algorithm using source codes from the Internet[2].

The first version is referred as *Sequential-Sliced Spiral Search-EMST*, all steps of the proposed EMST algorithm work sequentially in this version. The second version is referred as *Parallel-SlicedSpiralSearch-EMST-1*, it only incorporates the sliced spiral search step as a GPU Kernel while other steps are executed sequentially. The third version is referred as *Parallel-SlicedSpiralSearch-EMST-2*, it takes most advantages of parallel implementation as explained in this paper, like the GPU parallel sliced spiral search and two-direction BFS.

We test the proposed method on uniformly distributed data set with 20 instances and 20 benchmarks for Euclidean Traveling Salesman Problems (TSP) offered by TSPLIB [23]. These datasets simulate a large set of possible data distribution on 2D Euclidean space. For each test case, the average running time over 10 runs is reported. We implemented all algorithms with C++ and CUDA Toolkit v8.0. Code is compiled on laptop with CPU Intel(R) Core(TM) i7-4710HQ, 2.5 GHz running Windows and GPU card GeForce GTX 850M.

The algorithm finishes building EMST within very few iterations, for example 8 iterations for ch71009.tsp instance. Statistic results are shown in Figs. 7 and 8, EMSF visual examples are shown in Fig. 9. Running time includes time needed for data copy between CPU and GPU side when it is necessary.

From Fig. 7 and Fig. 8, we can see that the proposed sequential sliced spiral search runs faster than using Delaunay triangulation on 2D plane. Comparison between Sequential-SlicedSpiralSearch-EMST and Parallel-SlicedSpiralSearch-EMST-1 shows that the GPU parallel implementation of this slice spiral search gets huge acceleration over the sequential one. Comparison between Parallel-SlicedSpiralSearch-EMST-1 and Parallel- SpiralSearch-EMST-2 shows that the two-direction GPU parallel BFS can hardly get acceleration over sequential BFS implementation on our current GPU card.

6 Conclusion

This paper presents GPU implementation of Borůvka's EMSF/EMST algorithm starting only from input Euclidean points, we exploit GPU data parallelism with decentralized control for each necessary sub-step. It includes an improved spiral search methods and a GPU two-direction BFS implementation.

The proposed sliced spiral search further reduce time complexity of finding a point set's closest outgoing point from one vertex of this point set when following cellular partition-based Bentley spiral search algorithm. The proposed K-d search manner to access neighbor cells can be extended to 3D or 4D Euclidean spaces, which will be detailedly explained in our next paper. These algorithms

---

[2] Solving Euclidean Minimum Spanning Tree, source code from GitHub, https://github.com/hqythu/EMST.

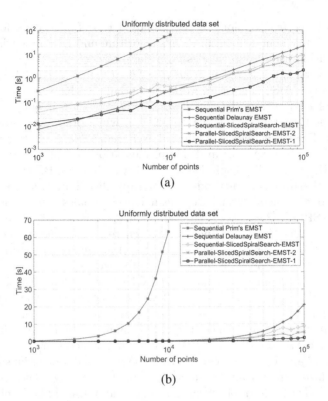

(a)

(b)

Fig. 7. Average running time of the five EMST algorithms on uniformly distributed data set. (a): exponential vertical coordinates. (b): general vertical coordinates.

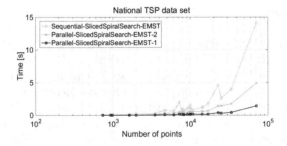

Fig. 8. Average running time of the proposed EMST algorithms on national TSP data set.

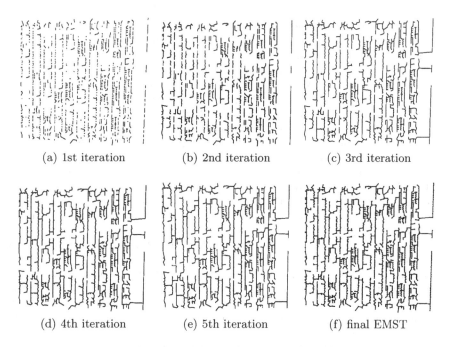

(a) 1st iteration   (b) 2nd iteration   (c) 3rd iteration

(d) 4th iteration   (e) 5th iteration   (f) final EMST

Fig. 9. Test of the proposed EMSF/EMST method working on rbw2481.tsp from TSPLIB. Images (a–e) show different EMSF built at previous five iterations, image (f) shows the final EMST built at the final iteration.

are attractive since their natural attributes to be parallelized with characteristics of global memory, decentralized control on GPU.

The GPU two direction BFS algorithm provides a GPU data parallel solution to traverse one spanning tree from any one of its vertex acting as a root. Characteristic is that it assigns one thread to treat one frontier vertex and allows communication between these simultaneously executed frontier vertexes. This solution provides basis for some higher level distributed or parallel optimization algorithms. Our experimental results were obtained allowing linear increase of memory and processors with problem size. With increasing quantity of physical cores in GPU hardwares, the proposed data parallel algorithm with decentralized control could yield to huge acceleration. Based on contribution of this work, application to many other Euclidean optimization problems, for example the well known traveling salesman problem or the hierarchical clusterings in an image, could be envisaged with high efficiency and within divide-and-conquer scheme.

References

1. Prim, R.C.: Shortest connection networks and some generalizations. Bell Syst. Tech. J. **36**(6), 1389–1401 (1957)
2. Kruskal, J.B.: On the shortest spanning subtree of a graph and the traveling salesman problem. Proc. Am. Math. Soc. **7**(1), 48–50 (1956)

3. Boruvka, O.: O jistém problému minimálním (1926)
4. Bader, D.A., Cong, G.: Fast shared-memory algorithms for computing the minimum spanning forest of sparse graphs. In: Proceedings of 18th International Parallel and Distributed Processing Symposium, p. 39. IEEE (2004)
5. Harish, P., Vineet, V., Narayanan, P.: Large graph algorithms for massively multithreaded architectures. Technical report, International Institute of Information Technology Hyderabad, IIIT/TR/2009/74 (2009)
6. Vineet, V., Harish, P., Patidar, S., Narayanan, P.: Fast minimum spanning tree for large graphs on the GPU. In: Proceedings of the Conference on High Performance Graphics, pp. 167–171. ACM, New York (2009)
7. Wang, W., Huang, Y., Guo, S.: Design and implementation of GPU-based prim's algorithm. Int. J. Mod. Educ. Comput. Sci. **3**(4), 55 (2011)
8. Ramaswamy, S.I., Patki, R.: Distributed minimum spanning trees (2015)
9. Lingas, A.: A linear-time construction of the relative neighborhood graph from the delaunay triangulation. Comput. Geom. **4**(4), 199–208 (1994)
10. Shamos, M.I., Hoey, D.: Closest-point problems. In: 16th Annual Symposium on Foundations of Computer Science, pp. 151–162. IEEE (1975)
11. March, W.B., Ram, P., Gray, A.G.: Fast euclidean minimum spanning tree: algorithm, analysis, and applications. In: Proceedings of the 16th ACM SIGKDD International Conference on Knowledge Discovery and Data Mining, pp. 603–612(2010)
12. Bentley, J.L.: Multidimensional binary search trees used for associative searching. Commun. ACM **18**(9), 509–517 (1975)
13. Rivest, R.L.: On the optimality of elia's algorithm for performing best-match searches. In: IFIP Congress, pp. 678–681 (1974)
14. Cleary, J.G.: Analysis of an algorithm for finding nearest neighbors in euclidean space. ACM Trans. Math. Softw. (TOMS) **5**(2), 183–192 (1979)
15. Zhou, K., Hou, Q., Wang, R., Guo, B.: Real-time kd-tree construction on graphics hardware. ACM Trans. Graph. (TOG) **27**(5), 126 (2008)
16. Qiu, D., May, S., Nüchter, A.: GPU-accelerated nearest neighbor search for 3D registration. In: Fritz, M., Schiele, B., Piater, J.H. (eds.) ICVS 2009. LNCS, vol. 5815, pp. 194–203. Springer, Heidelberg (2009). https://doi.org/10.1007/978-3-642-04667-4_20
17. Hu, L., Nooshabadi, S.: Massive parallelization of approximate nearest neighbor search on kd-tree for high-dimensional image descriptor matching. J. Vis. Commun. Image Represent. **44**, 106–115 (2017)
18. Bentley, J.L., Weide, B.W., Yao, A.C.: Optimal expected-time algorithms for closest point problems. ACM Trans. Math. Softw. (TOMS) **6**(4), 563–580 (1980)
19. Nvidia, C.: Programming guide (2010)
20. Robins, G., Salowe, J.S.: On the maximum degree of minimum spanning trees. In: Proceedings of the Tenth Annual Symposium on Computational Geometry, pp. 250–258. ACM, New York (1994)
21. Rajasekaran, S.: On the euclidean minimum spanning tree problem. Comput. Lett. **1**(1), 11–14 (2004)
22. Galler, B.A., Fisher, M.J.: An improved equivalence algorithm. Commun. ACM **7**(5), 301–303 (1964)
23. Reinelt, G.: Tsplib–a traveling salesman problem library. ORSA J. Comput. **3**(4), 376–384 (1991)

A Cooperative Learning Approach for the Quadratic Knapsack Problem

Eduardo Lalla-Ruiz[1,4], Eduardo Segredo[2,3], and Stefan Voß[1(✉)]

[1] Institute of Information Systems, University of Hamburg, Hamburg, Germany
{eduardo.lalla-ruiz,stefan.voss}@uni-hamburg.de
[2] School of Computing, Edinburgh Napier University, Edinburgh, UK
e.segredo@napier.ac.uk
[3] Dpto. de Ingeniería Informática y de Sistemas, Universidad de La Laguna,
San Cristóbal de La Laguna, Spain
esegredo@ull.edu.es
[4] Department of Industrial Engineering and Business Information Systems,
University of Twente, 7522NB Enschede, The Netherlands
e.a.lalla@utwente.nl

Abstract. The *Quadratic Knapsack Problem* (QKP) is a well-known optimization problem aimed to maximize a quadratic objective function subject to linear capacity constraints. It has several applications in different fields such as telecommunications, graph theory, logistics, hydrology and data allocation, among others. In this paper, we propose the application of a novel population-based metaheuristic referred to as Multi-leader Migrating Birds Optimization (MMBO), which exploits the concepts of cooperation and communication along the search leading to a collective learning, to solve a wide range of well-known QKP instances.

1 Introduction

The *Quadratic Knapsack Problem* (QKP) is a knapsack problem introduced by Gallo *et al.* [5] that includes the relationship among the items within a quadratic objective function. Formally, we are given a set of items $N = \{1, ..., n\}$, where each item i has a given weight $w_i > 0$, and an $n \times n$ profit matrix B with entries $b_{i,j}$ that indicates the benefit obtained when an item is packed with respect to itself and other items. In other words, if an item i is selected, then the profit obtained is equal to $b_{ii} + \sum_{j \in N \setminus \{i\}} b_{ij}$. Note that matrix B is symmetric, i.e., $b_{ij} = b_{ji}$. The goal of the QKP is thus determining the items to be packed in the knapsack taking into account the weight capacity c such that the total profit of the items packed is maximized. The selection of the items is ruled by the binary variable x_i which is equal to 1 if item i is selected and 0 otherwise. The formal definition of the QKP is as follows.

$$\max z(QKP) = \sum_{i \in N} \sum_{j \in N} b_{ij} x_i x_j \qquad (1)$$

© Springer Nature Switzerland AG 2019
R. Battiti et al. (Eds.): LION 12 2018, LNCS 11353, pp. 31–35, 2019.
https://doi.org/10.1007/978-3-030-05348-2_3

$$\sum_{i \in N} w_i x_i \leq c \qquad (2)$$

$$x_i \in \{0, 1\}, \quad i \in N \qquad (3)$$

Note that, if $b_{ij} = 0$ for $i \neq j$ then the QKP can be reduced to the *Knapsack Problem*. Moreover, the *Clique Problem* can be treated as a particular case of the QKP, where the *Max Clique* can be solved by a QKP algorithm by using a binary search [2].

In this work, we propose the application of a novel population-based metaheuristic, called *Multi-leader Migrating Birds Optimization* (MMBO) [6]. This approach exploits the communication among a population of individuals during the search, thus enabling cooperative learning. We test our method for the latest benchmark suite proposed for the QKP [3]. We attain competitive results, in terms of the objective function value obtained at the end of the runs in comparison to the best-performing state-of-the-art approaches used in those problem instances.

2   Multi-leader Migrating Birds Optimization

For solving the QKP, we propose the use of MMBO (Algorithm 1), which is a decentralized cooperative search approach inspired by the flight formation of migratory birds. In MMBO, the population is denoted as $P = \{1, 2, ..., p\}$, where p is the number of individuals representing solutions of the optimization problem at hand. During the search, individuals are distributed in a line formation, i.e., $(1, 2, ..., p)$, where individual 1 is directly connected to individual 2, and individual 2 is connected to individuals 1 and 3, and so on. Based on that line formation, a relationship structure is established according to a given *relationship criterion*, for instance, in terms of the objective function value associated with each member of the population. By means of that criterion, the role of each pair of individuals is determined, i.e., which individual provides and which individual receives information during the search.

Starting from each individual in P, k feasible neighbors are generated through a predefined neighborhood structure. In this work, two decision variables of a given individual are uniformly selected at random and their corresponding binary values are flipped in order to produce a novel neighbor. Depending on the relationship criterion and how information is shared among individuals, different roles arise:

- **Leader.** It is that individual with the best objective function value when compared to its adjacent individuals. Hence, it does not receive information from any individual, but shares it, in the form of δ neighbors, with its adjacent individuals. Each leader generates k neighbors. The set of leaders is denoted as P_L.

- **Follower**. It is that individual which explores the search space considering its own information and the information received from the individuals in front of it within the relationship structure. It generates $k - \delta$ neighbors and receives δ neighbors. The set of followers is denoted as P_F.
- **Independent**. It is that individual which is not included into any of the above categories because it has associated the same objective function value than its adjacent individuals. Hence, it does not exchange information with any other individual, but generates k neighbors. The set of independent individuals is denoted as P_I.

Algorithm 1 Pseudocode of MMBO

Require: p, K, k, and δ
1: Initialize the population P, which consists of p individuals generated at random
2: **while** (K neighbors have not been generated) **do**
3:     Determine the interaction among the individuals of P and establish the relationship structure

4:     **while** (the stopping criterion associated with the relationship structure is not met) **do**
5:         Generate k neighbors starting from each individual included within $P_L \cup P_I$
6:         Replace each individual included within $P_L \cup P_I$ by its fittest neighbor if the latter is fitter than the former
7:         Replace each individual included within P_I by its fittest neighbor
8:         **for all** (individual $f \in P_F$) **do**
9:             Generate $k - \delta$ neighbors starting from f
10:             Get the best unused δ neighbors from the previous individuals of f in the relationship structure
11:             Replace f by its fittest neighbor if the latter improves the former
12:         **end for**
13:     **end while**
14: **end while**
15: Return the fittest individual in P

---

3 Numerical Results

The proposed approach has been implemented in Java and executed on a computer equipped with an Intel i7 CPU 3.5 GHz and 16 GB of RAM. By preliminary experiments, we identified the following parameters $p = 20$, $\delta = 1$, and $k = 5$ with a stopping criterion $K = |n|^2$ neighbors. The problem instances used are those proposed in [3] by following the guidelines of previous works. The instances are generated for different density values, i.e., different percentage of non-zero cross benefits in $B = \{b_{ij} : i, j \in N, i < j\}$, where each b_{ij} is chosen from the interval $[1, 100]$. The knapsack capacity c is selected from $[50, \sum_{j=1}^{n} w_j]$.

Table 1 shows the computational results for instances with $n = 100$ items, which were generated by considering different density values (dst). The methods selected for comparison purposes are the best-performing ones in the related literature for the instances proposed in [3]. They are based on a *Dynamic Programming Heuristic* (DPH) and a *Non-Delayed Relax-and-Cut* (CSL) [4]. We should note that execution times are measured as integer values in [3]. Hence, some execution times are reported as 0 when the time invested is lower than 1 second. Therefore, we report the upper bound of the said times to avoid those cases. Our approach is able to reduce considerably the execution time with respect to the best approach in terms of the objective function value (CSL) while reporting a new best solution considering one of the instances ($id = 4$). Finally, it is worth mentioning that MMBO is able to reach the optimal solution in those problem instances proposed in [1] and solved to optimality.

4 Conclusions

In this work, we have proposed a MMBO approach for the Quadratic Knapsack Problem. Due to its self-organization and cooperation dynamics, it allows individuals to learn along the search process in a collective way. The collaborative relationship structure enhances the diversification of the search as individuals can be distributed over the search space. Intensification is addressed by the sum of efforts of the individuals belonging to the same group located in a particular region of the search space. Our algorithm reports high-quality solutions in terms of the objective function value while investing shorter computational times compared to state-of-the-art approaches. In this regard, we even attained a new best solution for one of the instances tested.

Table 1. Numerical results for instances of $n = 100$ items. Bold represents a new best solution

Instance			CSL		DPH		MMBO				
n	dst	id	Obj	Time	Obj	Time	Max	Avg.	Min.	σ^2	Time
100	25	1	53774	3	53757	2	53774	53759.6	53723	371.64	0.08
		2	7082	5	7076	1	7082	7079.6	7076	8.64	0.06
		3	60875	2	60875	1	60875	60875	60875	0.00	0.09
		4	18386	6	18386	2	18386	18273.4	18224	2651.24	0.04
		5	43014	4	43014	1	43014	43002.2	42896	1253.16	0.08
		6	50484	4	50484	1	50484	50479.6	50473	29.04	0.06
		7	21769	4	21769	2	21769	21755.4	21701	739.84	0.07
		8	30687	5	30687	1	30687	30652.4	30555	2724.84	0.06
		9	28719	6	28719	1	28719	28666.4	28585	2037.44	0.13
		10	5463	4	5421	2	5463	5461	5443	36.00	0.08
100	50	1	34653	5	34653	1	34653	34653	34653	0.00	0.08
		2	43178	5	43169	1	43178	43170.9	43156	70.69	0.06
		3	46243	6	46243	2	46243	46232.5	46138	992.25	0.09
		4	48894	5	48992	1	**49030**	49004.7	48894	1666.41	0.04
		5	41515	6	41515	1	41515	41515	41515	0.00	0.08
		6	71982	4	71982	2	71982	71982	71982	0.00	0.06
		7	69146	6	69177	1	69177	69091.8	68985	7414.56	0.07
		8	83085	3	83085	1	83085	83057.5	82948	2045.85	0.06
		9	9772	4	9772	2	9772	9772	9772	0.00	0.13
		10	62465	6	62407	1	62465	62415	62081	12918.4	0.08
Average			41559.30	4.65	41559.308	1.35	41567.65	41544.95	41483.75	1748	0.07

As future work, we aim to extend the results provided herein to analyze the different features provided by the approach to solve the QKP in more detail.

References

1. Billionnet, A., Soutif, É.: An exact method based on lagrangian decomposition for the 0–1 quadratic knapsack problem. Eur. J. Oper. Res. **157**(3), 565–575 (2004)
2. Caprara, A., Pisinger, D., Toth, P.: Exact solution of the quadratic knapsack problem. INFORMS J. Comput. **11**(2), 125–137 (1999)
3. Cunha, J.O., Simonetti, L., Lucena, A.: Lagrangian heuristics for the quadratic knapsack problem. Comput. Optim. Appl. **63**, 97–120 (2016)
4. Fomeni, F., Letchford, A.: A dynamic programming heuristic for the quadratic knapsack problem. INFORMS J. Comput. **26**(1), 173–182 (2013)
5. Gallo, G., Hammer, P.L., Simeone, B.: Quadratic knapsack problems. In: Padberg, M.W. (ed.) Combinatorial Optimization, pp. 132–149. Springer, Berlin (1980). https://doi.org/10.1007/BFb0120892
6. Lalla-Ruiz, E., de Armas, J., Expósito-Izquierdo, C., Melián-Batista, B., Moreno-Vega, J.M.: Multi-leader migrating birds optimisation: a novel nature-inspired metaheuristic for combinatorial problems. Int. J. Bio-Inspired Comput. **10**, 89–98 (2017)

An Improved BTK Algorithm Based on Cell-Like P System with Active Membranes

Linlin Jia, Laisheng Xiang, and Xiyu Liu[✉]

College of Management Science and Engineering, Shandong Normal University,
Jinan, Shandong, China
sdxyliu@163.com

Abstract. BTK algorithm is an efficient algorithm for mining top-rank-k frequent patterns. It proposes new tree and list structures to store the information, employs subsume index concept, early pruning strategy and threshold raising method to reduce the search space. In this paper, a fast BTK algorithm, called CP-BTK algorithm is proposed, which is based on cell-like P system with active membranes. Cell-like P system is new computing model inspired from biological cells, operations in cell-like P system are distributed and parallel, so it can save time and improve the efficiency of algorithm greatly. And finally a example is given to illustrate the practicability and effectiveness of the proposed algorithm.

Keywords: Pattern mining · Top-rank-k frequent patterns · Cell-like P system · Membrane computing

1 Introduction

Frequent itemsets mining is a fundamental field of data mining introduced by Agrawal et al. [1], dedicates to find itemsets with high frequency in large transaction databases. The number of algorithms for frequent itemsets mining are numerous [2–5], but there are still many shortcomings. First is difficult to set the threshold suitable. If it is set too low, the number of candidates will be too large, the operation hours and memory consumption will also grow. In addition, if the threshold is set too high, few itemsets will be discovered. The BTK algorithm provides solutions in these problems, first progress is that it puts forward a new TB-tree structure [6], which each node contains the item-name, count, parent-pointer, start-build and finish-build five fields, and these information can get by traversing the TB-tree once, better than the PPC-tree, an FP-tree like structure [7–9]. Then it proposes the strategies of raising threshold by the count of patterns [6], pruning by threshold early and an effective method to generate subsume index. Although, several improvements have been proposed in this algorithm, it remains computationally expensive, the consumption of time and memory is great.

© Springer Nature Switzerland AG 2019
R. Battiti et al. (Eds.): LION 12 2018, LNCS 11353, pp. 36–48, 2019.
https://doi.org/10.1007/978-3-030-05348-2_4

Membrane computing is a computational model abstracted from biological cells, it puts forward new ideas and methods to computer science. Cell-like P system consists of three parts: membrane structure, objects set and evolution rules. The important part of the membrane structure is not size or spatial layout but the nested relationship between the membranes [10,11]. The computing processes in different membranes run in parallel, while computing processes in the same membrane run also in parallel, parallel mechanism makes P system a powerful computing system [12–14]. In this study, the CP-BTK algorithm utilizes the parallel mechanism to improve the productiveness of the BTK algorithm. But to transform the complete BTK algorithm into P system is a difficult and complex work, the implementation process of the algorithm in cell-like P system will be described in subsequent chapters of the paper.

The remaining part of this paper is arranged as follows: Sect. 2 describes the top-k frequent patterns mining, BTK algorithm and the basic introduction for cell-like P system. Section 3 introduces two improvements to BTK algorithm, design of the cell-like P system with active membranes for BTK algorithm explanation of rules and the computing process. Section 4 applies an instance to explain how CP-BTK works. The conclusions are drawn in the Sect. 5.

2   Preliminaries

In this section, some basic definitions in top-k frequent patterns mining [15], BTK algorithm, and basic cell-like P system structure are introduced.

2.1   Top-Rank-k Frequent Patterns Mining

Let $I = \{I_1, I_2 \ldots I_n\}$ be a set of items. $DB = \{T_1, T_2 \ldots T_m\}$ is a transaction database with m transactions.

Definitions

(i) *Pattern*: a set of items $P \subseteq I$ is called a pattern or an itemset.

(ii) *Support count*: the number of transactions containing a certain pattern, represented as sup(P). And a pattern which count is equal to or larger than threshold is called frequent pattern.

(iii) *Rank of a pattern*: R_p is represented the rank of pattern $P \subseteq I$, defined as: $R_p = \{sup(X) | X \subseteq I \text{ and } sup(X) \geq sup(P)\}$

(iv) Top-rank-k frequent patterns: if the rank of a pattern is equal to or less than given threshold k: $R_p \leq k$, this pattern is called a top-rank-k frequent pattern.

(v) Subsume index: subsume(A) is represented the subsume index of 1-pattern A. The definition is: subsume(A) = $\{B \in I | g(A) \subseteq g(B)\}$

g(A): the transactions in database include item A.

Property: sup(A) = sup(A∪E)|∀ E ∈{subsets of subsume(A)} [16].

2.2   TB-Tree and B-List

TB-tree and B-list are new structures used to store information of frequent patterns [6].
Definitions:
(i) *TB-tree*: There are two requirements for building a TB-tree.
Firstly, the root of the tree is labeled as root, and then to create the child nodes of the tree, a set of item prefix subtrees.
Secondly, every children node in subtrees have five fields: item-name, count, parent-pointer, start-build and finish-build. The name of each node in the tree is stored in the item-name field. Count represents the number of occurrences of item in the path from root to this node. The parent-pointer points to parent node. Start-build indicates that the node is the s-th node of the tree from beginning. Finish-build means that the node is the f-th already constructed node of the tree.
Property: Two nodes A and B, if and only if A.start-build<B.start-build and A.finidh-build>B.finish-build, we call A is the ancestor of B.
(ii) *B-info-code*: B_n is represented the B-info-code of node N, stores the information (N.count,[N.start-build, N.finish-build]).
(iii) *B-list of a 1-pattern*: BL_p represents the B-list of 1-pattern P, consists of all the B-info-codes of nodes which name is P.
(iv) *B-list of a t-pattern*: P_X and Q_X are two t-patterns, $BL_{px} = \{B_{p1}B_{p2}...B_{pn}\}$ and $BL_{Qx} = \{B_{q1}, B_{q2}...B_{qn}\}$. Intersection of BL_{px} and BL_{Qx} generates the B-list of pattern PQX. First find nodes that have ancestor-descendant relationship, and if B_{pi} is an ancestor of B_{qj}, add (B_{qj}.count, [B_{pi}.start-build, B_{pi}.finish-build]) to the B-list of pattern PQX. And then merge all B-info-codes in BL_{PQX} which have same start-build and finish-build into one B-info-code. Finally update the count to total count of all merged nodes.

2.3   BTK Algorithm

In this section, the RSC(Raising Threshold by the Support of Candidates) and EP(Early Pruning by Threshold) strategies are presented first, then the BTK algorithm is discussed in detail [6].
RSC(Raising Threshold by the Support of Candidates): Tabk is a table to store the present frequent itemsets by descending order of rank. First, threshold is zero and the table is no entry. If there are patterns support no less than the threshold, insert them into Tabk and update threshold to the support of the k-th entry in Tabk every step.
EP(Early Pruning by Threshold): During the process of generating frequent t-patterns, the corresponding count will be subtracted if find nodes that do not have ancestor and descendant. And if there is a count not larger than the threshold, the intersection function will be stopped, because the pattern generated from two input itemsets is not a frequent pattern. The process is as follows:
(i) Compute the total counts of two input B-lists, separately represented by C_1 and C_2.

(ii) Every step in algorithm, find B-info-code B_w in each B-list that does not have ancestor-descendant relationship. And if B_w is in the first list, let $C_1 = C_1$ - B_w.count. If not, let $C_2 = C_2$ - B_w.count.

(iii) The intersection stops when there are counts not larger than threshold and return null.

The general operation of the algorithm can be described as follows (Table 1):

Table 1. Algorithm: BTK

Input. DB: a transactions database; K:a user-specified top rank value
(i) Scan the database to calculate the count of each item, get the frequent 1-patterns and the B-list and store them in P_1
(ii) Construct the subsume indexes of 1-patterns in P_1 and the TB-tree
(iii) Set the threshold to zero. The value of this variable is represented the current smallest support for patterns stored in Tabk
(iv) Cycle through each entry in P_1, and determine if the support is not less than the threshold. If yes, add the pattern to Tabk and the candidate list C_1(candidates list for looking higher level frequent patterns), and update threshold value
(v) loop to find higher level candidate until the procedure Candidate gen stops
(vi) When the final result Tabk has been determined, a loop is done over each pattern in Tabk. Group each itemset to the subset of subsume it contains, put the new pattern to result Tabk
Output. The top-rank-k frequent patterns

The program to generate candidate itemsets is as follows (Table 2):

Table 2. The candidate gen procedure

Input. frequent t-patterns
(i) First procedure loops over each pattern and checks the requirements for merging two itemsets: if they belong to the subsume index, if they are different by only one item, if their count larger than the threshold
(ii) If the above requirements are met, merges the two patterns to generate a pattern P. The support and the B-list of $P(BL_P)$ is generated by intersection function and the subsume is also updated
(iii) If the B-list of P is not null and all the (i-1)-pattern subsets of P are candidates, adds the pattern P to the set of patterns to be used for generating larger candidates and Tabk, updates the threshold value
Output. frequent (t+1)-patterns

2.4   Basic Cell-Like P System Structure

A traditional cell-like P system [17] is a construct:
$$\Pi = \{\mathbf{O}, \mathbf{T}, \mathbf{C}, \mu, W_1, W_2, \ldots W_n, R_1, R_2, \ldots R_n, \rho_1, \rho_2, \ldots \rho_n\}$$
Where:

(i) O is a non-empty alphabet that represents a collection of objects in the cell-like P system.

(ii) T is a collection of output objects from cell-like P system, and $T \subseteq O$.

(iii) $C \subseteq$ O-T is a catalyst. In the running process of P system, the objects in C will not change at all. Sometime only when the catalyst C exists, some specific rules in P system can be executed.

(iv) μ is a collection of membrane structures in system, labeled by 1, 2...m.

(v) W_i $(1 \leq i \leq n)$represents the set of all the objects placed in the membrane i.

(vi) $R_i (1 \leq i \leq n)$represents the collections of evolution rules in the regions 1, 2...m.

(vii) $\rho_i (1 \leq i \leq n)$represents the execution order of rules in membrane i.

3   CP-BTK Algorithm

This section begins with introducing two improvements of BTK algorithm, and then is the design of cell-like P system with active membranes [18], finally the explanation of rules and compute process are given.

3.1   Improvements to BTK Algorithm

The algorithm is modified in two aspects, one is to determine the threshold quickly, the other is to generate subsume index effectively.

Improvement 1(QDT: Quickly Determine the Threshold). This improvement adopts the Tabk that stores patterns by descending order of support. The threshold is set to the support of k-th entry in Tabk, if the number of entries are not enough to k first time, update the threshold in next Tabk till the threshold is determined. The user-specified top rank value K is not very large, and the threshold can usually be determined in first time. The threshold is checked only when a support count of patterns larger than the established threshold appears in Tabk. This method improves efficiency without updating the threshold in each step.

Improvement 2(VMGS: Vertical Method to Generate Subsume Index). First, the data is transformed from horizontal format to vertical format by scanning the database once. The support count of an item is the length of transactions set obtained from the first step. According to the definition of subsume index and the vertical format data, obtains subsume index of by intersecting transactions sets.

3.2   The CP-BTK Algorithm

First the initial P system for implementing BTK algorithm [19] is presented in Fig. 1. The Cell-like P system [20] based on the improved BTK algorithm is defined as follows:

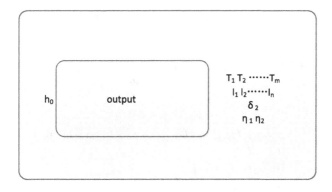

Fig. 1. Initial P system for implementing BTK algorithm

$$\Pi = \{\mathbf{O}, \ \mu; \ M_0, \ M_{h0}; \ R, \ \rho\}$$

The collection of all objects in the initial P system:

$O = \{T_1, T_2 \ldots T_m, \ I_1, I_2 \ldots I_n, \ \delta_2, \ \eta_1, \ \eta_2\}$

The initial structure of P system: $\mu = [_0[_{h0}]_{h0}]_0$

All objects in cell membrane 0:

$M_0 = \{T_1, T_2 \ldots T_M, \ I_1, I_2 \ldots I_n, \ \delta_2, \ \eta_1, \ \eta_2\}$

$M_{h0} = \{\lambda\}$ means that there is no object in the membrane h_0 in the initial state;

Evolution rules in cell membrane 0:

$R_1 = \{T_1 \rightarrow T_1 U_{I_1}[1]U_{I_2}[1]\ldots U_{I_n}[1]\}$

$\cup\{\delta_i T_i U_{I_1}[\text{i-1}]U_{I_2}[\text{i-1}]\ldots U_{I_n}[\text{i-1}] \rightarrow \delta_{i+1}T_i U_{I_1}[\text{i}]U_{I_2}[\text{i}]\ldots U_{I_n}[\text{i}], \ 2{\leq}i{\leq}\text{M-1}\}$

$\cup\{\delta_M T_M U_{I_1}[\text{M-1}]U_{I_2}[\text{M-1}]\ldots U_{I_n}[\text{M-1}] \rightarrow \delta_{M+1}T_M U_{I_1}U_{I_2}\ldots U_{I_n}\}$

$\cup\{\delta_{M+1}{\rightarrow}\lambda\}$

$R_2 = \{U_{I_1}U_{I_2}\ldots U_{I_n} \rightarrow U_{I_1}U_{I_2}\ldots U_{I_n}C_{I_1}C_{I_2}\ldots C_{I_n}\}$

R_3:

$r_{31} = \{C_{I_1}C_{I_2}\ldots C_{I_n} \rightarrow C^1 C^2 \ldots C^k \mathrm{H}, \ 1{\leq}k{\leq}K\}$

$r_{32} = \{\eta_1 C^1 C^2 \ldots C^k \rightarrow \eta_1 [C^k]_k, \ 1{\leq}k{\leq}K\}$

$R_4 = \{U_{I_1}U_{I_2}\ldots U_{I_n} \rightarrow X_{I_1}X_{I_2}\ldots X_{I_n}\}$

R_5:

$r_{51} = \{\mathrm{D} \rightarrow D_1\}$

$r_{52} = \{D_1 \rightarrow D_{11}D_{12}\ldots D_{1M}\}$

R_6:

$r_{61} = \{I_1 I_2 \ldots I_n D_{11}D_{12}\ldots D_{1M} \rightarrow V_{I_j}[\text{s}]D_2, \ 1{\leq}j{\leq}n\}$

$r_{62} = \{I_1 I_2 \ldots I_n D_{1M}D_{1,M-1}\ldots D_{11} \rightarrow D_2 W_{I_j}[\text{f}], \ 1{\leq}j{\leq}n\}$

$r_{63} = \{D_2 \rightarrow D_2^1 D_2^2 \ldots D_2^N, \ 1{\leq}N{\leq}n\}$

$r_{64} = \{D_2^1 D_2^2 \ldots D_2^N \rightarrow b_{I_j}[\text{<s,c>}], \ 1{\leq}j{\leq}n\}$

$r_{65} = \{V_{I_j}[\text{s}]W_{I_j}[\text{f}]I_j[\text{<s,c>}] \rightarrow B_{I_j}^1[\text{<s,f,c>}] \ B_{I_j}^2[\text{<s,f,c>}]\ldots B_{I_j}^n[\text{<s,f,c>}],$
$\quad 1{\leq}j{\leq}n\}$

$r_{66} = \{B_{I_j}^1[\text{<s,f,c>}]B_{I_j}^2[\text{<s,f,c>}]\ldots B_{I_j}^n[\text{<s,f,c>}] \rightarrow BL_{I_j}[B^1 B^2 \ldots B^n]C_{I_j},$
$\quad 1{\leq}j{\leq}n\}$

R_7:

$r_{71} = \{I_j \rightarrow I_j(X_{I_j}, I_j)_{in_k}, C_{I_j} = C^k, 1 \leq j \leq n, 1 \leq k \leq K\}$

$r_{72} = \{X_{I_j} I_j \rightarrow \lambda, C_{I_j} < H\}$

R_8:

$r_{81} = \{BL_{I_{j_1}} BL_{I_{j_2}} \rightarrow BL_{I_{j_1}} BL_{I_{j_2}} \theta_{I_{j_1 j_2}}, 1 \leq j_1 \leq j_2 \leq n\}$

$r_{82} = \{I_{j_1} I_{j_2} BL_{I_{j_1}} BL_{I_{j_2}} \rightarrow BL_{I_{j_1 j_2}} I_{j_1 j_2} | \neg \theta_{I_{j_1 j_2}} C_{I_{j_1 j_2}}, 1 \leq j_1 \leq j_2 \leq n\}$

$r_{83} = \{X_{I_{j_1}} X_{I_{j_2}} \rightarrow X_{I_{j_1 j_2}} | \neg \theta_{I_{j_1 j_2}}, 1 \leq j_1 \leq j_2 \leq n\}$

$r_{84} = \{I_{j_1 j_2} X_{I_{j_1 j_2}} \rightarrow \lambda, C_{I_{j_1 j_2}} < H, 1 \leq j_1 \leq j_2 \leq n\}$

$R_9 = \{I_{j_1 j_2} \rightarrow L_2(X_{j_1 j_2} I_{j_1 j_2})_{in_k}, C_{I_{j_1 j_2}} = C^k, 1 \leq j_1 \leq j_2 \leq n, 1 \leq k \leq K\}$

$R_{10} = \{L_2 \rightarrow L_2^1 L_2^2 \ldots L_2^h\}$

$R_{11} = \{\eta_1 L_2^i \rightarrow (\eta_1 L_2^i)_i, 1 \leq i \leq h\}$

$R_{12} = \{\eta_2 \rightarrow (BL_{I_{j_1 j_2}} X_{I_{j_1 j_2}})_{in_i}, 1 \leq i \leq h\}$

Evolution rules in cell membrane $i (1 \leq i \leq h)$:

$r_1' = \{L_t^i \rightarrow L_t^{i,1} L_t^{i,2} \ldots L_t^{i,x}, 1 \leq i \leq h\}$

$r_2' = \{L_t^{i,u} \rightarrow \lambda, y = 1, 2 \leq t, 1 \leq i \leq h, 1 \leq u \leq x\}$

$r_3' = \{BL_{I_{j_1 j_2 \ldots j_m}} BL_{I_{j_1 j_2 \ldots j_{m-1} j_{m+1}}} \rightarrow BL_{I_{j_1 j_2 \ldots j_m}} BL_{I_{j_1 j_2 \ldots j_{m-1} j_{m+1}}}$
$\theta_{I_{j_1 j_2 \ldots j_{m+1}}}, 1 \leq j_1 \leq \ldots \leq j_{m+1} \leq n\}$

$r_4' = \{BL_{I_{j_1 j_2 \ldots j_m}} BL_{I_{j_1 j_2 \ldots j_{m-1} j_{m+1}}} I_{j_1 j_2 \ldots j_m} I_{j_1 j_2 \ldots j_{m-1} j_{m+1}} \rightarrow$
$BL_{I_{j_1 j_2 \ldots j_{m+1}}} I_{j_1 j_2 \ldots j_{m+1}} | \neg \theta_{I_{j_1 j_2 \ldots j_{m+1}}} C_{I_{j_1 j_2 \ldots j_{m+1}}}, 1 \leq j_1 \leq \ldots \leq j_{m+1} \leq n\}$

$r_5' = \{X_{I_{j_1 j_2 \ldots j_m}} X_{I_{j_1 j_2 \ldots j_{m-1} j_{m+1}}} \rightarrow X_{I_{j_1 j_2 \ldots j_{m+1}}} | \neg \theta_{I_{j_1 j_2 \ldots j_{m+1}}},$
$1 \leq j_1 \leq \ldots \leq j_{m+1} \leq n\}$

$r_6' = \{I_{j_1 j_2 \ldots j_{m+1}} X_{I_{j_1 j_2 \ldots j_{m+1}}} \rightarrow \lambda, C_{I_{j_1 j_2 \ldots j_{m+1}}} < H,$
$1 \leq j_1 \leq \ldots \leq j_{m+1} \leq n\}$

$r_7' = \{I_{j_1 j_2 \ldots j_{m+1}} X_{I_{j_1 j_2 \ldots j_{m+1}}} \rightarrow L_{m+1}^i (I_{j_1 j_2 \ldots j_{m+1}} X_{I_{j_1 j_2 \ldots j_{m+1}}})_{in_k},$
$C_{I_{j_1 j_2 \ldots j_{m+1}}} = C^k\}$

Evolution rules in cell membrane $c^k (1 \leq k \leq K)$:

$r_1^k = \{I_{j_1} I_{j_1 j_2} \ldots I_{j_1 j_2 \ldots j_m} X_{I_{j_1}} X_{I_{j_1 j_2}} \ldots X_{I_{j_1 j_2 \ldots j_m}} \rightarrow I_{j_1 j_2} \ldots I_{j_1 j_2 \ldots j_m},$
$1 \leq j_1 \leq \ldots \leq j_{m+1} \leq n\}$

$r_2^k = \{I_{j_1} I_{j_1 j_2} \ldots I_{j_1 j_2 \ldots j_m} \rightarrow (P^k, out), 1 \leq j_1 \leq \ldots \leq j_m \leq n\}$

The order of evolution rules is as follows:

In cell membrane 0:

$\rho_0 = \{R_i > R_j, r_{ij_1} > r_{ij_2}, 1 \leq i < j \leq 12, 1 \leq j_1 < j_2 \leq 6\}$

In cell membrane $i (1 \leq i \leq h)$:

$\rho_i = \{r_i' > r_j', 1 \leq i < j \leq 8\}$

In cell membrane $k (1 \leq k \leq K)$:

$\rho_k = \{r_1^k > r_2^k\}$

After the algorithm is finished, the structure and state of the cell-like P system are presented in Fig. 2.

3.3 Computing Process

The cell-like P system designed to implement improved BTK algorithm has two membranes: 0 and h_0, and membrane 0 is the skin membrane, h_0 is the membrane to output final result.

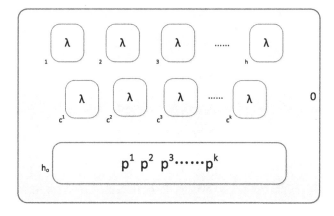

Fig. 2. The P system when algorithm finished

In the initial P system, rule R_1 in the skin membrane is executed first to traverse the database D, generate the object $U_{I_j}(1\leq j\leq n)$, and $U_{I_j}[i]$ represents the set of transactions T_i which include the item I_j. When the object δ_{M+1} appears, the database scanning process over. Rule R_2 is executed to calculate the number of transactions in the set $U_{I_j}(1\leq j\leq n)$, the count of the item I_j, recorded as C_{I_j}. Rule r_{31} is executed to arrange all C_{I_j} in descending order and get the first K counts recorded as $C^k(1\leq k\leq K)$, K is the user-specified top rank value, threshold H is equal to C^K. The object η_1 actives rule r_{32} to create the membrane 1, 2...K. Rule R_4 generates the subsume of 1-pattern according to the inclusion relation between $U_{I_j}(1\leq j\leq n)$. Rule r_{51} updates database D to D_1 which items and transactions sorted in descending order and rule r_{52} divides items in database to $D_{11}, D_{12}...D_{1m}$ by row. All rules in R_6 are used to generate B-lists of $I_j(1\leq j\leq n)$ denoted as BL_{I_j}. First label the items in D_{11} sequentially, then cycle to compare D_{1i} to $D_{1i-1}(1\leq i\leq m)$ from the first item and start marking from the first different item, finally get start-build value of each item in TB-tree. $V_{I_j}[s]$ represents start-build value sets of item $I_j(1\leq j\leq n)$. r_{62} is a rule that execute rule r_{61} in reverse order, first label the items in D_{1m} in reverse order, then cycle to compare D_{1i} to $D_{1i+1}(1\leq i\leq m)$ and start marking from the first different item, finally get the finish-build value of each item in TB-tree. $W_{I_j}[f]$ represents finish-build value sets of item $I_j(1\leq j\leq n)$. Moreover, rule r_{61} updates the database D_1 to D_2 which items are marked with start-build number, and then rule r_{63} divides database D_2 to $D_2^1, D_2^2...D_2^n$ by column. Rule r_{64} is executed to get the count of items which are marked in database D_2 and generate $b_{I_j}[<s,c>]$ which represents the set of two-tuples $<$start_build number, count$>$ of I_j. For each marked item in $D_2^1, D_2^2...D_2^n$, counting from itself until meets next marked item or the last item, get the count of I_j. Rule r_{65} is executed to get the B-info-code of each item in TB-tree, merge $b_{I_j}[<s,c>]$ and $W_{I_j}[f]$ with order to get $B_{I_j}[<s,f,c>]$. Rule r_{66} is executed to assemble all B_{I_j} to generate B-List of $I_j(1\leq j\leq n)$, recorded as BL_{I_j}, C_{I_j} represents the sum of counts of each B_{I_j}.

Table 3. The database of the illustrative instance

TID	Items
T100	I_2 I_4
T200	I_4 I_5 I_2
T300	I_3 I_4 I_5
T400	I_2 I_4 I_1
T500	I_5 I_3 I_2 I_4
T600	I_4 I_5
T700	I_2 I_1
T800	I_4 I_2 I_5 I_1

Next, rule R_7 is executed, if $C_{I_j}(1 \leq j \leq n)$ is equal to $C^k(1 \leq k \leq K)$, item I_j will be sent to membrane k, and if C_{I_j} is less than threshold H, item I_j and the subsume X_{I_j} will be deleted by rule r_{72}. Next, rule r_{81} is executed to check the B-Lists in membrane 0 through EP strategy. First find the B-info-code B_w that does not have ancestor-descendant relationship with any other nodes in two input B-Lists. And then if B_w is in the first list, let $C_{I_{j_1}} = C_{I_{j_1}} - B_w$.count. If not, let $C_{I_{j_2}} = C_{I_{j_2}} - B_w$.count. Object $\theta_{I_{j_1 j_2}}$ will be generated when there are C_{I_j} not larger than threshold H. Rule r_{82} is executed to generate $BL_{I_{j_1 j_2}}$ by intersecting B-Lists. Finding nodes that have ancestor-descendant relationship, if $B_{I_{j_1}}$ is an ancestor of $B_{I_{j_2}}$, add $([B_{I_{j_1}}.s, B_{I_{j_1}}.f], B_{I_{j_2}}.c)$ to $BL_{I_{j_1 j_2}}$. If there are B-info-codes in $BL_{I_{j_1 j_2}}$ which have same start-build and finish-build, turn them into one B-info-code. Object $\theta_{I_{j_1 j_2}}$ is the inhibitor of B-Lists intersection, and $C_{I_{j_1 j_2}}$ is the total count of $BL_{I_{j_1 j_2}}$. If $C_{I_{j_1 j_2}}(1 \leq j_1 < j_2 \leq n)$ is larger than the established threshold, check if it needs to be updated. If the threshold is updated, corresponding C^k and membrane c^k also need updating and rule R_7 will be executed in membrane k($1 \leq k \leq K$). The subsume is generated by rule r_{83}, then subsume and patterns which sum count less than threshold will be deleted by rule r_{84}. In rule R_9, if $C_{I_{j_1 j_2}}(1 \leq j \leq n)$ is equal to $C^k(1 \leq k \leq K)$, the itemset $I_{j_1 j_2}$ will be sent to membrane k, and the set of frequent 2-patterns L_2 will be generated. Rule r_{10} is executed to divide L_2 into L_2^1, $L_2^2 \ldots L_2^h$, assume the first item of L_2 has h species, and there is only one item different in each group. Object η_1 activates rule R_{11} to create h membranes and sends L_2^1, $L_2^2 \ldots L_2^h$ to corresponding membrane i($1 \leq i \leq h$). Meanwhile, object η_2 activates rule R_{12} sends $BL_{I_{j_1 j_2}}$ and $X_{I_{j_1 j_2}}$ $(1 \leq j_1 < j_2 \leq n)$ to Corresponding membrane i($1 \leq i \leq h$), activates the evolution rules in cell membrane i($1 \leq i \leq h$).

For the frequent t-patterns set $L_t^i(t \geq 2)$ in membrane i($1 \leq i \leq h$), rule r_1' is executed to divide L_t^i into $L_t^{i,1}$, $L_t^{i,2} \ldots L_t^{i,x}$(the number of groups is x, and the number of frequent t-patterns in each group is y). There is only one item different in each group, and they are not belong to subsume of each other. If there is only one pattern in group $L_t^{i,u}(1 \leq u \leq x)$, execute r_2' to eliminate this group. If there are more than one pattern, execute r_3' and r_4' to check and intersect B-Lists in each group just like the way in r_{81} and r_{82}, and get the B-Lists of $(t+1)$-patterns. If $C_{I_{j_1 j_2 \cdots j_{m+1}}}$ $(1 \leq j_1 < j_2 < j_{m+1} \leq n)$ is greater than the established threshold, check if it needs to be updated. If the threshold is updated, corresponding C^k and membrane c^k also need updating and rule r_6' and r_7' will be executed in membrane k$(1 \leq k \leq K)$. The subsume of $(t+1)$-patterns is generated by rule r_5', then subsume and patterns which total count less than threshold will be eliminated by rule r_6'. In rule r_7', if $C_{I_{j_1 j_2 \cdots j_{m+1}}}$ $(1 \leq j_1 \leq j_2 \leq j_{m+1} \leq n)$ is equal to $C^k(1 \leq k \leq K)$, the itemset $I_{j_1 j_2 \cdots j_{m+1}}$ will be sent to membrane k, and the set of frequent $(t+1)$-patterns L_{t+1} will be generated. When the number of groups in the membrane i is $0(x = 0)$, the object φ is generated to activate rules in membrane $c^k(1 \leq k \leq K)$.

Table 4. The groups in updated D_1

Groups	Items
D_{11}	$I_2\ I_1$
D_{12}	$I_4\ I_5\ I_3$
D_{13}	$I_4\ I_5$
D_{14}	$I_4\ I_2\ I_1$
D_{15}	$I_4\ I_2\ I_5\ I_3$
D_{16}	$I_4\ I_2\ I_5\ I_1$
D_{17}	$I_4\ I_2\ I_5$
D_{18}	$I_4\ I_2$

In membrane $c^k(1 \leq k \leq K)$, r_1^k is executed to find rest frequent patterns by subsume, combine patterns with all subsets generated from subsume of itself. Finally, rule r_2^k is executed to generate $P^k(1 \leq k \leq K)$ which contains all patterns in membrane $c^k(1 \leq k \leq K)$ and output them, the computation halts.

After the algorithm is finished, the final results, top-rank-k frequent patterns are stored in membrane h_0.

4 An Illustrative Example

To give a clear demonstration of how CP-BTK works, this section gives an instance to describe the execution of algorithm and verify its correctness and effectiveness. Table 3 shows the database used in example and given k = 4.

Table 5. The groups in updated D_2

Groups	Items
D_2^1	$I_2^2 I_4^3 I_4 I_4 I_4 I_4 I_4^{10}$
D_2^2	$I_1^1 I_5 I_5^4 I_2 I_2 I_2 I_2^9$
D_2^3	$I_3 I_1^5 I_5 I_5 I_5^8$
D_2^4	$I_3^6 I_1^7$

Table 6. The B-list of 1-pattern

1-pattern	B-list of 1-pattern	Count
I_1	<2,1,1> <7,5,1><10,7,1>	3
I_2	<1,2,1><6,9,5>	6
I_3	<5,3,1><9,6,1>	2
I_4	<3,10,7>	7
I_5	<4,4,2><8,8,3>	5

In the initial P system, Rule R_1 is executed first to traverse the database D, check if item I_j exists in transaction T_i, generate the objects: $U_{I_1} = \{T_4,T_7,T_8\}$, $U_{I_2} = \{T_1,T_2,T_4,T_5,T_7,T_8\}$, $U_{I_3} = \{T_3,T_5\}$, $U_{I_4} = \{T_1,T_2,T_3,T_4,T_5,T_6,T_8\}$, $U_{I_5} = \{T_2,T_3,T_5,T_6,T_8\}$. Rule R_2 records the count of each item through U_{I_j}, obtains $C_{I_1} = 3, C_{I_2} = 6, C_{I_3} = 2, C_{I_4} = 7, C_{I_5} = 5$. Through rule R_3 CP-BTK gets $C^1 = 7, C^2 = 6, C^3 = 5, C^4 = 3$, H = 3, and 4 membranes are created. According to the inclusion relation between U_{I_j} obtains the subsume of items, subsume(I_3) is I_4 and I_5, subsume(I_4) is I_5. Updated database by rule R_5 is shown in Table 4. Then R_6 is executed to obtain the B-List of items, $V_{I_1} = \{2,7,10\}$, $V_{I_2} = \{1,6\}$, $V_{I_3} = \{5,9\}$, $V_{I_4} = \{3\}$, $V_{I_5} = \{4,8\}$ and $W_{I_1} = \{1,5,7\}$, $W_{I_2} = \{2,9\}$, $W_{I_3} = \{3,6\}$, $W_{I_4} = \{10\}$, $W_{I_5} = \{4,8\}$, D_2 which divides by column is shown in Table 5. Rule r_{64} is executed to obtain $b_{I_1} = \{<2,1><7,1><10,1>\}$, $b_{I_2} = \{<1,1><6,5>\}$, $b_{I_3} = \{<5,1><9,1>\}$, $b_{I_4} = \{<3,7>\}$, $b_{I_5} = \{<4,2><8,3>\}$. Then $b_{I_j}[<s,c>]$ and $W_{I_j}[f]$ are merged with order and rule r_{66} is executed to assemble all the B-info-codes to generate B-List of $I_j(1 \le j \le n)$, the result is presented in Table 6.

Table 7. The B-list of 2-pattern

2-pattern	B-list of 2-pattern	Count
I_{12}	<1,2,1><6,9,2>	3
I_{14}	<3,10,1>	1
I_{15}	<8,8,1>	2
I_{24}	<3,10,5>	5
I_{25}	<6,9,3>	3

Rule r_{71} is executed to send I_4, I_2, I_5, I_1 to membrane c^1, c^2, c^3, c^4 respectively and send X_{I_4} to membrane c^1. I_3 and X_{I_3} are deleted by rule r_{72} because C_{I_3} is less than 3. Rule r_{81} is executed to check B-Lists in membrane 0 through EP strategy and generate object $\theta_{I_{j_1 j_2}}$ when there are counts not larger than 3. Rule r_{82} is executed to intersect B-Lists and generate B-lists of 2-patterns, the result is presented in Table 7. The subsume is generated by rule r_{83}, both $X_{I_{14}}$ and $X_{I_{24}}$ are I_5. $BL_{I_{14}}$, $BL_{I_{15}}$ and $X_{I_{14}}$ are deleted by rule r_{84}. Rule R_9 sends

I_{24} to membrane c^3, I_{12} and I_{25} to membrane c^4 and generates L_2 which contains all frequent 2-patterns. According to rule R_10, I_{24}, I_{12} and I_{25} are divided into one group L_2^1 and generate membrane 1. Then rules in membrane 1 are triggered.

Table 8. The final results stored in membrane h_0

k	Patterns	Count
1	I_4	7
2	I_2	6
3	$I_5\ I_{24}I_{45}$	5
4	$I_1I_{12}I_{25}I_{245}$	3

In membrane 1, rule r_1' is executed to divide L_2 to $L_2^{(1,1)} = \{I_{12}, I_{25}, I_{24}\}$(x = 1, y = 3). Rule r_3' and r_4' are executed to check and intersect B-Lists and get the B-Lists of 3-pattern, $BL_{I_{124}} = \{<3,10,2>\}$. The subsume of 3-patterns is generated by rule r_5', $X_{I_{124}} = I_5$. After,$BL_{I_{124}}$ and $X_{I_{124}}$ are deleted by rule r_6' because count is less than 3. At this time, the number of groups in membrane 1 is 0, object φ is generated to activate rules in the membrane $c^k(1{\leq}k{\leq}K)$.

In membrane c^3, r_1^3 is executed to generate I_{45} and I_{245}. Finally, rule r_2^3 is executed to generate object P^k which stores the final results and output them, the computation ends. The final results are presented in Table 8.

5   Conclusions

Membrane computing is initiated as a theoretical branch of natural computing, inspired from the structure and functioning of the biological cell. This paper introduces a cell-like P system with active membranes to implement BTK algorithm. The CP-BTK algorithm makes two improvements to mine subsume and determine threshold more efficient, creates h membranes to mine frequent t-patterns($1{<}t{\leq}n$) in parallel by rules in each membrane. Finally, an instance is given to illustrate the feasibility and effectiveness of the proposed algorithm based on cell-like P system, result shows that the CP-BTK algorithm is efficient and precise. Because the parallel mechanism used in membrane computing models, time consumption and algorithm efficiency are improved greatly. For further research, it can use some other P system to improve the BTK and other data mining algorithms.

Acknowledgments. Project is supported by National Natural Science Foundation of China(nos.61472231, 61502283, 61640201, 61170038), Social Science Foundation of Shandong Province, China (nos.16BGLJ06, 11CGLJ22), Ministry of Education of Humanities and Social Science Research Project, China (12YJA630152).

References

1. Agrawal, R., Imielinski, T., Swami, A.: Mining association rules between sets of items in large databases. SIGMOD Rec. **22**(2), 207–216 (1993)
2. Han, J.W., Pei, J., Yin, Y.W.: Mining frequent patterns without candidate generation:a frequent-pattern tree approach. Data Min. Knowl. Discov. **8**(1), 53–87 (2004)
3. Tasy, Y.J., Chiang, J.Y.: Cbar: an efficient method for mining association rules. Knowl.-Based Syst. **18**(2–3), 99–105 (2005)
4. Zaki, M.J., Gouda, K.: Fast vertical mining using diffsets. In: ACM SIGKDD International Conference on Knowledge Discovery and Data Mining, pp. 326–335 (2003)
5. Zaki, M.J., Parthasarathy, S., Ogihara, M., Li, W.: New algorithms for fast discovery of association rules. In: Proceedings of the Third International Conference on Knowledge Discovery and Data Mining, pp. 283–286 (1997)
6. Dam, T.L., Li, K., Fournier-Viger, P., et al.: An efficient algorithm for mining top-rank-k frequent patterns. Appl. Intell. **45**(1), 96–111 (2016)
7. Deng, Z.H.: Fast mining top-rank-k frequent patterns by using node-lists. Exp. Syst. Appl. **41**(4), 1763–1768 (2014)
8. Deng, Z.H.: Fast mining frequent itemsets using nodesets. Exp. Syst. Appl. **41**(10), 4505–4512 (2014)
9. Deng, Z.H., Wang, Z.H., Jiang, J.J.: A new algorithm for fast mining frequent itemsets using n-lists. Sci. China Inf. Sci. **55**(9), 2008–2030 (2012)
10. Păun, G.: Computing with membranes. Comput. Syst. Sci. **61**(1), 108–143 (2000)
11. Păun, G., Rozenberg, G., Salomaa, A.: The Oxford Handbook of Membrane Computing. Oxford University Press, Oxford (2010)
12. Wang, J., Shi, P., Peng, H.: Membrane computing model for iir filter design. Inf. Sci. **329**, 164–176 (2016)
13. Singh, G., Deep, K.: A new membrane algorithm using the rules of particle swarm optimization incorporated within the framework of cell-like p-systems to solve sudoku. Appl. Soft Comput. **45**, 27–39 (2016)
14. Zhang, G., Rong, H., Cheng, J., Qin, Y.: A population membrane-system-inspired evolutionary algorithm for distribution network reconfiguration. Chin. J. Electron. **23**(3), 437–441 (2014)
15. Deng, Z.H., Fang, G.D.: Mining top-rank-k frequent patterns. In: International Conference on Machine Learning and Cybernetics, pp. 851–856 (2007)
16. Song, W., Yang, B.R., Xu, Z.Y.: Index-bit table fi: an improved algorithm for mining frequent itemsets. Knowl.-Based Syst. **21**(6), 507–513 (2008)
17. Păun, G.: Membrane Computing. Springer, Berlin (2002). https://doi.org/10.1007/978-3-642-56196-2
18. Zandron, C., Ferretti, C., Mauri G.: Solving np-complete problems using p systems with active membranes. In: International Conference on Unconventional MODELS of Computation, pp. 289–301 (2000)
19. Xiyu, L., Yuzhen, Z., Minghe, S.: An improved apriori algorithm based on an evolution-communication tissue-like p system with promoters and inhibitors. Discret. Dyn. Nat. Soc. **2017**(1), 1–11 (2017)
20. Păun, G., Suzuki, Y., Tanaka, H.: On the power of membrane division in p systems. Theor. Comput. Sci. **324**(1), 61–85 (2004)

A Simple Algorithmic Proof of the Symmetric Lopsided Lovász Local Lemma

Lefteris Kirousis(iD) and John Livieratos[(⊠)](iD)

Department of Mathematics, National and Kapodistrian University of Athens,
Athens, Greece
{lkirousis,jlivier89}@math.uoa.gr

Abstract. We provide a simple algorithmic proof for the symmetric *Lopsided* Lovász Local Lemma, a variant of the classic Lovász Local Lemma, where, roughly, only the degree of the negatively correlated undesirable events counts. Our analysis refers to the algorithm by Moser (2009), however it is based on a simple application of the probabilistic method, rather than a counting argument, as are most of the analyses of algorithms for variants of the Lovász Local Lemma.

Introduction

The Lovász Local Lemma (LLL) first appeared in 1975 in a paper by Erdős and Lovász [1]. In its original form, LLL provides a sufficient condition for the possibility to avoid a number of undesirable events given that there is a small constant upper bound on the product of the maximum probability of the events with the maximum number of other events any given event depends on. This form is known as the *symmetric* LLL.

There is also a more general, *non-symmetric*, form where the sufficient condition asks for the existence of a family of numbers in $(0, 1)$, one for each event, such that for each event E its probability is bounded from above by a specified expression of the numbers that correspond to the neighbors of E. See e.g., the exposition paper by Szegedy [2] for the necessary background.

The proofs of these versions were non-constructive. The first algorithmic proof, that entailed an unexpectedly simple algorithm, was given much later by Moser [3] (work on constructive approaches was previously done by Alon [4], Beck [5], Srinivasan [6] and others). The initial proof of Moser was only for the application of LLL in the satisfiability problem. It was based on a counting argument that became known as the "entropic method": one bounds the entropy (cardinality) of structures that witness that the algorithm lasts for too many steps (see [7]). For an elegant presentation of this result based on a direct probabilistic proof, see Spencer [8]. Closely related to Spencer's proof is the proof in Giotis et al. [9].

The so called *lopsided* version of the Lovász Local Lemma (LLLL), where dependency of undesirable events is restricted to negative correlation, was proved, in a non-constructive way, by Erdős and Spencer in 1991 [10]. This

© Springer Nature Switzerland AG 2019
R. Battiti et al. (Eds.): LION 12 2018, LNCS 11353, pp. 49–63, 2019.
https://doi.org/10.1007/978-3-030-05348-2_5

version as well has a more general non-symmetric variant and has been used in, among others, the satisfiability problem: instead of declaring two clauses of a formula dependent on each other when they share a variable, one can consider clauses that are in conflict with each other (i.e. have opposing literals of the same variable). Berman et al. [11] and Gebauer et al. [12,13] have successfully used it to bound the number of occurrences per variable that can be allowed in a formula, while guaranteeing the existence of a satisfying assignment.

An algorithmic proof for the non-symmetric form, of both non-lopsided and lopsided versions, was given by Moser and Tardos [14]. The proof assumed that the events are expressed in terms of a family of independent variables. Specifically, it is assumed that each event depends on a subset of the independent variables that comprises the *scope* of the event; the dependency graph is given in terms of how the scopes of the events are related. This proof as well is based on the "entropic method".

An interesting application of both the symmetric and asymmetric LLL is in *covering arrays*, a problem closely related with software and hardware interaction testing. The objective is to find the *minimum* number N of rows for an $N \times k$ array A, with elements taken from a set Σ of cardinality $v \geq 2$, to exist, such that every $N \times t$ sub-array of A contains any element $x \in \Sigma^t$ as one of its rows. Sarkar and Colbourn [15] improve on known upper bounds for N, by using both of the aforementioned versions of the LLL, in conjunction with other known techniques. Notably, they also provide an algorithm that constructs an $N \times k$ array with the above properties, by using a variant of the Moser–Tardos algorithm [14].

Kolipaka and Szegedy [16] describe an algorithmic approach where the events are not assumed to be expressed in terms of independent variables. In their framework, they prove the most general form of the LLL, the Shearer version [17], which provides a necessary and sufficient condition for avoiding a number of undesirable events.

More recently, Harvey and Vondrák [18] provided an algorithmic proof of the general non-symmetric form of lopsided LLL, also without assuming that the events are expressed in terms of variables. Their proof was through the Shearer version.

It is worth pointing out that very recently Harris [19] claimed that the lopsided version of LLL in the variable framework can be stronger than the general Shearer version, if a suitable dependency graph is defined in the variable framework (this variable version was analyzed by Harris in [19], by the entropic method). The crucial point in their proof is that the dependency graph they define is essentially directed; directed dependency graphs cannot be defined in the framework without variables.

In this work, we deal with the lopsided LLL in the variable framework. We introduce a simple notion of *directed* dependency graph inspired by the undirected notion of lopsidependency of Gebauer et al. [13]. Our proof is based on the probabilistic method and not on a counting argument. Also it does not necessitate proving first the complex (but general) version of Shearer. However, we restricted our analysis to the symmetric case, where the product of the maximum probability over all events times the outgoing degree of the directed dependency graph (incremented by one) is assumed to be bounded by a constant. Although this version is less general than the non-symmetric versions, the lack of generality is compensated, we believe, by the simplicity of the analysis, which reflects the elegance of the algorithm of Moser [3] that we use. We consider this simplicity in comparison to some of the extant work as the primary contribution of this work; it is an outcome of using the probabilistic method (Spencer's first Durango lecture [20] vividly describes the power of this method). We also consider interesting the introduction of a simple variable-dependent directed notion of lopsidependency in Sect. 1.2; a notion that constitutes a further indication that for some applications, the variable framework might be advantageous over the no-variables framework of general probability spaces. Achlioptas and Iliopoulos [21] present a completely abstract framework for LLL, interestingly not in terms of probability spaces, that covers also the lopsided version with a directed dependency graph; for the proof they use counting arguments (entropic method).

Let us also point out that the difference of the algorithm by Moser [3], which we use, from the subsequent one of Moser and Tardos [14] is that the former, at each phase, resamples an occurring event that is dependent on the previously examined one, in contrast with the latter, which at each phase resamples an arbitrary event not necessarily depending on the previously examined one.

In Sect. 1.1, we formally present the variable framework. In Sect. 1.2, we formally present the various lopsidependency notions and show how they are related. Finally in Sect. 2, we present our main result.

1 Preliminaries

1.1 The Variable Framework

We will work in what is known as the *variable framework*, which was first used in [3,14]. Let X_i, $i = 1, ..., l$ be mutually independent random variables on a common probability space, taking values in the finite sets D_i, $i = 1, ..., l$, respectively. An assignment of values to the random variables is an l-ary vector $\alpha = (a_1, ..., a_l)$, where $a_i \in D_i$, for all $i \in \{1, ..., l\}$. Let $\Omega = \prod_{i=1}^{l} D_i$ be the probability space of all these assignments of values.

Let E_j, $j = 1, ..., m$, be events defined on the common probability space Ω. The *scope* $\mathrm{sc}(E_j)$ of an event E_j is the minimal subset of variables such that one can determine whether E_j is satisfied or not knowing only their values, i.e., event E_j depends only on the values of the variables of $\mathrm{sc}(E_j)$. Events are assumed to be ordered according to their index.

For each event E_j, we also define $\overline{E}_j := \Omega \setminus E_j$ to be the event that occurs *if and only if* E_j does not. These events are introduced only for notational purposes.

The events E_j are considered "undesirable", i.e., the objective is to design a randomized algorithm that will return an assignment of values α for which none of the events E_j holds.

1.2   Lopsidependent Events

We will first give a notion of dependency between a pair of events, that we call *Variable-dependent Directed Lopsidependency*, VDL in short (depending on the context, VDL may also stand for "Variable-dependent Directed Lopsidependent" –an adjective rather than a noun). Then, we will explain how this notion is related with the classical notion of *lopsidependency*, given my Moser and Tardos [14].

Definition 1. *Let E_i, E_j be events, $i, j \in \{1, ..., m\}$. We say that E_j is VDL on E_i if:*

1. *there exists an assignment α under which E_i and $\overline{E}_j$ occur and*
2. *there exists an assignment β that differs from α only in the values of variables in sc(E_i), under which E_j occurs.*

Intuitively, E_j is VDL on E_i if the effort to undo the undesirable E_i results in E_j occurring, although it did not before.

Notice that an event E_j can never be VDL on itself, nor on any event whose scope shares no variables with sc(E_j). The binary relation VDL defines a simple *directed* graph $G = (V, E)$, the *VDL graph* of events $E_1, ..., E_m$, where $V = \{1, ..., m\}$ and $E = \{(i, j) \mid E_j$ is VDL on $E_i\}$.

For $i = 1, ..., m$, let $\Gamma(E_i)$ be the outwards neighborhood of the event E_i in the VDL graph, i.e. $\Gamma(E_i) = \{E_j \mid E_j$ is VDL on $E_i\}$. The notion of VDL was inspired by the following *undirected* notion of Moser and Tardos [14]:

Definition 2 (Moser and Tardos [14]). *Let E_i, E_j be events, $i, j \in \{1, ..., m\}$. We say that E_i, E_j are lopsidependent if there exist two assignments α, β, that differ only on variables in sc(E_i) ∩ sc(E_j), such that:*

1. *α makes E_i occur and β makes E_j occur and*
2. *either $\overline{E}_i$ occurs under β or $\overline{E}_j$ occurs under α.*

Notice again that an event E_j is never lopsidependent on itself, nor on any event whose scope shares no variables with sc(E_j).

We now show, by means of Example 1, that the relation of VDL is indeed a refinement of that of lopsidependency.

Example 1. Suppose we have five *independent Bernoulli trials*, each represented by a random variable X_i, $i = 1, 2, 3, 4, 5$, with probability of success ($X_i = 1$) equal to $x \in [0, 1)$.

Consider the following events we want to avoid:

$$E_1 = \{(X_1 = 0 \vee X_3 = 1) \wedge X_2 = 0\},$$
$$E_2 = \{X_3 = 1 \wedge X_4 = 0\} \text{ and}$$
$$E_3 = \{X_4 = 1 \wedge X_5 = 0\}.$$

Consider the assignments $\alpha = (0, 0, 0, 0, 0)$ and $\beta = (0, 0, 1, 0, 0)$. Under α, E_1 occurs and E_2 does not. Also, β differs from α only in the value of X_3, which belongs in $\mathrm{sc}(E_1) \cap \mathrm{sc}(E_2)$ and, under it, E_2 occurs. By Definitions 1 and 2, E_2 is VDL on E_1 and E_1, E_2 are lopsidependent. By analogous arguments, it is easy to see that E_2 and E_3 are VDL on each other and lopsidependent.

On the other hand, E_1 is *not* VDL on E_2. To see this, we first restrict ourselves to the first four Bernoulli trials, since neither E_1 nor E_2 depend on X_5. Now, consider the four assignments under which E_2 occurs, namely:

$$(0,0,1,0), \ (0,1,1,0), \ (1,0,1,0), \ (1,1,0,0).$$

Under the first and third of the assignments above, E_1 occurs too. Thus, only the second and forth assignments satisfy the first condition of Definition 1. But, under both these assignments, $X_2 = 1$. Thus, under any assignment that differs from those only in variables in $\mathrm{sc}(E_2) = \{X_3, X_4\}$, $X_2 = 1$ and consequently, E_1 does not occur, failing the second condition of Definition 1.   ◇

We will now prove that two events are lopsidependent in the sense of Definition 2 if at least one is VDL on the other.

Claim. Two events E_i, E_j, $i, j \in \{1, ..., m\}$, are lopsidependent (in the sense of Definition 2), if and only if $E_j \in \Gamma(E_i)$ or $E_i \in \Gamma(E_j)$.

Proof. ($\Rightarrow$) By Definition 2, there exist assignments α, β that differ only on variables in $\mathrm{sc}(E_i) \cap \mathrm{sc}(E_j)$ such that E_i occurs under α, E_j occurs under β and either $\overline{E}_j$ occurs under α or $\overline{E}_i$ occurs under β.

It is immediate to see that if $\overline{E}_j$ occurs under α, $E_j \in \Gamma(E_i)$ and that, if $\overline{E}_i$ occurs under β, $E_i \in \Gamma(E_j)$.

($\Leftarrow$) Let $E_j \in \Gamma(E_i)$. Then, there are two assignments $\alpha = (a_1, ..., a_l), \beta = (b_1, ..., b_l)$ that differ only in $\mathrm{sc}(E_i)$, such that $E_i, \overline{E}_j$ occur under α and E_j occurs under β. If assignments α, β differed only in $\mathrm{sc}(E_i) \cap \mathrm{sc}(E_j)$, there would be nothing to prove.

Let the assignment $\beta' = (b'_1, ..., b'_l)$ be such that:

– $b'_r = a_r$, for all $r \in \{1, \ldots, l\}$ such that $X_r \notin \mathrm{sc}(E_i) \cap \mathrm{sc}(E_j)$ and
– $b'_r = b_r$, for all $r \in \{1, \ldots, l\}$ such that $X_r \in \mathrm{sc}(E_i) \cap \mathrm{sc}(E_j)$.

Since α differs from β only on variables in $\mathrm{sc}(E_i)$, it follows that β' differs from β only on variables in $\mathrm{sc}(E_i) \backslash \mathrm{sc}(E_j)$. Now, since E_j holds for β and does not depend on variables not in its scope, E_j holds under β' also. Thus, the assignments α, β' fulfill the requirements of Definition 2.

Similarly if $E_i \in \Gamma(E_j)$.   □

A direct consequence of the above claim, is that the lopsidependency undirected graph of the events under the binary relation of Definition 2 is the *undirected* graph that underlies their VDL graph.

Example 2. Consider the setting of Example 1. The leftmost graph below is the VDL graph of E_1, E_2, E_3 and the rightmost is their lopsidependency graph.

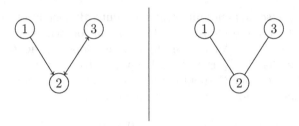

Note that, if $N(E_j)$ is the neighborhood of E_j in the lopsidependency graph (of Definition 2), $j = 1, 2, 3$, then $\Gamma(E_j) = N(E_j) = \{E_2\}$, $j = 1, 2$, but $\Gamma(E_2) = \{E_3\} \subseteq \{E_1, E_3\} = N(E_2)$.   ◇

Lastly, we consider the (non-directed) classical definition of lopsidependency formulated by Erdős and Spencer [10]:

Definition 3 (Erdős and Spencer [10]**).** *Let $E_1, \ldots, E_m$ be events in an arbitrary probability space, G a graph on the indices $1, \ldots, m$. We say that G is an Erdős–Spencer lopsidependency graph (for the events) if*

$$\Pr\left[E_j \mid \bigcap_{i \in I} \overline{E_i}\right] \leq \Pr[E_j]$$

for all j, I with $j \notin I$ and no $i \in I$ adjacent to j.

Observe that if two events are (strictly) negatively correlated, they should be connected by an edge in the Erdős–Spencer lopsidependency graph. Since the notion of negative correlation is symmetric, we conclude that it would not make sense to define a directed notion of lopsidependency in the general framework without variables.

We will now show that the underlying undirected graph of the VDL graph (in other words the lopsided undirected graph in the sense of Definition 2) is an Erdős–Spencer dependency graph. This fact shows that it might be advantageous for applications of LLLL to use the notion of VDL rather than the classical undirected Erdős–Spencer lopsidependency or the undirected lopsidependency of Moser and Tardos [14] (see Theorem 1 below).

Lemma 1. *For any event E_j, $j = 1, ..., m$, let I be a set of indices of events not in $\Gamma(E_j) \cup \{E_j\}$. Then, it holds that:*

$$Pr\left[E_j \mid \bigcap_{i \in I} \overline{E_i}\right] \leq Pr[E_j].$$

Proof. Let $E = \bigcap_{i \in I} \overline{E_i}$. Note that E is not necessarily some of the E_js (in fact if it is, then there is no assignment that avoids all the undesirable events). Since $E_i \notin \Gamma(E_j)$, for all $i \in I$, it holds that for any assignment α that makes E_j and E hold, there is no assignment β that differs from α only in $sc(E_j)$ that makes $\overline{E}$ hold.

Now, for the events E_j, E, it holds that:

$$Pr[E_j \mid E] \leq Pr[E_j] \Leftrightarrow \frac{Pr[E_j \cap E]}{Pr[E]} \leq Pr[E_j] \Leftrightarrow$$

$$Pr[E_j \cap E] \leq Pr[E_j] \cdot Pr[E].$$

Suppose now $\alpha = (a_1, ..., a_l), \beta = (b_1, ..., b_l)$ are two assignments obtained by independently sampling the random variables twice, once to get α and once to get β. It holds that:

$$Pr[\underbrace{(\alpha, \beta) \text{ are such that } \alpha \text{ makes } E_j, E \text{ happen}}_{\text{event } S}] = Pr[E_j \cap E].$$

Indeed, event S above imposes no restriction on β.

Let now $\alpha' = (a'_1, ..., a'_l), \beta' = (b'_1, ..., b'_l)$ be two assignments obtained by α, β by swapping values in variables in $\mathrm{sc}(E) \setminus \mathrm{sc}(E_j)$:

- $a'_i = b_i$, for all i such that $X_i \in \mathrm{sc}(E) \setminus \mathrm{sc}(E_j)$, $a'_i = a_i$ for the rest,
- $b'_i = a_i$, for all i such that $X_i \in \mathrm{sc}(E) \setminus \mathrm{sc}(E_j)$ and $b'_i = b_i$ for the rest.

Since $a'_i = a_i$ for all i such that $X_i \in \mathrm{sc}(E_j)$, α' makes E_j happen. Observe also that the coordinates of β' that correspond to variables in $\mathrm{sc}(E)$ and whose values are different than those of α, are all in $\mathrm{sc}(E_j) \cap \mathrm{sc}(E) \subseteq \mathrm{sc}(E_j)$. By the hypothesis, and since variables not in $\mathrm{sc}(E)$ do not influence E, if E_j and E hold under α, then E holds under β'.

Obviously α', β' are two independent samplings of all variables, since all individual variables were originally sampled independently, and we only changed the positioning of the individual variables. Thus, it holds that:

$$Pr[\underbrace{\alpha' \text{ makes } E_j \text{ happen and } \beta' \text{ makes } E \text{ happen}}_{\text{event } T}] = Pr[E_j]Pr[E].$$

Now, by the hypothesis and the construction of α', β', it also holds that S implies T. Thus:

$$Pr[S] \leq Pr[T] \Leftrightarrow Pr[E_j \cap E] \leq Pr[E_j]Pr[E].$$

$\square$

Suppose now that the size of the outwards neighborhood of any event in the VDL graph is bounded by $d \in \mathbb{N}$, i.e. $max\{|\Gamma(E_j)| \mid 1 \leq j \leq m\} \leq d$ and that $p \in [0, 1]$ is the maximum of the probabilities $Pr[E_j], j = 1, ..., m$. In the next section we will prove (e is the base of the natural logarithm):

Theorem 1 (Symmetric Lopsided Local Lemma for VDL). *If $ep(d+1) \leq 1$, then $Pr[\overline{E_1} \wedge \overline{E_2} \wedge \cdots \wedge \overline{E_m}] > 0$, i.e. there exists an assignment of values to the variables X_i for which none of the events E_j hold.*

We end this section by proving that, using the VDL notion of Definition 1 can be advantageous over using the lopsidependency notion of Definition 2.

Example 3. Recall the setting of Example 1. Each of the five Bernoulli trials succeeds with probability $x \in [0, 1)$, thus:

$$Pr[E_1] = ((1-x) + x^2) \cdot (1-x),$$
$$Pr[E_2] = Pr[E_3] = x(1-x).$$

Let $p \in [0, 1)$ and d as in Theorem 1 above. It is easy to see that:

$$p = Pr[E_1] = (-x^3 + 2x^2 - 2x + 1).$$

Now, since $d = 1$ by Example 2, Theorem 1 guarantees the existence of an assignment such that neither one of the events occur, for any $x > 0.778$.

Let d' be an upper bound in the size of the neighborhoods of the events in the lopsidependency graph of Example 2. The *symmetric lopsided* LLL states that, if $ep(d' + 1) \leq 1$, then a solution is guaranteed to exist. Thus, since $d' = 2$, this version of the LLL would guarantee us a solution only for $x > 0.861$.   ◇

2   Algorithmic Lopsided Lovász Local Lemma

2.1   The Algorithm

The notion of lopsidependency we use hereafter is the directed notion of Definition 1, which we called VDL. The outwards neighborhood operation of Γ is always in terms of VDL.

To prove Theorem 1 in our framework, we give a Moser-like algorithm, M-ALGORITHM 1 below, and find an upper bound to the probability that it lasts for at least n *rounds*, i.e. the number of recursive calls of RESAMPLE.

Algorithm 1 M-ALGORITHM.

1: Sample the variables X_i, $i = 1, ..., l$ and let α be the resulting assignment.
2: **while** there exists an event that occurs under the current assignment, let E_j be the least indexed such event and **do**
3:     RESAMPLE(E_j)
4: **end while**
5: Output current assignment α.

RESAMPLE(E_j)

1: Resample the variables in sc(E_j).
2: **while** some event in $\Gamma(E_j) \cup \{E_j\}$ occurs under the current assignment, let E_k be the least indexed such event and **do**
3:     RESAMPLE(E_k)
4: **end while**

It is trivial to see that if M-ALGORITHM ever stops, it returns an assignment for which none of the "undesirable" events $E_1, ..., E_m$ holds.

We first prove:

Lemma 2. *Consider an arbitrary call of Resample(E_j). Let $\mathcal{X}$ be the set of events that do not occur at the start of this call. Then, if and when this call terminates, all events in $\mathcal{X} \cup \{E_j\}$ do not occur.*

Proof. Let $E_k \in \mathcal{X} \cup \{E_j\}$, assume that RESAMPLE($E_j$) terminates and that E_k occurs under the current assignment α.

$E_k \neq E_j$, else the RESAMPLE(E_j) couldn't have exited the while-loop of line 2 and thus wouldn't have terminated.

Consequently, $E_k \in \mathcal{X}$ thus, under the assignment at the beginning of RESAMPLE(E_j), E_k did not occur. Then, it must be the case that at some point during this resample call, some resampling of the variables caused E_k to occur.

Let RESAMPLE(E_s) be the last time E_k became occurring, and thus remained occurring until the end of RESAMPLE(E_j). During this call, only the variables in sc(E_s) were resampled. Thus, it is easy to see that $E_k \in \Gamma(E_s)$.

Since E_k remains occurring, RESAMPLE(E_s) couldn't have exited the while-loop of line 2 and thus couldn't have terminated. Consequently, neither could RESAMPLE(E_j). Contradiction. □

A *root call* of RESAMPLE is any call of RESAMPLE made when executing line 3 of M-ALGORITHM and a *recursive call* of RESAMPLE is one made from line 2 of another RESAMPLE call. The time complexity of the algorithm will be given in terms of the number of rounds it will need to execute.

By Lemma 2, we know that the events of the root calls of Resample are pairwise distinct. Therefore:

Corollary 1. *There are at most m Resample root calls in any execution of* M-ALGORITHM.

2.2 Forests

To depict an execution of M-ALGORITHM, we will use *rooted forests*, i.e. forests of trees such that each tree has a special node designated as its root.

The nodes of rooted forests are labeled by the events E_j, $j = 1, ..., m$, with repetitions of the labels allowed and they are ordered as follows: children of the same node are ordered as their labels are; nodes in the same tree are ordered by preorder (respecting the ordering between siblings) and finally if the label on the root of a tree T_1 precedes the label of the root of T_2, all nodes of T_1 precede all nodes of T_2.

The number of nodes of a forest $\mathcal{F}$ is denoted by $|\mathcal{F}|$.

Definition 4. *A labeled rooted forest $\mathcal{F}$ is called* feasible *if the following conditions hold:*

1. *The labels of the roots of $\mathcal{F}$ are pairwise distinct.*
2. *If u, v are siblings (have common parent), then the labels of u, v are distinct.*
3. *Let E_i, E_j be the labels of nodes u, v respectively, where u is a child of v. Then, $E_i \in \Gamma(E_j) \cup \{E_j\}$.*

Now consider an execution of M-ALGORITHM that lasts for n rounds. We construct in a unique way, depending on the rounds, a feasible forest with n nodes as follows:

1. The forest under construction will have as many roots as root calls of RESAMPLE. These roots will be labeled by the event of the corresponding root call.
2. A tree that corresponds to a root call RESAMPLE(E_j) will have as many non-root nodes as the number of recursive calls of RESAMPLE within RESAMPLE(E_j). The non-root nodes will be labeled by the events of those recursive calls.
3. The non-root nodes are organized within the tree with root-label E_j so that a node that corresponds to a call RESAMPLE(E_k) is parent to a root that corresponds to a call RESAMPLE(E_l), if RESAMPLE(E_l) appears immediately on top of RESAMPLE(E_k) in the recursive stack that implements the root call RESAMPLE(E_j).

It is straightforward to verify, by inspecting the succession of steps of M-ALGORITHM 1 in an execution, and making use of Lemma 2, that a forest constructed as above from the consecutive rounds of the execution of M-Algorithm is indeed a feasible forest in the sense of Definition 4. It is not true however that every feasible forest corresponds to the consecutive rounds of some execution of M-ALGORITHM. For example consider a feasible forest where a node w has a child u and a descendent v, such that: (i) v is not a descendent of u and (ii) u and v have the same label (thus v cannot also be a child of w). By Lemma 2, this forest could never be constructed as above.

Definition 5. *The* witness forest *of an execution of* M-ALGORITHM *1 is the feasible forest constructed by the process described above. Given a feasible forest* $\mathcal{F}$ *with n nodes, we denote by $W_{\mathcal{F}}$ the event that in the first n rounds of* M-ALGORITHM, $\mathcal{F}$ *is constructed. This implies that* M-ALGORITHM *lasts for at least n rounds.*

We also set:

$$P_n := \Pr\left[\bigcup_{\mathcal{F}:|\mathcal{F}|=n} W_{\mathcal{F}} \right] = \sum_{\mathcal{F}:|\mathcal{F}|=n} \Pr\left[W_{\mathcal{F}} \right], \tag{1}$$

where the last equality holds, because the events $W_{\mathcal{F}}$ are disjoint. Obviously now:

$$Pr[\text{M-Algorithm lasts for at least } n \text{ rounds}] = P_n. \tag{2}$$

We will end this subsection with an example of a witness forest constructed by M-ALGORITHM.

Example 4. Consider the setting of Example 1. The following is an execution of M-ALGORITHM that results in an assignment of values such that neither of the events E_1, E_2, E_3 occurs. α_0 will denote the assignment after the initial sampling of the variables and α_i that of round i, $i = 1, 2, 3, 4$. The events are the least indexed occurring ones under the current assignment:

$$\alpha_0 = (0, 0, 1, 0, 0) \rightarrow E_1, \ \alpha_1 = (0, 1, 1, 0, 0) \rightarrow E_2,$$

$$\alpha_2 = (0, 0, 1, 1, 0) \rightarrow E_1, \ \alpha_3 = (1, 1, 0, 1, 0) \rightarrow E_3, \ \alpha_4 = (1, 1, 0, 0, 1).$$

The reader is encouraged to check that this is a valid execution of M-ALGORITHM that leads to a satisfying assignment.

Easily, the witness forest of the above execution is:

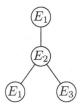

The nodes of the above tree are ordered as: E_1, E_2, E_1, E_3.

2.3 Validation Algorithm

To find an upper bound for P_n, consider the following algorithm, which we call the Validation Algorithm:

Algorithm 2 VALALG.

    Input: Feasible forest $\mathcal{F}$ with labels $E_{j_1}, \ldots, E_{j_n}$.

1: Sample the variables X_i, $i = 1, ..., l$.
2: **for** s=1,...,n **do**
3:     **if** E_{j_s} does not occur under the current assignment **then**
4:         **return failure** and exit.
5:     **else**
6:         Resample the variables in $\text{sc}(E_{j_s})$
7:     **end if**
8: **end for**
9: **return success.**

The validation algorithm VALALG, takes as input a feasible forest $\mathcal{F}$ with n nodes, labeled with the events $E_{j_1}, ..., E_{j_n}$ (ordered as their respective nodes) and outputs a Boolean value **success** or **failure**.

Intuitively, with input a feasible forest with labels $E_{j_1}, ..., E_{j_n}$, VALALG initially generates a random sampling of the variables, then checks the current event and, if it holds, resamples its variables and goes to the next event. If the algorithm manages to go through all events, it returns **success**, otherwise, at the first event that does not hold under the current assignment, it returns `failure` and stops. Note that whether there are occurring events under the last assignment generated by VALALG is *irrelevant* to its success of failure.

A round of VALALG is the duration of any **for** loop executed at lines 2–8.

We first give a result about the probability distribution of the assignment to the variables X_i during the execution of VALALG.

Lemma 3 (Randomness Lemma). *At the beginning of any round of the Validation Algorithm VALALG 2, the distribution of the current assignment of values to the variables X_i, $i = 1, ..., l$ is as if all variables have been sampled anew. Therefore the probability of any event occurring at such an instant is bounded from above by p.*

Proof. The result follows from the fact that if at the previous iteration of the **for** loop of line 2 the event E_{j_s} has been checked, only the values of the variables in $\mathrm{sc}(E_{j_s})$ have been exposed, and these are resampled anew.

Definition 6. *Given a feasible forest $\mathcal{F}$ with n nodes, we say that $\mathcal{F}$ is validated by VALALG if the latter returns* success *on input $\mathcal{F}$. The event of this happening is denoted by $V_{\mathcal{F}}$. We also set:*

$$\hat{P}_n = \sum_{\mathcal{F}: |\mathcal{F}|=n} \Pr[V_{\mathcal{F}}]. \tag{3}$$

We now claim:

Lemma 4. *For any feasible forest $\mathcal{F}$, the event $W_{\mathcal{F}}$ implies the event $V_{\mathcal{F}}$, therefore:*

$$P_n \leq \hat{P}_n. \tag{4}$$

Proof. Indeed, if the random choices made by an execution of M-ALGORITHM that produces as witness forest $\mathcal{F}$ are made by VALALG on input $\mathcal{F}$, then clearly VALALG will return success.

Lemma 5. *For any feasible forest $\mathcal{F}$ with n nodes, $\Pr[V_{\mathcal{F}}]$ is at most p^n.*

Proof. Immediate corollary of the Randomness Lemma 3.

Therefore to find an upper bound for $\hat{P}_n = \sum_{\mathcal{F}: |\mathcal{F}|=n} \Pr[V_{\mathcal{F}}]$, it suffices to find an upper bound on the number of feasible forests with n nodes, a fairly easy exercise in enumerative executed, whose solution we outline below.

First to any feasible forest $\mathcal{F}$ with n nodes we add new leaves to get a new labeled forest $\mathcal{F}'$ (perhaps not feasible anymore) comprising of m full $(d+1)$-ary trees whose internal nodes comprise the set of all nodes (internal or not) of $\mathcal{F}$ (recall m is the total number of events $E_1, \ldots, E_m$). Specifically, we perform all the following additions of leaves:

1. Add to $\mathcal{F}$ new trees, each consisting of a single root/leaf labeled with a suitable event, so that the set of labels of all roots of $\mathcal{F}'$ is equal to the the set of all events $E_1, \ldots, E_m$. In other words, the labels of the added roots/leaves are the events missing from the list of labels of the roots of $\mathcal{F}$.

2. Hang from all nodes (internal or not) of $\mathcal{F}$ new leaves, labeled with suitable events, so that the set of the labels of the children in $\mathcal{F}'$ of a node u of $\mathcal{F}$ labeled with E is equal to the set of the events in $\Gamma(E) \cup \{E\}$. In other words, the labels of the new leaves hanging from u are the events in $\Gamma(E) \cup \{E\}$ that were missing for the labels of the children of u in $\mathcal{F}$.
3. Add further leaves to all nodes u of $\mathcal{F}$ so that every node of $\mathcal{F}$ (internal or not in $\mathcal{F}$) has exactly $d + 1$ children in $\mathcal{F}'$; label all these leaves with the first $(d + 1) - |\Gamma(E) \cup \{E\}|$ events from the list $E_1, \ldots, E_m$, so that labels of siblings are distinct, and therefore siblings can be ordered by the index of their labels.

Notice first that indeed, the internal nodes of $\mathcal{F}'$ comprise the set of all nodes (internal or not) of $\mathcal{F}$.

Also, the labels of the nodes of a labeled forest $\mathcal{F}'$ obtained as above are uniquely determined from the rooted planar forest structure of $\mathcal{F}'$ when labels are ignored, but the ordering of the nodes imposed by them is retained (a forest comprised of rooted trees is called *rooted planar* if the roots are ordered and if the children of each internal node are ordered). Indeed, the roots of $\mathcal{F}'$ are labeled by $E_1, \ldots, E_m$; moreover, once the the label of an internal node of $\mathcal{F}'$ is given, the labels of its children are distinct and uniquely specified by steps (2) and (3) of the construction above.

Finally, distinct $\mathcal{F}$ give rise to distinct $\mathcal{F}'$.

By the above remarks, to find an upper bound to the number of feasible forests with n nodes (internal or not), it suffices to find an upper bound on the number of rooted planar forests with n internal nodes comprised of m full $(d + 1)$-ary rooted planar trees.

It is well known that the number t_n of full $(d+1)$-ary rooted planar trees with n internal nodes is equal to $\frac{1}{dn+1}\binom{(d+1)n}{n}$, see e.g. [22, Theorem 5.13]. Now by Stirling's approximation easily follows that for some constant $A > 1$, depending only on d, we have:

$$t_n < A\left(\left(1 + \frac{1}{d}\right)^d (d+1)\right)^n. \tag{5}$$

Also obviously the number f_n of rooted planar forests with n internal nodes that are comprised of m $(d+1)$-ary rooted planar trees is given by:

$$f_n = \sum_{\substack{n_1 + \cdots + n_m = n \\ n_1, \ldots, n_m \geq 0}} t_{n_1} \cdots t_{n_m}. \tag{6}$$

From (5) and (6) we get:

$$f_n < (An)^m \left(\left(1 + \frac{1}{d}\right)^d (d+1)\right)^n < (An)^m (e(d+1))^n. \tag{7}$$

So by Lemma 4, Eq. (3), Lemma 5 and Eq. (7) we get the following Theorem, which is essentially a detailed restatement of Theorem 1:

Theorem 2. *Assuming p and d are constants such that $\left(1 + \frac{1}{d}\right)^d p(d+1) < 1$, (and therefore if $ep(d+1) \leq 1$), there exists an integer N, which depends linearly on m, and a constant $c \in (0,1)$ (depending on p and d) such that if $n/\log n \geq N$ then the probability that M-ALGORITHM lasts for at least n rounds is $< c^n$.*

Clearly, when M-ALGORITHM stops we have found an assignment such that none of the events occurs. Since, by the above Theorem, this happens with probability close to 1 for large enough n, Theorem 1 follows.

References

1. Erdős, P., Lovász, L.: Problems and results on 3-chromatic hypergraphs and some related questions. Infin. Finite Sets **10**, 609–627 (1975)
2. Szegedy, M.: The Lovász local lemma – a survey. In: Bulatov, A.A., Shur, A.M. (eds.) CSR 2013. LNCS, vol. 7913, pp. 1–11. Springer, Heidelberg (2013). https://doi.org/10.1007/978-3-642-38536-0_1
3. Moser, R.A.: A constructive proof of the Lovász local lemma. In: Proceedings of the 41st annual ACM Symposium on Theory of Computing, pp. 343–350 (2009)
4. Alon, N.: A parallel algorithmic version of the local lemma. Random Struct. Algorithms **2**(4), 367–378 (1991)
5. Beck, J.: An algorithmic approach to the Lovász local lemma I. Random Struct. Algorithms **2**(4), 343–365 (1991)
6. Srivinsan, A.: Improved algorithmic versions of the Lovász local lemma. In: Proceedings of the Nineteenth Annual ACM-SIAM Symposium on Discrete Algorithms, pp. 611–620. Society for Industrial and Applied Mathematics (2008)
7. Tao, T.: Moser's entropy compression argument (2009). https://terrytao.wordpress.com/2009/08/05/mosers-entropy-compression-argument/
8. Spencer, J.: Robin Moser makes Lovász local lemma algorithmic! (2010). https://cs.nyu.edu/spencer/moserlovasz1.pdf
9. Giotis, I., Kirousis, L., Psaromiligkos, K.I., Thilikos, D.M.: On the algorithmic Lovász local lemma and acyclic edge coloring. In: Proceedings of the Twelfth Workshop on Analytic Algorithmics and Combinatorics. Society for Industrial and Applied Mathematics (2015). http://epubs.siam.org/doi/pdf/10.1137/1.9781611973761.2
10. Erdős, P., Spencer, J.: Lopsided Lovász local lemma and Latin transversals. Discret. Appl. Math. **30**(2–3), 151–154 (1991)
11. Berman, P.R., Scott, A., Karpinski, M.: Approximation hardness and satisfiability of bounded occurrence instances of SAT. ECCC 10(022) (2003)
12. Gebauer, H., Moser, R.A., Scheder, D., Welzl, E.: The Lovász local lemma and satisfiability. Efficient Algorithms, pp. 30–54. Springer, Berlin (2009). https://doi.org/10.1007/978-3-642-03456-5_3
13. Gebauer, H., Szabó, T., Tardos, G.: The local lemma is tight for SAT. In: Proceedings of the Twenty-Second Annual ACM-SIAM Symposium on Discrete Algorithms, pp. 664–674. SIAM (2011)
14. Moser, R.A., Tardos, G.: A constructive proof of the general Lovász local lemma. J. ACM (JACM) **57**(2), 11 (2010)
15. Sarkar, K., Colbourn, C.J.: Upper bounds on the size of covering arrays. SIAM J. Discret. Math. **31**(2), 1277–1293 (2017)

16. Kolipaka, K., Rao, B., Szegedy, M.: Moser and Tardos meet Lovász. In: Proceedings of the Forty-Third Annual ACM Symposium on Theory of Computing, pp. 235–244. ACM, New York (2011)
17. Shearer, J.B.: On a problem of Spencer. Combinatorica **5**(3), 241–245 (1985)
18. Harvey, N.J.A., Vondrák, J.: An algorithmic proof of the Lovász local lemma via resampling oracles. In: Proceedings 56th Annual Symposium on Foundations of Computer Science (FOCS), pp. 1327–1346. IEEE (2015)
19. Harris, D.G.: Lopsidependency in the Moser-Tardos framework: beyond the lopsided Lovász local lemma. In: Proceedings of the Twenty-Sixth Annual ACM-SIAM Symposium on Discrete Algorithms, pp. 1792–1808. SIAM (2015)
20. Spencer, J.: Ten Lectures on the Probabilistic Method, vol. 64. SIAM, Philadelphia (1994)
21. Achlioptas, D., Iliopoulos, F.: Random walks that find perfect objects and the Lovász local lemma. In: 2014 IEEE 55th Annual Symposium on Foundations of Computer Science (FOCS), pp. 494–503. IEEE (2014)
22. Sedgewick, R., Flajolet, P.: An Introduction to the Analysis of Algorithms. Addison-Wesley, Upper Saddle River (2013)

Creating a Multi-iterative-Priority-Rule for the Job Shop Scheduling Problem with Focus on Tardy Jobs via Genetic Programming

Georg E. A. Froehlich[1]([⊠]), Guenter Kiechle[1], and Karl F. Doerner[1,2]

[1] Faculty of Business Administration, University of Vienna, 1090 Vienna, Austria
georg.erwin.adrian.froehlich@univie.ac.at
[2] Data Science, University of Vienna, Vienna, Austria

Abstract. Genetic programming is used to create Priority Rules (PR) for the Job Shop Scheduling Problem with the aim of minimizing the weighted sum of tardy jobs. Four types of structures are used for the PR: Normal PR, Iterative-PR (IPR), Multi-PR (MPR), and Multi-Iterative-PR (MIPR). These are then compared among one another and with classical PR like shortest-processing-time. A modern metaheuristic based on local search using disjunct graphs and critical paths is used to solve the static problem as a benchmark. The results show that all types provide better results than classical PR and that with and without time limit the types from best to worst are: MIPR, MPR, IPR, and PR. The gaps to the metaheuristic are also reported.

1 Introduction

Real-world scheduling problems like the Job Shop Scheduling Problem (JSSP) often have to fare with disruptions of machines, additional jobs coming in during the day, or others being cancelled. This situation then frequently asks for fast procedures to adapt the schedule online. Priority Rules (PR) are one method to quickly generate good schedules and are also found commonly in practice. It is obvious that more sophisticated approaches like metaheuristics based on local search or population-based might generate far better results in comparison. But one of their drawbacks is that they are often very time consuming - getting worse the larger the instances grow, which can be quite large for real-world problems. Therefore, they might not be practicable during production, especially when also disruptions occur. Furthermore, such metaheuristics have to be adapted if new additional restrictions occur, which previously have not been considered. Compared to this PR can always be used, even if some restriction were not taken into account during their creation. Instead of creating good PR via manual composition of different PR or factors, which could be beneficial for certain criteria, tree based heuristics can be used to generate good PR. We apply Genetic Programming (GP) for the generation of these PR, which belongs

R. Battiti et al. (Eds.): LION 12 2018, LNCS 11353, pp. 64–77, 2019.
https://doi.org/10.1007/978-3-030-05348-2_6

to this class of heuristics and for which it has already been shown that better PR than "standard" PR can be created (e.g. [1–5]). While research has already been made for several objectives like makespan or tardiness, the weighted sum of tardy jobs has had far less attention (cf. the list of Branke et al. [6]). But sometimes it does not or barely matter how tardy a job is and instead only or mainly matter if it is tardy. There are several possible causes for this. A product could be needed up to a certain point in time and then loses its value for the customer like Halloween costumes. Customers expect deliveries until the due date and lose a lot of trust in the reliability of a company if the product is late even one day. Contracts might be designed so that a penalty only applies when the product is delivered late. The focus in this work, therefore, lies on this criterion and uses types of PR similar to these of Nguyen et al. [5]. In contrast to Nguyen et al. the different types of PR are compared, when all of them have the same runtime. Moreover, the PR are used in a manner to generate active schedules and not to generate non-delay schedules. Active schedules are characterized through the inability to prepone any operation on a machine, while keeping the remaining operations like they are, and the inability to exchange any two operations on a machine without worsening the solution quality. Non-delay schedules are a subset of active schedules. For them no machine may be idle, if an operation for the machine is available. For active schedules, it is guaranteed that they include all optimal schedules for regular criteria, which by finishing any job earlier, cannot become worse. However, for non-delay schedules this is not the case.

Section 2 gives a brief problem description. Section 3 explains GP and how the different types of PR were created. In Sect. 4 the results are given and finally Sect. 5 summarizes this paper once more and draws conclusions.

2   Problem Description

The JSSP belongs to the class of NP-hard problems. For a set of jobs J all their operations o_{jm} have to be scheduled on a set of machines M, where every machine may at most process a single operation at a time. Operations have got an associated processing time t_{jm} and release time a_{jm}. The completion time c_{jm} of an operation is the sum of its starting time s_{jm}, which may not be smaller than either the release time a_{jm} or the completion time of any preceding operations of the job, and its processing time t_{jm}. For every job there exists a sequence, in which his operations have to be fulfilled, a weight w_j indicating how important the job is, and a due date d_j, until which the last operation of the job should be done. Otherwise the job is tardy ($U_j = 1$) with a tardiness of $T_j = max(0, max_{m \in M} c_{jm} - d_j)$.

The objective to minimize the weighted sum of tardy jobs can be expressed by $\sum_{j \in J}(w_j \cdot U_j) \to min$, which belongs to the class of regular criteria.

In the following the related work, where GP was used to create PR for JSSP is presented. For a more detailed list see the surveys of Branke et al. [6] or Nguyen et al. [7]. Dimopoulos and Zalzala [1] created PR for the one-machine-setting with the aim of minimizing the tardiness. They compared their rule to other standard rules, where theirs proved superior. Miyashita [4] compared, if it would be better to have one PR for all machines or not. Three cases were compared: One PR for all. For every machine an individual PR. Two PR, one for machines, which were seen as bottleneck, and one for those, which were not, and additionally one function for differentiating whether a machine was a bottleneck or not. The case of a single PR for all was worse than the other two cases. Jakobovic and Budin [2] considered dynamic single- and multiple-machine-cases for tardiness and the number of tardy jobs. For the second case, they used a similar approach to Miyashita, where again the case with multiple rules yielded better results. A later work of Jakobovic et al. [3] looked at combinations of static and dynamic JSSP with and without setup times regarding makespan and weighted tardiness. Once looking at the makespan and once at the weighted tardiness Nguyen et al. [5] created different types of PR: A normal PR, an Iterative-PR (IPR), an IPR with variable neighborhood search (IPR-VNS), and an IPR-VNS with a lookahead component (IPR-VNS-L). For all types, except for the IPR-VNS-L non-delay schedules were created. Exactly this possibility of allowing idling was the lookahead component for the IPR-VNS-L. Furthermore, it was tested, whether the variable factors of the IPR should be determined by the GP or not, which seemed, however, rather unnecessary. Comparisons of the types showed that IPR-VNS and IPR-VNS-L were better than the rest. For makespan they were roughly similar, but regarding weighted tardiness IPR-VNS-L had a slight advantage. Nguyen et al. [8] created several scheduling policies for the multi-objective, dynamic JSSP case with the aim of minimizing makespan, normalized total weighted tardiness, and mean absolute percentage error of the flowtime estimation. The flowtime estimation error is created since they also try to assign due dates to incoming jobs. This is done in an individual tree, while a second tree is responsible for the assignment of operations. They notice that the three types of goals are conflicting and then compare their results to some existing PR and due date assignment rules, where their procedure proves superior.

The main differences of this work compared to Nguyen et al. [5] are that active schedules and not non-delay schedules are allowed in all steps, and that the objective is to minimize the number of tardy jobs, which is to the best of our knowledge not considered via iterative PR in other publications so far. Additionally, the different PR types were also compared with a fixed time limit and not only with a set amount of generations, since there are significantly different time requirements for the different types.

3   Solution Method

GP is a population based metaheuristic based on the idea of natural selection, in which every member of the current generation has a chance corresponding to its fitness (e.g. the objective value) to take part in the reproduction stage for the next generation. GP can be described in five main steps.

1. Generate the initial generation.
2. Evaluate the fitness of the current generation.

3. Decide based on the fitness, from which members of the generation the members of the next generation should be built. The selection can be done deterministically or randomly.
4. Create the next generation via crossover, mutation, and reproduction from the selected members.
5. If no stopping criteria is reached return to step 2. Otherwise return the best result.

When using GP PR are most often represented as trees. Figure 1 depicts the rule $(2 \cdot PT) + DD$. Nodes like $\cdot$ or $+$, which need further input are called function nodes, while nodes not needing input, e.g. processing time PT and due date DD, are terminal nodes.

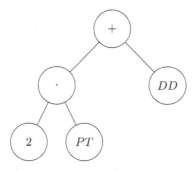

Fig. 1. Representation of PR as tree

This tree structure is used for mutation and crossover. For mutation, a number of branches are cut of the tree and randomly repaired. Crossover takes two trees. Then a branch is cut off from both trees, and both of them are exchanged. Care has to be taken that the resulting trees are still feasible. Should several branches be exchanged, this is called k-Crossover, where k is the number of exchanges. Reproduction is taking a tree and inserting it without any changes into the next generation.

The initial generation was created with a set of predetermined rules and the remaining rules being each with equal likelihood either of maximum size or smaller. The maximum size of a PR was reached, when every branch in the tree representation of the PR ended in the sixth and last layer. For creating further generations tournament selection was used. For this, a number of PR were selected by chance and the best according to the weighted sum of tardy jobs of them was selected. Due to this any scaling of the objective would not have changed the rank of the individual solution and was therefore not undertaken. For determining the weighted sum of tardy jobs the Giffler–Thompson algorithm [9] for generating active schedules was used.

Giffler–Thompson-Algorithm:

1. Let S be the set of currently possible operations and s^e_{jm} the earliest starting time for the corresponding operation o_{jm}.
2. Select from S an operation o_{kn}, for which the earliest completion time c_{kn} is minimal ($s^e_{kn} + t_{kn} <= s^e_{jm} + t_{jm} \forall o_{jm} \in S$).
3. Build a set of all operations o_{jn} taking place on machine n. These build the conflict set $C := \{o_{jn} \in S\}$.
4. Calculate with the PR for every operation in C the priority value and select the operation to schedule on machine n accordingly. (The generated PR are used here.)
5. Update the set S and the earliest starting times s^e_{jm}.
6. If set S is not empty, jump to step 2, otherwise done.

Different types of PR based on Nguyen et al. [5] were created. The first type was a simple PR, without any special features. IPR were the second type. For them iterative factors were included in the rule, which could change from run to run like completion time. The rule was performed on an instance as long as improvements took place. The idea behind IPR was that rules might determine, where problems existed during the last scheduling run and improve them in the next. The third type - Multi-PR (MPR) - consisted out of several simple PR, from which each one was performed once on an instance and the best result taken. MPR might yield improvements, since one type of PR can fare well with certain sets of instances, while it handles others rather badly. By then having one additional rule, which handles these instances better, improvements could take place. Multi-Iterative-PR (MIPR) combined these two concepts. Instead of having multiple simple PR, they had multiple IPR. The iterative factors were not reset, when the next IPR of a MIPR was performed. Only when all IPR were performed without any of them leading to an improvement, MIPR stopped and gave back the best result.

Table 1 lists the terminal- and functionsets for all the different types of PR. The first set of terminals and functions is used by every type, while the second set of terminals is used additionally for IPR and MIPR, while the second set of functions is used additionally for MPR and MIPR. While $+$, $-$, and $\cdot$ are self-explanatory, it should be noted that $\div$ stands for the protected division and returns 1, if it is divided by 0. The operators $||$, max, and min return the absolute value, the largest value, and the smallest value, respectively. The operators $<$ and $=$ compare their first two inputs and return their third input, should the first input be smaller for $<$ or both equal for $=$, and otherwise their forth. The MPR-base-node is just for connecting the different PR/IPR for MPR/MIPR. Most of the terminals are very basic and easy to understand like due date d_j, processing time of the operation t_{jm}, weight w_j, release date a_{jm}, total processing time of the job $\sum_{m \in M} t_{jm}$. Remaining operations of the job denotes the number of operations, which have not been performed (o_{jm} incomplete), and remaining processing time of the job is the sum of all their processing times $\sum_{m \in M | o_{jm} incomplete} t_{jm}$. Work in next queue for an operation o_{jm} is the sum of the processing times of all operations waiting at the machine, where the next operation of job j would take place. Recorded completion time of the job c_j, recorded waiting time of the job $c_j - a_j$, and recorded waiting time of the operation $c_{jm} - max(a_{jm}, c_{pred(o_{jm})})$, with $pred(o_{jm})$ being the predecessing operation of o_{jm}, are each dependent on just the last execution result of the PR. Recorded delays, however, depends on all previous results for this instance. Every time a job is delayed, recorded delay for it becomes increased by 1. Otherwise recorded delay is decreased by 1, if the resulting value would not be smaller than 0.

Table 1. Terminal- and functionsets

Terminals: (All)	Due date (DD), remaining processing time of the job (RPT), processing time of the operation (PT), weight (W), release date (RlD), total processing time of the job (TPT), work in next queue (WNQ), remaining operations of the job (ROp), constant
Functions: (All)	$+, -, \cdot, \div, ||, max, min, <, =$
Terminals: (IPR & MIPR only)	Recorded completion time of the job (RcC), recorded waiting time of the job (RWJ), recorded waiting time of the operation (RWO), recorded delays (RcD)
Functions: (MPR & MIPR only)	MPR-base-node (MPRF)

Besides the different types of PR two types of pruning have been tried as well. Pruning in GP means that some branches of the tree are cut off. If this is performed on branches, which do not yield any additional value regarding the fitness, the remaining parts of the trees have a higher probability of being strongly presented in the next generations and might cause better results. Additionally, a smaller tree needs less time to be evaluated, which in the best case might be a saving larger than the additional time needed for the pruning process.

For the first version, a tree was taken and its fitness determined. Then for every node, which was a direct successor of the root node, the root node with all its branches was replaced by just the branch of the successor. If at least one of the successor led to a better or equal result, the best of these results was accepted and the procedure repeated from this node onwards. Otherwise the procedure continued one layer below. This method was tested with different probabilities of how likely it is that a PR is pruned after a generation. However, this procedure led to clearly worse results. Further examination of the PR showed that of the earlier rather large PR only very small parts were remaining, suggesting that this caused a far too low genetic variety.

The second version only pruned obviously redundant parts of the PR, which were:

1. *max/min*: If the successors are identical terminals, replace the branch with just a single of these terminals.

2. <: If the first two successors are identical terminals, replace the branch with the fourth branch, which is the branch executed if the result of the first branch is not smaller than the result of the second branch.
3. =: If the first two successors are identical terminals, replace the branch with the third branch, which is the branch executed if the results of the first two branches is identical.
4. ||: For some terminals, which cannot be negative replace || with just them.

Compared to the first approach this type of pruning needs almost no computation time, since it was not necessary to evaluate the new rules. The results were neither significantly better nor worse than the ones without pruning.

4   Results

This approach was programmed in C#. Several classes from HeuristicLab [10] supplying the base code for a GP have been taken. The program was run on the Vienna Scientific Cluster with the compiler Mono. For evaluating the GP the Taillard instances [11] were chosen, even though they were not created with any due dates. Due dates were later added by Mati et al. [12] in a procedure similar to Singer and Pinedo [13] with three differently tightness factors f - 1.3, 1.5, 1.6 - for each instance. The due dates are determined according to the tightness factor and are equal to the release time a_j plus the total processing time for the job times the tightness factor ($d_j = a_j + f \sum_{m \in M} t_{jm}$). Every job has exactly one operation on every machine and always ten instances are of the same size with the smallest instances having 15 jobs and machines each and the largest having 30 jobs and 20 machines.

For the three trainings sets, used to generate the PR, tai01 to tai05 and tai11 to tai15 were used with the first five having 15 jobs and machines each and the other having 20 jobs and 15 machines. Instances tai06 to tai10 and tai16 to tai20, which were of similar sizes, were used for the first three test sets. Furthermore, tai41 to tai50 built the later three test sets with larger instances having 30 jobs and 20 machines. Sets using the same instances have different tightness factors. Of three sets the first one has 1.3 as a tightness factor, while the second and third have 1.5 and 1.6 respectively. Parameter tuning for number of elite solutions, mutation, predetermined rules, and tournament size was performed first. Table 2 lists the different settings for each of the parameters. Not all possible combinations were tested. Instead, from one base setting exactly one parameter was changed with the basic setting being elite 5%, mutation 10%, 50 predetermined rules, and a tournament size of 7 (cf. bold, underlined number in Table 2). Population and generations were set to 400 and 25 respectively for these runs. For each setting 25 runs were performed.

It could be deduced that the mutation should be larger 0%, since it yielded a clearly worse average. The other rates did not yield very different averages, but 10% seemed better in two of three trainings sets. The predetermined rules in the initial generation did not lead to any significant differences after 25 generations. For trainings set 2 and 3 also the changes in between any generation

Table 2. Setting for parameters

Parameter	Settings
Elite	0%, 1%, **5 %**, 10%
Mutation	0%, 5%, **10 %**, 15%, 20%
Predetermined rules	0, 10, 25, **50**, 100
Tournament size	4, 5, **7**, 9

Note: Bold, underlined values are the basic setting

were marginal. Only for trainings set 1 there was a very noticeable difference for the first 20 generations. At first 0 predetermined rules led to worse results, then to better results, but after 25 generations the results were again rather similar. The parameter variation for the tournament size suggested that a tournament size of 7 seemed preferable, since sizes 4 and 5 led to worse results than 7 and 9, and 9 also to worse than 7. For the elite rate, it seemed preferable to have an elite larger 0%, since the results for trainings set 1 and 3 were clearly worse without.

Following this results the setting was set to elite 1%, mutation 10%, predetermined rules 50, and tournament size 7 for further runs. The second test focused on increasing the amount of PR that are generated during one run of the GP. While for the parameter tuning 10,000 PR were generated not including the initial generation the number was raised to 20,000 PR and different combinations of population and generations compared. The combinations were populations with 200, 400, 625, and 800 individuals and 100, 50, 32, and 25 generations respectively. 15 repetitions were done for each setting. From the results could be taken that the setting 200×100 was not only worse than the other settings coming up to 20,000 PR, but also worse than 400×25. However, compared to all other settings 400×25 fared worse. Since for trainings set 1 400×50, 625×32, and 800×25 were the best, as were 400×50 and 800×25 for trainings set 3, while 400×50 showed itself as superior on trainings set 2, the setting 400×50 seemed to be the best.

For the third step the population and generation were set to 400 and 50 and for IPR, MPR, and MIPR 20 replications made, for which the results are in Fig. 2. The graphs show the average of the best results - regarding the weighted number of tardy jobs with identical settings. The x-axis denotes the generations with the initial generation starting at 0, while the y-axis denotes the averages. It can be seen that for all instance the rules are from best to worst MIPR, MPR, IPR, and PR with some rather large gaps compared to the ones from the parameter tuning and generation and population test.

Table 3 lists the final results for all tested settings and the coefficients of variation with the best value of every test being printed in bold. The rather high coefficients of variation for several settings point towards either the GP being not stable enough or not enough time for convergence. Since the first tests only took the amount of PR created into account and not the time needed, a further

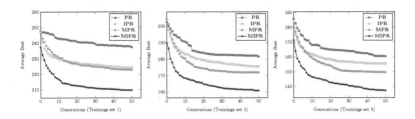

Fig. 2. Results for different PR-types

test with a run time limit of 36 h and 40 replications was done. Especially for IPR, MPR, and MIPR more time should be needed than for simple PR.

From the last eight lines of Table 3, which list the results for the different types of PR with and without time limit, it can be seen that not only did the results for the simple PR get better by roughly up to 2.2%, but also did the coefficients of variation drop from the two-digits to below 3.05%. The absolute gap between IPR and MPR also closed by a bit, even though both of them improved. Only the MIPR worsened its results. These changes can be attributed to some runs of MIPR having less than 50 generations, while every other rule had every run clearly more than 50 generations. The simple PR has on average even five times the amount of generations. Table 4 lists the average amount of generations, the coefficient of variation, and the minimum and maximum amount of generations when a time limit of 36 h is used.

For comparing the results of the via GP generated PR the percentage gaps ($\frac{Obj(GP)-Obj(PR)}{Obj(PR)}$) between them and classical PR were calculated (cf. Table 5). As benchmarks the best result of Shortest-Processing-Time (SPT), Shortest-Remaining-Processing-Time, Slack, Earliest-Due-Date (EDD), Modified-Due-Date, Work-In-Next-Queue (WINQ), PT+WINQ+Slack, Release, PT+Slack, Critical-Ratio+SPT, Longest-Processing-Time (LPT), LPT+EDD were used. Table 5 lists the PR, which yielded the best result, the set, the result, and the gaps for the generated PR. Since the quality loss attributed to PR, which were generated for trainings sets with different tightness factors than the test sets, should be examined as well, test sets 1–3 show up twice. LR stands for Limited Rules, meaning only rules from trainings sets with an identical tightness factor were used. AR in contrast stands for All Rules, meaning the rules from all trainings sets are used. The results in these two cases are on average rather similar. While they got worse for test set 1 and 3, they improved for test set 2. Changing the size of the instance, however, led to worse results in almost all cases and a by roughly 1.5 percentage points smaller percentage gap. But even then, the percentage gap of all the generated rules with −17.14, −17.29%, and −20.74% are still a significant improvement.

Table 3. Results after fixed number of generations

Setting	Trainings set 1		Trainings set 2		Trainings set 3	
	Avg. Best	Cov	Avg. Best	CoV	Avg. Best	CoV
Elite 0%	238.68	2.98%	187.35	7.66%	166.35	9.73%
Elite 1%	**237.35**	5.35%	**187.05**	8.63%	164.3	9.35%
Elite 5%	238.9	4.34%	188.2	8.65%	164.3	6.41%
Elite 10%	239.15	6.76%	188	10.82%	**163.45**	6.96%
Mutation 0%	243.85	5.04%	188.8	22.96%	166.55	14.36%
Mutation 5%	238.2	6.07%	187.3	7.37%	164.1	8.03%
Mutation 10%	239.4	7.23%	**184.7**	10.38%	**162.85**	14.06%
Mutation 15%	**238.05**	3.60%	185.45	12.69%	165.8	8.56%
Mutation 20%	240.7	6.34%	185.85	13.75%	163.85	6.99%
Pr. Rules 0	**237.9**	10.88%	185	12.67%	163.8	12.47%
Pr. Rules 10	238.8	10.58%	186.8	10.43%	164.4	11.66%
Pr. Rules 25	238.3	2.44%	**184.1**	4.88%	163.4	20.32%
Pr. Rules 50	239.4	7.23%	184.7	10.38%	**162.85**	14.06%
Pr. Rules 100	239.8	8.37%	184.65	5.87%	165.05	9.77%
Tournament Size 4	240	3.78%	187.55	5.55%	164.85	9.14%
Tournament Size 5	239.55	3.81%	186.95	8.58%	164.65	9.02%
Tournament Size 7	**239.4**	7.23%	**184.7**	10.38%	**162.85**	14.06%
Tournament Size 9	239.7	4.82%	185.5	10.89%	163.35	10.90%
Pop. 200, Gen. 100	239.27	3.73%	184.87	15.31%	166.27	7.36%
Pop. 400, Gen. 50	**237.4**	11.72%	**181.5**	16.27%	**160.4**	14.75%
Pop. 625, Gen. 32	237.53	3.3%	184.07	11.29%	162.6	8.68%
Pop. 800, Gen. 25	237.73	6.82%	184.4	7.34%	160.67	5.57%
PR	237, 4	11.72%	181.5	16.27%	160.4	14.75%
IPR	224.25	1.52%	175.55	1.66%	155.5	2.8%
MPR	222.95	1.38%	171.85	2.03%	149.6	1.58%
MIPR	**209.7**	1.48%	**160.8**	3.57%	**137.2**	6.74%
PR-TL	232.25	2.10%	178.68	2.71%	157.4	3.05%
IPR-TL	223.65	1.12%	172.53	2.72%	152.35	2.5%
MPR-TL	220.28	1.94%	170.5	1.84%	148.7	2.14%
MIPR-TL	**211.7**	1.28%	**162.38**	2.89%	**140.38**	3.56%

Note: For "Pop. x, Gen. y" y generations. Otherwise 50 generations.

As a comparison of the generated PR to a modern metaheuristic MDL by Mati et al. [12] was used. MDL is a general neighborhood search developed for JSSP for any regular type of criterion. Neighborhoods are defined on a disjunctive graph and generated by swapping critical arcs. The first step is improving

Table 4. Generations within the time limit of 36 h

		PR	IPR	MPR	MIPR
Trainings set 1	Average	272.95	145.53	122.15	58.03
Trainings set 1	CoV	18.52%	24.95%	10.82%	9.46%
Trainings set 1	Minimum	181	77	90	45
Trainings set 1	Maximum	413	208	148	68
Trainings set 2	Average	250.73	134.58	111.8	51.7
Trainings set 2	CoV	17.30%	17.59%	9.81%	11.55%
Trainings set 2	Minimum	178	75	94	41
Trainings set 2	Maximum	337	177	135	64
Trainings set 3	Average	228.35	123.65	101.68	51.35
Trainings set 3	CoV	17.38%	16.67%	8.24%	12.01%
Trainings set 3	Minimum	145	85	89	41
Trainings set 3	Maximum	347	172	128	66

the primary criterion (e.g. the number of tardy jobs). If no improvement can be found an auxiliary criterion is introduced (e.g. the total tardiness). Should improvements regarding the primary criterion be made the method returns the the first step. Should no improvements be possible for the auxiliary criterion a third step is initiated, which performs random swaps and is intended for diversification. The comparison with MDL showed that the results are still very far off from near optimal results, with even the gaps of MIPR except for one case being between 42.74% and 307.44%. Only for test set 4 the differences were rather small with 9.60%, which, most likely, is attributed to the difficulty of not violating the due dates.

To illustrate the for the JSSP via GP generated PR, the best trees of MIPR for the training sets without time limit are given in Fig. 3. The first, second, and third tree were created for the first, second, and third training set respectively and all the trees have already been pruned. The branches following the $MPRF$-nodes are the different IPR, which the MIPR uses.

In contrast to the first intuition that the iterative terminals RcC, RWJ, RWO, and RcD would be used in most rules for prioritizing jobs in the next run, which were late/too late, it seems to be the other way around. The reddish branches in Fig. 3 like $W \div RWJ$ in the third tree increase the priority either for a larger absolute value of W or a smaller RWJ.

Within these three trees not a single branch can be found, where later completion or tardiness, yields a higher priority in the next run. Visual inspection of more trees showed that this is not just a random occurrence, but the case for almost all trees.

Table 5. Gaps compared to classical PR

Classical PR	Set	Result	IPR - Gap	MPR - Gap	MIPR-Gap
EDD	Trainings set 1	298	−24.95%	−26.08%	−28.96%
EDD	Trainings set 2	261	−33.90%	−34.67%	−37.79%
SRPT	Trainings set 3	222	−31.37%	−33.02%	−36.77%
SRPT	Test set 1 - LR	299	−14.46%	−16.17%	−20.04%
SRPT	Test set 2 - LR	256	−18.17%	−20.20%	−22.47%
SRPT	Test set 3 - LR	233	−20.61%	−21.33%	−26.81%
SRPT	Test set 1 - AR	299	−13.99%	−14.55%	−19.33%
SRPT	Test set 2 - AR	256	−18.51%	−20.21%	−24.19%
SRPT	Test set 3 - AR	233	−19.51%	−21.48%	−26.53%
SRPT	Test set 4	547	−14.29%	−13.86%	−17.85%
SRPT	Test set 5	487	−18.06%	−18.70%	−21.74%
SRPT	Test set 6	452	−19.06%	−19.32%	−22.62%

5 Conclusion

This paper focused on JSSP and GP with the aim to minimize the weighted number of tardy jobs. For this, four different types of PR were created. There were simple PR, IPR, MPR, MIPR. IPR had several iterative factors, which could change between every evaluation of an instance and thereby prioritize a job more or less, depending on the last run. This should enable the rule itself to learn. MPR did not try to learn and instead offered multiple PR. This multitude should enable MPR to solve more instances adequately. MIPR combined these two concepts by having several IPR, which all shared the values of the iterative factors. Before these types of rules were tested with and without time limit, parameter tuning of elite, mutation, predetermined rules, tournament size, and the generation amount and population size took place.

The results without time limit showed that the types got on average better from PR to IPR to MPR up to MIPR. Even when including a time limit this did not change, even though the margins between them diminished a bit.

Compared to classical PR all generated PR proved themselves to yield an on average 30% smaller weighted sum of tardy jobs on the trainings sets. Further examination on test sets showed that this gap got smaller, when test sets either had differently tight due dates or were of a larger size. Still, the gap for the best rules was on average −21.24%. A comparison to a modern metaheuristic based on a local search procedure showed, however, that there were still gaps between 42.74% and 307.44% - not including the one test set, which yielded a surprisingly small gap of 9.60%. Visual analysis of the rules implied that contrary to the first assumption the iterative factors of IPR and MIPR favored jobs, which were finished early in the last scheduling run and had rather low waiting times.

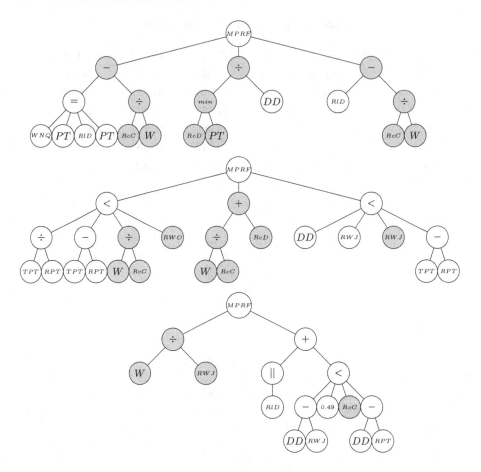

Fig. 3. Best trees for trainings sets without time limit

References

1. Dimopoulos, C., Zalzala, A.M.S.: Investigating the use of genetic programming for a classic one-machine scheduling problem. Adv. Eng. Softw. **32**, 489–498 (2001)
2. Jakobović, D., Budin, L.: Dynamic scheduling with genetic programming. In: Collet, P., Tomassini, M., Ebner, M., Gustafson, S., Ekárt, A. (eds.) EuroGP 2006. LNCS, vol. 3905, pp. 73–84. Springer, Heidelberg (2006). https://doi.org/10.1007/11729976_7
3. Jakobović, D., Jelenković, L., Budin, L.: Genetic programming heuristics for multiple machine scheduling. In: Ebner, M., O'Neill, M., Ekárt, A., Vanneschi, L., Esparcia-Alcázar, A.I. (eds.) EuroGP 2007. LNCS, vol. 4445, pp. 321–330. Springer, Heidelberg (2007). https://doi.org/10.1007/978-3-540-71605-1_30
4. Miyashita, K.: Job-shop scheduling with genetic programming. In: Proceedings of the 2nd Annual Conference on Genetic and Evolutionary Computation, pp. 505–512 (2000)

5. Nguyen, S., Zhang, M., Johnston, M., Chen Tan, K.: Learning iterative dispatching rules for job shop scheduling with genetic programming. Int. J. Adv. Manuf. Technol. **67**, 85–100 (2013)
6. Branke, J., Nguyen, S., Pickardt, C.W., Mengjie, Z.: Automated design of production scheduling heuristics: a review. IEEE Trans. Evol. Comput. **20**(1), 110–124 (2016)
7. Nguyen, S., Zhang, M., Johnston, M., Tan, K.C.: A computational study of representations in genetic programming to evolve dispatching rules for the job shop scheduling problem. IEEE Trans. Evol. Comput. **17**(5), 621–639 (2013)
8. Nguyen, S., Zhang, M., Johnston, M., Tan, K.C.: Automatic design of scheduling policies for dynamic multi-objective job shop scheduling via cooperative coevoluation genetic programming. IEEE Trans. Evol. Compu. **18**(2), 193–208 (2014)
9. Giffler, B., Thompson, G.L.: Algorithms for solving production-scheduling problems. Oper. Res. **8**, 487–503 (1959)
10. Wagner S. et al.: Architecture and Design of the HeuristicLab Optimization Environment. In: Klempous R., Nikodem J., Jacak W., Chaczko Z. (eds) Advanced Methods and Applications in Computational Intelligence. Topics in Intelligent Engineering and Informatics, vol 6. Springer, Heidelberg (2014)
11. Taillard, E.: Benchmarks for basic scheduling problems. Euro. J. Oper. Res. **64**(2), 278–285 (1993)
12. Mati, Y., Dauzere-Peres, S., Lahlou, C.: A general approach for optimizing regular criteria in the job-shop scheduling problem. Euro. J. Oper. Res. **212**, 33–42 (2011)
13. Singer, M., Pinedo, M.: Computational study of branch and bound techniques for minimizing the total weighted tardiness in job shops. IIE Trans. **29**, 109–119 (1998)

A Global Optimization Algorithm for Non-Convex Mixed-Integer Problems

Victor Gergel, Konstantin Barkalov$^{(\boxtimes)}$, and Ilya Lebedev

Lobachevsky State University of Nizhny Novgorod, Nizhny Novgorod, Russia
konstantin.barkalov@itmm.unn.ru

Abstract. In the present paper, the mixed-integer global optimization problems are considered. A novel deterministic algorithm for solving the problems of this class based on the information-statistical approach to solving the continuous global optimization problems has been proposed. The comparison of this algorithm with known analogs demonstrating the efficiency of the developed approach has been conducted. The stable operation of the algorithm was confirmed also by solving a series of several hundred mixed-integer global optimization problems.

Keywords: Global optimization · Non-convex constraints · Mixed-integer problems

1 Introduction

In the present paper, the global optimization problems and the method of solving these ones are considered. The global optimization problems are the computation-costly ones since the global optimum is an integral characteristic of the problem being solved and requires the investigation of the whole search domain. The problems, in which some parameters can take the integer values only (mixed-integer global optimization problems) are of special interest because for these ones it is more difficult to build the estimates of the optimum as compared to the continuous problems.

A lot of publication have been devoted to the methods of solving the mixed-integer problems (see, for example, the reviews [1,2]). The well known deterministic methods of solving the problems of this class are based, as a rule, on the Branch-and-Bound approach or on the Branch-and-Reduce one. Also, a number of the metaheuristic and genetic algorithms are known, which are based one way or another on the random search concept.

In the present study, we proposed a novel deterministic method for solving the mixed-integer problems based on the information-statistical approach to solving the global optimization problems [3,4]. The paper text reflecting the results of preliminary studies is composed as follows. First, an approach to solving the problems with continuous parameters has been considered, then its generalization for the mixed-integer problems was proposed. In the final section, the results of comparison of the proposed method with the known analogs are presented.

© Springer Nature Switzerland AG 2019
R. Battiti et al. (Eds.): LION 12 2018, LNCS 11353, pp. 78–81, 2019.
https://doi.org/10.1007/978-3-030-05348-2_7

2   Global Search Algorithm and Dimension Reduction

Let us consider a multiextremal optimization problem in the form

$$\varphi(y^*) = \min \{\varphi(y) : y \in D, \ g_i(y) \le 0, \ 1 \le i \le m\}, \tag{1}$$
$$D = \{y \in R^N : a_i \le y_i \le b_i, 1 \le i \le N\}.$$

Suppose, that the objective function $\varphi(y)$ (henceforth denoted by $g_{m+1}(y)$) and the left-hand sides $g_i(y)$, $1 \le i \le m$, of the constraints satisfy the Lipschitz condition and may be multiextremal.

Using a continuous single-valued mapping $y(x)$ (Peano-type space-filling curve, or *evolvent*) of the interval $[0, 1]$ onto D, a multidimensional problem (1) can be reduced to a one-dimensional problem

$$\min \{g_{m+1}(y(x)) : x \in [0, 1], \ g_i(y(x)) \le 0, \ 1 \le i \le m\}. \tag{2}$$

The reduction of dimensionality matches the multidimensional problem with a Lipschitzian objective function and constraints with a one-dimensional problem where the respective functions satisfy the uniform Hölder condition (see [3]).

An efficient global search algorithm (GSA) for solving the constrained optimization problem (2) has been developed at University of Nizhni Novgorod. A detailed description of the algorithm and the corresponding convergence theory are presented in [3].

3   Global Search Algorithm for Mixed-Integer Problems

Now, let us consider a method of adaptation of GSA for solving the mixed-integer global optimization problems

$$\min \{g_{m+1}(y) : y \in D, \ g_i(y) \le 0, \ 1 \le i \le m\}, \tag{3}$$
$$D = \{a_i \le y_i \le b_i, \ 1 \le i \le N, \ y_j \in Z, \ j \in J, \ y_i \in R, \ i \notin J\}.$$

Let us formulate a set of problems based on the problem (3)

$$\min \{g_{m+1}^s(y) : y \in D, \ g_i^s(y) \le 0, \ 1 \le i \le m\}, \tag{4}$$
$$D = \{a_i \le y_i \le b_i, \ y_i \in R, \ i \notin J\}, \ s \in \{1, ..., S\}.$$

Each of these problems corresponds to the initial problem (3) with some combination of integer parameters. The number of problems S will correspond to the number of all possible combinations of the integer parameters.

Using the dimensionality reduction scheme utilizing the evolvent $y(x)$ one can compose a united problem

$$\min \{g_{m+1}^s(y^s(x)) : x \in [0, S], \ g_i^s(y^s(x)) \le 0, \ 1 \le i \le m\}, \tag{5}$$

where the mapping $y^s(x)$ is formed on the basis of initial evolvent $y(x)$ as follows

$$y^s(x) = y(x - s), \ x \in [s - 1, s], \ s \in \{1, ..., S\}.$$

The functions $g_i^s(y^s(x))$ can have the breaks at the points $s \in \{1, ..., S-1\}$, therefore, these points were considered as the punctured ones, the values of the objective function and constraints at these points remain undefined.

As an illustration, Fig. 1 presents the plots of the functions corresponding to a problem with one continuous parameter and one binary parameter.

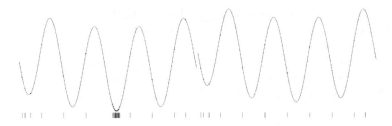

Fig. 1. Reduced mixed-integer global optimization problem

Applying the global search algorithm to solving the problem (5), we will find the solution of the problem (3). In this case the major part of trials will conducted in the subproblem, the solving of which corresponds to the solving of the initial problem (3). In the rest subproblems, a only minor part of trials will be performed since the solutions of these subproblems are the locally optimal ones. All the above is confirmed by Fig. 1, where the points of trials executed in the course of solving this problem are denoted by the dashes.

Thus, we have constructed the *Mixed-Integer Global Search Algorithm* (MIGSA) based on the reduction of the mixed-integer non-convex optimization problem to the non-convex optimization problem.

4   Results of Experiments

Let us compare the proposed MIGSA method with a genetic algorithm for solving the mixed-integer optimization problems implemented in Matlab Global Optimization Toolbox [6]. In Table 1, the numbers of trials required for solving the known test mixed-integer problems by these methods are presented. For both methods, the same accuracy of search 10^{-2} were used. All numerical experiments were conducted on a computer with Intel Core i5-7300 2.5 GHz processor and 8 Gb RAM under MS Windows 10. The results of experiments have demonstrated the advantage to the MIGSA method in the number of iterations as well as in the execution time.

In order to demonstrate the reliability of the MIGSA method, we have solved two series of 100 multiextremal mixed-integer problems each constructed on the basis of modified problems of *Simple* and *Hard* classes generated by GKLS generator. This generator of multiextremal functions is often used for the investigations of the global optimization algorithms [8–11]. In the problems generated in

Table 1. Comparison of MIGSA and GA

Test problem	GA		MIGSA	
	k	t	k	t
Problem 2 [5]	481	0.0601	417	0.04
Problem 6 [5]	641	0.0510	118	0.001
Problem 1 [7]	481	0.1378	66	0.0007
Problem 2 [7]	481	0.0473	57	0.0006
Problem 7 [7]	841	0.0736	372	0.017

our experiments, there were 5 integer and 3 continuous parameters. The accuracy of the search was equal to 10^{-2}. All problems of the series have been solved successfully and for solving the problems based on the *Simple* class 11988 trials in average were required whereas for solving the problems based on the *Hard* class 24750 trials were required.

Acknowledgments. This study was supported by the Russian Science Foundation, project No 16-11-10150.

References

1. Burer, S., Letchford, A.N.: Non-convex mixed-integer nonlinear programming: a survey. Surv. Oper. Res. Manag. Sci. **17**, 97–106 (2012)
2. Boukouvala, F., Misener, R., Floudas, C.A.: Global optimization advances in Mixed-Integer Nonlinear Programming, MINLP, and Constrained Derivative-Free Optimization CDFO. Eur. J. Oper. Res. **252**, 701–727 (2016)
3. Strongin, R.G., Sergeyev, Y.D.: Global Optimization with Non-convex Constraints. Sequential and Parallel Algorithms. Kluwer Academic Publishers, Dordrecht (2000)
4. Sergeyev, Ya.D., Strongin, R.G., Lera, D.: Introduction to Global Optimization Exploiting Space-Filling Curves. Springer (2013)
5. Floudas, C.A., Pardalos, P.M.: Handbook of Test Problems in Local and Global Optimization. Springer (1999)
6. https://www.mathworks.com/help/gads/mixed-integer-optimization.html
7. Deep, K., Singh, K.P., Kansal, M.L., Mohan, C.: A real coded genetic algorithm for solving integer and mixed integer optimization problems. Appl. Math. Comput. **212**(2), 505–518 (2009)
8. Paulavičius, R., Sergeyev, Y., Kvasov, D., Žilinskas, J.: Globally-biased DISIMPL algorithm for expensive global optimization. J. Glob. Optim. **59**(2–3), 545–567 (2014)
9. Sergeyev, Y.D., Kvasov, D.E.: A deterministic global optimization using smooth diagonal auxiliary functions. Commun. Nonlinear. Sci. Numer. Simul. **21**(1–3), 99–111 (2015)
10. Lebedev, I., Gergel, V.: Heterogeneous parallel computations for solving global optimization problems. Procedia Comput. Sci. **66**, 53–62 (2015)
11. Gergel, V., Sidorov, S.: A two-level parallel global search algorithm for solution of computationally intensive multiextremal optimization problems. Lect. Notes Comput. Sci. **9251**, 505–515 (2015)

Massive 2-opt and 3-opt Moves with High Performance GPU Local Search to Large-Scale Traveling Salesman Problem

Wen-Bao Qiao$^{(\boxtimes)}$ and Jean-Charles Créput

Le2i, CNRS, Arts et Métiers, University Bourgogne Franche-Comté,
Besançon, France
rapidbao@outlook.com

Abstract. 2-opt, 3-opt or $k-$opt heuristics are classical local search algorithms for traveling salesman problems (TSP) in combinatorial optimization area. This paper introduces a judicious decision making methodology of offloading which part of the $k-$opt heuristic works in parallel on Graphics Processing Unit (GPU) while which part remains sequential, called "multiple $k-$opt evaluation, multiple $k-$opt moves", in order to simultaneously execute, without interference, massive 2-/3-opt moves that are globally found on the same TSP tour or the same Euclidean space for many edges, as well as keep high performance for GPU massive $k-$opt evaluation. We prove the methodology is judicious and valuable because of our originally proposed sequential non-interacted 2-/3-exchange set partition algorithm taking linear time complexity and a new TSP tour representation, array of ordered coordinates-index, in order unveil how to use GPU on-chip shared memory to achieve the same goal as using doubly linked list and array of ordered coordinates for parallel $k-$opt implementation. We test this methodology on 22 national TSP instances with up to 71009 cities and with brute initial tour solution. Average maximum 997 non-interacted 2-opt moves are found and executed on the same tour of ch71009.tsp instance in one iteration of our proposed method. Experimental comparisons show that our proposed methodology gets huge acceleration over both classical sequential and a possible current fastest state-of-the-art GPU parallel 2-opt implementation.

Keywords: Massive 2-opt moves · Parallel 2-opt · Optimization
High performance GPU computing

1 Introduction

The traveling salesman problem (TSP) [5,7] is one of the well-known combinatorial optimization problems that have been classified as NP-hard [2]. One solution to a TSP instance is one permutation of input N cities, while different possible permutations form various solutions. According to the permutation, the

© Springer Nature Switzerland AG 2019
R. Battiti et al. (Eds.): LION 12 2018, LNCS 11353, pp. 82–97, 2019.
https://doi.org/10.1007/978-3-030-05348-2_8

salesman travels each city once and returns to its starting city. Symmetric TSP forms undirected graph where distance between every two nodes is same. Asymmetric TSP forms directed graph where paths may not exist or the distances in opposite direction might be different between two cities. To optimize a given initial TSP solution, various TSP heuristics are used to constitute combinatorial optimization mataheuristics like TABU search [3,4] and variable neighborhood search (VNS) [10].

One of the most commonly used heuristic methods to approach TSP is repeating a series of local search algorithm called 2-opt [1], 3-opt and $k-$opt, which will often result in a tour with length less than 5% (2-opt) or 3% (3-opt) above the Held-Karp bound [5,9]. The $k-$opt algorithm includes two sub-steps: search and execute $k-$opt. **The classical complete sequential 2-opt search** that aims to search the best 2-opt move for the whole tour often indicates that every pair of edges have been checked once, which leads to totally $\frac{N*(N-1)}{2}$ 2-opt checks needed to be done in **one 2-opt run**. Similarly, a complete 3-opt search leads to totally $\frac{N*(N-1)*(N-2)}{6}$ checks in one 3-opt run in order to find the best 3-opt move for the whole tour. However, this classical $k-$opt implementation is obviously time-consuming when treating an initial brute large-scale TSP solution that needs to iterate millions of 2-opt or 3-opt moves before getting local optimal. Besides, implementing a $k-$opt algorithm involves reversing permutation order of cities in some segments of the original tour, which should be carefully treated in order to keep the tour valid.

Though it exists various sequential or parallel strategies to accelerate iterative $k-$opt local search in literature, we do not find a fast 2-opt implementation that simultaneously executes massive 2-opt moves globally found on the same tour or in the same Euclidean space for many edges except our previous work in [12]. Related works either execute multiple 2-opt moves in separate disjoint partitions of the original tour [18] or partitions of the Euclidean space [6], or execute one 2-opt move after a whole procedure of parallel evaluation [16,17], or take quadratic time complexity to select non-interacted 2-opt moves [15].

In this paper, we propose a judicious decision making methodology of offloading which part of the $k-$opt heuristic work in parallel on GPU while which part remains sequentially, in order to simultaneously execute, without interference, massive 2-/3-exchanges that are globally found on the same TSP tour for many edges as well as keeping high performance on GPU side. It consists of two parts as shown in Algorithm 1, various GPU parallel $k-$opt local search implementations evaluate candidate $k-$opt moves along current global tour for many (or for all) edges, but only these non-interacted $k-$opt moves can be sequentially selected within linear time in order to be executed simultaneously in one iteration, which is simplified as "multiple $k-$opt evaluation, multiple $k-$opt moves".

This methodology is judicious since intervention of a sequential tour reversal operation taking $O(N)$ time for each 2-/3-opt move is unavoidable when using arrays as TSP tour data structure for high performance GPU computing that considers coalesced memory access and usage of limited on-chip shared memory.

This methodology is valuable because of an originally proposed sequential $O(N)$ time complexity non-interacted 2-exchange set partition algorithm and a new TSP data structure, *array of ordered coordinates-index*, in order to unveil how to use GPU on-chip shared memory to achieve the same goal as using doubly linked list and array of ordered coordinates for parallel $k-$opt implementation. Besides executing massive 2-/3-opt moves, these two innovations further reduce total amount of costly tour reversal operations and data transmission between host and GPU sides for iterative 2-/3-opt implementations.

Algorithm 1 Methodology of massive 2-/3-opt moves along the global tour with high performance GPU local search, named as *"multiple k-opt evaluation, multiple k−opt moves"*.

1: **while** before termination condition **do**
2:      Prepare array of ordered coordinates-index for GPU parallel computing;
3:      Data copy to GPU;
4:      Various parallel $k-$opt implementations that can find candidate exchanges for many (or for all) edges along current global TSP tour;
5:      Transfer information of these candidate exchanges to CPU;
6:      Sequentially select non-interacted $k-$opt exchanges with linear time complexity;
7:      Concurrently execute these selected non-interacted $k-$opt moves;
8: **end while**

The following paper is organized as this: Sect. 2 gives a brief introduction of the 2-/3-opt algorithms and the TSP formulation used in this paper. Section 3 presents related sequential and parallel work on accelerating iterative 2-/3-opt implementations. Section 4 presents a new TSP tour data structure for GPU high performance parallel 2-/3-opt local search in order to obtain massive candidate 2-/3-opt moves. Section 5 presents our proposed algorithms about "multiple $k-$opt evaluation, multiple $k-$opt moves" methodology. Section 6 shows experimental results and comparisons, while Sect. 7 concludes this paper.

2   TSP Formulation and the Iterative 2-/3-opt

Given a symmetric Euclidean TSP instance, let V be a set of N cities in 2D plane and $d(u, v) \in \Re$ a distance for each pair of cities $u, v \in V$. Let S be a set of all possible permutation solutions of a complete undirected graph $(V, V \times V)$ and $s \in S$ a solution. Let $\pi : \{0, ..., k, k+1, ..., N-1\} \rightarrow V$ be a bijection. A TSP tour solution $s = (\pi_0, \pi_1, \pi_2, ..., \pi_k, \pi_{(k+1)}, ..., \pi_{(N-1)})$ indicates that the input N cities have been permuted with an increasing permutation order from 0 to $(N-1)$ along current tour direction. Edge e_k indicates the salesman travels from city π_k to π_{k+1} in order to avoid traveling one city twice, $(k+1)modN = 0$ in case of $k = (N-1)$. The goal is to find a permutation solution $s \in S$ that minimize the cost function $f(s) = \sum_{k=0}^{N-1} d(e_k)$.

Consider two neighbor TSP tour solutions $s_i, s_{i+1} \in S$, if $s_{(i+1)}$ would be shorter than s_i by removing two edges and inserting two edges, this transition is defined as **a 2-opt exchange (or 2-opt move)** $\mathcal{N}_2$ [8]. Searching one 2-opt move should consider related cities' tour order. Given two oriented edges (e_k, e_p) and four related cities $i = \pi_k, i + 1 = \pi_{k+1}, j = \pi_p, j + 1 = \pi_{p+1}$, the algorithm should check whether $d(i, i+1) + d(j, j+1) > d(i, j) + d(i+1, j+1)$ to remain the tour valid, which is called one **2-opt check** or **2-opt evaluation**. Removing two edges from the original tour produces two **sub-tours**, while inserting two new edges will reverse original permutation order of cities in one of the two sub-tours as shown in Fig. 1.

3-opt $\mathcal{N}_3$ or $k-$opt $\mathcal{N}_k$ works in similar fashion with 2-opt. Removing three edges from the original tour produces three sub-tours. After executing a correct 3-exchange, one or two sub-tours' original permutation order have been inversed.

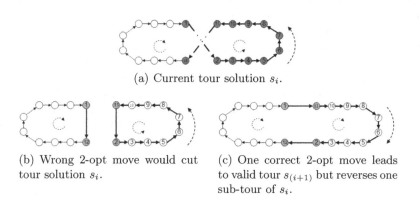

(a) Current tour solution s_i.

(b) Wrong 2-opt move would cut tour solution s_i.

(c) One correct 2-opt move leads to valid tour $s_{(i+1)}$ but reverses one sub-tour of s_i.

Fig. 1. Tour order plays an important role for iterative 2-opt implementation. ① represents one city's tour order on current tour solution s_i.

3   Related Works

Both sequential and parallel strategies have been studied to accelerate performance of the classical iterative 2-opt local search implementation.

Various sequential speeding up strategies exist to improve efficiency of the classical sequential iterative 2-/3-opt implementation. For example, instead of executing complete $\frac{N*(N-1)}{2}$ 2-opt checks to find the best 2-opt move for the integral tour, the first strategy executes $(N - 1)$ 2-opt checks along the global tour to find the best 2-opt move for one edge, which is the well-known "best improvement for one edge". The second strategy further decreases local search ranges for one edge's optimization, for example, manually keep a fix m nearest neighbor for one edge's optimization, or the well-known "first improvement for one edge" strategy that prunes the k-opt search if the first optimization for one edge is found. After one edge is optimized, these algorithms begin next edge's optimization by repeating the same procedure until termination condition is met.

TSP tour data structure is an important factor since it influences the efficiency of tour reversal operation, efficiency of multiple 2-/3-opt moves, and efficiency of parallel local search on GPU CUDA architecture. For example, in order to re-make an integral TSP tour, it is necessary to take $O(N)$ time complexity for the tour reversal operation after each $k-$opt move when using an array of ordered coordinates as TSP tour data structure. This is bad for brute initial TSP solutions that might need millions of selected $k-$opt moves before getting to local optimal. However, when working on doubly linked list where each node has two links pointing to its two neighbor cities, this reversal operation only involves changing links of related cities and does not take $O(N)$ time to re-make an integral tour. When considering parallel computing on GPU with coalesced memory access and usage of the limited size of on-chip shared memory, array of ordered coordinates is often preferred.

Parallelism of 2-opt implementations have been studied since a long time ago [6,12,13,15–18]. Nevertheless, these parallel $k-$opt implementations can not execute simultaneously massive 2-opt or 3-opt moves that are globally found along the same tour or in the integral Euclidean space for many edges, since the limitation of parallel hardwares or the necessary tour reversal operation that needs communication between $k-$opt moves. For example, parallel 2-opt implementations based on data decomposition like these works [6,13,18] execute multiple 2-opt moves in independent disjoint partitions by partitioning the integral tour or Euclidean space; parallel 2-opt implementations like Rocki et al. [16,17] distribute complete $\frac{N*(N-1)}{2}$ 2-opt checks to GPU threads to select the best 2-opt move in current tour, while they need CPU-aided tour reversal operation after each 2-opt move to prepare the TSP array of ordered coordinates for next iteration of high performance parallel $\frac{N*(N-1)}{2}$ 2-opt checks. They achieve a high performance GPU local search implementation because each thread responds to certain quantity of 2-opt checks on GPU on-chip shared memory with coalesced memory access.

Our previous work in [12] had mentioned this "multiple $k-$opt evaluation, multiple $k-$opt moves" methodology and the $O(N)$ time complexity noninteracted 2-opt set partition algorithm, in order to execute multiple 2-opt moves found in the same Euclidean space, while much less details were explained due to the page limitation. Rios et al. [15] propose a distributed variable neighborhood descent metaheuristic on multi-GPU system, while their method verifies conflict between a candidate move and the remaining moves [15], so that each verification goes over all the remaining moves and takes quadratic time complexity to judge conflict between multiple moves.

In general, two main researching branches on multiple 2-/3-opt moves can be categorized: one is parallel local search on disjoint partitions of the same tour [18] or Euclidean space [6], the other one is our proposed massive 2-/3-opt moves that are found globally on the same tour or the integral Euclidean space.

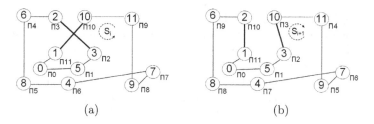

(a)   (b)

Fig. 2. (a) Tour solution s_i possessing two candidate 2-exchanges; (b) New tour after executing the 2-exchange $\mathcal{N}_2^{2,10}$ on tour s_i. ① represents one city' input index in the input array of coordinate, π_1 indicate the city's current tour order.

4   TSP Tour Data Structure for GPU Massive $k-$opt Evaluation

TSP tour data structure for GPU massive $k-$opt evaluation should concern two main aspects: tour reversal operation and coalesced memory access for GPU high performance computing. In the proposed methodology show in Algorithm 1, we profit from doubly linked list data structure to execute massive selected non-interacted 2-opt moves, while this data structure can not support high performance GPU parallel computing that often prefer arrays of ordered coordinates.

Array of ordered coordinates shown in Fig. 3, where the order of cities in the array also represents current tour's permutation order at any time [11,17] has two advantages for GPU $k-$opt implementation: coalesced memory access and it can be split into segments to be partially reloaded on GPU on-chip shared memory when the input TSP size is too large to be entirely reloaded by on-chip shared memory [17]. However, considering the necessary reversal operation during $k-$opt implementation, too much cities will be involved in relocation for each exchange when using this data structure, as one case shown in Fig. 3 for single 2-exchange $\mathcal{N}_2^{2,10}$ shown in Fig. 2. Also, changing one node in the array of ordered coordinate will influence two oriented edges in current tour solution.

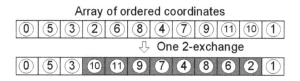

Fig. 3. The way the 2-opt move $\mathcal{N}_2^{2,10}$ in Fig. 2(a) happens on TSP tour represented by array of ordered coordinates.

Doubly linked list (DLL) shown in Fig. 4, where each city has two links pointing to its two neighbor cities in current TSP tour, is suitable for doing this tour reversal job because an exchange on DLL only needs to change links of related cities. Also, changing one of the two links possessed by a node in DLL only influences one oriented tour edge. However, memory access pattern is scattered when using DLL, and DLL can not work on GPU on-chip shared memory when the size of TSP instance exceeds the size of on-chip shared memory, because DLL should necessarily have global access to the input array of coordinates.

We propose a new TSP tour data structure called **Array of Ordered Coordinates-Index.** Each element of this array consists of one city's Euclidean coordinates (usually two float values) and an index of this city. This index is identifier of a city on the array of doubly linked list, which is same with the index of input array of city coordinate and is used for later registering candidate k-exchanges' information during the parallel $k-$opt local search. This array is also an ordered array where order of a city in the array also represents its permutation tour order. Figure 5 is an example to represent the tour shown in Fig. 2(a).

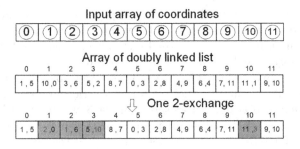

Fig. 4. The way the 2-opt move $\mathscr{N}_2^{2,10}$ in Fig. 2 happens on TSP tour represented by doubly linked list.

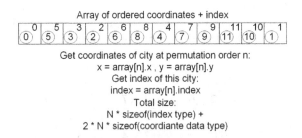

Fig. 5. Array of ordered coordinates-index for GPU parallel $k-$opt evaluation that combines the advantages of doubly linked list and array of ordered coordinates.

5   Multiple $k-$opt Evaluation, Multiple $k-$opt Moves

In order to obtain high performance GPU $k-$opt evaluation, it is a better choice to use ordered arrays as TSP tour data structure, which leads to a problem that tour reversal operation is unavoidable for each 2-/3-opt move and usually happens sequentially taking $O(N)$ time complexity. This naturally serialize the execution of multiple 2-/3-opt moves.

One question appears to be "can massive 2-opt moves that are found globally along the same tour or in the same Euclidean space be executed simultaneously?" The answer is yes, while the necessary tour reversal operation of $k-$opt implementation makes massively and concurrently global $k-$opt moves become complex. We take the basic 2-opt as an example to explain this complexity:

- One 2-opt check needs to consider tour ordering to avoid cutting the tour, as shown in Fig. 1;
- Executing massive correct 2-opt moves that are found globally along the same tour for many edges may also cut the tour as shown in Fig. 6(a, b).

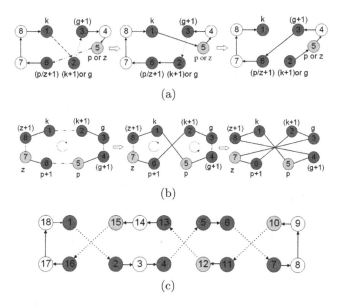

Fig. 6. Cases of massive 2-opt moves found during one run of parallel 2-opt local search. (a, b) Cases of multiple 2-opt moves interacting with each other: in (a) 2-opt move $\mathcal{N}_2^{1,5} \rightarrow (e_1, e_5)$ interacts with $\mathcal{N}_2^{2,5} \rightarrow (e_2, e_5)$; in (b) 2-opt move $\mathcal{N}_2^{1,5} \rightarrow (e_1, e_5)$ interacts with $\mathcal{N}_2^{3,7} \rightarrow (e_3, e_7)$; execution of the two 2-opt moves in (a,b) will cut the original integral tour. (c) One case that multiple 2-opt moves do not interact with each other. ① represents one city's tour order.

However, when working on doubly linked list TSP data structure, these massively *non-interacted 2-opt moves* shown in Fig. 6(c) can be executed in parallel without cutting the tour or doing tour reversal operation after each exchange. Definition of non-interacted 2-opt moves mainly profits from distribution of the special sub-tour $\pi_{k+1} \rightarrow \pi_p$ of each oriented 2-exchange $\mathcal{N}_2^{k,p}$ along the same

TSP tour direction, as shown in Fig. 7. We introduce below the two main steps to answer the question posed at beginning of this section by using the proposed methodology of "multiple k−opt evaluation, multiple k−opt moves" in Algorithm 1.

5.1   Multiple k−opt Evaluation on GPU

A certain parallel k−opt implementation evaluates (but not executes) many edges' or all edge's k−opt optimization on the same global tour, which mainly takes advantages of parallel reading ability on GPU device memory. As this step is not main contribution of this paper, we will simply extend it in the experimental section. However, various parallel k−opt Evaluation operators should satisfy following two requirements for later selecting multiple non-interacted k−opt moves, we take 2-opt as an example:

- Firstly, each 2-opt check $\mathcal{N}_2^{kp} \to \{e_k, e_p\}$ is oriented along the same tour direction, edge $e_p = (\pi_p, \pi_{p+1})$ is visited after edge $e_k = (\pi_k, \pi_{k+1})$ in current oriented tour solution $s = (\pi_0, \pi_1, ..., \pi_{(N-1)})$.
- Secondly, for each 2-opt check between two edges $\{e_k, e_p\}$ that optimizes the original tour, the parallel 2-opt local search implementation has to memorize index of city π_p only to city π_k. This is information of candidate 2-exchanges for later selecting and executing non-interacted 2-opt moves $\mathcal{N}_2^{kp}$ only by visiting the city π_k.

To satisfy these two requirements as well as keep high performance of GPU parallel local search that considers coalesced memory access to the GPU shared memories, the proposed data structure, array of ordered coordinates-index, is an ideal choice for TSP tour representation.

5.2   Non-interacted 2-opt Moves Set Partition with Linear Time

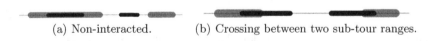

(a) Non-interacted.   (b) Crossing between two sub-tour ranges.

Fig. 7. Possible distribution of sub-tours (red and black) produced by two candidate 2-exchanges along the same TSP tour.

We propose an originally sequential algorithm to select these non-interacted 2-opt moves on the same TSP solution $s_i \in S$, which takes $O(N)$ time complexity. The algorithm is sequential since these candidate 2-opt moves as well as their related sub-tour ranges are randomly distributed along the original tour, while this operation needs communication among massive candidate 2-opt moves.

This sequential selecting algorithm selects non-interacted 2-opt moves while only traverses the tour solution s_i once, as shown in Algorithm 2. It accesses each city of s_i along the same tour direction in which these massive candidate exchanges are found. The function **JudgeNonInteracted** in Algorithm 2 judges whether a candidate 2-exchange interacts with the previously selected one according to whether their sub-tours $\pi_{k+1} \to \pi_p$ are partially crossed as shown in Fig. 7(b). It takes advantage of two vacant stacks to decrease complexity for judging non-interacted exchanges and only compare the newly met 2-exchange with the last selected 2-exchange whose information is stored at top of the Stacks. The algorithm steps through each city along the original tour solution, if the newly met city plays a role of π_p of a selected 2-exchange $\mathcal{N}_2^{kp}$, the algorithm POPs top element π_k from stack A as well as StackB and semaphores city π_k as possessing a selected non-interacted 2-exchange $\mathcal{N}_2^{kp}$. Then the algorithm steps into next city. If the newly met city possesses a 2-exchange $\mathcal{N}_2^{k_{new},p_{new}}$ that does not interact with last selected one, city $\pi_{k_{new}}$ is pushed into stack A, city $\pi_{p_{new}}$ is pushed into stack B.

Algorithm 2 Sequentially select *non-interacted 2-opt moves* along current TSP solution s_i with linear time complexity.

Require: From s_i's starting city π_0, traverse each city π_k according to current permutation order of s_i. Consider a candidate 2-exchange $\mathcal{N}_2^{kp}, p \neq (k+1), (k, p+1) mod N$, stack A is for city π_k , stack B is for city π_p.

1: **for** each city π_k from the starting city π_0 **do**
2:    **if** A.size > 0 && π_k is in Stack B **then** //(*City π_k has been occupied by previously selected 2-exchange.)*
3:        Mark A.top possesses a selected non-interacted 2-exchange $\mathcal{N}_2^{kp}$;
4:        Pop A.top;
5:        Pop B.top;
6:    **else if** π_k possesses a $\mathcal{N}_2^{kp}$ **then**
7:        Extract π_p from π_k's information package;
8:        **if** A.size == 0 && B.size == 0 **then**
9:            Mark first selected 2-exchange;
10:            Push π_k to stack A; Push π_p to stack B;
11:            Mark π_p is in stack B;
12:        **else** JudgeNonInteracted($\mathcal{N}_2^{kp}$, A.top, B.top) //*(Compare the new $\mathcal{N}_2^{kp}$ with the top 2-opt move stored in the stacks.)*
13:            **if** Non-interacted **then**
14:                Push π_k to stack A; Push π_p to stack B;
15:                Mark π_p is in stack B;
16:            **end if**
17:        **end if**
18:    **else**
19:        Continue next city $\pi_{(k+1)}$ along s_i; //*(One edge only participate in one 2-exchange.)*
20:
21:    **end if**
22: **end for**

Execute Massive 2-Exchanges. For the reason that information of each candidate 2-exchange $\mathcal{N}_2^{kp}$ has been memorized to its starting city π_k during the parallel 2-opt evaluation step, and all non-interacted 2-opt moves have been marked in the sequential selection step, they can be executed independently in arbitrary order on the doubly linked list data structure.

5.3   Extension to Massive $k-$opt Moves

With following four building blocks: high performance GPU parallel $k-$opt local search along the same tour direction, doubly linked edge list, array of ordered coordinates-index and the proposed sequential non-interacted k-exchange set partition algorithm, the above explanation for executing massive 2-opt moves can be extended to execute 3-opt or $k-$opt moves within the methodology of "multiple $k-$opt evaluation, multiple $k-$opt moves".

For example, one GPU parallel 3-opt evaluation $\mathcal{N}_3^{kpq}$ should memorize city π_p, π_q to city π_k during GPU parallel 3-opt evaluation; the sequential selection algorithm needs three stacks(stack A for π_k, stack B for π_p, stack C for π_q) to judge interference between a newly met 3-exchange and the last selected one, namely judge interference between the special sub-tours of the two 3-exchanges $\mathcal{N}_3^{kpq}$. An oriented 2-exchange $\mathcal{N}_2^{kp}$ produces one special sub-tour $\pi_{k+1} \rightarrow \pi_p$ while an oriented 3-exchange produces two special sub-tours $\pi_{k+1} \rightarrow \pi_p$ and $\pi_{p+1} \rightarrow \pi_q$.

6   Experiments

Various GPU parallel local search can be integrated into this multiple 2-/3-opt moves methodology once they support the two requirements (explained in Sect. 5.1) for the non-interacted k-exchange set partition algorithm. Key characteristic lies in that massive processors search many edges' optimization without using various disjoint tour or area partitions. In the literature, it exists following strategies that satisfy the key characteristic: brute GPU parallel implementation of classical sequential "best/first improvement for one edge" running simultaneously for many edges on GPU; Rocki' implementation [17] that distributes total $\frac{N*(N-1)}{2}$ 2-opt checks to GPU threads to select the best 2-opt move in one run of classical complete 2-opt search; our previous work [12] that searches N edges' 2-opt improvement among their separate Euclidean neighborhood edges using local spiral search. Considering efficiency of GPU computing, the parallel framework proposed by Rocki et al. [17] gets higher efficiency on GPU to search all edges' 2-opt optimization along the global tour.

Rocki's parallel framework can be briefly explained with Fig. 8. The total $\frac{N*(N-1)}{2}$ 2-opt checks for finding the best 2-opt move for the whole tour are distributed in the color boxes. Every pair of numbers in each box indicate current tour order of two cities π_k and π_p for $\mathcal{N}_2^{kp}$ separately. Due to the algorithm works on array of ordered coordinates data structure, one number in the box actually represent one oriented edge along current tour. One run of these $\frac{N*(N-1)}{2}$ 2-opt checks will produce massive candidate 2-exchanges along the global tour for many edges. For example the two red boxes marked in Fig. 8 will find the two

2-opt moves shown in Fig. 2(a), while Rocki's implementation has to perform two times of this parallel 2-opt runs and tour reversal operations to search and execute these two 2-opt moves in Fig. 2(a).

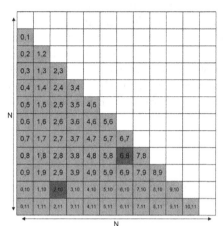

i,j : permutation order on array of order coordinates

Fig. 8. GPU parallel 2-opt local search scheme proposed by Rocki [17], in order to perform total $\frac{n*(n-1)}{2}$ 2-opt checks in one run of classical complete 2-opt search.

In our experiments, we build the well-known sequential iterative 2-opt with "first improvement along the global tour for one edge" which is simplified as "Serial-First-AT", and "best improvement along the global tour for one edge", simplified as "Serial-Best-AT". These two sequential algorithms work on doubly linked list data structure to avoid tour reversal operation after each exchange. We implement Rocki's implementation, marked as "Rocki-original". Our proposed 2-opt implementation is marked as "multiple 2-opt evaluation, multiple 2-opt moves" using adaptation to Rocki' parallel 2-opt evaluation strategy to achieve massive 2-opt moves. Tests are executed on National TSP instances from TSPLIB [14].

We implemented all algorithms with C++ and CUDA Toolkit v8.0. Code is compiled on laptop with CPU Intel(R) Core(TM) i7-4710HQ, 2.5 GHz running Windows and GPU card GeForce GTX 850M. GPU CUDA configuration is set as $<<<N/512 + 1, 512>>>$.

One test starts from the same initial solution whose permutation order is the same with the order of cities in original input files, ends with a status where no optimization can be found within a single iteration. The computing time includes necessary data copy between GPU and CPU. Figure 9 shows convergence speed of three algorithms on two TSP instances, lu980.tsp and sw24978.tsp. In Fig. 9, the vertical coordinates show percentage deviation compared with the optimum TSP tour solution provided by TSPLIB (%PD) after each iteration of three algorithms, the horizontal coordinate shows the accumulated time taken by previous

2-opt iterations before the end of one test. Figure 10 shows percentage deviation from the optimum of the mean results over 10 tests on national TSP instances, namely %PDM. Figures 11 and 12 show the average running time of one test. The TSP result of the proposed massive 2-opt methodology is slightly worse than other implementations, while the running time is greatly economized. For the largest "ch71009.tsp", as shown in Fig. 12, our proposed methodology only runs average 362.54 s in one test due to the reason that the methodology only iterates average 786 parallel 2-opt runs (iterations) to execute average total 306631 2-opt moves, namely average 392 non-interacted 2-opt moves are found and executed on the same tour in each iteration, while the classical sequential algorithms take more than 2 days per test to approach TSP instances bm33708 and ch71009 on our personal laptop.

7 Conclusion

In this paper, we propose detailed algorithms to simultaneously execute, without interference, massive 2-/3-opt moves that are globally found on the same

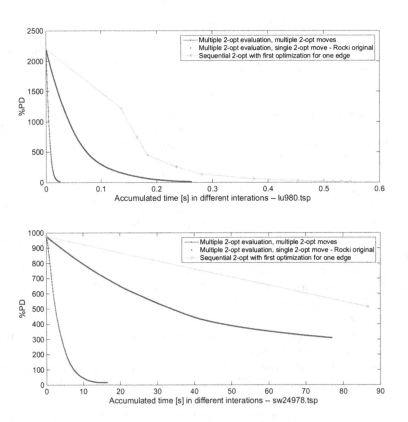

Fig. 9. Changes of TSP result's quality over time. The figure compares three iterative 2-opt implementations on two TSP instances lu980.tsp and sw24978.tsp separately.

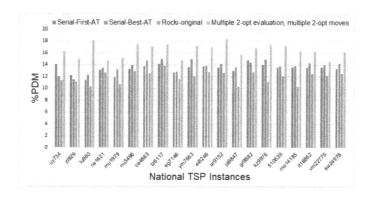

Fig. 10. PDM of TSP results using different iterative 2-opt implementations.

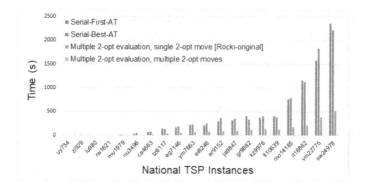

Fig. 11. Average running time of one test using different iterative 2-opt implementations on national TSP instances.

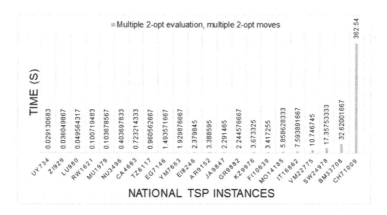

Fig. 12. Average running time of one test using our proposed "multiple 2-opt evaluation, multiple 2-opt moves" methodology on national TSP instances.

TSP tour, as well as to keep high performance both on host and GPU side. Experiments results show that our proposed methodology could possibly achieve the fastest convergence speed to get 2-/3-opt local optima for large scale TSP instances. Profiting from this methodology and continuous improvement in GPU hardware, very large instances, which could not be solved before, can now be solved in an acceptable time using the k−opt heuristic method.

References

1. Croes, G.A.: A method for solving traveling-salesman problems. Oper. Res. **6**(6), 791–812 (1958)
2. Garey, M.R., Johnson, D.S.: A Guide to the Theory of NP-Completeness, p. 70. WH Freemann, New York (1979)
3. Gendreau, M., Hertz, A., Laporte, G.: A tabu search heuristic for the vehicle routing problem. Manag. Sci. **40**(10), 1276–1290 (1994)
4. Glover, F.: Tabu search–part i. ORSA J. Comput. **1**(3), 190–206 (1989)
5. Johnson, D.S., McGeoch, L.A.: The traveling salesman problem: a case study in local optimization. Local Search Comb. Optim. **1**, 215–310 (1997)
6. Karp, R.M.: Probabilistic analysis of partitioning algorithms for the traveling-salesman problem in the plane. Math. Oper. Res. **2**(3), 209–224 (1977)
7. Laporte, G.: The traveling salesman problem: an overview of exact and approximate algorithms. Euro. J. Oper. Res. **59**(2), 231–247 (1992)
8. Lin, S.: Computer solutions of the traveling salesman problem. Bell Syst. Tech. J. **44**(10), 2245–2269 (1965)
9. Lourenço, H.R., Martin, O.C., Stützle, T.: Iterated local search. Handbook of Metaheuristics, pp. 320–353. Springer, Boston (2003). https://doi.org/10.1007/0-306-48056-5_11
10. Mladenović, N., Hansen, P.: Variable neighborhood search. Comput. Oper. Res. **24**(11), 1097–1100 (1997)
11. Mulder, S.A., Wunsch, D.C.: Million city traveling salesman problem solution by divide and conquer clustering with adaptive resonance neural networks. Neural Netw. **16**(5), 827–832 (2003)
12. Qiao, W., Créput, J.: Massive parallel self-organizing map and 2-opt on GPU to large scale TSP. In: Rojas, I., Joya, G., Catala, A. (eds.) IWANN 2017. LNCS, vol. 10305, pp. 471–482. Springer, Cham (2017). https://doi.org/10.1007/978-3-319-59153-7_41
13. Qiao, W.B., Créput, J.C.: Parallel 2-opt local search on GPU. World Acad. Sci. Eng. Technol. Int. J. Electr. Comput. Energ. Electron. Commun. Eng. **11**(3), 281–285 (2017)
14. Reinelt, G.: Tsplib–a traveling salesman problem library. ORSA J. Comput. **3**(4), 376–384 (1991)
15. Rios, E., Ochi, L.S., Boeres, C., Coelho, V.N., Coelho, I.M., Farias, R.: Exploring parallel multi-GPU local search strategies in a metaheuristic framework. J. Parallel Distrib. Comput. **111**, 39–55 (2018)
16. Rocki, K., Suda, R.: Accelerating 2-opt and 3-opt local search using GPU in the travelling salesman problem. In: High Performance Computing and Simulation, pp. 489–495. IEEE (2012)
17. Rocki, K., Suda, R.: High performance GPU accelerated local optimization in TSP. In: 2013 IEEE 27th International Parallel and Distributed Processing Symposium Workshops & Ph.D. forum (IPDPSW), pp. 1788–1796. IEEE (2013)

18. Verhoeven, M., Aarts, E.H., Swinkels, P.: A parallel 2-opt algorithm for the traveling salesman problem. Future Gener. Comput. Syst. **11**(2), 175–182 (1995)

Instance-Specific Selection of AOS Methods for Solving Combinatorial Optimisation Problems via Neural Networks

Teck-Hou Teng[2(✉)], Hoong Chuin Lau[1(✉)], and Aldy Gunawan[2(✉)]

[1] School of Information Systems, Singapore Management University, Singapore, Singapore
hclau@smu.edu.sg

[2] Fujitsu-SMU Urban Computing and Engineering Corporate Laboratory, Singapore Management University, Singapore, Singapore
{thteng,aldygunawan}@smu.edu.sg

Abstract. Solving combinatorial optimization problems using a fixed set of operators has been known to produce poor quality solutions. Thus, adaptive operator selection (AOS) methods have been proposed. But, despite such effort, challenges such as the choice of suitable AOS method and configuring it correctly for given specific problem instances remain. To overcome these challenges, this work proposes a novel approach known as I-AOS-DOE to perform Instance-specific selection of AOS methods prior to evolutionary search. Furthermore, to configure the AOS methods for the respective problem instances, we apply a Design of Experiment (DOE) technique to determine promising regions of parameter values and to pick the best parameter values from those regions. Our main contribution lies in the use a self-organizing neural network as the offline-trained AOS selection mechanism. This work trains a variant of FALCON known as FL-FALCON using performance data of applying AOS methods on training instances. The performance data comprises derived fitness landscape features, choices of AOS methods and feedback signals. The hypothesis is that a trained FL-FALCON is capable of selecting suitable AOS methods for unknown problem instances. Experiments are conducted to test this hypothesis and compare I-AOS-DOE with existing approaches. Experiment results reveal that I-AOS-DOE can indeed yield the best performance outcome for a sample set of quadratic assignment problem (QAP) instances.

1 Introduction

Evolutionary search is commonly used to solve combinatorial optimization problems. During evolutionary search, adaptive operator selection (AOS) is used to select operators adaptively that will hopefully lead the search toward optimality. Though it may be highly probable for seasoned practitioners to gain specialized knowledge of performance characteristics for a subset of problem instances, it

© Springer Nature Switzerland AG 2019
R. Battiti et al. (Eds.): LION 12 2018, LNCS 11353, pp. 98–114, 2019.
https://doi.org/10.1007/978-3-030-05348-2_9

is a non-trivial endeavor to possess such specialized knowledge for all problem instances. Moreover, with the availability of more sophisticated machine learning techniques, such specialized knowledge may have become obsolete.

Following on a work that investigates varying operator settings [18], there has been interest on designing better AOS mechanism. Evidences of such interest include improving bandit-based operator selection mechanisms by [4,5]. There is also a study on the use of adaptive pursuit in neuro-evolution algorithm [8]. These works contributed to the state-of-the-art by extending specific AOS methods. To our knowledge, there has been no attempt to achieve a generic method for selecting AOS methods at the problem instance level. Also of interest to this work is on the topic of fitness landscape analysis [12], the use of fitness landscape for selecting differential evolution operators [13] and algorithm selection problem [2]. This work is also encouraged by a neural network-based AOS method using fitness landscape features for selecting crossover operators [16].

This work aims to overcome the problem of selecting suitable AOS methods and configuring it to parameter values correct for specific problem instances [18]. Thus, we propose a novel approach, known as I-AOS-DOE, integrating self-organizing neural networks and design of experiment (DOE) techniques for instance-specific selection of AOS methods. Using DOE [6], our proposed approach treats the parameter values and AOS operators as decision variables. DOE was initially applied to identify the important parameters and determine ranges of their values. Further analysis is performed to identify the best parameter values in those ranges. A class of self-organizing neural networks known as Fusion Architecture for Learning and Cognition (FALCON) [14] is used as an adaptive selector of AOS methods. A specific variant of FALCON known as FL-FALCON [17] is trained using performance data of AOS methods on the problem instances. The performance data comprises derived landscape features, the selected AOS, the AOS parameter set and feedback signals indicating the solution quality. The hypothesis of this work is that the trained FL-FALCON can select suitable AOS methods for unknown problem instances. To test this hypothesis, experiments were conducted to evaluate the performance of the proposed approach using Quadratic Assignment Problem (QAP) instances. Comparing the performance of I-AOS-DOE with existing approaches, experiment results reveal that I-AOS-DOE can indeed yield the best performance outcome for a sample set of QAP instances.

The presentation of this work continues in Sect. 2 where several related works are reviewed. The problem addressed through this work is defined in Sect. 3. The proposed approach is presented in Sect. 4. This is followed by the presentation of the experiments and the results in Sect. 5. Last but not least, Sect. 6 summarizes and conclude this work.

2 Related Works

SATzilla was introduced as an algorithm portfolio selection methodology to solve SAT problem [20]. This is achieved by using a number of problem-specific features for a given SAT instance. The goal is to provide a runtime prediction model for SAT solvers. Later, SATzilla was enhanced to include explicit cost-sensitive loss function [21]. Earlier, Hydra was proposed as using parameter tuning to address algorithm portfolios [19]. Hydra tunes the solver using ParamILS and add parameterization to a SATzilla portfolio. In addition, ISAC was introduced as an instance-specific algorithm configurator for solving MaxSAT problems [7]. It overcomes the problem of proper parameterizing of algorithms by clustering training instances produced by those algorithms on MaxSAT problems. Later, ISAC++ generalizes tuning of individual solvers and combine multiple solvers into a solver portfolio [1]. ISAC++ takes an additional step of using any algorithm selector to choose one of the parameterizations.

SNAP-NEAT incorporates adaptive pursuit as the AOS mechanism for solving fractured problems [8]. To do so, SNAP-NEAT makes continuous updates and initial estimation of the operator values and probability. Evaluated using several problem instances, the experiment results reveal that SNAP-NEAT can select the best operators intelligently for the problem instances. Thus, SNAP-NEAT has demonstrated the ability to combine the strengths of NEAT, RBF-NEAT and Cascade-NEAT effect for solving reactive control and high-level problems.

A dynamic variant of the Multi-Armed Bandit (MAB) upper confidence bound algorithm was introduced as an AOS mechanism [5]. It uses a sliding window to update operator quality estimates, discard ancient events while preserving the recent ones. The MAB-based AOS mechanism was evaluated using artificial and real optimization problems for operator quality distribution. There is also an MAB-based AOS mechanism using fitness landscape analysis to better describe the resultant population during evolutionary search [4]. An online learning algorithm known as dynamic weighted majority (DWM) is used to model concept drifts. It was evaluated using several problem instances and compared with MAEN*-II. Other AOS mechanisms using fitness landscape analysis includes a landscape-based AOS mechanism for differential evolution (LSAOS-DE) for selecting multiple DE operators [13]. LSAOS-DE uses problem landscape information and performance histories of operators in an adaptive operator selection mechanism. LSAOS-DE is evaluated using single-objective optimization functions from CEC2014 and CEC2015 competitions. It is also compared with other heuristic-based selection method and state-of-the-art algorithms.

One-sided support vector regression was introduced to address the algorithm selection problem [2]. It employs exploratory landscape analysis to discover landscape features. That proposed approach is evaluated using several BBOB functions. The BBOB functions are divided into four classes according to modality, separability and global structure. Results from the experiments shows that one-sided support vector regression can generalize better than the selected benchmarks. It has also provided better insights on how landscape features can be mapped to efficient algorithms. There is also a meta-learning framework for addressing algorithm selection problem in continuous optimization problems [10]. The independent variables are the landscape features and algorithm parameters while the dependent variable is the algorithm performance. Performance of the proposed neural network-based approach is evaluated using the CEC2005 benchmarks. The predicted rankings made by the proposed approach are compared with the actual rankings and random ranking.

3   Problem Statement

This work addresses the problem of solving quadratic assignment problem (QAP) instances unknown to the solver. Landscape features are derived using fitness values from a different and unknown solver. Only the derived landscape features are presented as input to the solver for solving QAP problem instances. The generated solution is evaluated using fitness function on the QAP problem instances. In this work, there are several QAP instances, AOS methods and possible values for the AOS parameters. The performance of the AOS methods across the problem instances can be rather different [20]. Thus, the challenge here is to perform instance-specific selection of AOS methods for specific problem instances.

The application of solver to QAP problem instance j results in a fitness landscape comprising a list of objective values. A QAP problem instance has also an objective value of a best-known solution. This work subtracts the objective value of best-known solution from the objective values in the fitness landscape to obtain a list of objective value deviations. A set of fitness landscape features $\mathcal{F}^j$ for problem instance j is derived using the objective value deviations. The proposed solution is to recommend the choice of AOS method aos_i and possible values for AOS parameters ap_i^h for solving unknown problem instance j. Clearly, the resultant fitness landscape and solution is expected to be better than the quality of the initial fitness landscape $\mathcal{F}^j$. The quality of the fitness landscape is quantified using mean value and minimum value of the objective value deviation.

4   Instance-Specific Selection of AOS Methods

This paper proposed a framework, illustrated using Fig. 1, to address the problem of solving combinatorial optimization problems. Known as I-AOS-DOE, it performs instance-specific selection of AOS methods prior to conducting evolutionary search. I-AOS-DOE uses an artificial neural network to make good

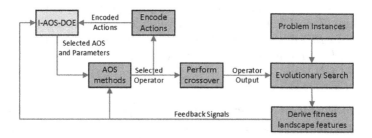

Fig. 1. Illustration of the proposed I-AOS-DOE framework for selecting AOS methods to unknown problem instances

recommendations of AOS methods and parameter values for problem instances adaptively. The statistical technique known as Design of Experiment (DOE) is used to identify good parameter values for the AOS methods.

4.1 Neural Network-Based Approach

I-AOS-DOE uses a class of ART-based self-organizing neural network known as FALCON [14] for selecting AOS methods. FALCON is favored for the ability to learn incrementally in real time. Based on the adaptive resonance theory (ART) [3], FALCON has been demonstrated in several prior works [14–17] to be capable of striking a delicate balance between specificity and generalizability. A specific variant of FALCON known as FL-FALCON [17] is used in this work.

Structure and Operating Modes. Seen in Fig. 2, FL-FALCON has a two-layer architecture, comprising an input/output (IO) layer and a knowledge layer. The IO layer has a sensory field F_1^{c1} for accepting state vector **S**, an action field F_1^{c2} for accepting action vector **A**, and a reward field F_1^{c3} for accepting reward vector **R**. The category field F_2^c at the knowledge layer stores the committed and uncommitted cognitive nodes. Each cognitive node j has template weights $\mathbf{w}^{ck}$ for $k = \{1, 2, 3\}$.

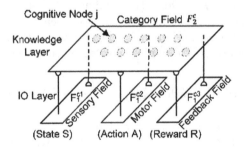

Fig. 2. The FALCON architecture.

FL-FALCON operates in one of the following operating modes. In *PER-FORM* mode, FL-FALCON selects cognitive node J for deriving action choice a for state s. In *LEARN* mode, FL-FALCON learns the effect of action choice a on state s. In *INSERT* mode, domain or external knowledge can be assimilated into FL-FALCON [15].

Insertion of Performance Data. This work prepares FL-FALCON for selecting AOS methods by inserting performance data of AOS methods on problem instances into it. This step prepares FL-FALCON for selecting AOS methods. Performance characteristics of AOS methods on problem instances are inserted into FL-FALCON in two stages. The first stage is to encode input pattern $\mathbf{x}^{ck}$

where derived fitness landscape features are encoded as state vector $\mathbf{x}^{c1}$, decision variables are encoded as action vector $\mathbf{x}^{c2}$ and feedback signal r are encoded as reward vector $\mathbf{x}^{c3}$, i.e., $\mathbf{x}^{ck} = \{\mathbf{x}^{c1}, \mathbf{x}^{c2}, \mathbf{x}^{c3}\}$. The second stage is to insert the encoded performance data into FL-FALCON. The insertion operation is performed by selecting a winning cognitive node J for input pattern $\mathbf{x}^{ck}$ using the code selection steps. Template learning is applied on winning cognitive node J to learn input pattern $\mathbf{x}^{ck}$.

The decision variables encoded in action vector $\mathbf{x}^{c2}$ are the identity of problem instance, the choice of AOS method and parameters of the selected AOS method. Different AOS methods have different number of parameters. To account for such differences, action vector $\mathbf{x}^{c2}$ is defined to be capable of holding the largest number of parameters. The continuous values of AOS parameters are between 0.0 and 1.0. A parameter having value x is encoded into action vector $\mathbf{x}^{c2}$ as $\{x, 1.0 - x\}$. Identity of problem instance and AOS method are nominal decision variables. To encode such information into action vector $\mathbf{x}^{c2}$, a numerical index v_i is used for decision variable i. After that, it is divided by available number of choices D_i to give a continuous numerical value d_i between 0.0 and 1.0. After that, numerical equivalent of decision variable i d_i is encoded into $\mathbf{x}^{c2}$ as $\{d_i, 1.0 - d_i\}$. This way of encoding a numerical value using two elements is known as complement coding.

Code Selection. Code selection is performed to identify winning cognitive node J for deriving action choice a for state s in *PERFORM* mode. A winning cognitive node J is identified in *LEARN* mode to learn the mapping of state s to action choice a with effect r. In *INSERT* mode, a winning cognitive node J is identified to learn an unit of domain knowledge.

A winning cognitive node J is identified in two stages. The first stage is known as code activation. This stage derive choice function T_j^c for the committed cognitive node j with respect to input pattern $\mathbf{x}^{ck}$ for $k = \{1, 2, 3\}$ using

$$T_j^c = \sum_{k=1}^{3} \gamma^{ck} \frac{|\mathbf{x}^{ck} \wedge \mathbf{w}_j^{ck}|}{\alpha^{ck} + |\mathbf{w}_j^{ck}|} \tag{1}$$

where the fuzzy AND operation $(\mathbf{p} \wedge \mathbf{q})_i \equiv min(p_i, q_i)$, the norm $|.|$ is defined by $|\mathbf{p}| \equiv \sum_i p_i$ for vectors $\mathbf{p}$ and $\mathbf{q}$, $\alpha^{ck} \in [0, 1]$ is the choice parameters, $\gamma^{ck} \in [0, 1]$ is the contribution parameters and $k = \{1, 2, 3\}$. This is followed by a competition among the cognitive nodes using their choice function T_j^c to find winning cognitive node J using

$$J = \arg \max_j \{T_j^c : \text{for all } F_2^c \text{ node } j\} \tag{2}$$

This is followed by the second stage where the match functions m_j^{ck} of winning cognitive node J with respect to input pattern $\mathbf{x}^{ck}$ is derived and checked against vigilance parameters ρ^{ck} using

$$m_J^{ck} = \frac{|\mathbf{x}^{ck} \wedge \mathbf{w}_J^{ck}|}{|\mathbf{x}^{ck}|} \geq \rho^{ck} \tag{3}$$

A winning cognitive node J is only confirmed when it satisfies the vigilance criterion as expressed using (3). Using winning cognitive node J, FL-FALCON performs operation specific to the operating mode. In *LEARN* and *INSERT* modes, FL-FALCON executes the template learning operation. In *PERFORM* mode, FL-FALCON executes the activity readout operation using $\mathbf{w}_J^{c2}$.

Template Learning. This step is executed to learn an input pattern $\mathbf{x}^{ck}$ using a winning cognitive node J identified using code selection. Winning cognitive node J can either be a committed or uncommitted cognitive node. A committed cognitive node j contains a previously learned pattern while an uncommitted cognitive node j has not learned an input pattern.

Template learning takes place by modifying weight $\mathbf{w}_J^{ck}$ of winning cognitive node J using

$$\mathbf{w}_J^{ck(\text{new})} = (1 - \beta^{ck})\mathbf{w}_J^{ck(\text{old})} + \beta^{ck}(\mathbf{x}^{ck} \wedge \mathbf{w}_J^{ck(\text{old})}) \tag{4}$$

where $\beta^{ck} \in [0,1]$ is the learning rate. After that, cognitive node J becomes a part of the knowledge layer. It will participate in subsequent rounds of code selection.

Activity Readout. This is the process of acquiring action choice a from action field $\mathbf{w}_J^{c2}$ of winning cognitive node J. Previous works encode a single decision variable in $\mathbf{w}_J^{c2}$. By doing so, action choice a can be read out using

$$\mathbf{x}^{c2(\text{new})} = \mathbf{x}^{c2(\text{old})} \wedge \mathbf{w}_J^{c2} \tag{5}$$

In this work, decision variables are encoded as action vector $\mathbf{x}^{c2}$. The readout of the decision variables from action field $\mathbf{w}_J^{c2}$ of winning cognitive node J is performed in reverse order to the encoding scheme.

4.2   Input Pattern

Input pattern $\mathbf{x}^{ck}$ to FL-FALCON comprises state vector $\mathbf{x}^{c1}$, action vector $\mathbf{x}^{c2}$ and reward vector $\mathbf{x}^{c3}$. State vector $\mathbf{x}^{c1}$ encodes the fitness landscape features. Action vector $\mathbf{x}^{c2}$ encodes the decision variables. Reward vector $\mathbf{x}^{c3}$ encodes the feedback signal.

Fitness Landscape Features. Combinatorial optimization problem instances are solved using evolutionary search. Solution s_i^j generated at search iteration i for problem instance j has an objective value $o(s_i^j)$. Problem instance j has an optimal solution os^j with objective value $o(os^j)$. Objective value deviation δ_i^j of solution s_i^j at search iteration i to optimal solution os^j is derived using

$$\delta_i^j = \frac{|o(s_i^j) - o(os^j)|}{o(os^j)}. \tag{6}$$

Evolutionary search is conducted for $\mathcal{N}$ search iterations. Fitness landscape $\mathcal{F}^j$ is formed using $\mathcal{N}$ values of objective value deviation. Fitness landscape analysis (FLA) [12] and descriptive statistics techniques are applied on fitness landscape $\mathcal{F}^j$ to derive 10 landscape-based features comprising the minimum value (min), the maximum value (max), the median (med), the standard deviation ($std-dev$), the coefficient of variation (cv), range (r), skewness (sk), kurtosis (kt), proportion of upticks ($uptick$) and proportion of downticks ($downticks$).

For minimizing optimization problem j, min feature represents an objective value deviation δ_i^j of solution s_i^j having an objective value $o(s_i^j)$ closest to objective value $o(os^j)$ of optimal solution os^j. Similarly, max feature represents an objective value deviation δ_k^j of solution s_k^j having an objective value $o(s_i^j)$ furthest away from objective value $o(os^j)$ of optimal solution os^j. The difference between $o(s_i^j)$ and $o(s_k^j)$ gives the range of objective value deviation of fitness landscape $\mathcal{F}^j$. The standard deviation $std - dev$ feature measures the average amount of quantitative deviation from the mean value of fitness landscape $\mathcal{F}^j$. The coefficient of variation cv measures the dispersion of the objective value deviation δ_i^j making up fitness landscape $\mathcal{F}^j$. For this work, skewness (sk) is a measure of symmetry of fitness landscape $\mathcal{F}^j$. Kurtosis (kt) is a measure of *peakedness* of fitness landscape $\mathcal{F}^j$. With respect to search i, an uptick represents a worse solution is found at search iteration $i+1$. The *uptick* feature is the proportion of upticks with respect to $\mathcal{N}$ search iterations. Similarly, a downtick represents a better solution is found at search $i + 1$. The *downtick* feature is the proportion of downticks with respect to $\mathcal{N}$ search iterations. The derived landscape features are normalized and encoded using complement coding.

Decision Variables. The decision variables comprises the problem instance, the AOS method and values of AOS parameters. The choices of problem instance and AOS method have nominal values while the parameters are normalized continuous values. Thus, the problem instance and AOS method choices are encoded as part of the action vector by representing them numerically as indices. The problem index and AOS index are then normalized using the total number of problem instances and AOS methods respectively. The AOS methods have different number of tunable parameters. Thus, a fixed-length action vector is defined to accommodate the largest number of AOS parameters. Similar to the state vector, complement coding is used to encode the decision variables into the action vector. A compatible decoding scheme is used to retrieve the decision variables from the action vector.

Feedback Signal. This is a continuous value indicating the *quality* of fitness landscape for problem instance j. The quality of fitness landscape is measured with respect to the objective value of its optimal solution. Feedback signal f^j is derived as an average of the mean value $m(\delta^j)$ and the minimum value $min(\delta_i^j)$ of objective value deviations. The mean value gives an aggregated perception of *proximity* of raw fitness values to the fitness value of the optimal solution for a problem instance. Thus, smaller mean value indicates fitness landscape with better quality. The minimum value of objective value deviation represents the best possible solution for a problem instance. Thus, smaller minimum value of objective value deviation indicates the presence of better quality solution. Due to such significance, the feedback signal f^j is derived using $\frac{m(\delta^j)+min(\delta_i^j)}{2}$.

4.3 Select AOS Methods

Properly parameterized AOS methods are selected in two stages. The first stage is to insert performance data into FL-FALCON. This is following by the second stage of selecting AOS methods using the trained FL-FALCON. The steps for inserting performance data into FL-FALCON is outlined using Algorithm 1. It entails the derivation of fitness landscape features $\mathcal{F}^j$ for problem instance j. The derived fitness landscape features $\mathcal{F}^j$ are then encoded as state vector $\mathbf{x}^{c1}$. The decision variables and feedback signals are encoded as action vector $\mathbf{x}^{c2}$ and reward vector $\mathbf{x}^{c3}$ respectively. Together, they form the input pattern $\mathbf{x}^{ck}$ to be inserted into FL-FALCON.

Algorithm 1 Insertion of Performance Data into FL-FALCON

Require: performance data for applying AOS methods on problem instances
Ensure: FL-FALCON is initialize to appropriate structure and parameter values.
1: **while** has more performance data **do**
2:     derive fitness landscape features from raw fitness landscape
3:     encode derived fitness landscape features $\mathcal{F}^j$ as state vector $\mathbf{x}^{c1}$
4:     encode decision variables as action vector $\mathbf{x}^{c2}$
5:     encode feedback signal as reward vector $\mathbf{x}^{c3}$
6:     insert input pattern $\mathbf{x}^{ck} = \{\mathbf{x}^{c1}, \mathbf{x}^{c2}, \mathbf{x}^{c3}\}$ into FL-FALCON operating in *INSERT* mode
7: **end while**
8: **return** trained FL-FALCON

The trained FL-FALCON selects the AOS methods along with parameter values using the steps outlined in Algorithm 2. The proposed approach has two loops. Given the derived fitness landscape features $\mathcal{F}$, the inner loop search for the best response from FL-FALCON. The winning cognitive node J is inhibited after each iteration of the inner loop to allow other cognitive nodes to win. The chosen response comprising AOS method and parameter values are returned to the outer loop where it is used in evolutionary search. After that, a new set of derived fitness landscape features are obtained using the collected objective values. Together with the chosen response and feedback signals, it is presented to FL-FALCON for further learning. The updated FL-FALCON is used in subsequent rounds of search for better AOS methods and parameter sets. The outer loop terminates after meeting the closing criteria.

4.4 Design of Experiment

Design of Experiment (DOE) is a well-studied statistical technique for determining key parameters of particular process [9] and best parameter values for target algorithms [6]. In this paper, we focus on defining promising regions for the

Algorithm 2 Solving combinatorial optimization problem using I-AOS-DOE
***

Require: trained FL-FALCON
Require: derived fitness landscape features $\mathcal{F}$
Require: State vigilance threshold ζ^{c1} and state vigilance stepsize $\delta\zeta^{c1}$
Ensure: best reward $r* = 0.0$
  1: **while** not end of search **do**
  2:     encode derived fitness landscape features $\mathcal{F}$ as state vector $\mathbf{x}^{c1}$
  3:     set action vector $\mathbf{x}^{c2}$ and reward vector $\mathbf{x}^{c3}$ to uncommitted patterns
  4:     **while** $\rho_p^{c1} > \zeta^{c1}$ **do**
  5:         present input pattern $\mathbf{x}^{ck}$ to FL-FALCON for action selection
  6:         **if** $w_J^{c3}(0) > r*$ **then**
  7:             $r* = w_J^{c3}(0)$
  8:             readout AOS methods and parameter values from action vector $\mathbf{w}_J^{c2}$
  9:         **end if**
 10:         inhibit winning cognitive node J
 11:         reduce state vigilance ρ_p^{c1} using $\delta\zeta$
 12:     **end while**
 13:     clear inhibition for all cognitive nodes
 14:     **if** no valid response **then**
 15:         select AOS method and parameter values randomly
 16:     **end if**
 17:     apply AOS method configured with DOE-derived parameter values to solve problem instance j
 18:     remember best solution to problem instance j
 19:     derive fitness landscape features $\mathcal{F}^j$ from a new set of objective values
 20:     encode derived fitness landscape features $\mathcal{F}$ as state vector $\mathbf{x}^{c1}$
 21:     encode selected response as action vector $\mathbf{x}^{c2}$
 22:     encode feedback signal as reward vector $\mathbf{x}^{c3}$
 23:     present input pattern $\mathbf{x}^{ck} = \{\mathbf{x}^{c1}, \mathbf{x}^{c2}, \mathbf{x}^{c3}\}$ to FL-FALCON for learning
 24: **end while**
 25: **return** best solution to problem instance j
***

parameter configurations and define the best parameter values for AOS methods using DOE technique. We briefly explain the idea of this approach.

Given a set of parameters K, it is assumed that the initial range value of each parameter $k \in K$ is known and bounded by a numerical interval $[LB_k, UB_k]$. A $2^{|K|}$ factorial design is applied to screen and rank the parameters. A complete design requires $rep \times 2^{|K|}$ observations where rep represents the number of replicates for a set of parameter values. Screening is performed to determine the parameters that are statistically significant. This step reduces the number of parameters to tune. The following example provides an illustration of how the screening phase is applied.

Suppose we wish to study the effect of two parameters k_1 and k_2. The 2^2 factorial design would consist of four experimental units where each unit is run rep times:

unit (1): set k_1 at LB_{k_1} and k_2 at LB_{k_2} unit (k_1): set k_1 at LB_{k_1} and k_2 at UB_{k_2}
unit (k_2): set k_1 at UB_{k_1} and k_2 at LB_{k_2} unit (k_1k_2): set k_1 at UB_{k_1} and k_2 at UB_{k_2}

A factorial experiment is then analyzed using Analysis of Variance (ANOVA) to estimate the main effect for a particular parameter. The test of significance of the main effect of the parameters with a significance level (e.g. $\alpha = 5\%$) is conducted for determining the importance of parameters. The ranking of the important parameters is then done by comparing the absolute values of the main effects of those parameters. Each non-significant parameter is then set to a constant value. After that, we find promising ranges for m important parameters for $m \leq |K|$. This process begins by exploring a larger space where the linear relationship holds allowing for the application of standard approach for linear model checking and diagnosis [9]. The AOS is run with respect to the parameter configuration space which contains $(2^m + 1)$ possible parameter configurations with an additional setting defined by the centre points of the m important parameters. Centre points are added to protect against curvature.

The interaction and curvature tests are used for checking model adequacy. The planar model can still be applied as long as either one of them is not statistically significant. Otherwise, the region of planar local optimality has been reached and the promising region has been found. The process is continued by applying the steepest descent, in order to bring the parameter to the vicinity of the optimum values. Once the region of the optimum values has been found (e.g. one of two statistical tests is statistically significant), the planar model becomes invalid. The important parameters are assumed to be in their promising range. The best parameter value is taken by discretizing the range and pick the value that provides the best result. Details on the DOE technique are in [6].

5   Performance Evaluation

Experiments are conducted to evaluate the performance of I-AOS-DOE and compared it with selected AOS methods and crossover operators. To do so, it is necessary to inform the readers of the necessary details of the experiments such as the choice of AOS methods and crossover operators in Sect. 5.1, the default parameters and DOE-derived parameters in Sect. 5.2 and the selected QAP problem instances in Sect. 5.3. After that, the results are presented and analysed in Sect. 5.4.

5.1   AOS and Crossover Operators

The AOS methods and crossover operators used in this work are presented in the following paragraphs.

Adaptive Operator Selection. As mentioned in [4], adaptive operator selection is composed of credit assignment and operator selection sub-tasks. The credit assignment sub-task allocates a credit value signifying the effect of the selected operator on the quality of the solution. The operator selection sub-task decides on the choice of operators based on the knowledge of credit value for the operators. Several AOS methods implementing different approaches for the sub-tasks are known. In this work, the AOS methods that can be selected using I-AOS-DOE are probabilistic matching (PM), adaptive pursuit (AP), multi-armed bandit (MAB), reinforcement learning (RL) and a self-organizing artificial neural network (NN) known as FL-FALCON [17].

Crossover Operators. The AOS methods are implemented to select crossover operators during the search iterations. Crossover operator picks parts of the parent chromosomes and use it to form a child chromosome. In this way, the child chromosome may possess characteristics of the parent chromosomes. Different crossover operators implement the crossover operation differently. The crossover operators used in this work are the cycle crossover (CX), distance-preserving crossover (DPX), partially-mapped crossover (PMX) and order crossover (OX). Details on the crossover operators are available in [16].

5.2   Design of Experiment

The steps of DOE are presented in Sect. 4.4. Table 1 summarizes the initial ranges for parameters of each AOS method. One column is used to present the best parameter values for each AOS when solving the training instances. For example, the best parameters for Reinforcement Learning are $\alpha = 0.2$ and $\delta = 1.0$. The last column contains the default parameters.

Table 1. Parameter values of AOS

AOS	par	$[LB_{par}, UB_{par}]$	Best value	Default value [16]
Probability matching	α	$[0.0, 1.0]$	0.5	0.3
	p_{min}	$[0.0, 0.2]$	0.1	0.05
Adaptive pursuit	α	$[0.0, 1.0]$	0.5	0.3
	β	$[0.0, 1.0]$	1.0	0.3
	p_{min}	$[0.0, 0.2]$	0.2	0.05
Multi-armed bandit	γ	$[0.0, 1.0]$	0.5	1.0
Reinforcement learning	α	$[0.0, 0.2]$	0.2	0.03
	δ	$[0.0, 1.0]$	1.0	0.9
Neural network	ρ_p^{c1}	$[0.9, 1.0]$	1.0	0.97
	ρ_p^{c2}	$[0.9, 1.0]$	0.95	0.0
	ρ_p^{c3}	$[0.0, 0.1]$	0.05	0.03

5.3   Problem Instances

This work focuses on solving selected Quadratic Assignment Problem (QAP) problem instances for demonstrating the efficacy of I-AOS-DOE. The QAP problem instances are selected using prior knowledge of the difficulty levels. The primary source of such knowledge is the work on the use of self-organizing neural network as an AOS method in [16]. It is discovered that the AOS methods and crossover operators have rather broad range of performance profile on 18 QAP instances. Thus, the interest here is to propose a robust approach capable of stabilizing the performance on the following 18 QAP problem instances.

chr15a, chr15c, chr18a, chr20a, chr20b, chr20c, chr22a, chr22b, chr25a,
kra30a, kra30b, kra32, lipa40a, nug17, nug24, nug28, nug30, rou20,

5.4   Result Analysis

The results used to illustrate the performance of the AOS methods and crossover
operators are seen in Figs. 3 and 4. The results are presented in the form of ranks
among the various approaches. These approaches are ranked according to their
best objective values for solving the selected QAP problem instances.

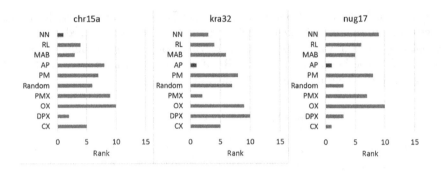

Fig. 3. Illustration of the differences in performance of AOS methods on different
problem instances

It is known from [16] that there are problem instances where the AOS methods have rather diverse outcome. Figure 3 is meant to illustrate this observation.
From Fig. 3, NN is observed having the best performance for chr15a problem
instance whereas AP is observed performing better than NN for kra32 and nug17
problem instances. In addition, the other AOS methods have rather different performance outcome among these problem instances as well. Thus, I-AOS-DOE is
designed to exploit the strength of AOS methods for different problem instances.

From Fig. 4, I-AOS-DOE is showing rather robust and stable performance. It
has the best rank among the approaches in both experiments. At the other end
of the spectrum, order crossover operator (OX) has the worst performance aggregated over the selected problem instances. The worst performing AOS method
configured using the default parameters is AP. The worst performance AOS
method configured using DOE-derived parameters is RL.

In the second experiment, only the parameters of the AOS methods are
derived using DOE. Thus, the effect of DOE on the performance of AOS methods can be observed objectively using Fig. 5. The objective value difference seen
in Fig. 5 is obtained by subtracting the objective value of the approach whose
parameter values are derived using DOE from that of the same approach configured with the default parameter values. This approach is taken because of
the expectation that DOE-derived parameters can help the AOS methods to

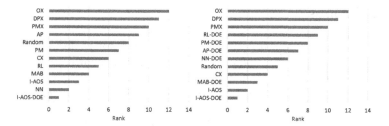

Fig. 4. Illustration of the ranking obtained by AOS methods configured using DOE-derived parameters (left) and default parameters (right) and the crossover operators.

improve on their performance. Thus, if the application of DOE has helped, the objective value difference will be positive and vice avers. The approaches seen in Fig. 5 are ranked in ascending order with respect to their objective value difference. Positive objective value difference indicates the application of DOE has helped in improving the performance of the AOS method for solving combinatorial optimisation problem using evolutionary search.

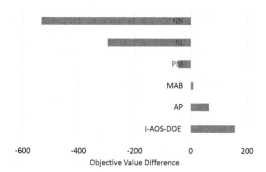

Fig. 5. Illustrations of the ranking obtained on the effect of DOE on the performance of the AOS methods

I-AOS-DOE is only affected by the application of DOE indirectly. This is because the parameters of I-AOS-DOE are not tuned using DOE. Rather, it is the parameters of AOS methods selected by I-AOS-DOE that are derived using DOE. From Fig. 5, I-AOS-DOE, MAB and AP are observed having positive objective value differences. This means the performance of these three approaches are helped by the application of DOE. Given that the parameters of I-AOS-DOE are not tuned using DOE, its performance improvement can be attributed to the selection of AOS methods helped by the application of DOE.

6   Conclusions

This work proposed a neural network-based approach for performing instance-specific selection of adaptive operator selection methods known as I-AOS-DOE. Using a class of self-organizing neural network known as FALCON, I-AOS-DOE is trained to select AOS methods for solving specific QAP problem instances. Training of a specific variant of FALCON known as FL-FALCON is carried out by inserting performance data of AOS methods on QAP problem instances. The performance data comprises a set of derived fitness landscape features as the state features, a set of decision variables comprising identity of problem instance, the choice of AOS method and the set of AOS parameter values as features of the action space. A feedback signal aggregating the average objective value difference and the best objective value difference to the objective value of a known optimal solution is encoded as the reward vector.

The trained FL-FALCON selects a set of decision variables using a set of derived fitness landscape features as inputs. Without knowing the identity of the problem instance, the trained FL-FALCON is to recommend a AOS method along with the appropriate parameter set for solving the problem instance using evolutionary search. Experiments were conducted using 18 QAP problem instances, 5 AOS methods and 4 crossover operators. Experiment results are obtained for AOS methods configured using the default parameter values and DOE-derived parameter values. The aggregated effect of these two sets of AOS methods over the selected QAP problem instances are compared and contrasted. From the results, it is observed that I-AOS-DOE is most effective and robust in solving the selected QAP instances. In addition, it is shown that DOE is effective towards certain AOS methods in solving the selected QAP instances using evolutionary search. Consequentially, the use of DOE turns out to have positive effect on the performance of I-AOS-DOE.

This work can be extended in several dimensions. In particular, we might incorporate I-AOS-DOE into the Hyflex framework [11]. By being in this Hyflex framework, I-AOS-DOE can be evaluated on problem domains such as maximum satisfiability problem (SAT), the flow shop sequencing problem, the bin packing problem and the personnel scheduling. It may also be possible to compare against other approaches for solving these benchmark problems. Beyond that, there are also many fitness landscape analysis (FLA) techniques that can be exploited to better analyse and characterise the raw fitness landscape of problem instances. The derivation of fitness landscape features agnostic to approaches used for solving problem instances can help I-AOS-DOE learn a set of generalizable policies for selecting AOS methods for unknown problem instances.

Acknowledgments. This research project is funded by National Research Foundation Singapore under its Corp Lab @ University scheme and Fujitsu Limited.

References

1. Ansotegui, C., Gabas, J., Malitsky, Y., Sellmann, M.: MaxSAT by improved instance-specific algorithm configuration. Artif. Intell. **235**, 26–39 (2016)
2. Bischl, B., Mersmann, O., Trautmann, H., Preuß, M.: Algorithm selection based on exploratory landscape analysis and cost-sensitive learning. In: Proceedings of the 14th Annual Conference on Genetic and Evolutionary Computation, pp. 313–320. ACM (2012)
3. Carpenter, G.A., Grossberg, S.: A massively parallel architecture for a self-organizing neural pattern recognition machine. Comput. Vis. Graph. Image Process. **37**(1), 54–115 (1987)
4. Consoli, P.A., Mei, Y., Minku, L.L., Yao, X.: Dynamic selection of evolutionary operators based on online learning and fitness landscape analysis. Soft Comput. **20**(10), 3889–3914 (2016)
5. Fialho, Á., Da Costa, L., Schoenauer, M., Sebag, M.: Analyzing bandit-based adaptive operator selection mechanisms. Ann. Math. Artif. Intell. **60**(1), 25–64 (2010)
6. Gunawan, A., Lau, H.C.: Lindawati: fine-tuning algorithm parameters using the design of experiments approach. In: Coello Coello, C. (ed.) LION'11, pp. 278–292. Springer, LNCS (2011)
7. Kadioglu, S., Malitsky, Y., Sellmann, M., Tierney, K.: ISAC-instance-specific algorithm configuration. In: Proceedings of the 2010 Conference on ECAI 2010: 19th European Conference on Artificial Intelligence, pp. 751–756 (2010)
8. Kohl, N., Miikkulainen, R.: An integrated neuroevolutionary approach to reactive control and high-level strategy. IEEE Trans. Evol. Comput. **16**(4), 472–488 (2012)
9. Montgomery, D.: Design and Analysis of Expeirments, 6th edn. Wiley, Inc., New Jercy (2005)
10. Muñoz, M.A., Kirley, M., Halgamuge, S.K.: A meta-learning prediction model of algorithm performance for continuous optimization problems. In: Coello Coello, C.A., Cutello, V., Deb, K., Forrest, S., Nicosia, G., Pavone, M. (eds.) PPSN 2012. LNCS, vol. 7491, pp. 226–235. Springer, Heidelberg (2012). https://doi.org/10.1007/978-3-642-32937-1_23
11. Ochoa, G., et al.: HyFlex: A Benchmark Framework for Cross-Domain Heuristic Search. In: Hao, J.-K., Middendorf, M. (eds.) EvoCOP 2012. LNCS, vol. 7245, pp. 136–147. Springer, Heidelberg (2012). https://doi.org/10.1007/978-3-642-29124-1_12
12. Pitzer, E., Affenzeller, M.: A comprehensive survey on fitness landscape analysis. Recent Adv. Intell. Eng. Syst. **378**, 161–191 (2012)
13. Sallam, K.M., Elsayed, S.M., Sarker, R.A., Essam, D.L.: Landscape-based adaptive operator selection mechanism for differential evolution. Inf. Sci. **418**, 383–404 (2017)
14. Tan, A.H.: FALCON: a fusion architecture for learning, cognition, and navigation. In: Proceedings of the IJCNN, pp. 3297–3302 (2004)
15. Teng, T.H., Tan, A.H., Zurada, J.M.: Self-organizing neural networks integrating domain knowledge and reinforcement learning. IEEE Trans. Neural Netw. Learn. Syst. **26**(5), 889–902 (2015)
16. Teng, T.-H., Handoko, S.D., Lau, H.C.: Self-organizing neural network for adaptive operator selection in evolutionary search. In: Festa, P., Sellmann, M., Vanschoren, J. (eds.) LION 2016. LNCS, vol. 10079, pp. 187–202. Springer, Cham (2016). https://doi.org/10.1007/978-3-319-50349-3_13

17. Teng, T.H., Tan, A.H.: Fast reinforcement learning under uncertainties with self-organizing neural networks. In: Proceedings of IAT, pp. 51–58 (2015)
18. Tuson, A., Ross, P.: Adapting operator settings in genetic algorithms. Evol. Comput. **6**(2), 161–184 (1998). https://doi.org/10.1162/evco.1998.6.2.161
19. Xu, L., Hoos, H.H., Leyton-Brown, K.: Hydra: automatically configuring algorithms for portfolio-based selection. In: Proceedings of the 24th AAAI Conference on Artificial Intelligence, pp. 210–216 (2010)
20. Xu, L., Hutter, F., Hoos, H.H., Leyton-Brown, K.: SATzilla: portfolio-based algorithm selection for SAT. J. Artif. Intell. Res. **32**(1), 565–606 (2008)
21. Xu, L., Hutter, F., Shen, J., Hoos, H.H., Leyton-Brown, K.: Satzilla 2012: improved algorithm selection based on cost-sensitive classification models. In: Proceedings of SAT Challenge, pp. 57–58 (2012)

CAVE: Configuration Assessment, Visualization and Evaluation

André Biedenkapp$^{(\boxtimes)}$, Joshua Marben, Marius Lindauer, and Frank Hutter

University of Freiburg, Freiburg, Germany
{biedenka,marbenj,lindauer,fh}@cs.uni-freiburg.de

Abstract. To achieve peak performance of an algorithm (in particular for problems in AI), algorithm configuration is often necessary to determine a well-performing parameter configuration. So far, most studies in algorithm configuration focused on proposing better algorithm configuration procedures or on improving a particular algorithm's performance. In contrast, we use all the collected empirical performance data gathered during algorithm configuration runs to generate extensive insights into an algorithm, given problem instances and the used configurator. To this end, we provide a tool, called *CAVE*, that automatically generates comprehensive reports and insightful figures from all available empirical data. *CAVE* aims to help algorithm and configurator developers to better understand their experimental setup in an automated fashion. We showcase its use by thoroughly analyzing the well studied SAT solver *spear* on a benchmark of software verification instances and by empirically verifying two long-standing assumptions in algorithm configuration and parameter importance: (i) Parameter importance changes depending on the instance set at hand and (ii) Local and global parameter importance analysis do not necessarily agree with each other.

1 Introduction

In the AI community, it is well known that the algorithm parameters have to be tuned to achieve peak performance. Since manual parameter tuning is a tedious and error-prone task, several methods were proposed in recent years to automatically optimize parameter configurations of arbitrary algorithms [1–5]. This led to performance improvements in many AI domains, such as propositional satisfiability solving [6], AI planning [7], the traveling salesperson problem [8], set covering [9], mixed-integer programming [10], hyper-parameter optimization of machine learning algorithms [11] and architecture search for deep neural networks [12]. These studies focus either on proposing more efficient automated parameter optimization methods or on improving the performance of a particular algorithm (the so-called *target algorithm*).

To determine a well-performing parameter configuration of a given algorithm on a given instance set, algorithm configuration procedures (in short *configurators*) have to collect a lot of empirical performance data. This entails running the algorithm at hand with different parameter configurations on several problem

© Springer Nature Switzerland AG 2019
R. Battiti et al. (Eds.): LION 12 2018, LNCS 11353, pp. 115–130, 2019.
https://doi.org/10.1007/978-3-030-05348-2_10

instances to measure its performance. Only the best performing configuration is returned in the end and all the collected performance data is typically not used further, although it was very expensive to collect.

In this paper, we reuse this data to further analyze all parts involved in the configuration process. This includes the target algorithm, its parameters, the used instances as well as the configurator. Hence, users will not only obtain a well-performing parameter configuration by using our methods in combination with a configurator but several additional insights.

A potential use-case is that algorithm developers implemented a new algorithm and want to empirically study that algorithm thoroughly as part of a publication. In a first step, they run a configurator to optimize their algorithm's parameters on a given instance set to achieve peak performance. As the next step, our automatic analysis tool *CAVE* (Configuration Assessment, Visualization and Evaluation) generates figures that can be directly used in a publication.

In *CAVE*, we build on existing methods to analyze algorithms and instances, reaching from traditional visualization approaches (such as scatter plots) over exploratory data analysis (used, e.g., for algorithm selection [13,14]) to recent parameter importance analysis methods [15–18]. We combine all these methods into a comprehensive analysis tool tailored to algorithm development and configuration studies, and propose two new approaches to complement their insights. Specifically, the contributions of our paper are:

1. We give an overview of different approaches to analyze a target algorithm, a given instance set and the configurator's behavior based on collected empirical data. We use these for an exemplary analysis of the SAT solver *spear* [19] on a benchmark of SAT-encoded software verification instances.
2. We propose a new qualitative analysis of configurator footprints based on a recently proposed similarity metric of configurations [20].
3. We propose a new parameter importance analysis by studying the impact on performance when changing one parameter at a time, thus exploring the immediate neighborhood of the best found configuration.
4. We provide a ready-to-use toolkit, called *CAVE*[1], for such analyses, which can be directly used in combination with the configurator *SMAC* [3].
5. We show the value of our tool and the need of such comprehensive analyses by verifying two common assumptions for algorithm configuration:
   (a) Parameter importance changes depending on the instance set at hand.
   (b) Local and global parameter importance analysis do not necessarily agree with each other and hence complement each other.

2   Related Work

Empirical evaluation of algorithms is as old as computer science. One of the first systematic approaches for ensuring reproducibility and insights in comparing a set of algorithms is the PAVER service [21,22]. The tool is primarily tailored to

---

[1] https://github.com/automl/CAVE

mixed integer programming solvers and provides some tables and visualization of the performance of algorithms. In contrast to PAVER, our tool also considers parameters of an algorithm and is designed to be used on arbitrary algorithms.

In the context of algorithm selection [23], a lot of performance data of different algorithms is collected, which can be used to analyze algorithms and instance sets. An example of an exploratory data analysis for algorithm selection is part of the algorithm selection library ASlib [13] that provides some simple performance and distribution tables and corresponding plots, e.g., scatter and box plots.

The first system that included an automatic analysis of algorithm performance in the context of algorithm configuration was the high-performance algorithm laboratory *HAL* [24]. Its main purpose was to help algorithm developers to apply automated algorithm design methods (such as algorithm configuration and selection) in a uniform framework by handling all interfaces, data storage, running experiments and some basic aggregation and visualizations of results, such as scatter and performance distribution plots. In contrast to *HAL*, *CAVE* focuses on the analysis part and provides a far more extensive analysis.

The tool *SpySMAC* [25] followed a similar approach as *HAL* but was specifically tailored to the needs of the propositional satisfiability (SAT) community. It provided an easy-to-use interface to run the configurator *SMAC* [3] for optimizing parameter configurations of a SAT solver. Our approach is inspired by *SpySMAC* but with the focus on the analysis part and extends it substantially. Furthermore, our new approach is no longer specific to SAT and *SMAC*, but it can be applied to any algorithm and configurator.

In the context of black-box optimization and in particular for hyperparameter optimization of machine learning algorithms, Golovin et al. [26] proposed *Google Vizier*. Similar to *HAL* and *SpySMAC*, it is a service to run optimization benchmarks and also provides some visualizations, such as performance-over-time of the target function or parallel coordinate visualizations [27]. Since *Google Vizier* focuses on black-box optimization, it does not have a concept of instances, which are an integral part of algorithm configuration.

Lloyd et al. [28] proposed automatically constructed natural language reports of regression models, giving raise to the automatic statistician tool. Although we have the same goal (providing automatically constructed reports to help users to get more out of their data), our goal is not to provide a natural language report, but to leave the interpretation of the results to the users.

3   Generation of Algorithm Configuration Data

In this section, we describe the general work-flow of generating algorithm configuration data (see Fig. 1), which will be the input for *CAVE*'s analyses in the next section. The typical inputs of configurators are[2]:

---

[2] We ignore in this simplified view that several budgets have to be defined, such as, the configuration budget (e.g., time budget or maximal number of algorithm calls) and resource limits of the target algorithm runs (e.g., runtime and memory limits).

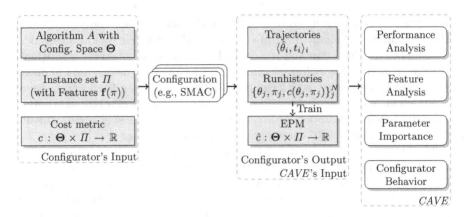

Fig. 1. Work-flow of algorithm configuration (AC) and analysis.

- *a target algorithm A* with a description of the parameter configuration space Θ, i.e., all possible parameter configurations,
- *an instance set Π* drawn from a distribution $\mathcal{D}_\Pi$ over all possible instances[3],
- *a cost metric $c : \Theta \times \Pi \to \mathbb{R}$* to be optimized (e.g., the runtime of an algorithm, the quality of a plan for an AI planning problem instance or the error of a machine learning algorithm).

A configurator's task is to find $\theta^* \in \Theta$ such that the cost c of running A using θ^* across Π is optimized, i.e., $\theta^* \in \arg\min_{\theta \in \Theta} \sum_{\pi \in \Pi} c(\theta, \pi)$. Since a configurator iteratively improves the currently best configuration $\hat{\theta}$ (called *incumbent*), its *trajectory* includes the incumbent $\hat{\theta}_i$ at each time point t_i.

In this process, the configurator follows a strategy to explore the joint space of configurations and instances, e.g., local search [1], genetic algorithms [2] or model-based [3] and it collects empirical cost data. Internally, a configurator keeps track of a set $\{\theta_j, \pi_j, c_j\}_{j=1}^N$ of all evaluated configurations θ_j on instance π_j with the corresponding cost c_j, called the *runhistory*. The runhistory is an optional output, but it is crucial for *CAVE*'s analyses. If several runs of a configurator were performed, *CAVE* can use all resulting outputs as input.

To guide the search of configurators, a recent approach is to use *empirical performance models (EPMs) $\hat{c} : \Theta \times \Pi \to \mathbb{R}$* that use the observed empirical data from the runhistory to predict the cost of new configuration-instance pairs [29], e.g., using random forests [30]. Based on these predictions, a configurator can decide whether to explore new regions in the configuration space or to exploit knowledge of the presumably well-performing regions in the configuration space. Model-based configurators [3,4] can return these EPMs directly, and for model-free configurators [1,5], such a model can be subsequently learned based on the

---

[3] Typically, the instance set is split into a training and a test set. On the training set, the target algorithm is optimized and on the test set, an unbiased cost estimate of the optimized parameter configuration is obtained.

returned runhistories. To this end, an optional input to configurators are instance features providing numerical descriptions of the instances at hand [29,31]. Hutter et al. [29] showed that predictions of new cost data is fairly accurate if enough training data is provided. Further, Eggensperger et al. [32] showed that EPMs trained on runhistory data from configurator runs are good surrogates for real target algorithm runs. Thus, we also use EPMs trained on the union of all runhistories for our analyses, e.g., to impute missing cost data of configurations that were evaluated only on some but not all instances.

Our tool *CAVE* analyzes all this data as described in the next section. Thereby we use an extended version of the output format defined for the second edition of the algorithm configuration library AClib [33] such that *CAVE* can be in principle used with any configurator. Right now, we have a ready-to-use implementation in combination with the configurator *SMAC* [3].

4   Analyzing Algorithm Configuration Data

In this section, we give a brief overview of all components of our analysis report generated based on the trajectory, runhistory data and EPMs described in the last section. A detailed description of the individual elements of *CAVE* can be found in the online appendix[4]. As a running example, we show figures for studying the SAT solver *spear* [19] on SAT instances encoding software verification problems based on three 2-day SMAC runs.[5] In addition to the data generated by SMAC, we validated all incumbent configurations to decrease the uncertainty of our EPM in the important regions of the configuration space.

4.1   Performance Analysis

The performance analysis of *CAVE* mainly supports different ways of analyzing the final incumbent and the performance of the algorithm's default parameter configuration (the results obtained by downloading the algorithm and running it with its default parameter settings). In particular, the performance analysis part of *CAVE* consists of a qualitative tabular analysis providing aggregated performance values across all instances, scatter plots showing default performance vs. optimized performance for each instance (Fig. 2a), empirical cumulative performance distribution (eCDF) plots across the instance set (Fig. 2b) and algorithm footprint plots [14] (Fig. 3a).

What have we learned about spear? Figure 2 shows the scatter plot and the eCDF for *spear* on software verification instances. From these plots, we learn that the performance of *spear* was not improved on all instances, but on many of them, with large speedups on some. The optimized configuration solved all instances in at most 20 seconds, while the default led to many timeouts. Based on the

---

[4] http://ml.informatik.uni-freiburg.de/papers/18-LION12-CAVE.pdf

[5] The complete generated report can be found at
http://ml.informatik.uni-freiburg.de/~biedenka/cave.html

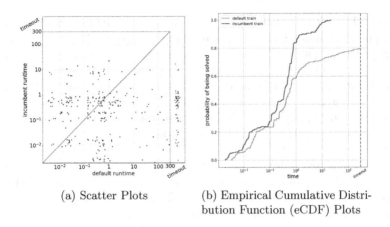

(a) Scatter Plots

(b) Empirical Cumulative Distribution Function (eCDF) Plots

Fig. 2. Comparison of the empirical performance of the default and final incumbent.

eCDF plot, the optimized configuration solved all instances whereas the default configuration solved only 80% of the instances. Furthermore, if the cutoff time of *spear* is larger than 0.8 seconds, the optimized configuration performed better; the blue curve is above the red curve consistently. The algorithm footprint plot (Fig. 3a) shows that the incumbent performed well on different types of instances. Compared to the scatter plot, we expected to see more red points in the footprint plots (instances where the default outperforms the incumbent). Looking more deeply into the data revealed that several of these are overlapping each other because the instances features are missing for these instances.

4.2   Instance and Instance Feature Analysis

If instances are characterized by instance features (as done for model-based configurators), these features can be studied to better understand an instance set. To obtain instance features for the SAT instances at hand, we used the instance feature generator accompanied by the algorithm selection tool *SATzilla* [29,34]. In particular, the feature analysis part of *CAVE* consists of box and violin plots for each instance feature, clustering plots based on a PCA into the 2-dimensional feature space, correlation heatmaps for each pair of features (Fig. 3b), and feature importance plots based on greedy forward selection [15].

What have we learned about the software verification instances? Based on these plots, we learned that there are at least three instance clusters mixed together; knowing the source of these instances (also reflected in the instance names), we can verify that software verification for four different software tools were encoded in this instance set and that clustering approximately recovered the sources (merging two sources into one cluster). Since the PCA plot and the footprint plot (Fig. 3a) indicate that the instance set is heterogeneous [35], using algorithm configuration on each individual software tool or per-instance algorithm

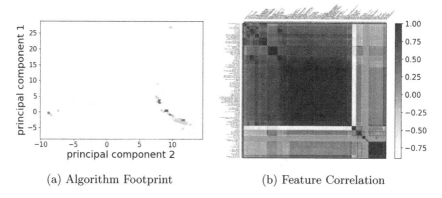

(a) Algorithm Footprint   (b) Feature Correlation

Fig. 3. (a) Green dots indicate instances where the incumbent configuration performed well i.e., at most 5% worse than the oracle performance of default and incumbent. All other instances are plotted as red dots, indicating that the default configuration performed well. Instances might be mapped closely together in the 2D reduced space. (b) The correlation matrix for all features is shown.

configuration [36, 37] could improve the performance of *spear* even further. From the feature correlation plot, we see that roughly half of the features are highly correlated and some of the features could be dropped potentially.

4.3   Configurator Behavior

Besides insights into the algorithm and instances at hand, the trajectory and the runhistory returned by a configurator also allow for insights into how the configurator tried to find a well-performing configuration. This may lead to insights into how the optimization process could be adjusted. In particular, the configurator behavior analysis consists of a plot showing the performance of the target algorithm over the time spend by the configurator (Fig. 4a) and parallel coordinate plots showing the interactions between parameter settings and performance of the target algorithm [26] (Fig. 4c).

Configurator Footprint. As a novel approach, we propose in this paper to study how a configurator iteratively sampled parameter configurations, i.e., all configurations in the runhistory, see Algorithm 1 and exemplary Fig. 4b. It is based on a similarity metric for parameter configurations[6] which is used in a multi-dimensional scaling (MDS) [38] based on the *SMACOF* algorithm [39] to obtain a non-linear mapping into 2-dimensions [20]. We extend this analysis by highlighting incumbent configurations in the trajectory and by scaling the dots (parameter configurations) wrt. the number of instances they were evaluated

---

[6] In contrast to Xu et al. [20], we normalize the relabelling cost of continuous parameters to $[0, 1]$ since otherwise relabelling of continuous parameters would dominate the similarity metric compared to relabelling of discrete parameters.

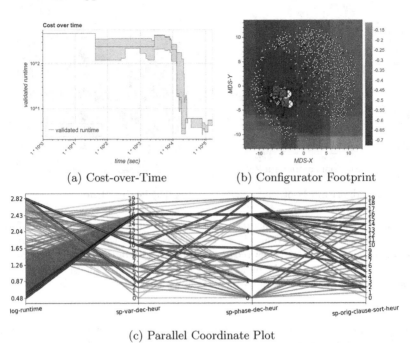

(a) Cost-over-Time   (b) Configurator Footprint

(c) Parallel Coordinate Plot

Fig. 4. (a) depicts the predicted performance over time. The blue area gives the first
and third quantiles over three configurator runs. (b) Dots represent sampled configurations, where the size represents the amount of times a configuration was evaluated. The
background shows the predicted performance in the 2D space using an MDS. Incumbents are plotted as red squares, the default as orange triangle and the final incumbent as red inverted triangle. Configurations sampled from an acquisition function are
marked with an X, all other configurations were purely randomly sampled. (c) Parallel
Coordinate Plot for the three most important parameters with a subsampled set of
500 configurations. The best performing configuration has the brightest shade of red,
whereas the darkest shade of blue depicts the worst observed configuration.

on. For racing-based configurators, this corresponds to how well a configuration
performed compared to the current best found configuration. Furthermore, we
use an EPM in the 2D space to highlight promising parts of the configuration
space. Finally, the figures in the html-report also have a mouse-over-effect that
allows to see the parameter configuration corresponding to each dot in the figure.

What have we learned about the configurator and spear? From the cost-over-time
plot (Fig. 4a), we learned that *SMAC* already converged after 40,000 seconds
and investing further time did not improve *spear*'s performance further. From
the configurator footprints (Fig. 4b), we see that *SMAC* covered most parts of
the space (because every second parameter configuration evaluated by *SMAC* is
a random configuration) but it also focused on promising areas in the space. The
fraction of good configurations is fairly large, which also explains why *SMAC* was

---

Algorithm 1: Configurator Footprint (Visualization of a runhistory)

---

1 **Input:** Runhistory $\mathcal{H} = \{\theta_j, \pi_j, c_j\}_j^N$; trajectory $\mathcal{T} = \langle \hat{\theta}_k, t_k \rangle_k$; Instance set Π

2 For each pair $\langle \theta_i, \theta_j \rangle$, compute similarity $s(\theta_i, \theta_j)$ [20];

3 Fit 2D MDS based on similarities $s(\theta_i, \theta_j)$;

4 Replace each θ in $\mathcal{H}$ by 2D projection $MDS(\theta)$;

5 Plot each θ in 2D space $MDS(\theta)$ with size proportional to #entries in $\mathcal{H}$;

6 Highlight incumbents $\hat{\theta}$ of trajectory $\mathcal{T}$;

7 Fit EPM $\hat{c} : \mathbb{R}^2 \times \Pi \to \mathbb{R}$ based on $\mathcal{H}$;

8 Plot heatmap in background based on $\frac{1}{|\Pi|} \sum_{\pi \in \Pi} \hat{c}(MDS(\theta), \pi)$

---

able to find a well-performing configuration early on. The parallel coordinate plot (Fig. 4c) reveals that sp-var-dec-heur should be set to 16 for peak performance; however, the runtime also depends on other parameters such as sp-phase-dec-heur.

4.4 Parameter Importance

Besides obtaining a well-performing parameter configuration, the most frequently asked question by developers is which parameters are important to achieve better performance. This question is addressed in the field of parameter importance. Existing approaches for parameter importance analysis (used in AC) include greedy forward selection [15], ablation analysis [17,18] and functional ANOVA [16].

Parameter	fANOVA	Ablation	LPI
sp-var-dec-heur	62.052	93.224	93.351
sp-phase-dec-heur	7.535	< 0.01	0.508
sp-orig-clause-sort-heur	0.764	< 0.01	< 0.01
sp-learned-clause-sort-heur	0.239	< 0.01	< 0.01
sp-restart-inc	< 0.01	4.239	3.473
sp-clause-activity-inc	< 0.01	2.406	< 0.01
sp-clause-decay	< 0.01	1.097	< 0.01
sp-variable-decay	< 0.01	< 0.01	1.005

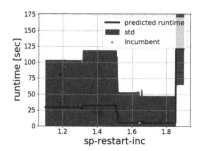

(a) Parameter Importance   (b) LPI on sp-restart-inc

Fig. 5. (a) Relative parameter importance values for the most important parameters of each method. The parameters are ordered by fANOVA's importance values. (b) Exemplary LPI plot. The shaded area is the model uncertainty.

Local Parameter Importance (LPI). In addition to the existing approaches, we propose a new parameter importance analysis approach, which we dub LPI. It is inspired by the human strategy to look for further improved parameter configurations or to understand the importance of parameter changes in the neighborhood of a parameter configuration. For example, most users are interested in understanding which parameters in optimized parameter configurations are crucial for the achieved performance.

Using an EPM, we study performance changes of a configuration along each parameter. To quantify the importance of a parameter value θ_p, we compute the variance of all cost values by changing θ_p and then compute the fraction of all variances. Given the parameter space of the target algorithm Θ, a set of parameters P and an EPM $\hat{c}\colon \Theta \to \mathbb{R}$ marginalized over Π, the local importance of parameter $p \in P$ with domain Θ_p is given by the fraction of variance caused by p over the sum of all variances

$$LPI(p \mid \theta) = \frac{\text{Var}_{v \in \Theta_p} \hat{c}(\theta\,[\theta_p = v])}{\sum_{p' \in P} \text{Var}_{w \in \Theta_{p'}} \hat{c}(\theta\,[\theta_{p'} = w])}$$

Compared to an ablation analysis, our new analysis is even more local since it solely focuses on one parameter configuration. Nevertheless, it is also similar to fANOVA since we quantify the importance of a parameter by studying the variance of performance changes. However the marginalization across all other parameter settings is not a part of LPI.

What have we learned about spear? Figure 5a gives an example of LPI estimated parameter importance and contrasts it to fANOVA and ablation results. The ablation analysis reveals that the most important parameter change was setting *sp-var-dec-heur* to 16 instead of the default of 0. Only a few of the 25 other parameters were important for the performance improvement. Overall, fANOVA and LPI agree with this assessment. However, LPI and ablation give a much larger importance to *sp-var-dec-heur* than fANOVA does; this is in part due to the fact that fANOVA (in contrast to LPI and Ablation) considers higher-order interaction effects as important, but also in part due to fANOVA being a global method and LPI and Ablation being local methods.

5   Exemplary Study of Parameter Importance using *CAVE*

In this section we will show that *CAVE* enables comprehensive studies. We will study two assumptions in the field of algorithm configuration regarding parameter importance:

Q1 Does the set of important parameters change depending on the instance set? If that is false, parameter importance of a given algorithm has only to be studied once and warmstarting methods of configurators [40] should perform quite well in general. Alternatively, parameter importance studies would be required on each new instance set again and warmstarting methods should perform quite poorly.

Q2 Do local and global parameter importance approaches agree on the set of important parameters? The common assumption is that local and global parameter importance analysis are complementary. A globally important parameter may not be important in a local analysis since it might already be set to a well-suited parameter value in a specific configuration.

Setup: To study these two questions, we run (capped) fANOVA [16][7], ablation analysis [18] and our newly proposed LPI analysis, through *CAVE* for different algorithms from different domains on several instance sets, see Table 1. All these benchmarks are part of the algorithm configuration library [33][8]. We note that *clasp* and *probSAT* were the respective winners on these instance sets in the configurable SAT solver challenge 2014 [6]. To collect the cost data for training the required EPMs, we ran *SMAC* (v3 0.7.1)[9] ten times on each of the benchmarks using a compute cluster with nodes equipped with two Intel Xeon E5-2630v4 and 128GB memory running CentOS 7.

Table 1. Algorithm Configuration Library benchmarks, with #P the number of parameters of each algorithm.

Algorithm	Domain	#P	Instance sets
lpg[41]	AI plan.	65	satellite, zenotravel, depots
clasp(-ASP)[42]	ASP	98	ricochet, riposte, weighted-sequence
CPLEX	MIP	74	cls, corlat, rcw2, regions200
SATenstein[43]	SAT	49	cmbc, factoring, hgen2-small, k3-r4 26-v600, qcp, swgcp
clasp(-HAND)	SAT	75	GI, LABS, queens
clasp(-RAND)	SAT	75	3cnf-v350, K3, unsat-unif-k5
probSAT[44]	SAT	9	3SAT1k, 5SAT500, 7SAT90

Metric: For both questions, we are interested in the sets of important parameters. To this end, we define a parameter to be important if it is responsible for at least 5% cost improvement (ablation) or explains at least 5% of the variance in the cost space (fANOVA, LPI). To compare the sets of important parameters for two instance sets (Q1) or for ablation, fANOVA and LPI (Q2), we report the fraction of the intersection and the union of the two sets. For example, if both sets are disjoint, this metric would return 0%; and if they are identical, the score would be 100%. In Tables 2 & 3, we show the averaged results for each solver across all instance sets; for Q1 we averaged over all pairs of instance sets and for Q2 we averaged over all instance sets.

---

[7] In capped fANOVA, all cost values to train a marginalized EPM are capped at the cost of the default configuration θ_{def}: $c(\theta) := \min\left(c(\theta_{\mathrm{def}}), c(\theta)\right)$.

[8] http://aclib.net/

[9] https://github.com/automl/SMAC3

Table 2. Q1: Comparison of fANOVA/ablation/LPI results across different instance sets. The values show the percentage of how often a method determined the same parameters to be important on a pair of instance sets.

Algorithm	ablation		fANOVA		fANOVA_c		LPI	
	μ	σ	μ	σ	μ	σ	μ	σ
clasp(-ASP)	8.33	5.89	41.67	18.00	21.92	27.95	31.30	4.54
clasp(-HAND)	0.00	0.00	50.00	13.61	13.61	11.34	24.94	10.24
clasp(-RAND)	13.61	4.83	11.11	15.71	2.02	1.89	27.51	4.86
CPLEX	4.03	6.29	15.83	20.56	0.00	0.00	36.37	8.44
lpg	16.19	11.97	30.00	14.14	33.33	47.14	37.63	12.44
probSAT	46.67	37.71	31.67	13.12	30.95	14.68	60.65	19.12
SATenstein	14.90	28.18	26.33	13.06	15.45	27.98	26.99	16.85

Table 3. Q2: Comparison of results obtained with different importance methods, on the same instance sets. The values show the percentage of how often two methods agreed on the set of most important parameters.

Algorithm	fANOVA				fANOVA_c				ablation	
	vs. ablation		vs. LPI		vs. ablation		vs. LPI		vs. LPI	
	μ	σ	μ	σ	μ	σ	μ	σ	μ	σ
clasp(-ASP)	8.33	11.79	5.87	1.54	3.48	2.47	25.71	19.87	12.36	4.02
clasp(-HAND)	6.67	9.43	9.98	4.06	4.88	6.90	20.23	7.53	21.97	19.15
clasp(-RAND)	38.10	44.16	13.35	3.16	35.86	45.46	13.89	7.88	31.60	15.92
CPLEX	6.86	8.59	6.76	2.64	0.82	1.64	7.12	14.25	12.95	6.18
lpg	42.86	10.10	37.90	3.90	27.78	20.79	20.97	11.76	38.91	2.83
probSAT	4.17	5.89	21.94	8.62	23.81	33.67	33.06	23.96	32.26	21.51
SATenstein	11.58	10.42	12.96	6.22	15.63	18.56	13.89	15.94	34.38	15.51

Q1: Parameter Importance across different Instance Sets. As shown in the right part of Table 2, the overlap of important parameters on pairs of instance sets is often quite small. Hence it depends on the instance set whether a parameter is important or not. Surprisingly, the results of ablation and fANOVA are similar in this respect. This indicates that for some algorithms (e.g. *probSAT*), a subset of the important parameters is constant across all considered instance sets. This supports the results on warmstarting of configurators [40], but also shows that warmstarting will potentially fail for some algorithms, e.g., *clasp*-(RAND), *CPLEX* and *SATenstein*.

Q2: Comparison of Local and Global Parameter Importance. As shown in Table 3, fANOVA and Ablation do not agree on the set of important parameters for most algorithms. Only for *lpg* and *clasp*(-RAND), both parameter importance approaches return some overlapping parameters, i.e., more than a third of the parameters are on average important according to both approaches. LPI results tend to agree more with ablation results, but there is also some overlap with capped fANOVA. Thus, local and global parameter importance analysis are not redundant and indeed provide a different view on the importance of parameters.

6 Discussion and Conclusion

Algorithm configurators generate plenty of data that is full of potential to learn more about the algorithm or instance set at hand as well as the configurator itself. However, this potential so far remains largely untapped. *CAVE* provides users with the opportunity to broaden their understanding of the algorithm they want to inspect, by automatically generating comprehensive reports as well as insightful figures. We also introduced two new analysis approaches: configurator footprints and local parameter importance analysis.

We demonstrated the usefulness of such an automatic tool by using it to verify the assumption that local and global parameter importance are complementary and to demonstrate that important parameters depend on the examined set of instances.

CAVE could be further extended in many ways. In particular, we plan to analyze instance sets for their homogeneity [35]; to this extent, *CAVE* could recommend users to use per-instance algorithm configuration methods [36,45] instead of conventional configurators if the instance set is strongly heterogeneous. We also plan to improve the uncertainty estimates of *CAVE*'s EPM by replacing the random forest models by quantile regression forests [46] as shown by Eggensperger et al. [32].

Acknowledgments. The authors acknowledge support by the state of Baden-Württemberg through bwHPC and the German Research Foundation (DFG) through grant no INST 39/963-1 FUGG and the Emmy Noether grant HU 1900/2-1.

References

1. Hutter, F., Hoos, H., Leyton-Brown, K., Stützle, T.: ParamILS: an automatic algorithm configuration framework. JAIR **36**, 267–306 (2009)
2. Ansótegui, C., Sellmann, M., Tierney, K.: A gender-based genetic algorithm for the automatic configuration of algorithms. In: Gent, I.P. (ed.) CP 2009. LNCS, vol. 5732, pp. 142–157. Springer, Heidelberg (2009). https://doi.org/10.1007/978-3-642-04244-7_14
3. Hutter, F., Hoos, H.H., Leyton-Brown, K.: Sequential model-based optimization for general algorithm configuration. In: Coello Coello, C.A. (ed.) LION 2011. LNCS, vol. 6683, pp. 507–523. Springer, Heidelberg (2011). https://doi.org/10.1007/978-3-642-25566-3_40
4. Ansótegui, C., Malitsky, Y., Sellmann, M., Tierney, K.: Model-based genetic algorithms for algorithm configuration. In: Yang, Q., Wooldridge, M. (eds.) Proceedings of IJCAI'15, pp. 733–739 (2015)

5. López-Ibáñez, M., Dubois-Lacoste, J., Caceres, L.P., Birattari, M., Stützle, T.: The irace package: iterated racing for automatic algorithm configuration. Oper. Res. Perspect. **3**, 43–58 (2016)
6. Hutter, F., Lindauer, M., Balint, A., Bayless, S., Hoos, H., Leyton-Brown, K.: The configurable SAT solver challenge (CSSC). AIJ **243**, 1–25 (2017)
7. Fawcett, C., Helmert, M., Hoos, H., Karpas, E., Roger, G., Seipp, J.: Fd-autotune: domain-specific configuration using fast-downward. In: Helmert, M., Edelkamp, S. (eds.) Proceedings of ICAPS'11 (2011)
8. Mu, Z., Hoos, H.H., Stützle, T.: The impact of automated algorithm configuration on the scaling behaviour of state-of-the-Art inexact TSP solvers. In: Festa, P., Sellmann, M., Vanschoren, J. (eds.) LION 2016. LNCS, vol. 10079, pp. 157–172. Springer, Cham (2016). https://doi.org/10.1007/978-3-319-50349-3_11
9. Wagner, M., Friedrich, T., Lindauer, M.: Improving local search in a minimum vertex cover solver for classes of networks. In: Proceedings of IEEE CEC, pp. 1704–1711. IEEE (2017)
10. Hutter, F., Hoos, H.H., Leyton-Brown, K.: Automated configuration of mixed integer programming solvers. In: Lodi, A., Milano, M., Toth, P. (eds.) CPAIOR 2010. LNCS, vol. 6140, pp. 186–202. Springer, Heidelberg (2010). https://doi.org/10.1007/978-3-642-13520-0_23
11. Snoek, J., Larochelle, H., Adams, R.P.: Practical Bayesian optimization of machine learning algorithms. In: Proceedings of NIPS'12, pp. 2960–2968 (2012)
12. Zoph, B., Le, Q.V.: Neural architecture search with reinforcement learning. In: Proceedings of ICLR'17 (2017)
13. Bischl, B., et al.: ASlib: a benchmark library for algorithm selection. AIJ 41–58 (2016)
14. Smith-Miles, K., Baatar, D., Wreford, B., Lewis, R.: Towards objective measures of algorithm performance across instance space. Comput. OR **45**, 12–24 (2014)
15. Hutter, F., Hoos, H., Leyton-Brown, K.: Identifying key algorithm parameters and instance features using forward selection. In: Proceedings of LION'13, pp. 364–381 (2013)
16. Hutter, F., Hoos, H., Leyton-Brown, K.: An efficient approach for assessing hyperparameter importance. In: Proceedings of ICML'14, pp. 754–762 (2014)
17. Fawcett, C., Hoos, H.: Analysing differences between algorithm configurations through ablation. J. Heuristics **22**(4), 431–458 (2016)
18. Biedenkapp, A., Lindauer, M., Eggensperger, K., Fawcett, C., Hoos, H., Hutter, F.: Efficient parameter importance analysis via ablation with surrogates. In: Proceedings of AAAI'17, pp. 773–779 (2017)
19. Babić, D., Hutter, F.: Spear theorem prover. Solver description. SAT Competition (2007)
20. Xu, L., KhudaBukhsh, A.R., Hoos, H.H., Leyton-Brown, K.: Quantifying the similarity of algorithm configurations. In: Festa, P., Sellmann, M., Vanschoren, J. (eds.) LION 2016. LNCS, vol. 10079, pp. 203–217. Springer, Cham (2016). https://doi.org/10.1007/978-3-319-50349-3_14
21. Bussieck, M., Drud, A.S., Meeraus, A., Pruessner, A.: Quality assurance and global optimization. In Bliek, C., Jermann, C., Neumaier, A. (eds.) Proceedings of GOCOS. Lecture Notes in Computer Science, vol. 2861. Springer (2003) 223–238
22. Bussieck, M., Dirkse, S., Vigerske, S.: PAVER 2.0: an open source environment for automated performance analysis of benchmarking data. J. Glob. Optim. **59**(2–3), 259–275 (2014)
23. Rice, J.: The algorithm selection problem. Adv. Comput. **15**, 65–118 (1976)

24. Nell, C., Fawcett, C., Hoos, H.H., Leyton-Brown, K.: HAL: a framework for the automated analysis and design of high-performance algorithms. In: Coello Coello, C.A. (ed.) LION 2011. LNCS, vol. 6683, pp. 600–615. Springer, Heidelberg (2011). https://doi.org/10.1007/978-3-642-25566-3_47
25. Falkner, S., Lindauer, M., Hutter, F.: SpySMAC: automated configuration and performance analysis of SAT solvers. In: Proceedings of SAT'15, pp. 1–8 (2015)
26. Golovin, D., Solnik, B., Moitra, S., Kochanski, G., Karro, J., Sculley, D.: Google vizier: a service for black-box optimization. In: Proceedings of KDD, pp. 1487–1495. ACM (2017)
27. Heinrich, J., Weiskopf, D.: State of the art of parallel coordinates. In: Proceedings of Eurographics, Eurographics Association, pp. 95–116 (2013)
28. Lloyd, J., Duvenaud, D., Grosse, R., Tenenbaum, J., Ghahramani, Z.: Automatic construction and natural-language description of nonparametric regression models. In: Proceedings of AAAI'14, pp. 1242–1250 (2014)
29. Hutter, F., Xu, L., Hoos, H., Leyton-Brown, K.: Algorithm runtime prediction: methods and evaluation. AIJ **206**, 79–111 (2014)
30. Breimann, L.: Random forests. MLJ **45**, 5–32 (2001)
31. Nudelman, E., Leyton-Brown, K., Hoos, H.H., Devkar, A., Shoham, Y.: Understanding random sat: beyond the clauses-to-variables ratio. In: Wallace, M. (ed.) CP 2004. LNCS, vol. 3258, pp. 438–452. Springer, Heidelberg (2004). https://doi.org/10.1007/978-3-540-30201-8_33
32. Eggensperger, K., Lindauer, M., Hoos, H., Hutter, F., Leyton-Brown, K.: Efficient benchmarking of algorithm configuration procedures via model-based surrogates. Mach. Learn. (2018) (To appear)
33. Hutter, F., et al.: AClib: a benchmark library for algorithm configuration. In: Pardalos, P.M., Resende, M.G.C., Vogiatzis, C., Walteros, J.L. (eds.) LION 2014. LNCS, vol. 8426, pp. 36–40. Springer, Cham (2014). https://doi.org/10.1007/978-3-319-09584-4_4
34. Xu, L., Hutter, F., Hoos, H., Leyton-Brown, K.: SATzilla: Portfolio-based algorithm selection for SAT. JAIR **32**, 565–606 (2008)
35. Schneider, M., Hoos, H.H.: Quantifying homogeneity of instance sets for algorithm configuration. In: Hamadi, Y., Schoenauer, M. (eds.) LION 2012. LNCS, pp. 190–204. Springer, Heidelberg (2012). https://doi.org/10.1007/978-3-642-34413-8_14
36. Xu, L., Hoos, H., Leyton-Brown, K.: Hydra: automatically configuring algorithms for portfolio-based selection. In: Proceedings of AAAI'10, pp. 210–216 (2010)
37. Kadioglu, S., Malitsky, Y., Sellmann, M., Tierney, K.: ISAC - instance-specific algorithm configuration. In: Proceedings of ECAI'10, pp. 751–756 (2010)
38. Kruskal, J.: Multidimensional scaling by optimizing goodness of fit to a nonmetric hypothesis. Psychometrika **29**(1), 1–27 (1964)
39. Groenen, P., van de Velden, M.: Multidimensional scaling by majorization: A review. J. Stat. Softw. 73(8) (2016)
40. Lindauer, M., Hutter, F.: Warmstarting of model-based algorithm configuration. In: Proceedings of the AAAI conference (2018) (To appear)
41. Gerevini, A., Serina, I.: LPG: a planner based on local search for planning graphs with action costs. In: Proceedings of AIPS'02, pp. 13–22 (2002)
42. Gebser, M., Kaufmann, B., Schaub, T.: Conflict-driven answer set solving: from theory to practice. AI **187–188**, 52–89 (2012)
43. KhudaBukhsh, A., Xu, L., Hoos, H., Leyton-Brown, K.: SATenstein: automatically building local search SAT solvers from components. In: Proceedings of IJCAI'09, pp. 517–524 (2009)

44. Balint, A., Schöning, U.: Choosing probability distributions for stochastic local search and the role of make versus break. In: Cimatti, A., Sebastiani, R. (eds.) SAT 2012. LNCS, vol. 7317, pp. 16–29. Springer, Heidelberg (2012). https://doi.org/10.1007/978-3-642-31612-8_3
45. Kadioglu, S., Malitsky, Y., Sabharwal, A., Samulowitz, H., Sellmann, M.: Algorithm selection and scheduling. In: Lee, J. (ed.) CP 2011. LNCS, vol. 6876, pp. 454–469. Springer, Heidelberg (2011). https://doi.org/10.1007/978-3-642-23786-7_35
46. Meinshausen, N.: Quantile regression forests. JMLR **7**, 983–999 (2006)

The Accuracy of One Polynomial Algorithm for the Convergecast Scheduling Problem on a Square Grid with Rectangular Obstacles

Adil Erzin[1,2(⊠)] and Roman Plotnikov[1]

[1] Sobolev Institute of Mathematics, 4 Koptyug avenue, 630090 Novosibirsk, Russia
[2] Novosibirsk State University, 1 Pirogova str, 630090 Novosibirsk, Russia
{adilerzin,prv}@math.nsc.ru

Abstract. In the Convergecast Scheduling Problem, it is required to find in the communication graph an oriented spanning aggregation tree with a root in a base station and the arcs oriented to the root and to build a conflict-free min-length schedule for aggregating data along the arcs of the aggregation tree. This problem is NP-hard in general, however, if the communication graph is a unit square grid in each node of which there is a sensor and in which a data packet is transmitted along any edge during a one-time slot, the problem is polynomially solvable. In this paper, we consider a communication graph in the form of a square grid with rectangular obstacles impenetrable by the messages. In our previous paper, we proposed a polynomial algorithm for constructing a feasible schedule and intensive numerical experiment allowed us to make a hypothesis that the algorithm constructs an *optimal* solution. In this paper, we present a counterexample and prove that the proposed algorithm constructs a schedule of length at most one time round longer than the optimal schedule.

1 Introduction

When designing a wireless sensor network (WSN), three main optimization problems are solved consistently, allowing rational use of sensor's energy, and thereby prolonging the network lifetime. The first problem is to build an energy-efficient cover [2, 3, 16]. After solving it, the placement of sensors becomes known and it is necessary to solve the second problem – the synthesis of an energy efficient network [1, 8]. After solving the second problem, the communication graph becomes known, along the edges of which the information collected by the sensors is transmitted. Since the communication graph is built on the basis of the energy saving criterion of the transmitters, it is highly sparse, and the information between the vertices is usually transmitted by transit through other vertices [4, 10, 11, 15]. Within the framework of the TDMA (Time Division Multiple Access) standards, time is divided into the rounds (slots) in such a way that the duration of one round is sufficient to transmit a data packet along any edge of the graph.

© Springer Nature Switzerland AG 2019
R. Battiti et al. (Eds.): LION 12 2018, LNCS 11353, pp. 131–140, 2019.
https://doi.org/10.1007/978-3-030-05348-2_11

When *aggregating* data, the information received by each vertex is analyzed and then a packet with aggregated data is sent until the necessary information reaches the base station – the selected vertex of the graph (the sink). After that, the data aggregation session ends and a new session can be started. Minimizing time for the aggregated convergecast is equivalent to minimizing the number of time slots required for all packets to reach the sink. The length of the schedule is the number of time rounds during which the aggregated data is collected at the base station. The smaller the length of the schedule, the faster the system can react to events. In systems related, for example, to security, it is required to find a min-length schedule for aggregating data.

In wireless networks, as a rule, radio communication is used for data transmission. In this case, the transmitters use a limited number of channels (radio frequencies) [12], often one channel [4,10,11]. This leads to the conflicts that increase the aggregation time. Firstly, if more than one transmitter is operating in the receiving zone sharing one frequency, the recipient can not receive the data packet intended for it because of a phenomenon such as *interference* of radio waves. Secondly, in half-duplex systems, a sensor can either receive or transmit a packet at a time. Thirdly, a sensor can receive or transmit no more than one packet during one time round. And, fourthly, for the energy efficiency purposes, each sensor sends the packet only once during the data aggregation session. This means that in the communication graph it is necessary to find an oriented spanning *aggregation tree* (AT) with a root in the base station and the arcs directed to the root [10]. Then each non-pendent vertex must first receive the data packets from all its children and only then send the aggregated packet to its parent node in the AT.

We call a schedule *feasible* if it satisfies all four of the above requirements. The *Convergecast Scheduling Problem* (CSP) is to build the AT and to find a min-length feasible schedule for aggregating the data. The CSP is NP-hard in general case [4] and even in the case of a given AT [6]. A few polynomially solvable cases of the CSP are known. For example, if CSP is considered on a unit square grid, in each node of which there is a sensor, then the problem is polynomially solvable when the transmission distance is 1 [9] and when the transmission distance is 2 [7].

In [5], the CSP for a unit square grid with rectangular disjoint obstacles, which are impermeable to radio waves in the case when the transmission range of each sensor is equal to 1, was first considered. In this case, the sensors are located in all nodes of the grid except the interior of the obstacles. In [5], the algorithm LGA (Lexicographic Greedy Algorithm) is proposed for constructing a feasible schedule for a given AT and the notions of "vertical tree" (VT) and "horizontal tree" (HT) are introduced. Both trees VT and HT are spanning, and each vertex is connected to the base station by a path with the smallest possible number of edges. That is, both VT and HT are shortest path trees (SPT). When constructing the AT, if the vertex has adjacent vertices both from below and from the left, then the lower adjacent vertex is selected as the parent vertex in the VT, and the left one in the HT. In [13] it is shown that in an arbitrary communication graph the optimal schedule of conflict-free data aggregation is not necessarily based on the SPT. However, if the communication graph is a unit square grid, then it is sufficient to search the optimal schedule on the SPT [9]. In [5], a numerical experiment was conducted in which the obstacles of different sizes were accidentally placed in the grids of different dimensions. In 100% of the random examples, the LGA constructed the *optimal* schedule for at least one of the trees VT or HT, and the length of the constructed schedule coincided with the lower bound – the distance from the base station to the farthest vertex. This allowed us to formulate a hypothesis that the LGA *always* constructs an *optimal* schedule for at least one of the trees VT or HT [5].

In this paper, we refuted the above hypothesis by constructing a special counterexample, and proved that the LGA builds an *approximate* schedule on any of the trees VT or HT whose length exceeds the lower bound by not more than 1. As a result, the complexity of the problem remains an open question.

The rest of the paper is organized as follows. The mathematical formulation of the problem is given in Sect. 2. In Sect. 3, the properties of the solution of CSP are considered. In Sect. 4, we present the counterexample, in Sect. 5 we prove our main result, and the paper is concluded in Sect. 6.

2   Problem Formulation

The communication graph $G = (V, E)$ is given as a unit square grid of dimension $(n+1) \times (m+1)$ with a set of disjoint (tangency is allowed) rectangular obstacles $\{O_i, \ i = 1, \ldots, q\}$ with sides parallel to the axes. At each vertex $(x, y) \in V$, $x = 0, 1, \ldots, n$, $y = 0, 1, \ldots, m$, which is not the interior point of some obstacle O_i, except the origin $(0, 0)$, a sensor is placed. A sink (base station) is placed at the origin. Two vertices (x_1, y_1) and (x_2, y_2) are connected by an edge $e \in E$, if the Manhatten distance between them is $|x_1 - x_2| + |y_1 - y_2| = 1$ and there is no obstacle between them. During one time round, every vertex $i \in V$ can transmit a packet only along one edge $(i, j) \in E$ incident to it. Obviously, the graph G is connected. Figure 1a shows an example of a grid with obstacles.

A *schedule* is a function that assigns to each vertex of the communication graph a positive integer, the time round, when this vertex sends the packet. We call a schedule *feasible* if it satisfies the following conditions:

- it does not contain interference conflicts;
- during one time round, every vertex may either transmit or receive;
- during one time round, every vertex can receive or transmit at most one packet;
- each vertex sends a packet only once during the aggregation session.

In the CSP, it is required to find an aggregation tree and a feasible min-length schedule for data aggregation. Let's call the considered problem on the unit square grid with obstacles as a *Convergecast Scheduling Problem on a Grid with Obstacles* (CSPGO).

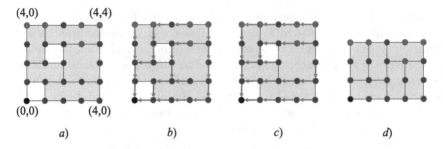

Fig. 1. *a*) The instance of a grid-graph with obstacles (grey rectangles); *b*) The VT; *c*) The HT; *d*) The counterexample.

3   Properties of the CSPGO

We call a *layer* a subset of the vertices of a grid with coinciding ordinates, and a subset of vertices with the same abscissas will be called a *column*. Below are some simple remarks and properties for the CSPGO.

Remark 1. If there are no obstacles between the top and the previous layer, then in the optimal solution all vertices of the top layer (vertices with maximal ordinates) can send packets simultaneously down to the corresponding vertices of the previous layer.

Remark 2. If there are no obstacles between the right and the previous column, then in the optimal solution all vertices of the right column (vertices with maximal abscissas) can send packets simultaneously to the left to the corresponding vertices of the previous column.

Considering Remark 1 and Remark 2, we can reduce the dimension of the grid and further assume that there are at least one obstacle with the side coinciding with the upper boundary of the grid and at least one obstacle with the side coinciding with the right boundary of the grid.

We call the *distance* and denote it as $d(i, j)$ between two vertices i and j the minimum number of edges in the path connecting these vertices. In the CSPGO, we will be interested in the distance $d(i, 0)$ from any vertex i to the base station. For any graph the following properties hold.

Property 1. If at least two vertices are at a distance of R from the base station, then the schedule length cannot be less than $R + 1$.

The following property is not used further, but it helps to understand the structure of the solution, so we decided to formulate it.

Property 2. If in the square grid there are q vertices at equal distance from the base station, then at least $\lceil q/2 \rceil$ of them can send messages simultaneously without any conflict.

In the LGA [5], at each step, a set S_t of vertices is searched, which send messages during the time round t. In this set, the leaves of the AT are included in the decreasing order of the distances from the sink, starting from the most remote one, if this does not lead to a conflict. The vertices that transmit the messages are then deleted together with the incident edges, and the procedure is repeated during the next time round.

To illustrate the operation of the LGA, refer to the Fig. 2, in which the green vertices transmit packets, and the red vertices are forbidden to transmit due to a conflict.

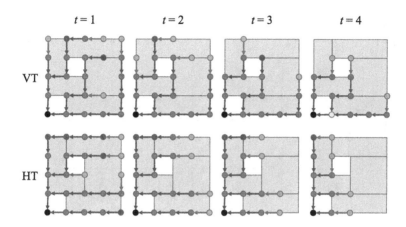

Fig. 2. 4 iterations of LGA.

In this example, in the VT two outermost vertices are transmitted during the first time round, as well as two leaves at a distance 4 from the base station. A red vertex at a distance 6 cannot transmit, otherwise, it would lead to the conflict with the transmission of the most remote vertex. During the second time round, all the pendent vertices of the VT transmit. During the third time round, one leaf (a red one) of the VT at a distance 5 cannot transmit. As a result, after the third time round, we have the two most remote vertices (they are circled in red). This means that at the yellow node the packets from the left and from the top vertices will arrive simultaneously, but they will have to transfer in turn and as a result, the aggregation time will increase by 1 (Property 1). In this example the algorithm constructs an optimal schedule on the HT, the length of which is equal to the distance to the most remote vertex.

Corollary 1. To match the length of the schedule with the lower bound $D = n + m$, it is necessary that at each step of the LGA there was only one most remote vertex. Hence, all the vertices that are the closest to remote, except the one to which the most remote vertex transmits, must also transmit.

In order for the length of the schedule to coincide with D, all vertices at a distance k from the base station must transmit no later than during the time round $D - k$. This means that not all conflicts increase the length of the schedule. We call the conflict *critical* if it leads to an increase in the schedule length.

4   Counterexample

As noted above, in [5], we formulated the hypothesis that the LGA constructs an *optimal* schedule on at least one of the trees VT or HT. Figure 1*d* shows an example that refutes this hypothesis. The illustration of this is shown in the Figs. 3 and 4.

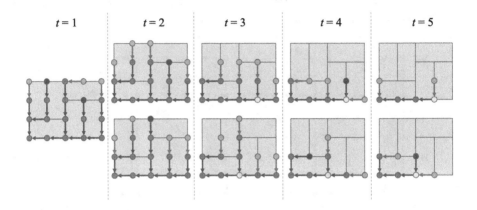

Fig. 3. 5 iterations of LGA on the VT. The length of the schedule is 8.

Figure 3 shows all the options for scheduling during the first 5 rounds on the VT. At the third iteration, we have the two most remote vertices, which are circled in red. Such a situation inevitably leads to a critical conflict in the yellow vertex.

Figure 4 shows all the options for scheduling during the first 5 rounds in the HT. At the fifth iteration of the algorithm, we have the two most remote vertices, which are circled in red, and such a situation leads to a critical conflict in the yellow vertex.

In all cases, the conflict is inevitable, and the length of the schedule is 8. For this example, the length of the optimal schedule coincides with the lower bound and is equal to 7 (Fig. 5). It can be seen that in this case there are conflicts during the first two rounds, but they are not critical, because the red vertex at the distance $D - 3$ is silent only during two first rounds.

So, in [5], we assumed that the LGA constructs an optimal schedule, of length D on at least one tree VT or HT, and a representative numerical experiment has strengthened our assumption. But the above example refutes our hypothesis, therefore in the next section, we prove a weaker statement.

5   Accuracy of the LGA

We will formulate the main result for the VT, although it is also true for the HT.

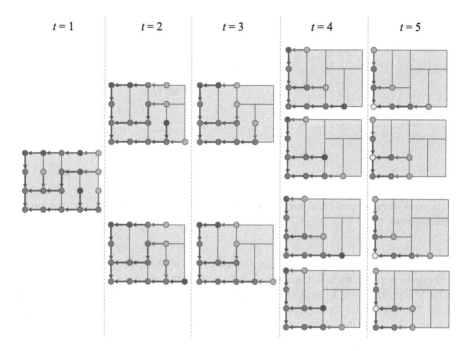

Fig. 4. 5 iterations of LGA on the HT. The length of the schedule is 8.

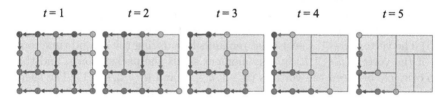

Fig. 5. 5 iterations of LGA on the optimal tree. The length of the schedule is 7.

Theorem 1. *The LGA builds a schedule on a VT, the length of which exceeds the lower bound D by not more than 1.*

Proof. We prove the statement by induction on the number of layers $k = 1, \ldots, m$.

(1) Let $k = 1$. In this case, it is easy to verify that, for any arrangement of the obstacles, the LGA constructs an *optimal* schedule on the VT.

(2) Suppose the statement is true for any number of layers $1, \ldots, k - 1$.

(3) Let us prove the statement for k layers. Note that in the VT, the path from the outermost vertex goes first along the right boundary, and then along the lower boundary of the grid. We denote this path by P_1. If we exclude the P_1, we obtain the number of layers equal to $k - 1$ and, by assumption (2), the theorem is proved. Hence, we need to consider the path P_1. The distance to the

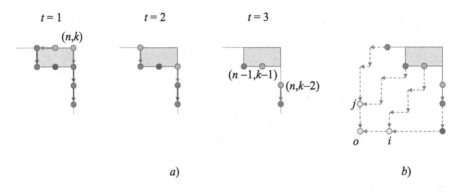

Fig. 6. *a*) 3 iterations of LGA on the instance of the VT; *b*) Illustration to the proof of the theorem.

vertex (n, k) (the length of the P_1) is D and suppose that the statement is false, i.e. there is a path with a transmission time greater than $D + 1$. In the LGA on the VT vertices (n, k) and $(n - 1, k)$ transmit messages during the first time round. The vertex (n, k) sends the packet down, and the vertex $(n - 1, k)$ sends either down, or (if there is an obstacle) to the left. In either case (since obstacles are rectangular), the vertex $(n - 1, k - 1)$ cannot transmit the packet during the first time round, because it will either interfere with the transmission by the vertex (n, k), or be the addressee of the vertex $(n - 1, k)$.

Let's consider two cases: (i) the transmission time along the path P_1 is greater than D, (ii) the transmission time along the path P_1 is D.

(i) The vertex $(n - 1, k - 1)$ must remain silent during the second time round only if the vertex $(n - 2, k)$ at this moment sends the packet down, and there is no obstacle between the vertices $(n - 1, k - 1)$ and $(n - 2, k - 1)$. The silence of the vertex $(n - 1, k - 1)$ during the first two rounds leads to the conflict, because after two time rounds at a distance of $D - 2$ there are two vertices $(n - 1, k - 1)$ and $(n, k - 2)$ (Fig. 6*a*). This results in an increase of the schedule length by 1 (Property 1), because the messages from the vertices $(n-1, k-1)$ and $(n-2, k-1)$ will come simultaneously to some vertex i (Fig. 6*b*), and one packet will have to wait. Let's consider further a path P_1 to some vertex i (Fig. 6*b*), the transmission along which turned out to be longer for 1 time round than its length. Can there still be a delay in the continuation of this path to the base station? Another conflict will occur if at some vertex j (Fig. 6*b*) the last data packet arrives at the same time round $d((n, k), i) + 1$ as in the vertex i. But, as it is easy to see, none of the paths to the vertex j can be longer than $d((n, k), i) - 1$. Considering the fact that at the vertex i the packets arrive 1 moment later than the distance $d((n, k), i)$, there should be a delay of 2 on the path to the vertex j. But the vertex j is above the vertex i, therefore, by assumption (2), on the path to the vertex j there cannot be more than one conflict increasing the packet transmission time by no more than 1. This contradicts the assumption that there is a path with a transmission time greater than $D + 1$.

(ii) If the transmission time along the path P_1 is D, then there is a vertex o belonging to the path P_1, which also belongs to another path P_2 higher than the path P_1, and the arrival time of the packet along the P_2 is the same as the arrival time of the packet to the vertex o along the P_1 (Fig. 6b). But by assumption (2), the delay along the path P_2 cannot be more than 1. Moreover, the path P_2 is shorter than the path P_1. Consequently, such a situation is excluded.

The theorem is proved.

Without loss of generality, let's assume that $n \leq m$. In fact, VT is constructed with complexity $O(nm)$. The LGA performs no more than $O(n + m) = O(m)$ iterations. At each iteration, it is necessary to find in the current tree the max-cardinality subset of pendent vertices at some distance from the sink that can send the packets without conflicts. This can be done with $O(n)$–time complexity [5]. Consequently, the total complexity of LGA is $O(nm)$.

6   Conclusion

In this paper, a special case of the problem of minimizing the time of conflict-free data aggregation in a single-channel wireless sensor network, known as the convergecast scheduling problem, is considered. In the case considered, the sensors are located in the nodes of a square grid of dimensions $(n + 1) \times (m + 1)$ with rectangular disjoint obstacles that are impermeable to the radio waves. The transmission range of each sensor is 1, and the base station is at the origin $(0, 0)$.

A polynomial algorithm, the LGA with complexity $O(nm)$ (proposed in [5]), that constructs a feasible schedule for a given aggregation tree is considered. It is obvious that the length of the schedule depends on the aggregation tree, but it cannot be less than the distance $n+m$ to the most remote vertex of the grid. We disproved the hypothesis expressed in [5] that at least one schedule constructed by the LGA on "horizontal tree" (HT) or "vertical tree" (VT) is optimal.

In this paper, we proved that the LGA constructs an approximate schedule on the VT, whose length exceeds the lower bound $n + m$ by not more than 1.

Acknowledgments. The research of A.I. Erzin is partly supported by the Russian Foundation for Basic Research, Projects 16-07-00552 and by the program of fundamental scientific researches of the SB RAS No. I.5.1. (project 0314-2016-0014). The research of R.V. Plotnikov is partly supported by the Russian Foundation for Basic Research, Project 16-37-60006.

References

1. Calinescu, G., Mandoiu, I.I., Zelikovsky, A.: Symmetric connectivity with minimum power consumption in radio networks. In: Baeza-Yates, R.A., Montanari, U., Santoro, N. (eds.) Proc. 2nd IFIP International Conference on Theoretical Computer Science. IFIP Conference Proceedings, vol. 223, pp. 119–130. Kluwer, Dordrecht (2002)

2. Cardei, M., Wu, J.: Energy-efficient coverage problems in wireless ad-hoc sensor networks. Comput. Commun. **29**, 413–420 (2006)
3. Carle, J., Simplot, D.: Energy-efficient area monitoring by sensor networks. IEEE Comput. **37**, 40–46 (2004)
4. Chen, X., Hu, X., Zhu, J.: Minimum data aggregation time problem in wireless sensor networks. In: Jia, X., Wu, J., He, Y. (eds.) MSN 2005. LNCS, vol. 3794, pp. 133–142. Springer, Heidelberg (2005). https://doi.org/10.1007/11599463_14
5. Erzin, A., Plotnikov, R.: Efficient algorithm for convergecast scheduling problem on a square grid with obstacles. In: Proceedings of the OPTIMA-2017, Petrovac, Montenegro. CEUR-WS, vol. 1987, pp. 187–193 (2017)
6. Erzin, A., Pyatkin, A.: Convergecast scheduling problem in case of given aggregation tree: the complexity status and some special cases. In: 10th International Symposium on Communication Systems, Networks and Digital Signal Processing, article 16. IEEE-Xplore, Prague (2016)
7. Erzin, A.: Solution of the convergecast scheduling problem on a square unit grid when the transmission range is 2. In: Battiti, R., Kvasov, D.E., Sergeyev, Y.D. (eds.) LION 2017. LNCS, vol. 10556, pp. 50–63. Springer, Cham (2017). https://doi.org/10.1007/978-3-319-69404-7_4
8. Erzin, A.I., Mladenovich, N., Plotnikov, R.V.: Variable neighborhood search variants for min-power symmetric connectivity problem. Comput. Oper. Res. **78**, 557–563 (2017)
9. Gagnon, J., Narayanan, L.: Minimum latency aggregation scheduling in wireless sensor networks. In: Gao, J., Efrat, A., Fekete, S.P., Zhang, Y. (eds.) ALGOSENSORS 2014. LNCS, vol. 8847, pp. 152–168. Springer, Heidelberg (2015). https://doi.org/10.1007/978-3-662-46018-4_10
10. Incel, O.D., Ghosh, A., Krishnamachari, B., Chintalapudi, K.: Fast data collection in tree-based wireless sensor networks. IEEE Trans. Mob. Comput. **11**(1), 86–99 (2012)
11. Malhotra, B., Nikolaidis, I., Nascimento, M.A.: Aggregation convergecast scheduling in wireless sensor networks. Wirel. Netw. **17**, 319–335 (2011)
12. Ghods, F., Yousefi, H., Mohammad, A., Hemmatyar, A., Movaghar, A.: MC-MLAS: multi-channel minimum latency aggregation scheduling in wireless sensor networks. Comput. Netw. **57**, 3812–3825 (2013)
13. Tian, C.: Neither shortest path nor dominating set: aggregation scheduling by greedy growing tree in multihop wireless sensor networks. IEEE Trans. Veh. Ttchnolohy. **60**(7), 3462–3472 (2011)
14. Wang, P., He, Y., Huang, L.: Near optimal scheduling of data aggregation in wireless sensor networks. Ad Hoc Netw. **11**, 1287–1296 (2013)
15. Xu, X., Li, X.-Y., Mao, X., Tang, S., Wang, S.: A delay-efficient algorithm for data aggregation in multihop wireless sensor networks. IEEE Trans. Parallel Distrib. Syst. **22**, 163–175 (2011)
16. Zalyubovskiy, V., Erzin, A., Astrakov, S., Choo, H.: Energy-efficient area coverage by sensors with adjustable ranges. Sensors **9**, 2446–2460 (2009)

An Effective Heuristic for a Single-Machine Scheduling Problem with Family Setups and Resource Constraints

Júlio C. S. N. Pinheiro, José E. C. Arroyo$^{(\boxtimes)}$, and Ricardo G. Tavares

Department of Computer Science, Universidade Federal de Viçosa, Viçosa, MG
36570-900, Brazil
{julio.pinheiro,ricardo.tavares}@ufv.br, jarroyo@dpi.ufv.br

Abstract. This paper presents a simple and effective iterated greedy heuristic to minimize the total tardiness in a single-machine scheduling problem. In this problem the jobs are classified in families and setup times are required between the processing of two jobs of different families. Each job requires a certain amount of resource that is supplied through upstream processes. The total resource consumed must not exceed the resource supply up. Therefore, jobs may have to wait and the machine has to be idle due to an insufficient availability of the resource. The iterated greedy heuristic is tested over an extensive computational experience on benchmark of instances from the literature and randomly generated in this work. Results show that the developed heuristic significantly outperforms a state-of-the-art heuristic in terms of solution quality.

Keywords: Single machine scheduling · Family setup-times · Resource constraints · Total tardiness · Meta-heuristics

1 Introduction

This paper considers a single-machine scheduling problem with the goal of minimizing the total tardiness of the jobs. For each job is given a processing time and a due date, and it belongs to a given family. The machine processes only one job at a time without preemption. A setup time is required between the processing of two jobs of different families and during this time the machine cannot process any other job. A resource constraint is also considered in the problem. Each job requires a certain amount of a common resource provided by the machine, which is supplied by upstream processes. At any time, the cumulative resource consumption must not overcome the cumulative resource supply. Thus, jobs may have to wait and the machine has to be idle due to an insufficient availability of the resource.

This studied problem has practical applications in many industries. For example, it arises in the continuous casting stage of steel production (Herr and Goel, [10]). Ladles with liquid steel of certain grades are fed to a continuous casting

© Springer Nature Switzerland AG 2019
R. Battiti et al. (Eds.): LION 12 2018, LNCS 11353, pp. 141–153, 2019.
https://doi.org/10.1007/978-3-030-05348-2_12

machine, and each ladle contains allocated orders to it that determine a due date. Whenever two ladles of similar steel grades (within the same setup family) are consecutively processed no setup work is needed. Still, if a change of ladles of different steel grades occurs, a setup process is required. The liquid steel is supplied by the blast furnace at a constant rate. At any time of the casting is not allowed to consume more liquid steel than the supplied by the blast furnace.

Similar situations also occur in many multi-stage production processes where upstream work systems supply a common resource that is consumed in downstream tasks executed on a machine, as in assembly lines where different products are assembled by their component parts provided by an upstream stage.

Since this problem has not been thoroughly exploited, our contribution with this work is providing a distinct and effective heuristic method for solving it.

The remainder of this paper is organized as follows. Section 2 gives a literature review of related works. Section 3 describes the problem under study. Our solution approach based on an iterated greedy (IG) algorithm is described in Sect. 4. Section 5 presents the implementation of a heuristic algorithm proposed in the literature for the same problem. In Sect. 6, the computational experiments and results are showed. Finally, the conclusions of this paper are presented in Sect. 7.

2   Literature Review

The problem addressed is this paper comprises three main streams of study in literature. First, single-machine scheduling problems with the goal of minimizing the total tardiness. Second, single-machine scheduling with setup regards. Third, scheduling problems with resource constraints.

Without setup considerations, the single-machine scheduling with total tardiness minimization is proved to be NP-hard (Du and Leung, [5]). Scheduling problems with tardiness minimization are studied by Koulamas [11] and Sen et al. [20]. For single-machine problems with total tardiness minimization, Emmons [6] described conditions that when fulfilled would result in an optimal schedule. Lawler [12] presented a decomposition approach that separates the problem into two mutually exclusive sub-problems using the longest job as a separator.

Surveys of Allahverdi et al. [1] and Potts and Kovalyov [15] present the main studies on scheduling problems with setup times considerations. Many meta-heuristics approaches were applied for minimizing the total tardiness on single-machine scheduling problems with sequence-dependent setup times. Gupta and Smith [9] presented a greedy randomized adaptive search procedure (GRASP) heuristic as well as a space-based local search procedure; Liao and Juan [13] introduced an ant colony optimization (ACO) based algorithm; Lin and Ying [14] proposed a simulated annealing and a tabu search algorithm; Ying et al. [25] developed a local search based iterated greedy heuristic; and Sioud et al. [21] used a hybrid genetic algorithm. An exact branch and bound algorithm was presented by Bigras et al. [2]. For a variant of the problem where the goal is to minimize the weighted tardiness of the jobs, an exact method based on successive sublimation dynamic programming and an iterated local search (ILS) heuristic were presented by Tanaka and Araki [23] and Subramanian et al. [22], respectively.

Some authors addressed scheduling problems with family setup-times and total tardiness objective. Gupta and Chantaravarapan [8] studied the problem under consideration of the group technology assumption (GTA) (see e.g. Potts and Van Wassenhove, [16]). They presented a mixed-integer programming (MIP) formulation that solves small problem instances as well as a heuristic algorithm for larger instances. Based on the properties described by Emmons [6], Schaller [18] developed two branch and bound procedures for the family scheduling problem with and without GTA. Moreover, Schaller and Gupta [19] proposed exact and heuristic methods, with and without GTA, for a single-machine scheduling problem with family setups. Furthermore, Jacob and Arroyo [24] presented iterated local search heuristics for a single-machine scheduling problem with sequence-dependent family setup times to minimize total tardiness.

Briskorn et al. [3] studied a single-machine scheduling problem where the unavailability of the required resource prevents a job of being processed. They considered the minimizing of the maximum lateness and the number of tardy jobs. Briskorn et al. [4] studied the single-machine problem with inventory constraints to minimize the total weighted completion times. Based on properties derived for an optimal solution, they presented a branch and bound and a dynamic programming method for solving the problem. They concluded that even for small problem instances with 20 jobs, exact approaches are not efficient for solving the problem, then heuristics methods are required.

The formulation of the problem studied in this paper is given by Herr and Goel [10]. They presented a MIP model as well as an ILS heuristic for solving the problem. Furthermore, they considered two variants of the problem: first, setup times are only considered between jobs of different families; second, setup times are also considered between jobs of the same family. In this paper we address the first variant of the problem.

3   Problem Description

A complete MIP model for problem under study is presented by Herr and Goel [10]. The problem can be described as follows. Let J the set of n jobs to be processed on a single machine. Each job j has a processing time p_j, a due date d_j and a quantity q_j representing the amount of resource required from the machine when the job j is processed. It is assumed that the resource is consumed at a constant rate of q_j/p_j. Also, each job belongs to a setup family f_j. For any pair of jobs $i, j \in J$, if i is scheduled before j and $f_i \neq f_j$, a setup time of duration $s_{ij} > 0$ is required between processing the jobs. Otherwise, $s_{ij} = 0$.

The single machine can process only one job at a time without preemption. Initially the machine has a resource amount r^* and it is supplied at a constant rate r per unit of time. In order to prevent that the cumulative demand of a job surpasses the cumulative supply of the machine, the machine has to remain idle and the starting of the job has to be delayed. This idle time must be sufficient so the delayed job can be processed without resource constraints. Whether in case of setup time or idle time, the machine cannot process any other job. The goal of the problem is to find a schedule (processing sequence of the jobs) that minimizes the total tardiness.

Following the MIP model of Herr and Goel [10], we assume the processing of a job may be delayed for any period of time due unavailability of resource. For a job j in a sequence, let Q_j denotes the cumulative demand of resources of all jobs up to j. The tardiness of a job j is computed as follows.

The completion time of the first job i in the sequence is

$$C_i = \max\{p_i, \frac{Q_i - r^*}{r}\} \tag{1}$$

The completion times of the other jobs j is computed as

$$C_j = \max\{C_i + s_{ij} + p_j, \frac{Q_j - r^*}{r}\} \tag{2}$$

where i is the job processed before j. Then, the tardiness of any job j is

$$T_j = \max\{0, C_j - d_j\} \tag{3}$$

Therefore, the total tardiness of all the jobs is $TT = \sum_{j \in J} T_j$.

4   Solution Approach

Herr and Goel [10] showed that solving instances of the problem under study with a large number of jobs using a MIP solver is inefficient in terms of computational time. Even if all the jobs belonged to a same family and the initial amount of resource was sufficiently large the problem would be reduced to the problem studied by Du and Leung [5], which is proven to be NP-hard. Therefore, the problem studied in this paper is also NP-hard.

In this study we developed an iterated greedy (IG) algorithm to tackle the problem. The IG algorithm was first used in scheduling problems by Ruiz and Stützle [17] to solve a permutation flow shop scheduling problem, and since then has been applied to others types of scheduling problems (see e.g. Ying et al. [25], and Fanjul-Peyro and Ruiz, [7]). A typical IG algorithm has few control parameters and is simple to implement.

The general scheme of the proposed IG heuristic is presented in Algorithm 1. Our algorithm has two input parameters: α (destruction level) and $nIter$ (maximum number of consecutive iterations of the algorithm without improvements of the best solution). The algorithm begins generating an initial solution (Step 1). This initial solution is generated by using the earliest due date (EDD) greedy rule. That is, an initial sequence is obtained by sorting the jobs in non decreasing order of their due dates.

Algorithm 1 Iterated Greedy $(\alpha, nIter)$

1: $S_{best} \leftarrow$ InitialSolution()
2: $S \leftarrow S_{best}$
3: $k \leftarrow 0$
4: **while** $k < nIter$ **do**
5:     $S_{partial} \leftarrow$ Destruction(S, α)
6:     $S \leftarrow$ Reconstruction$(S_{partial})$
7:     $S \leftarrow$ LocalSearch$(S, nIter)$
8:     **if** $TT(S) < TT(S_{best})$ **then**
9:         $S_{best} \leftarrow S$
10:        $k \leftarrow 0$
11:    **else**
12:        $k \leftarrow k + 1$
13:    **end if**
14: **end while**
15: **return** S_{best}

The iterations of the algorithm are computed in Steps 4–14 until the maximum number of consecutive iterations without improvements is reached. In the destruction phase (Step 5), $\alpha \times n$ jobs are randomly selected and removed from the sequence. In the reconstruction phase, the jobs are greedily reinserted (Step 6). For each job removed, it is tentatively reinserted in all positions of the sequence and then put in the position that gives the lowest total tardiness of the partial solution. Afterwards, a local search (LS) procedure is performed in order to improve the new constructed solution (Step 7). A counter k is used to get the maximum number of consecutive iterations without improvements. If the solution returned by the local search procedure improves the current best solution, the best solution obtained so far is updated (Step 9), and the counter k is set to zero (Step 10). If the best solution is not improved, the counter k is incremented (Step 12). The IG algorithm returns the best solution found (Step 15).

The next subsection provides an explanation of the local search procedure used in the IG algorithm.

4.1   Local Search Procedure

A local search (LS) procedure based on six neighborhood operators is performed to improve the solution given by the reconstruction phase. At each iteration of the LS procedure, a random neighbor solution is generated from the general current solution, then its total tardiness is computed. If neighbor solution improves the general solution, the general solution is updated and a counter variable is reset. Otherwise, the counter variable is incremented. The LS procedure ends when the current solution is not improved during $nIter$ consecutive iterations.

The pseudocode of the LS procedure is given in Algorithm 2. The used neighborhood operators are described in next subsection.

Algorithm 2 Local Search $(S, nIter)$

1: $k \leftarrow 0$
2: **while** $k < nIter$ **do**
3:     Select at random a neighborhood operator
4:     $S_{\text{neighbor}} \leftarrow \text{RandomNeighbor}(S)$
5:     **if** $TT(S_{\text{neighbor}}) < TT(S)$ **then**
6:         $S \leftarrow S_{\text{neighbor}}$
7:         $k \leftarrow 0$
8:     **else**
9:         $k \leftarrow k + 1$
10:     **end if**
11: **end while**
12: **return** S

4.2 Neighborhood Operators

The neighborhood operators are algorithmic objects for modifying a solution (schedule or sequence). Each operator modifies a sequence in a specific way. The operators used in our heuristic are the same six operators used by Herr and Goel [10]. Some of those operators regards the fact that in family scheduling problems a subset of jobs within a same family processed consecutively in a sequence can be grouped as a batch. Such batch-based operators have the advantage of maintaining some structural properties of the solution and avoiding unnecessary setup times (Herr and Goel, [10]).

The Batch Move and Job Move operators select a batch and a job, respectively, and move them to another position in the schedule. The Batch Exchange and Job Exchange operators select two different batches and two different jobs, respectively, and swap them. The Batch Break operator selects a batch, sorts it by the the due dates of its jobs and breaks it in the position with the largest due date difference. After that, the two batches are inserted in different positions in the schedule. The Batch Combine operator selects two batches of the same family and bind them, then insert the combined batch in the schedule.

5   Implemented Algorithms from Literature

To the best of our knowledge, there is only one paper that studies the addressed problem. Herr and Goel [10] proposed an iterated local search (ILS) heuristic to solve the problem. In this work, we re-implemented this ILS heuristic according to the original paper in order to evaluate the efficiency and effectiveness of our IG algorithm.

The ILS implementation can be described as follows. Each neighborhood operator described in Sect. 4.2 was implemented as a first-improvement local search procedure. An initial solution is generated by using the EDD rule. Then, at each iteration the local search procedures are randomly sorted and consecutively applied to the incumbent solution. The stopping criterion of each local search operator is whether a local minimum was found.

The perturbation approach used in the ILS algorithm calculates a promising position for each job in the sequence in terms of total tardiness minimisation (Herr and Goel, [10]). The promising position is calculated based on the lateness of each job. Positive lateness suggests a shifting of the job towards the beginning of the sequence, whereas negative lateness suggests a shifting of the job towards the ending of the sequence. As in Herr and Goel [10], the maximum number of iterations used in the ILS heuristic was 3.

6   Computational Experiments

In this section we describe the computational experiments carried out in order to study the performance of our IG algorithm. We compare the IG algorithm against the ILS algorithm proposed by Herr and Goel [10]. First, we ensure that the results of our ILS implementation are at least as good as the results presented by Herr and Goel [10]. We also compare several configurations of the IG algorithm with different parameters settings.

All the heuristic algorithms were coded in C++ and executed on a PC machine IntelR Core TM i7-4790K CPU @ 4.00GHz x 8 with 32GB RAM, running Ubuntu 14.04 LTS 64 bits. The quality of the solutions were measured by the relative percentage deviation (RPD) from the best known solution (reference solution). The RPD measure represents the percentage of how much experimental results differs from a reference value, and is calculated following the equation:

$$RPD(\%) = 100 \times \frac{TT_{\text{experimental}} - TT_{\text{reference}}}{TT_{\text{reference}}} \tag{4}$$

where $TT_{\text{reference}}$ is the value of the reference solution, and $TT_{\text{experimental}}$ is the obtained solution by a given heuristic.

The performance of the heuristic algorithms are tested on 80 small instances provided by Herr and Goel [10], and 30 large instances randomly generated in this paper according to (Herr and Goel, [10]). The factors of the small instances are: number of jobs $n \in \{8, 10, 12, 15, 20, 30, 40, 50\}$ and number of families $\in \{1, 2, 3, 4, 5\}$. The factors of the large instances are: number of jobs $n \in \{100, 150, 200\}$ and number of families $\in \{5, 6, 7\}$.

6.1   ILS Effectiveness

In this subsection we analyze the performance of the ILS algorithm of Herr and Goel [10]. We denote by ILS_I the ILS algorithm implemented in this paper, and by ILS_L the ILS algorithm implemented by Herr and Goel [10].

The ILS_I algorithm was ran on the instances provided by Herr and Goel [10] and the obtained results are compared with the results of the ILS_L algorithm available in the literature. The performance of each ILS algorithm is measured by RPD determined by Eq. (4), where $TT_{reference}$ is the minimum TT value found for the instance, and $TT_{experimental}$ is the minimum TT value found by a given ILS algorithm. Therefore, we are able to determine whether ILS_I is able to found results at least as low as the presented in the literature.

The results of the experiment were analyzed by means of Kruskal-Wallis statistical test by using the RPD measure as response variable. In this test two hypothesis are tested: the null hypothesis states that the performance of all the implementations tested are statically similar; and the alternative hypothesis states at least one implementation performance is significantly different. Figure 1 shows the mean plot and confidence intervals at a 95% confidence level. Overlapping intervals suggest that could be no difference between implementations. Since there are no overlapping intervals, the ILS_I presents the lowest mean, and the calculated P-value = $0.000 \leq 0.05$ suggests the null hypothesis can be rejected. We can conclude that the implemented ILS_I is effective enough to represent the implemented one in literature. Table 1 shows how the ILS implementations perform as the number of jobs increases.

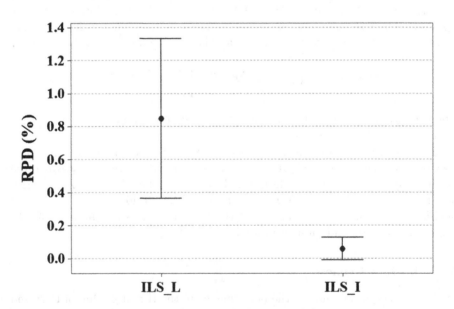

Fig. 1. Mean plot and confidence intervals at 95% confidence level for ILS_L versus ILS_I experiment.

Table 1. RPD of ILS implementations on literature instances per number of jobs.

Implementation	Number of jobs							
	8	10	12	15	20	30	40	50
ILS_L	0.00	0.00	0.00	0.42	0.24	2.32	0.86	2.95
ILS_I	0.00	0.17	0.00	0.31	0.00	0.00	0.00	0.00

6.2 IG Parameter Setting

The developed IG depends only on two parameters: the number of jobs to be randomly removed in the destruction phase ($\alpha \times n$), and the maximum number of iterations of the algorithm ($nIter$). Preliminary tests were conducted to determine the parameters calibrations. Each parameter was tested at three levels: $\alpha \in \{0.10, 0.15, 0.20\}$ and $nIter \in \{100, 200, 300\}$. We carried out a full factorial experiment with the levels to be tested, resulting in 9 versions of our IG algorithm. Each instance was solved 30 times by all the 9 versions of the algorithm.

The results were analyzed by means of Kruskal-Wallis using the RPD as response variable. Following equation (4), $TT_{\text{experimental}}$ was made as the average TT value found for the instance by a given IG version, and $TT_{\text{reference}}$ was made as smallest TT value found for the instance among all the versions. Figure 2 shows the mean plot and confidence intervals at a 95% confidence level. We can see that almost every version interval overlaps with one another as the P-value = 0.085 > 0.05 from the test indicates that the null hypothesis assumption is the most likely to be correct.

Although no significant difference level is indicated by the test, some conclusions can be derived from Fig. 2 about the IG parameters. We can see that as the number of iterations $nIter$ grows the means turn lower, and as the α value increases the intervals turn wider. Thus, we are prone to chose an algorithm version with a large $nIter$. Among the largest $nIter$ values, the IG with $\alpha = 0.15$ and $nIter = 300$ outstands for bearing the lowest RPD. Therefore, this configuration is the chosen one to be used in the next experiments.

6.3 ILS Versus IG

In this section we present the comparison of solution quality generated by the algorithms ILS_I and IG. Two experiments are performed. In the first experiment, the algorithms are compared on the 80 small instances provided by Herr and Goel [10]. In the second experiment, the algorithms are compared on the 30 large instances generated in this paper. The performance of each algorithm is also measured by RPD determined by Eq. (4), where $TT_{\text{reference}}$ is the best known solution obtained among the two algorithms.

Results on Small Instances: These experimental results were analyzed by means of Kruskal-Wallis test. The obtained P-value = 0.034 ≤ 0.05 indicates that

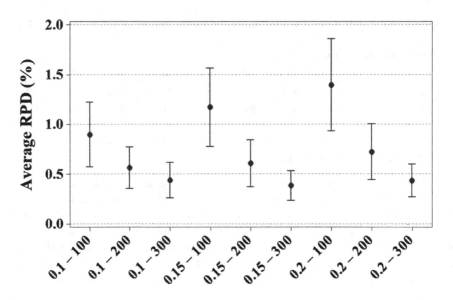

Fig. 2. Mean plot and confidence intervals at 95% confidence level for IG calibration experiment.

the null hypothesis can be rejected, i.e., one algorithm is significant different. Figure 3 shows, for the two heuristics, the mean plot and confidence intervals with 95% confidence level. Since there are no overlapping intervals and the IG holds the lowest RPD, it is safe to conclude that the IG outperforms the ILS on small instances. Table 2 shows how the heuristic algorithms perform as the number of jobs increases.

Table 2. Average RPD of the heuristics on small instances per number of jobs.

Heuristic	Number of jobs (n)							
	8	10	12	15	20	30	40	50
ILS	0.00	0.17	0.00	0.31	0.48	0.99	1.48	2.36
IG	0.00	0.09	0.00	0.00	0.01	0.00	0.05	0.64

Results on Large Instances: The results for large instances are presented in Table 3 and Fig. 4. Figure 4 shows the means plot and confidence intervals at 95% confidence level. Overlapping intervals indicates that no statistically significant difference exists among the overlapped means, i.e. the algorithms are statistically equivalent. Although, submitting the results for testing by means of Kruskal-Wallis test, the calculated P-value $= 0.012 \leq 0.05$ and the IG lowest RPD prones us to conclude that the IG heuristic also outperforms the ILS on large instances in this experiment. Table 3 shows how the heuristics perform as the number of jobs increases.

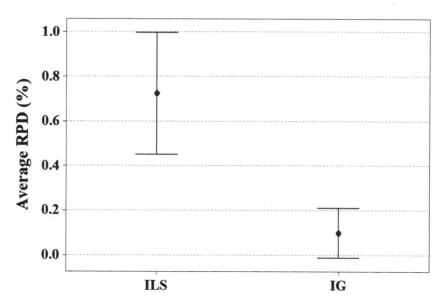

Fig. 3. Mean plot and confidence intervals at 95% confidence level for ILS versus IG experiment on small instances.

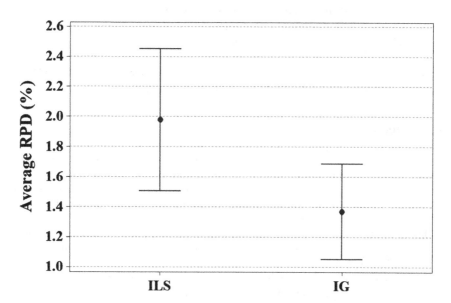

Fig. 4. Mean plot and confidence intervals at 95% confidence level for ILS versus IG experiment on large instances.

Table 3. Average RPD of the heuristics on large instances per number of jobs.

Heuristic	Number of jobs		
	100	150	200
ILS	2.61	1.50	1.82
IG	1.31	1.12	1.68

7   Conclusions

In this study we consider a single-machine scheduling problem with family setups and resource constraints with the goal of minimizing the total tardiness. For the best of our knowledge this problem has not been fully exploited in the literature, consequently, methods for solving this problem are so far in shortage. Since the problem is NP-hard and exact methods are time inefficient, heuristic methods are required to solve the problem. We presented a simple and effective iterated greedy algorithm which uses a local search procedure based on six neighborhood operators. Two set of instances were used to analyze the performance of the IG algorithm: small instances provided by Herr and Goel [10], and large instances generated in this paper. Extensive computational experiments have shown that the performance of our IG algorithm increases as the number of iterations increases as well. Computational experiments demonstrated the effectiveness of our IG algorithm. It outperforms the ILS algorithm of Herr and Goel [10]. The obtained results have been statistically tested.

Future research is to extend our approach for other manufacturing environments, such as parallel processing machines.

Acknowledgments. The authors thanks the financial support of FAPEMIG, CAPES and CNPq, Brazilian research agencies.

References

1. Allahverdi, A., Ng, C., Cheng, T.E., Kovalyov, M.Y.: A survey of scheduling problems with setup times or costs. Eur. J. Oper. Res. **187**(3), 985–1032 (2008)
2. Bigras, L.-P., Gamache, M., Savard, G.: The time-dependent traveling salesman problem and single machine scheduling problems with sequence dependent setup times. Discret. Optim. **5**(4), 685–699 (2008)
3. Briskorn, D., Choi, B.-C., Lee, K., Leung, J., Pinedo, M.: Complexity of single machine scheduling subject to nonnegative inventory constraints. Eur. J. Oper. Res. **207**(2), 605–619 (2010)
4. Briskorn, D., Jaehn, F., Pesch, E.: Exact algorithms for inventory constrained scheduling on a single machine. J. Sched. **16**(1), 105–115 (2013)
5. Du, J., Leung, J.Y.-T.: Minimizing total tardiness on one machine is np-hard. Math. Oper. Res. **15**(3), 483–495 (1990)
6. Emmons, H.: One-machine sequencing to minimize certain functions of job tardiness. Oper. Res. **17**(4), 701–715 (1969)

7. Fanjul-Peyro, L., Ruiz, R.: Iterated greedy local search methods for unrelated parallel machine scheduling. Eur. J. Oper. Res. **207**(1), 55–69 (2010)
8. Gupta, J.N., Chantaravarapan, S.: Single machine group scheduling with family setups to minimize total tardiness. Int. J. Prod. Res. **46**(6), 1707–1722 (2008)
9. Gupta, S.R., Smith, J.S.: Algorithms for single machine total tardiness scheduling with sequence dependent setups. Eur. J. Oper. Res. **175**(2), 722–739 (2006)
10. Herr, O., Goel, A.: Minimising total tardiness for a single machine scheduling problem with family setups and resource constraints. Eur. J. Oper. Res. **248**(1), 123–135 (2016)
11. Koulamas, C.: The single-machine total tardiness scheduling problem: review and extensions. Eur. J. Oper. Res. **202**(1), 1–7 (2010)
12. Lawler, E.L.: A "pseudopolynomial" algorithm for sequencing jobs to minimize total tardiness. Ann. Discret. Math. **1**, 331–342 (1977)
13. Liao, C.-J., Juan, H.-C.: An ant colony optimization for single-machine tardiness scheduling with sequence-dependent setups. Comput. Oper. Res. **34**(7), 1899–1909 (2007)
14. Lin, S., Ying, K.: A hybrid approach for single-machine tardiness problems with sequence-dependent setup times. J. Oper. Res. Soc. **59**(8), 1109–1119 (2008)
15. Potts, C.N., Kovalyov, M.Y.: Scheduling with batching: a review. Eur. J. Oper. Res. **120**(2), 228–249 (2000)
16. Potts, C.N., Van Wassenhove, L.N.: Integrating scheduling with batching and lot-sizing: a review of algorithms and complexity. J. Oper. Res. Soc. **43**(5), 395–406 (1992)
17. Ruiz, R., Stützle, T.: A simple and effective iterated greedy algorithm for the permutation flowshop scheduling problem. Eur. J. Oper. Res. **177**(3), 2033–2049 (2007)
18. Schaller, J.: Scheduling on a single machine with family setups to minimize total tardiness. Int. J. Prod. Econ. **105**(2), 329–344 (2007)
19. Schaller, J.E., Gupta, J.N.: Single machine scheduling with family setups to minimize total earliness and tardiness. Eur. J. Oper. Res. **187**(3), 1050–1068 (2008)
20. Sen, T., Sulek, J.M., Dileepan, P.: Static scheduling research to minimize weighted and unweighted tardiness: a state-of-the-art survey. Int. J. Prod. Econ. **83**(1), 1–12 (2003)
21. Sioud, A., Gravel, M., Gagné, C.: A hybrid genetic algorithm for the single machine scheduling problem with sequence-dependent setup times. Comput. Oper. Res. **39**(10), 2415–2424 (2012)
22. Subramanian, A., Battarra, M., Potts, C.N.: An iterated local search heuristic for the single machine total weighted tardiness scheduling problem with sequence-dependent setup times. Int. J. Prod. Res. **52**(9), 2729–2742 (2014)
23. Tanaka, S., Araki, M.: An exact algorithm for the single-machine total weighted tardiness problem with sequence-dependent setup times. Comput. Oper. Res. **40**(1), 344–352 (2013)
24. Jacob, V.V., Arroyo, J.E.C.: ILS heuristics for the single-machine scheduling problem with sequence-dependent family setup times to minimize total tardiness. J. Appl. Math. **2016** (2016)
25. Ying, K.-C., Lin, S.-W., Huang, C.-Y.: Sequencing single-machine tardiness problems with sequence dependent setup times using an iterated greedy heuristic. Exp. Syst. Appl. **36**(3), 7087–7092 (2009)

Learning the Quality of Dispatch Heuristics Generated by Automated Programming

Andrew J. Parkes[✉], Neema Beglou, and Ender Özcan

School of Computer Science, University of Nottingham, NG8 1BB, Nottingham, UK
Andrew.Parkes@nottingham.ac.uk

Abstract. One of the challenges within the area of optimisation, and AI in general, is to be able to support the automated creation of the heuristics that are often needed within effective algorithms. Such an example of automated programming may be performed by search within a space of heuristics that will be applied to a target domain. In this, brief proof-of-concept, paper, we consider the case of online bin-packing as the target domain, and consider the potential for machine learning methods to aid the associated automated programming problem. Simple numerical 'policy matrices' are used to represent heuristics, or 'dispatch policies', controlling the placement of item into bins as they arrive. We report on an initial investigation of the potential for neural nets to analyse and classify the resulting 'policy matrices', and find strong evidence that simple nets can be trained to learn to predict which heuristics, expressed as policy matrices, exhibit better or worse fitness. This gives the potential for them to be used as a surrogate fitness function to enhance the usage of search algorithms for finding heuristics. It also supports the prospect of using machine learning to extract the patterns that lead to successful heuristics, and so generate explanations and understanding of machine-generated heuristics.

1 Introduction

The effective solution of many real-world combinatorial optimisation problems, such as scheduling, timetabling, resource assignment, etc., requires good heuristics to be used within the heart of many algorithms. Usually, these heuristics are generated by human experts - however, the rise of powerful methods in search and AI holds the potential for them to be created automatically. To further this aim, in this paper, we consider the specific and simple case of online bin packing: items arrive one at a time, and immediately, and irrevocably need to be assigned to a bin, either an existing bin with sufficient space, or by opening a new bin, and the goal is simply to minimise the number of bins are used. The decision process then requires a 'dispatch policy' or 'heuristic' to decide which bin should be used. The quality of a heuristic can be evaluated by simply using it to pack many items and then taking the average fullness of the bins

© Springer Nature Switzerland AG 2019
R. Battiti et al. (Eds.): LION 12 2018, LNCS 11353, pp. 154–158, 2019.
https://doi.org/10.1007/978-3-030-05348-2_13

- usually expressed as a percentage, and higher is better. A natural way to do this is with an 'index policy' in which a score is given to each packing option based on: the size s of the item under consideration, and the remaining space, r, within the bin, and then the highest scoring option is taken. In the 'CHAMP' approach [1,3], Parkes and Özcan introduced a "policy matrix" representation, denoted as $M[s,r]$ to store the scores, and so a matrix constitutes a heuristic. Then standard search methods (evolutionary algorithms or metaheuristics) are used to find good values for those matrices. (Note that the associated search problem is in the space of heuristics and not the direct space of solutions to the target domain.) Earlier work [2] in this area used genetic programming (GP) to build a numeric score function $f(s, r)$ as an arithmetic tree of the relevant inputs; however, the automatically-generated matrix policies heuristics significantly improved upon the GP approaches. A surprise of the CHAMP work was that the resulting matrices were rather 'spiky', and did not have the smooth structure that was usually (implicitly) assumed in human-generated heuristics or those arising in the GP approaches. Accordingly, the table-lookup representation seems to offer the potential for search-based methods to create new kinds of heuristics. However, a disadvantage is that the resulting policy matrices can be hard to interpret – it is not immediately evident which matrices are good, and which are bad. It is then natural to ask whether machine learning methods can identify the structures or patterns in such "machine-discovered heuristics", and so give insight into what properties a heuristic ought to have. Given that the policy matrices are two-dimensional matrices, and so similar to 2-d pictures, it is particularly natural to ask whether systems used in machine vision might be applicable, and in particular to try neural nets. Accordingly, in this (short proof-of-concept) paper, we show that indeed neural nets can be trained to distinguish between good and bad matrices. Furthermore, even simple nets do well, and this gives evidence for a potential for use of neural nets (or machine learning in general) to be used for the automated generation and understanding of heuristics.

2 Experimental Results

In this paper, we will use an example of online bin-packing as in previous work [3]; "UBP(20,5,10)", with bins of capacity 20, and item sizes selected uniformly at random from the range [5,...,10]. See [1,3] for explicit examples of some resulting matrices. For this paper, the relevant entries in each matrix can be simply regarded as a set of integers (the size of the set happens to be 57). Furthermore, we restrict the value of each entry to be in $\{1, 2, 3\}$; this restriction still allows matrices of high fitness. The set of such numbers then define the heuristic. The standard heuristics, such as "best-fit" and "first-fit" can easily be represented as matrices and give a bin fullness of around 92%; however, using a GA much better matrices can be found with a fullness of up to about 98%. The aim here is to determine whether machine learning methods can be used to predict and aid understanding the fitness of such matrices.

For the study, we used a combination of random generation and local search to generate a diverse dataset of over 150k matrices, each with the associated fitness[1].

Table 1. Single layer MLP using classes N:[92-94]–P:[94-100].

g-mean	Predicted	Predicted
86.4%	P:[94-100]	N:[92-94]
Actual	982	298
P:[94-100]	76.7%	23.3%
Actual	300	12063
N:[92-94]	2.4%	97.6%

Table 2. Single layer MLP using classes N:[92-93]–P:[94-100].

g-mean	Predicted	Predicted
93.3%	P:[94-100]	N:[92-93]
Actual	1190	106
P:[94-100]	91.8%	8.2%
Actual	161	8781
N:[92-93]	1.8%	98.2%

In order to train a neural net there were two primary decision: how to give the matrices as input to the net, and the structure of the net. Firstly, the input is done using a standard 'one-hot' scheme - for each matrix entry, each of the possible values from $\{1, 2, 3\}$ is selected using a set of 3 $\{0, 1\}$ inputs. Hence, each matrix is converted to $3*57 = 171$ binary inputs. (Other schemes were tried but were less effective.) Secondly, for the neural net structure, we used a simple feed-forward Multi-layer Perceptron (MLPs), with one hidden layer. We followed a common heuristic that each layer should be about half the size of the previous one, so the hidden layer was give $171/2 = 85$ nodes After some initial experiments we decided on sigmoid activation functions for the hidden layer. The system was implemented using DeepLearning4J[2] and the training parameters were fairly standard e.g. using stochastic gradient descent. Again other options were tried, but were no more effective.

This proof-of-concept paper just aims to give evidence that neural nets have the ability to detect patterns in the matrices that are correlated with their fitness. Hence, instead of doing regression we just did binary classification. It is usual to discuss results in terms of "Positive (P)" and "Negative (N)" classes, and we take the P class to correspond to the better matrices. (The net is trained so that the output gives the class memberships.) As usual, we will report confusion matrices. The positive class will usually have fewer instances, and so we need to also use performance measures that are appropriate when the class sizes are imbalanced [4]; the accuracy would be a poor measure, hence we also report the standard geometric mean 'g-mean' of the true positive and true negative rates.

Due to lack of space, we just give representative results using two pairs of classes. Since we are mostly interesting in good matrices then it is reasonable to restrict to instances with fitness of at least 92%; Table 1 shows results when

---

[1] The data will available via http://www.cs.nott.ac.uk/~pszajp/CHAMP/.

[2] https://deeplearning4j.org/.

the two classes are then formed from a threshold at fitness 94%. This case is arguably challenging, as the range of fitness values is small, and there will be fewer of the very bad matrices that are (presumably) easier to spot. However, the performance is still good, with 77% of those in the higher-fitness class being recognised. This case has 'adjacent classes' with a boundary at 94%, and so small differences in the fitness could lead matrices to be placed in separate classes; such boundary cases are likely to be more difficult to learn, and so results could be unduly pessimistic. Accordingly, we put a small gap between the two classes by excluding the fitness range [92%-93%], giving a second case with two classes N:[92-93]–P:[94-100]. The results in Table 2 indicate that the classification performance did then improve significantly; now 92% of the higher fitness class was recognised, again with a very good value for the g-mean.

3   Conclusions and Future Work

We considered the issue of finding and identifying good heuristic dispatch policies for the online bin-packing problem. The heuristics are represented numerically by an integer matrix, with the pertinent entries being scores that are used during the decision process of where to allocate items as they arrive. We investigated whether or not a neural net can predict the quality of the matrices. On converting to binary classification problems, the results clearly indicate that nets with just one hidden layer can achieve high discriminating and predictive power. This indicates that the properties of the matrix that lead to high fitness are indeed susceptible to being learned. The first natural potential impact is that such learned model could be used as a surrogate fitness value during a search in the space of matrices; being used to select matrices more likely to perform well when evaluating their true fitness.

Arguably more importantly, a potential impact is using a learned model in order to generate understanding of the space of heuristics (matrices). Inspections of the weights in the trained nets should be at least able to indicate which of the entries of the matrix are considered most important, and deeper analysis might also reveal the conditions needed on the relevant values. For this, it is particularly interesting that even just a single hidden layer gives good results; as such simple nets are likely to be much easier to analyse. Future work will look at other methods, such as SVM (with appropriate kernels) and decision trees, as this might well give results that are easier to interpret. Such understanding will hopefully also lead to better insight of how to search for good heuristics.

Overall, the main novelty and contribution of this study is to show that (standard) machine learning can be applied in an intriguing way to study the space of 'special purpose' heuristic functions that are used in dispatch policies, or in optimisation algorithms.

References

1. Asta, S., Özcan, E., Parkes, A.J.: CHAMP: creating heuristics via many parameters. Expert. Syst. Appl. **63**, 208–221 (2016)
2. Burke, E.K., Hyde, M.R., Kendall, G., Woodward, J.R.: The scalability of evolved on line bin packing heuristics. In: 2007 IEEE Congress on Evolutionary Computation, pp. 2530–2537 (2007). https://doi.org/10.1109/CEC.2007.4424789
3. Özcan, E., Parkes, A.J.: Policy matrix evolution for generation of heuristics. In: Proceedings of the 13th Annual Conference on Genetic and Evolutionary Computation (GECCO) (2011)
4. López, V., Fernández, A., Garca, S., Palade, V., Herrera, F.: An insight into classification with imbalanced data: empirical results and current trends on using data intrinsic characteristics. Inf. Sci. **250**, 113–141 (2013). https://doi.org/10.1016/j.ins.2013.07.007

Explaining Heuristic Performance Differences for Vehicle Routing Problems with Time windows

Jeroen Corstjens[✉], An Caris, and Benoît Depaire

UHasselt, Research Group Logistics, Agoralaan, 3590 Diepenbeek, Belgium
`jeroen.corstjens@uhasselt.be`

Abstract. Heuristic algorithms are most commonly applied in a competitive context in which the algorithm is tested on well-known benchmarks of some problem application with the objective of obtaining better performance results than the state-of-the-art. Focusing on characterising heuristic algorithm behaviour to acquire insight and knowledge of how these solution procedures operate given a certain problem application, is a rarely applied research context. In this paper we strive to obtain a better understanding of heuristic performance. Based on an exploratory analysis of a large neighbourhood search algorithm applied on instances of the vehicle routing problem with time windows, we perform a detailed study on one of the detected patterns and seek to explain it. We learn that a regret operator functions best when it can take into account many and good alternatives, which is not the case when removing geographical clusters of customers. In the latter case some customers become isolated and have no feasible insertion option in one of the existing routes at the start of the repair phase. Their insertion is therefore postponed, but we show that it is beneficial for performance to assign them a higher priority through the creation of individual routes.

1 Introduction

For many years metaheuristics have been successfully applied to solve computationally challenging optimisation problems. These general solutions procedures are typically tested on benchmark problem sets and their performance evaluated against state-of-the-art methods. The objective is to present a better performing method [18] – 'better' in terms of solution quality, computation speed or their trade-off. However, this approach does not provide much insight into heuristic algorithm behaviour [1]. A detailed study of how metaheuristic performance is established is rarely the focus of a publication in operations research literature.

In recent years, there is increased attention on identifying the algorithm elements that are most relevant to performance [5,11,12]. A next step in explaining heuristic algorithm behaviour is to understand why two configurations – or more generally two algorithms – perform differently. What can be learned about the differences in components, strategies or implementations that result in an either

© Springer Nature Switzerland AG 2019
R. Battiti et al. (Eds.): LION 12 2018, LNCS 11353, pp. 159–174, 2019.
https://doi.org/10.1007/978-3-030-05348-2_14

better or worse performing heuristic method. This is the question we aim to address in this paper. We start from an exploratory analysis of the parameters' impact on performance and try to explain patterns we observe through iterative experimentation. Specifically, we seek to explain the performance difference observed in [4] between two configurations of a large neighbourhood search algorithm applied on the vehicle routing problem with time windows. This study indicated that iteratively removing customers at random is expected to perform better than iteratively removing clusters of customers when in both cases these customers are reinserted based on a difficulty (i.e., regret) measure. The counterintuitivity of the observation that a simpler (i.e., random) removal method performs better than one using a more sophisticated logic (i.e., related removal) has sparked our interest to further investigate it and find an explanation.

Before going into the follow-up experimental analysis we summarise the applied methodology and case study in Sect. 2. Next, the set-up of the new experiments is described in Sect. 3, followed by the search for explanations in Sect. 4 relying on regression analyses. A conclusion is finally given in Sect. 5.

2   Related Work

A clear understanding of metaheuristic performance provides insights and knowledge that is useful when designing, optimising and comparing heuristic algorithms. Gaining such an understanding implies decomposing the solution procedures as they consist of several interacting heuristic mechanisms and subsequently analysing their impact on performance following the principles of design of experiments. The analysis concerns the parameters controlling which heuristic mechanisms are activated, how they behave and interact with other mechanisms. The choices made for these parameters substantially impact the efficacy with which a heuristic algorithm solves a given (class of) problem instance(s). The challenge is finding a set of parameter values that optimises the performance of the heuristic algorithm. This problem is formalised in the algorithm configuration problem [8]. It was initially solved manually without applying a formal tool or statistical methodology. Given the performance impact, more rigorous ways were sought for determining parameter values and resulted in the development of automatic algorithm configurators [2]. The latter's success has inspired a software design paradigm called Programming by Optimisation that encourages to parameterise design choices as much as possible instead of committing to one design alternative. It leads to a rich design space from which an automated configurator can then determine the best design alternative for a specific problem [9].

Still, these configurators – relying heavily on the computational power of computers – are used in the context of achieving an algorithm that is able to beat the competition. A focus on explaining how a heuristic algorithm obtains its performance remains a rare research objective, showing that experimenters most frequently focus on development goals rather than research goals [7,15]. Which elements contribute to the performance of the heuristic method? How does the contribution of these elements vary when considering different problem instances? What is the reason for any performance differences between two (configurations of the same) heuristic methods? Such questions are not commonly addressed when discussing experimental results of heuristics [5,11,12]. A competitive evaluation methodology is useful when the aim is to develop the fastest possible procedure for a specific environment. When the goal is to understand how performance is obtained, to discover which elements in the algorithm contribute to its performance and to draw conclusions that are valid beyond the specific problem instances chosen, one has to rely on a statistical evaluation method [15].

In previous work [4] we extensively promote performing experiments with heuristic algorithms that are focused on understanding the results the algorithm produces. A heuristic evaluation methodology is suggested, based on well-known design of experiments principles, for analysing heuristic algorithm performance with the aim of obtaining a better understanding of which (combination of) elements operating within the algorithm work well and which do not and what influence the specific problem instance to be solved has. The methodology is summarised in Sect. 2.1, followed by a discussion in Sect. 2.2 of the experimental analysis performed in [4] that sparked the research question for this paper.

2.1 A Multilevel Evaluation Methodology

The methodology relies on multilevel models to efficiently analyse the numerous combinations of parameter settings and problem instances. Multilevel implies the presence of a hierarchical structure in the data. In the context of heuristic experimentation it means testing several different parameter settings on a single problem instance such that it is clear that any performance differences observed are due to the algorithm parameters and not the problem instance. Doing this for multiple problem instances, enables the exposure of the problem instance influence. This data structure is created by first randomly sampling problem instances and then randomly defining a number of parameter settings per problem instance. The efficiency gain compared to single level design is that less combinations need to be analysed while not losing any statistical power [6]. The analysis of the multilevel design is performed by relying on multilevel regression models since this type of data structure violates the assumption of independent error terms made by traditional regression analysis. If classical regression is used, statistical estimations might be inaccurate due to biased (i.e., underestimated) standard errors that result in spuriously significant results [10].

2.2 Case Study: Observing Patterns in the Data

In [4] we apply the methodology on a data set in which a number of instances of the vehicle routing problem with time windows (VRPTW) are solved using a large neighbourhood search (LNS) algorithm [13]. Large neighbourhood search is a popular metaheuristic framework that is successfully applied to multiple vehicle routing variants. It has many algorithm parameter choices, making it an interesting algorithm for performance analysis. It constructs an initial solution that is gradually improved by iteratively removing and reinserting customers until some stopping criterion is met. This process is executed using a destroy and repair operator.

The experiment performed in [4] investigates the correlations between algorithm parameters and between algorithm parameters and problem instance characteristics. These correlations are analysed by fitting a multilevel regression model with random effects for all algorithm parameters, i.e. the performance impact of an algorithm parameter is allowed to vary with the problem instance characteristics. The interaction effects are kept fixed for all problem instances.

The experimental goal is to observe patterns, which are to be further investigated in new experiments. One pattern concerns the relative performance of some combinations of operators employed within the algorithm. The results predict certain combinations to perform better than others. Yet, they do not provide indications as to why they perform differently. More specifically, it is observed that iteratively removing customers at random (i.e., random removal) and reinserting them using a difficulty measure (i.e., regret-2) results in a better predicted performance than iteratively removing geographical clusters of customers (i.e., related removal). This seems odd since the latter involves applying a logic for selecting these customers that is aimed to work better than pure random selection. Hence, the study spurred new research questions that are to be answered in consecutive experiments. This is referred to as the iterative inductive-deductive process [3]: experimenting is an iterative learning process, each time gaining knowledge as well as raising new questions. This paper focuses on one of those new questions. We want to explain why removing customers at random works better than removing clusters of customers given the use of a certain repair logic.

Therefore, the destroy and repair operators are of interest in this follow-up research. We consider two destroy operators – random and related removal – and three variations on the regret-k operator, all used in the implementation of [13]. Regret-2 is the only regret variant used in [4]. In this paper regret-3 and regret-4 are included to find out whether the observed pattern also prevails when looking further ahead in the repair process. First, the definitions of the relevant destroy and repair operators are discussed.

Random Removal. This destroy operator randomly selects q customers to remove. The idea is to diversify the search towards a better solution.

Related Removal. This destroy operator removes q 'related' customers. The experimenter can choose how to define the relatedness. Shaw [17] introduced the strategy in order to obtain solution improvements and the measure should therefore provide good insert opportunities to reach this objective. One definition is based on distance [13]. The idea is that customers close to each other can more easily switch routes and/or positions, while more distanced customers have a higher chance of being inserted back in their original position [17]. Other definitions are whether two customers are in the same route or have similar time windows [17]. In [16] the measure is based on four terms that account for distance, time, capacity and which vehicles can serve both customers.

Regret-k. The repair operator prioritises 'difficult' customers [14]. The difficulty of an insertion is determined as the difference between a customer's cheapest insertion route (i.e., insertion in this route adds the smallest distance to the overall travelled distance) and its second and third, ... up until its k-th cheapest insertion route. This is formulated in equation (1).

$$\sum_{n=2}^{k}(f_i^k - f_i^1) \ . \tag{1}$$

The term f_i^1 represents the change in the objective function value when inserting customer i at its best position in its cheapest route, while the term f_i^k indicates the change when inserting customer i at its best position in its k-cheapest route. Customers having high regret values are considered difficult to insert, in the sense that the additional cost incurred of not choosing the cheapest route for insertion is high. These customers should therefore be prioritised and inserted in their cheapest route at their minimum cost position. If there are any ties in the calculated value, these are broken by choosing the customer with the lowest insertion cost.

3   Experimental Set-up

Since the focus in this paper is on analysing individual operators and not a complete metaheuristic, the operators are extracted from the metaheuristic framework and we measure performance after a single destroy and repair iteration. This will allow detailed analysis of the destroy and repair process.

A data set of 10 000 observations is generated following the same multilevel experimental design as employed in [4]. It consists of 200 artificial VRPTW instances and 50 random parameter settings tested per problem instance. A parameter setting in this experiment is defined as a combination of a single destroy operator – either random or related removal – with a single repair operator – either regret-2, regret-3 or regret-4. Information on the problem instance characteristics and algorithm parameters can be found in [4]. The analysis relies on multilevel regression models, fitted using the R packages *lme4* and *brms*.

4   Analysis of Results

Since each scenario only performs one iteration and customers are to be served by a limited number of vehicles — i.e., at most the number of vehicles used in the initial solution —, it can occur that during the repair phase the final customer(s) cannot be feasibly reinserted and are placed in a request bank with a penalty cost of 100 000 per customer. This is the case for 3239 scenarios. Such occurrences can be accounted for in through a request bank variable, but it still proved to be problematic to obtain a model that complies with the underlying assumptions of independence, normality, and homoscedasticity of the errors. Therefore, we only consider the scenarios that were able to reinsert all removed customers. This is also the only relevant type of solution since it corresponds with a regular LNS solution. So, a truncated data set of 6761 observations is used.

Several regression analyses are performed on the experimental results. Section 4.1 verifies whether the operator pattern observed in [4] is also present in the new experimental data. In Sect. 4.2 the search for explanations starts with an analysis of the customer difficulty measure used to reinsert customers. This leads to an investigation of the priorities assigned to the removed customers (Sect. 4.3).

4.1 Verifying Observed Operator Pattern

A first analysis verifies whether the pattern observed in [4] is confirmed in this new experiment. A multilevel regression model is formulated in equations (2) to (4) predicting the total cost of a solution in terms of the total distance travelled (Y_i). The factors affecting this performance measure are the choice of destroy and repair operator (variables *Related, Regret-3 and Regret-4*), the percentage of customers removed (*Percentage removed*) and the total number of customers to be served (*Customers*). Variables for random removal and regret-2 are not included because of multicollinearity issues[1]. They are the baseline operator choices for which the effect is accounted in the regression intercept and the effects of all other operators are relative to this baseline operator scenario. All continuous variables are centred around their mean value to facilitate interpretation of effect estimates. Random effects are considered up to two-way interaction effects.

$$
\log Y_i = \alpha_{j[i]} + \beta_{1j[i]} Related_i + \beta_{2j[i]} Regret3_i + \beta_{3j[i]} Regret4_i +
$$
$$
\beta_{4j[i]} \log Request\_bank_i + \beta_{5j[i]} Percentage\_removed_i
$$
$$
+... + \beta_{17j[i]} Related_i \times Regret4_i \times Percentage\_removed_i + \epsilon_i \ . \ (2)
$$
$$
\alpha_j = \mu_0^\alpha + \mu_1^\alpha Customers_j + \eta_j^\alpha \ . \tag{3}
$$
$$
\beta_{kj} = \mu_0^{\beta_k} + \mu_1^{\beta_k} Customers_j + \eta_j^{\beta_k} \ . \tag{4}
$$

with

$i \in I$ scenario, a combination of a problem instance with a parameter setting
$j \in J$ problem instance
$k \in K$ algorithm parameter
$j[i]$ index variable to code problem instance membership $(j[i] = j)$, e.g., $j[10] = 5$ means the 10th scenario solves problem instance 5
Y_i objective function value of scenario i

---

[1] Multicollinearity may lead to unstable coefficient estimates. This makes it difficult to interpret results. Including all destroy or repair variables would lead to perfect multicollinearity. One variable of each needs to be left out as a reference value.

$\alpha_{j[i]}$ varying regression intercept: the objective function value given scenario i
 and problem instance j when the value for all k parameters is 0

$\beta_{kj[i]}$ varying effect of parameter k given scenario i and problem instance j

$\mu_0^{\beta_k}$ mean problem effect on the coefficient β of algorithm parameter k

η_j error at the problem instance level and is assumed to be $\sim N(0,\sigma^2)$

ϵ_i error at the parameter setting level and is assumed to be $\sim N(0,\sigma_e^2)$

The results confirm the previous findings in [4]: random removal performs
on average better than related removal when combined with regret-k. Figure 1
plots the predicted performance for the various operator combinations given a
problem instance with an average number of customers. Regret-3 and regret-4
perform a bit worse than regret-2 with random removal, while they perform a
bit better with related removal. This is also clear from the coefficient estimates.
The estimates for *Regret-3* and *Regret-4* are positive with random removal, but
become negative when accounting for the estimates for the interaction effects
Related × *Regret-3* (−0.25) and *Related* × *Regret-4* (−0.24). These significant
estimates are given in Table 1. Having verified the destroy operator performance
difference in the new experimental data, the explanatory study can begin.

Table 1. Significant effects model pattern verification

Dependent variable: $\sqrt{\text{total cost}}$	Estimate	Std. Error	95% Confidence interval
Intercept	174.60***	1.14	[172.32, 176.80]
Related	1.56***	0.07	[1.42, 1.71]
Regret-3	0.08**	0.05	[−0.01, 0.17]
Regret-4	0.08**	0.05	[−0.01, 0.17]
Percentage removed	−0.09***	0.003	[−0.10, −0.09]
Customers	0.40***	0.01	[0.38, 0.43]
Related × Regret-3	−0.25***	0.08	[−0.40, −0.09]
Related × Regret-4	−0.24***	0.08	[−0.40, −0.08]
Related × Customers	0.01***	0.001	[0.005, 0.01]
Regret-3 × Customers	−0.001*	0.0004	[−0.001, 0.0002]
Regret-4 × Customers	0.0003*	0.0004	[−0.0005, 0.001]
Percentage removed × Customers	−0.0001***	0.000	[−0.0002, −0.0001]
Related × Percentage removed	0.09***	0.01	[0.08, 0.10]
Related × Percentage removed × Customers	0.0002***	0.0000	[0.0002, 0.0003]
Related × Regret-3 × Percentage removed	−0.02**	0.01	[−0.03, −0.003]
Related × Regret-4 × Percentage removed	−0.02**	0.01	[−0.03, −0.003]
Observations	6761		
Num. groups: problem instances	200		

Note: *p < 0.1; **p < 0.05; ***p < 0.01

4.2   Investigating Customer Difficulty

In the search for explanations for this counterintuitive result, the process of
destroying and repairing a solution is decomposed and illustrated for a problem

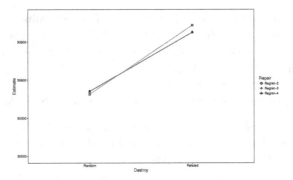

Fig. 1. Predictions for a problem instance with an average number of customers.

instance with 100 customers in Figs. 2 and 3. The white customer nodes in Fig. 2 are a random sample identified for removal, while in Fig. 3 they are all related customers. We define relatedness in terms of distance, so this operator will remove a number of geographically nearby customers. Figures 2c and 3c show the repaired solution relying on regret-3. Since the regret value is the key measure used in the repair phase and applied in both removal scenarios, a first investigation focuses on this measure.

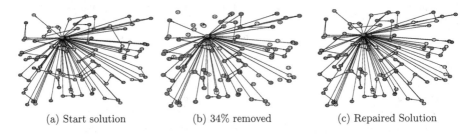

(a) Start solution   (b) 34% removed   (c) Repaired Solution

Fig. 2. Destroy and repair iteration when customers are removed at random.

It is observed in Fig. 2c that reinserting randomly removed customers does not lead to major changes in the overall solution structure, it remains largely the same to the initial solution (cf. Fig. 2a). During the repair process only small adjustments will be made that potentially result in a better final solution. When removing a geographical cluster of customers, on the other hand, a (large) part of the solution structure is completely destroyed and has to be rebuilt from scratch. This is shown in the upper left corner of Fig. 3b where there is not a single route left. For many of the removed customers, there are not a lot of existing routes nearby that are potential candidates to insert a customer. Consequently, for these customers the number of good alternative routes for their cheapest insertion route is expected to be small. This means that the cost difference between inserting a customer in its cheapest route and – depending

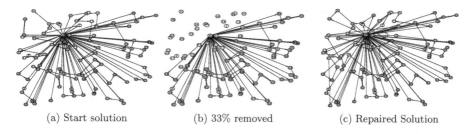

(a) Start solution   (b) 33% removed   (c) Repaired Solution

Fig. 3. Destroy and repair iteration when related customers are removed.

on the k value – its second, third and fourth cheapest route is large. A randomly removed customer, on the other hand, has more existing routes nearby and thus better alternatives for the cheapest insertion route, resulting in a lower regret value. Hence, the expectation is that a customer, part of a cluster of removed customers that are geographically nearby, has fewer feasible alternatives compared to a randomly removed customer. Further, due to the difference in number of feasible alternatives, it is expected that the average regret value of the customer selected for insertion is higher for scenarios removing a cluster of customers. Statistical evidence for these expectations is sought such that the following two null hypotheses posing no difference can be be rejected.

Hypothesis 1 (H1):   *The average number of feasible route options each removed customer can be inserted in is no different when this customer is selected at random or when it is part of a geographical cluster of removed customers.*

Hypothesis 2 (H2):   *The average (maximum) regret value of the customer selected for insertion is no different when the removed customer is selected at random or when it is part of a geographical cluster of removed customers.*

The first hypothesis is validated by fitting a multilevel regression model predicting the average number of feasible insertions each removed customer has based on factors listed in Table 2. Given a scenario with an average number of customers (i.e., about 184), removing an average percentage of customers (i.e., about 28%) and using random removal and regret-2, each removed customer is expected to have 3.10 feasible insertion routes on average – this is the Intercept value in Table 2. If related removal is used instead of random removal, the number decreases to 2.53. The findings are similar for the combination with regret-3 and regret-4, but with slightly lower averages in both removal scenarios. The decrease is statistically significant and the regression results therefore provide statistical evidence to reject Hypothesis 1. Intuitively, a rejection is logical since removing a geographical cluster customers most likely results in the removal of one or multiple entire routes and thus it is not surprising that in such a scenario the removed customers have less feasible options for insertion than in other scenarios. This regression analysis provides a statistical confirmation.

Hypothesis 2 is validated in a model predicting the average regret value. The dependent variable is log transformed to comply with the typical regres-

Table 2. Regression table hypothesis 1

Dependent variable: $\sqrt{\text{avg} \# \text{ of feasible insertions}}$	Estimate	Std. Error	95% Confidence interval
Intercept	1.76^{***}	0.01	[1.74, 1.78]
Related	-0.17^{***}	0.01	$[-0.19, -0.14]$
Regret-3	-0.01	0.01	$[-0.03, 0.003]$
Regret-4	-0.01^{*}	0.01	$[-0.03, 0.002]$
Percentage removed	0.004^{***}	0.001	[0.003, 0.005]
Customers	0.0013^{***}	0.0001	[0.0011, 0.0015]
Related × Regret-3	-0.06^{***}	0.01	$[-0.089, -0.034]$
Related × Regret-4	-0.06^{***}	0.01	$[-0.088, -0.034]$
Related × Percentage removed	-0.01^{***}	0.001	$[-0.007, -0.004]$
Regret-3 × Percentage removed	0.0003	0.001	$[-0.002, 0.001]$
Regret-4 × Percentage removed	-0.001	0.001	$[-0.002, 0.0008]$
Related × Customers	-0.001^{***}	0.0001	$[-0.0009, -0.0007]$
Regret-3 × Customers	-0.0002^{***}	0.0001	$[-0.0003, -0.0001]$
Regret-4 × Customers	-0.0004^{***}	0.0001	$[-0.0005, -0.0002]$
Related × Regret-3 × Percentage removed	-0.001	0.001	$[-0.003, 0.001]$
Related × Regret-4 × Percentage removed	-0.0004	0.001	$[-0.003, 0.002]$
Observations	6761		
Num. groups: problem instances	200		

Note: *p < 0.1; **p < 0.05; ***p < 0.01

sion assumptions. The values are back-transformed by taking the exponential function. For example, the average[2] regret value when using random removal and regret-2 is $\exp(6.61) = 742.48$ while it is $\exp(6.61 - 0.22) = 595.86$ when using related removal. For the combination with regret-3 the predicted values are respectively 5884.05 and 8866.19, while for the combination with regret-4 they are 13 359.73 and 17 676.65. The output in Table 3 indicates that this average regret measure is significantly higher for scenarios using related removal, but for the combination with regret-2 the regression model indicates the measure to be significantly lower when using related removal. This suggests that the second best alternative each removed customer has is on average of a better quality (meaning the difference in insertion cost is smaller) when removing a geographical cluster of customer than when removing randomly dispersed customers.

The fact that each customer has less feasible alternatives with related removal also explains the higher average regret value measured since a large cost (i.e., 9999) is assigned to every infeasible insertion point within a route. This is most notable in the scenarios relying on regret-3 and regret-4. The large cost is assigned to prevent customers with more feasible alternatives having a larger regret value and thus higher priority than customers with less feasible alternatives.

---

[2] The antilog of the arithmetic mean of log-transformed values is the geometric mean. For positively skewed data, like our experimental data, the geometric mean will be less than the arithmetic mean and often a good estimate of the original median.

Table 3. Regression Table Hypothesis 2

Dependent variable: Log(avg regret value)	Estimate	Std. Error	95% Confidence interval
Intercept	6.61***	0.04	[6.54, 6.69]
Related	−0.22***	0.05	[−0.31, −0.13]
Regret-3	2.07***	0.03	[2.00, 2.13]
Regret-4	2.89***	0.03	[2.82, 2.96]
Percentage removed	−0.056***	0.001	[−0.06, −0.05]
Customers	−0.004***	0.0003	[−0.005, −0.003]
Related × Regret-3	0.41***	0.05	[0.31, 0.50]
Related × Regret-4	0.28***	0.05	[0.18, 0.36]
Related × Percentage removed	0.02***	0.002	[0.016, 0.025]
Regret-3 × Percentage removed	0.036***	0.002	[0.033, 0.039]
Regret-4 × Percentage removed	0.045***	0.002	[0.041, 0.047]
Related × Customers	0.0007*	0.0004	[−0.00016, 0.0015]
Regret-3 × Customers	0.0025***	0.0003	[0.0019, 0.0031]
Regret-4 × Customers	0.0033***	0.0003	[0.0028, 0.0039]
Related × Regret-3 × Customers	0.001*	0.0004	[−0.00002, 0.0017]
Related × Regret-4 × Customers	0.0001	0.0004	[−0.00079, 0.00095]
Related × Regret-3 × Percentage removed	−0.003	0.003	[−0.0083, 0.0023]
Related × Regret-4 × Percentage removed	−0.01***	0.003	[−0.016, −0.0044]
Observations	6761		
Num. groups: problem instances	200		

Note: *p < 0.1; **p < 0.05; ***p < 0.01

4.3 Customer Prioritisation

We proceed by investigating the sequence in which customers are reinserted back in the solution to see which customers are considered difficult and thus prioritised. For the problem instance plotted in Fig. 3b customer 90 is inserted first, followed by customers 11, 25, 67, ... and finally customers 48 and 30. In this sequence the majority of customers considered first are almost all in near proximity of an existing route, while the more "isolated" customers (nodes 48 and 30 in Fig. 3b) are postponed to the final insertions. Characterising for these "isolated" customers is that they all have not one feasible insertion possibility in one of the existing routes. This means that the calculated regret value for these customers is zero and they are thus assigned the lowest priority of all removed customers. Their single alternative is to create a route straight from the depot to the customer and back. In the regret-k implementation used in this study, such an alternative is assigned a regret value of zero since there is no immediate cost loss of postponing the insertion for that customer. The customer has no other alternatives, so its insertion is not urgent. Secondly, as the solution of routing problems typically not only strive to minimise the total distance travelled, but doing so with the fewest number of vehicles, the current implementation does not favour the creation of additional routes. Yet, this observation raises the question what the impact would be if these isolated customers are taken into account at the start of the repair phase instead of initially being ignored. Perhaps their prioritisation might benefit other removed customers as this adds (better) alternative routes in an area with few or no routes at all. Therefore, we expect that scenarios removing clusters of customers lead to more isolated customers than scenarios that remove customers at random. Secondly, we wish to analyse the solution quality and how it differs between both removal strategies if isolated customers are prioritised instead of postponing their insertion.

Hypothesis 3 (H3): *The number of removed customers which have no other feasible insertion possibility than an individual route from the depot to the customer and back is no different when these customers are selected at random or when they belong to a geographically clustered group of removed customers.*

Hypothesis 4 (H4): *After performing a single destroy and repair iteration, there is no difference in the average solution quality for scenarios removing geographical clusters of customers if priority is given to customers which have no other feasible insertion possibility than an individual route from the depot to the customer and back or if these customers are ignored until the final insertions.*

Hypothesis 3 is validated by a model predicting the number of isolated customers at the start of the repair phase. Since this model involves predicting a (discrete) count variable, a negative binomial regression model is more suitable rather than the standard Gaussian model. The output in Table 4 shows that scenarios using random removal have on average $exp(0.42) = 1.52$ isolated customers while scenarios using related removal have a significantly higher average of $exp(0.42 + 1.76) = 8.85$ isolated customers, regardless of the regret variant.

Table 4. Regression Table Hypothesis 3

Dependent variable: $\sqrt{\text{isolated customers}}$	Estimate	Std. Error	95% Confidence interval
Intercept	0.42***	0.05	[0.33, 0.50]
Related	1.76***	0.04	[1.68, 1.84]
Percentage removed	0.05***	0.002	[0.04, 0.05]
Customers	0.002***	0.0004	[0.001, 0.003]
Related × Customers	0.003***	0.0004	[0.002, 0.004]
Percentage removed × Customers	−0.00002	0.00001	[−0.00004, 0.00001]
Related × Percentage removed	0.01***	0.002	[0.01, 0.02]
Observations	6761		
Num. groups: problem instances	200		

Note: *p < 0.1; **p < 0.05; ***p < 0.01

Hypothesis 4 is validated by changing the prioritisation mechanism in regret-k such that the most isolated customers are now assigned a higher priority. Their regret value is calculated as the difference between a large cost value of 9999 and twice the distance from the customer to the depot. The 10 000 scenarios are run again using the modified prioritisation. In 1687 scenarios one or multiple customers failed to be reinserted and remain in the request bank, which is almost half the number observed in the previous experiment. Hence, the truncated data set that will be analysed has 8313 observations. Customers 48 and 30 in Fig. 3 are now inserted a lot sooner resulting in a predicted total cost of 19 282.93 compared to 19 821.75 in the original implementation. The predicted total cost for the scenario using random removal (plotted in Fig. 2) remains at 19 335.85. A statistical analysis is performed by fitting the regression model formulated in (2) to (4) on the new performance results. The significant effects are listed in Table 5 and the average performance is plotted in Fig. 4. Note that, compared to Fig. 1, the average performance values are higher for all operator scenarios, but this is not a completely fair comparison since the new experimental data includes 1552 observations more, which might explain the higher averages. The important observation in Fig. 4 is that the majority of the performance difference between random and related removal is reduced. A significant improvement is even observed for the combination with regret-3 for problem instance sizes up to 222 customers. The combination of related removal with regret-2 or regret-4 performs significantly better on the smaller problem sizes (i.e., 25 to 65 customers), but still performs significantly worse than random removal for the larger instances (i.e., 208 customers or more). Nonetheless, the performance deterioration in this problem size range is largely reduced.

The previous analysis is based on a single destroy and repair iteration performed. LNS typically runs a large number of iterations and returns the best found solution across these iterations. Whether the smaller performance gap is also observed when running multiple iterations is verified in a final experiment similar to the one performed in [4]. A data set of 4000 scenarios is generated (200 problem instances with 20 parameter settings per instance). A detailed description of all parameters can be found in [4]. A parameter setting still has only one destroy and one repair operator. Two separate LNS experiments are run using the same inputs: one gives low priority to isolated customers while the second one assigns a high priority. The results show an improvement in solution quality if isolated customers are not ignored. Furthermore, their prioritisation makes the LNS also more efficient on those instances where both experiments obtain the same best solution: less iterations are required to reach this best found solution.

What is clear from this experimental study is that repairing a destroyed solution, applying a measure that accounts for cost increases of postponing customer insertions, works best when the repair operator has many and good alternatives available to make the best possible choice. Removing geographical clusters of customers with the underlying idea that these customers can more easily shuffle within and between routes, leading to a better performance, has to be nuanced. The removal of clusters will probably also remove entire routes, thereby removing feasible alternatives for the removed customers when they are to be reinserted.

Table 5. Significant effects regression model Hypothesis 4

Dependent variable: Log(total cost)	Estimate	Std. Error	95% Confidence interval
Intercept	178.77***	1.36	[176.17, 181.52]
Related	0.11**	0.05	[0.003, 0.21]
Regret-3	0.06	0.05	[−0.03, 0.15]
Regret-4	0.06	0.05	[−0.03, 0.15]
Percentage removed	−0.12***	0.003	[−0.13, −0.11]
Customers	0.40***	0.01	[0.38, 0.43]
Related × Regret-3	−0.23***	0.07	[−0.36, −0.09]
Related × Regret-4	−0.06	0.07	[−0.20, 0.07]
Related × Percentage removed	0.03***	0.004	[0.02, 0.04]
Regret-3 × Percentage removed	0.01	0.004	[−0.00, 0.01]
Regret-4 × Percentage removed	0.003	0.004	[−0.005, 0.01]
Related × Customers	0.002***	0.0001	[0.001, 0.003]
Regret-3 × Customers	−0.001	0.0004	[−0.001, 0.0002]
Regret-4 × Customers	−0.0001	0.0004	[−0.001, 0.001]
Percentage removed × Customers	−0.0002***	0.00	[−0.0002, −0.0001]
Related × Regret-3 × Customers	−0.001*	0.001	[−0.002, 0.0001]
Related × Regret-4 × Customers	−0.0005	0.001	[−0.002, 0.001]
Related × Percentage removed × Customers	0.0001***	0.00	[0.0001, 0.0002]
Regret-3 × Percentage removed × Customers	−0.00*	0.00	[−0.0001, 0.00]
Regret-4 × Percentage removed × Customers	−0.00	0.00	[−0.0001, 0.00]
Related × Regret-3 × Percentage removed	−0.01*	0.01	[−0.02, 0.0002]
Related × Regret-4 × Percentage removed	−0.01	0.01	[−0.02, 0.01]
Observations	8313		
Num. groups: problem instances	200		

Note: *p < 0.1; **p < 0.05; ***p < 0.01

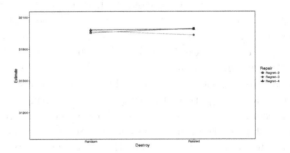

Fig. 4. Predictions for a problem instance with an average number of customers.

5   Conclusion

In this research work we perform an experimental study of heuristic performance with the focus on gaining a better understanding of how the observed performance is established. Putting the spotlight on insight and understanding is no common practice when experimenting with heuristic algorithms. Practitioners more typically focus on the competitive aspect of being able to develop a method that performs better than other ones on specific cases. Yet, studying how heuristic performance is characterised might provide knowledge that facilitates making a method competitive, but is not limited to specific cases.

This research work sought to explain why two destroy operators perform differently when combined with the same type of repair operator. It is found that removing geographical clusters of customers reduces the number of insertion alternatives during the repair phase. Several customers do not even have a single feasible insertion option in one of the existing routes and can therefore be considered isolated cases. Postponing the insertion of these customers is found to have a detrimental impact on the solution quality. It is tested what the effect is of assigning these customers a higher priority by allowing their insertion in an individual route, an option that was previously considered as a last alternative. Permitting these individual routes to be created sooner in the repair process adds good insertion alternatives for many other removed customers and thus enables the regret operator to make better choices. Hence, a regret operator will make a better estimation of customer difficulty and consequently a better prioritisation.

We explain the majority of the performance gap between both destroy scenarios. A small significant difference still remains for which an explanation might be found by further improving how customers are prioritised or by using a preparatory step in the regret operator.

Acknowledgments. The computational resources and services used in this work were provided by the VSC (Flemish Supercomputer Center), funded by the Research Foundation Flanders (FWO) and the Flemish Government - department EWI.

References

1. Bartz-Beielstein, T., Preuss, M.: The future of experimental research. Experimental Methods for the Analysis of Optimization Algorithms, pp. 17–49. Springer, Berlin (2010)
2. Birattari, M.: Tuning Metaheuristics. Springer, Berlin (2009)
3. Box, G.E.P., Hunter, J.S., Hunter, W.G.: Statistics for Experimenters: Design, Innovation, and Discovery. Wiley-Interscience, New Jersey (2005)
4. Corstjens, J., Depaire, B., Caris, A., Sörensen, K.: A multilevel evaluation method for heuristics with an application to the VRPTW. Sumbitted for publication (2017)
5. Fawcett, C., Hoos, H.H.: Analysing differences between algorithm configurations through ablation. J. Heuristics **22**(4), 431–458 (2015)
6. Gelman, A., Hill, J.: Data Analysis Using Regression and Multilevel/Hierarchical Models. Cambridge University Press, Cambridge (2006)
7. Hooker, J.N.: Testing heuristics: we have it all wrong. J. Heuristics **1**(1), 33–42 (1995)
8. Hoos, H.H.: Automated algorithm configuration and parameter tuning. In: Hamadi, Y., Monfroy, E., Saubion, F. (eds.) Autonomous Search, pp. 37–71. Springer, Berlin (2011)
9. Hoos, H.H.: Programming by optimization. Commun. ACM **55**(2), 70–80 (2012)
10. Hox, J.J., Moerbeek, M.: Schoot, R.v.d.: Multilevel Analysis: Techniques and Applications, 2nd edn, Routledge, Abingdon (2010)

11. Hutter, F., Hoos, H., Leyton-Brown, K.: An efficient approach for assessing hyper-parameter importance. In: ICML, pp. 754–762 (2014)
12. Hutter, F., Hoos, H.H., Leyton-Brown, K.: Identifying key algorithm parameters and instance features using forward selection. In: LION 7. Lecture Notes in Computer Science, vol. 7997, pp. 364–381. Springer, Berlin (2013)
13. Pisinger, D., Ropke, S.: A general heuristic for vehicle routing problems. Comput. Oper. Res. **34**(8), 2403–2435 (2007)
14. Potvin, J.Y., Rousseau, J.M.: A parallel route building algorithm for the vehicle routing and scheduling problem with time windows. Eur. J. Oper. Res. **66**(3), 331–340 (1993)
15. Rardin, R.L., Uzsoy, R.: Experimental evaluation of heuristic optimization algorithms: a tutorial. J. Heuristics **7**(3), 261–304 (2001)
16. Ropke, S., Pisinger, D.: An adaptive large neighborhood search heuristic for the pickup and delivery problem with time windows. Trans. Sci. **40**(4), 455–472 (2006)
17. Shaw, P.: Using constraint programming and local search methods to solve vehicle routing problems. In: Maher, M., Puget, J.F. (eds.) Principles and Practice of Constraint Programming CP98. Lecture Notes in Computer Science, vol. 1520, pp. 417–431. Springer, Berlin (1998)
18. Smith-Miles, K., Bowly, S.: Generating new test instances by evolving in instance space. Comput. Oper. Res. **63**, 102–113 (2015)

Targeting Well-Balanced Solutions in Multi-Objective Bayesian Optimization Under a Restricted Budget

D. Gaudrie[1(✉)], R. Le Riche[2], V. Picheny[3], B. Enaux[1], and V. Herbert[1]

[1] Groupe PSA, Vélizy-Villacoublay, France
david.gaudrie@mpsa.com
[2] CNRS LIMOS, École Nationale Supérieure des Mines de Saint-Étienne, Saint-Étienne, France
[3] Institut National de la Recherche Agronomique, MIAT, Toulouse, France

Abstract. Multi-objective optimization aims at finding trade-off solutions to conflicting objectives. These constitute the Pareto optimal set. In the context of expensive-to-evaluate functions, it is impossible and often non-informative to look for the entire set. As an end-user would typically prefer solutions with equilibrated trade-offs between the objectives, we define a Pareto front center. We then modify the Bayesian multi-objective optimization algorithm which uses Gaussian Processes to maximize the expected hypervolume improvement, to restrict the search to the Pareto front center. The cumulated effects of the Gaussian Processes and the center targeting strategy lead to a particularly efficient convergence to a critical part of the Pareto set.

Keywords: Gaussian processes · Parsimonious optimization
Computer experiments · Preference-based optimization

1 Introduction

Multi-objective optimization aims at minimizing m objectives simultaneously: $\min_{\mathbf{x}\in X\subset\mathbb{R}^d}(f_1(\mathbf{x}),..,f_m(\mathbf{x}))$. As these objectives are generally competing, optimal trade-off solutions $\mathbf{x}_1^*,..,\mathbf{x}_q^*$ known as the Pareto optimal set $\mathcal{P}_\mathcal{X}$ are sought. These solutions are *non-dominated* (ND): it is not possible to improve one objective without worsening another ($\forall i = 1,..,q, \nexists \mathbf{z} \preceq \mathbf{x}_i^*$). The image of $\mathcal{P}_\mathcal{X}$ in the objective space is called the Pareto front, $\mathcal{P}_\mathcal{Y} = \{f(\mathbf{x}), \mathbf{x} \in \mathcal{P}_\mathcal{X}\}$. The Ideal and the Nadir points bound the Pareto front and are defined respectively as $\mathbf{I} = (\min_{\mathbf{y}\in\mathcal{P}_\mathcal{Y}} y_1,.., \min_{\mathbf{y}\in\mathcal{P}_\mathcal{Y}} y_m)$ and $\mathbf{N} = (\max_{\mathbf{y}\in\mathcal{P}_\mathcal{Y}} y_1,.., \max_{\mathbf{y}\in\mathcal{P}_\mathcal{Y}} y_m)$.

Multi-objective optimization algorithms aim at constructing the best approximation to $\mathcal{P}_\mathcal{Y}$, called the *empirical* Pareto front $\widehat{\mathcal{P}_\mathcal{Y}}$ which is made of non-dominated observations. At the end of the search, $\widehat{\mathcal{P}_\mathcal{Y}}$ is delivered to a Decision Maker (DM) who will choose the solution he/she prefers.

© Springer Nature Switzerland AG 2019
R. Battiti et al. (Eds.): LION 12 2018, LNCS 11353, pp. 175–179, 2019.
https://doi.org/10.1007/978-3-030-05348-2_15

However, when dealing with expensive computer codes, only a few designs $\mathbf{x}$ can be evaluated. In Bayesian optimization, a surrogate is built for each objective upon all past evaluations using Gaussian Process (GP) regression [4]. Information given by these metamodels is used in order to sequentially evaluate new promising inputs with the aim of reaching the Pareto front. As the latter encompasses a large number of solutions when *many* objectives are considered, it may be impossible to compute an accurate approximation within the restricted computational budget. That approximation being used afterwards by a DM, it may anyway be irrelevant to provide the whole Pareto front because it will contain many uninteresting solutions from the DM's point of view.

Without additional knowledge about the preferences of the DM, we argue that "well-balanced" solutions, in a sense that will be defined hereafter as the central part of the Pareto front, are the most interesting ones. Therefore, we show how to estimate the center of the Pareto front and how classical infill criteria used in Bayesian optimization can be tailored to intensify the search towards it.

2   Center of the Pareto Front: Definition and Estimation

Definition. We define the center, $\mathbf{C}$, of the Pareto front as the projection (intersection in case of a continuous front) of the closest ND point on the Ideal-Nadir line (in the Euclidean objective space). An example of Pareto front center can be seen in Fig. 1. This center corresponds visually to an equilibrium among all objectives. Alternative definitions involving e.g. the barycenter of the Pareto front, are likely to be harder to calculate in high-dimensional spaces. Furthermore, this center has the property of being insensitive to a linear scaling of the objectives in a bi-objective case[1]. $\mathbf{C}$ is also very little sensitive to perturbations of the Ideal or the Nadir point: under mild regularity conditions on the Pareto front, $|\frac{\partial C_i}{\partial I_j}|$ and $|\frac{\partial C_i}{\partial N_j}| < 1$, $i, j = 1, \ldots, m$.

Estimation. As the Ideal and the Nadir of the empirical Pareto front will sometimes be weak substitutes for the real ones (leading to a biased estimated center), those two points have to be truly estimated for the purpose of computing the center. The probabilistic nature of the metamodels (GPs) allows to simulate possible responses of the objective functions. Conditional GP simulations are thus performed to create possible Pareto fronts, each of which defines a sample for $\mathbf{I}$ and $\mathbf{N}$. The estimated Ideal and Nadir are the medians of the samples. The intersection between the line $\widehat{\mathcal{L}}$ joining those points and the empirical Pareto front (or the projection if there is no intersection) is the estimated center $\widehat{\mathbf{C}}$.

3   Targeting Infill Criteria for Bayesian Optimization

Articulating preferences has already been addressed in evolutionary multi-objective optimization [1]. In Bayesian multi-objective optimization, fitted to

---

[1] Non-sensitivity to a linear scaling of the objectives is true when the Pareto front intersects the Ideal-Nadir line. Without intersection, exceptions may occur for $m \geq 3$.

costly objectives, new points are sequentially added using an infill criterion whose purpose is to guide the search towards the Pareto front. After having been evaluated, that point is used to update the metamodel. The Expected Hypervolume Improvement (EHI, [3]) is a commonly used multi-objective infill criterion. EHI chooses the input $\mathbf{x}$ that maximizes the expected growth of the hypervolume dominated by $\widehat{\mathcal{P}_y}$ up to a reference point $\mathbf{R}$. Classically, $\mathbf{R}$ is taken beyond the observed Nadir, e.g. [5], in order to cover the entire front. WHI [2] is a variant of EHI for targeting particular regions of the objective space through a user-defined weighting function.

Our approach targets the central part of the Pareto front with EHI by solely controlling the reference point $\mathbf{R}$. Indeed, the choice of $\mathbf{R}$ is instrumental in deciding the combination of objectives for which *Improvement* occurs: $\mathcal{I}_{\mathbf{R}} = \{\mathbf{y} \in Y : \mathbf{y} \preceq \mathbf{R}\}$. Positioning $\mathbf{R}$ at the (estimated) center of the Pareto front $\widehat{\mathbf{C}}$ will favour EHI's search at objective vectors belonging to $\mathcal{I}_{\widehat{\mathbf{C}}}$.

When the objectives are modeled by independent GPs and the used reference point $\mathbf{R} \not\preceq \widehat{\mathcal{P}_y}$, one has $\text{EHI}(\cdot; \mathbf{R}) = \text{mEI}(\cdot; \mathbf{R})$, where mEI stands for the product of the famous mono-objective Expected Improvement (EI) [4] considering $\mathbf{R}$ as the observed minimum in each objective, $\text{mEI}(\mathbf{x}; \mathbf{R}) = \prod_{i=1}^{m} \text{EI}_i(\mathbf{x}; R_i)$. This observation is particularly appealing from a computational point of view: EHI requires the computation of m-dimensional hypervolumes involving expensive Monte-Carlo estimations in a many-objectives case. For this reason, in our algorithm where $\mathbf{R} \equiv \widehat{\mathbf{C}}$ is not dominated, EHI is replaced by mEI.

Targeting a particular (here, central) part of the Pareto front leads to a fast local convergence. Once $\mathbf{R}$ is on the real Pareto front, the algorithm will try to improve non-improvable values (see left of Fig. 2). To avoid wasting costly evaluations, the convergence has to be checked. To this aim, we estimate $p(\mathbf{y})$, the probability of dominating $\mathbf{y} \in Y$, simulating Pareto fronts through conditional GPs. $p(\mathbf{y})(1 - p(\mathbf{y}))$ is a measure of domination uncertainty, which tends to 0 as $p(\mathbf{y})$ tends to 0 or 1. We assume local convergence when the *line-uncertainty*, $\int_{\widehat{\mathcal{L}}} p(\mathbf{y})(1 - p(\mathbf{y})) d\mathbf{y}$, is small enough. The complete algorithm, called C-EHI, is summarized below. Figure 1 shows a convergence of $\mathbf{R} \equiv \widehat{\mathbf{C}}$ to the true center of the Pareto front, $\mathbf{C}$.

4   Optimization Results

We apply the proposed methodology to a benchmark built from real-world airfoil aerodynamic data. Figure 2 shows that, compared with standard techniques, the proposed methodology leads to a faster and a more precise convergence to the central part of the Pareto front at the cost of a narrower covering of the front. The results are shown at the iteration which triggers the convergence criterion: only marginal gains would indeed be obtained continuing targeting the same region.

**Center-Targeting
Algorithm (C-EHI)**

Create an initial DoE and
initialize m GPs for each
objective $f_i, i = 1, .., m$;
$t = 0$;
line-uncertainty$=+\infty$;

while
*(line-uncertainty$> \varepsilon$) **and**
(t $\leq$ budget) **do**
   estimate **I**, **N** and **C**;
   $\mathbf{x}^{(t+1)} =$
   $\arg\max_{\mathbf{x}\in X} \mathrm{mEI}(\mathbf{x}; \mathbf{C})$;
   evaluate $f_i(\mathbf{x}^{(t+1)})$
   and update the GPs;
   compute
   line-uncertainty(GPs,
   I, **N**);
   $t = t + 1$;
end

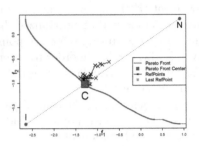

Fig. 1. Reference points
R successively used for
directing the search. They
lie close to the dashed
Ideal-Nadir line (**IN**) and
lead the algorithm to the
center of the Pareto front
(**C**).

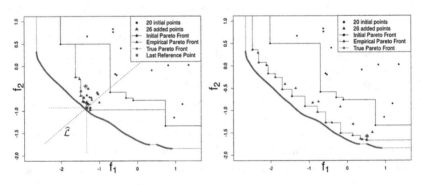

Fig. 2. Two objectives optimization with C-EHI (left). The initial approximation
(black) has mainly been improved around the center. Compared with a standard EHI
(right), the proposed methodology achieves convergence to the central part of the front.
EHI considers more compromises between objectives, but cannot converge within the
given budget (46 evaluations).

Further work will consider the continuation of the search to a broader but
central part of the Pareto front. That newly targeted region, controlled again
through **R**, will be as large as possible while being still attainable within the
remaining budget.

References

1. Bechikh, S., Kessentini, M., Said, L.B., Ghédira, K.: Chap. 4: Preference incorporation in evolutionary multiobjective optimization: a survey of the state-of-the-art. Adv. Comput. **98**, 141–207 (2015)
2. Brockhoff, D., Bader, J., Thiele, L., Zitzler, E.: Directed multiobjective optimization based on the weighted hypervolume indicator. J. Multi-Criteria Decis. Anal. **20**(5–6), 291–317 (2013)
3. Emmerich, M.T., Deutz, A.H., Klinkenberg, J.W.: Hypervolume-based expected improvement: monotonicity properties and exact computation. In: 2011 IEEE Congress on Evolutionary Computation (CEC), pp. 2147–2154. IEEE (2011)
4. Jones, D.R., Schonlau, M., Welch, W.J.: Efficient global optimization of expensive black-box functions. J. Glob. Optim. **13**(4), 455–492 (1998)
5. Ponweiser, W., Wagner, T., Biermann, D., Vincze, M.: Multiobjective optimization on a limited budget of evaluations using model-assisted s-metric selection. In: International Conference on Parallel Problem Solving from Nature, pp. 784–794. Springer (2008)

How *Grossone* Can Be Helpful to Iteratively Compute Negative Curvature Directions

Renato De Leone[1] , Giovanni Fasano[2(✉)] , Massimo Roma[3] ,
and Yaroslav D. Sergeyev[4]

[1] Scuola di Scienze e Tecnologie, Università di Camerino, Camerino, Italy
`renato.deleone@unicam.it`
[2] Dipartimento di Management, Università Ca' Foscari Venezia, Venice, Italy
`fasano@unive.it`
[3] Dipartimento di Ingegneria Informatica, Automatica e Gestionale 'A. Ruberti',
SAPIENZA, Università di Roma, Rome, Italy
`roma@dis.uniroma1.it`
[4] Dipartimento di Ingegneria Informatica, Modellistica, Elettronica e Sistemistica,
Università della Calabria, Rende, Italy
`yaro@dimes.unical.it`

Abstract. We consider an iterative computation of negative curvature directions, in large scale optimization frameworks. We show that to the latter purpose, borrowing the ideas in [1,3] and [4], we can fruitfully pair the Conjugate Gradient (CG) method with a recently introduced numerical approach involving the use of grossone [5]. In particular, though in principle the CG method is well-posed only on positive definite linear systems, the use of grossone can enhance the performance of the CG, allowing the computation of negative curvature directions, too. The overall method in our proposal significantly generalizes the theory proposed for [1] and [3], and straightforwardly allows the use of a CG-based method on indefinite Newton's equations.

Keywords: Negative curvature directions
Second order necessary optimality conditions · Grossone
Conjugate gradient method

1 Introduction

When considering the solution of the unconstrained optimization problem

$$\min_{x \in \mathbb{R}^n} f(x), \tag{1}$$

where $f : \mathbb{R}^n \to \mathbb{R}$ is a nonlinear smooth function and n is large, specific methods should be applied in case stationary points satisfying second order necessary optimality conditions are sought. Moreover, additional cares on the algorithms

R. Battiti et al. (Eds.): LION 12 2018, LNCS 11353, pp. 180–183, 2019.
https://doi.org/10.1007/978-3-030-05348-2_16

adopted are definitely mandatory, since standard stationarity conditions in general do not ensure the convexity of the objective function in a neighborhood of critical points. In this regard, the computation of negative curvature directions for the objective function is an essential tool, to guarantee convergence to second order points.

Observe that convergence towards second order stationary points, where the Hessian matrix is positive semidefinite, requires to efficiently explore the eigenspaces of the Hessian matrix at the solution. Thus, the optimization method adopted to solve (1) should be able to cope also with nonconvexities of the objective function. In particular, as showed in [3], this can be accomplished by suitable Krylov-subspace methods, such that at each iteration j, a pair of directions (s_j, d_j) is computed satisfying specific properties. The vector s_j must be a direction which approximately solves Newton's equation $\nabla^2 f(x_j)\, s = -\nabla f(x_j)$ at x_j. Its purpose is essentially to ensure convergence to stationary points. On the other hand, the direction d_j is a so called *negative curvature direction*, which is used to possibly force convergence to stationary points which satisfy second order necessary optimality conditions [9]. This implies that the sequence $\{d_j\}$ is expected to satisfy the next assumption (which might be possibly weakened).

Assumption 1. *The vectors in the sequence $\{d_j\}$ are bounded and*

$$d_j^T \nabla^2 f(x_j) d_j < 0$$

$$d_j^T \nabla^2 f(x_j) d_j \longrightarrow 0 \qquad \Longrightarrow \qquad \min\left[0,\ \lambda^{\,min}(\nabla^2 f(x_j))\right] \longrightarrow 0,$$

being $\lambda^{\,min}(\nabla^2 f(x_j))$ the smallest eigenvalue of the Hessian matrix $\nabla^2 f(x_j)$.

Roughly speaking, the conditions in Assumption 1 imply that the negative curvature directions $\{d_j\}$ need to eventually approximate an eigenvector associated with the smallest negative eigenvalue of the Hessian matrix. In [3] indications on the computation of the pair (s_j, d_j) were given, though the computation of d_j involved the use of Planar-CG methods, which impose an heavy computation in case some (so called) planar iterations are performed. This approach proved to be effective but required a complex analysis involving different articulated subcases. Here, we aim at describing a strong simplification in the computation the directions $\{d_j\}$, by using a novel approach which extends some ideas in [4].

2   Our Proposal

To this purpose, we use the Krylov-subspace method in [1], for indefinite Newton's equations. This method alternates standard CG iterations and *planar* iterations, each of them being equivalent to a double CG iteration. This method is used to satisfy the next lemma (the proof follows from Theorem 3.2 in [3]).

Lemma 1. *Given problem (1), suppose at iteration n (being n the space dimension) of the Krylov-subspace method used to solve Newton's equation $\nabla^2 f(x_j)s = \nabla f(x_j)$, the decompositions*

$$\nabla^2 f(x_j) R_j = R_j T_j, \qquad T_j = L_j B_j L_j^T$$

are available, where $R_j \in \mathbb{R}^{n \times n}$ is orthogonal, $T_j \in \mathbb{R}^{n \times n}$ has the same eigenvalues of $\nabla^2 f(x_j)$, with at least one negative eigenvalue, and $L_j, B_j \in \mathbb{R}^{n \times n}$ are nonsingular. Let z be the unit eigenvector corresponding to the smallest eigenvalue of B_j, and $\bar{y} \in \mathbb{R}^n$ be the bounded solution of the linear system $L_j^T y = z$. Then, the vector $d_j = R_j \bar{y}$ satisfies Assumption 1.

The main drawback of the latter approach is that the eigenvector z of B_j and the solution of the linear system $L_n^T y = z$ should be of easy computation, which is hardly guaranteed uniquely using the instruments in [1] and [3]. To fill this gap, let us consider the following matrices, obtained applying the method in [1]:

$$
L_n =
\begin{pmatrix}
1 & & & & & & & & \\
-\sqrt{\beta_1} & \ddots & & & & & 0 & & \\
& \ddots & 1 & & & & & & \\
& & -\sqrt{\beta_{k-1}} & 1 & & & & & \\
& & & 0 & 1 & & & & \\
& & & -\sqrt{\beta_k \beta_{k+1}} & 0 & 1 & & & \\
& 0 & & & & -\sqrt{\beta_{k+2}} & \ddots & & \\
& & & & & & \ddots & 1 & \\
& & & & & & & -\sqrt{\beta_{n-1}} & 1
\end{pmatrix},
$$

$$
B_n =
\begin{pmatrix}
1/a_1 & & & & & \\
& \ddots & & & 0 & \\
& & 1/a_{k-1} & 0 & \sqrt{\beta_k} & \\
& & 0 & \sqrt{\beta_k} & e_{k+1} & \\
& 0 & & & \ddots & \\
& & & & & 1/a_n
\end{pmatrix},
$$

where $\{a_i\}$, $\{\beta_i\}$, e_{k+1} are suitable scalars, and we assume (for the sake of simplicity) that the method performed all CG iterations, with the exception of only one planar iteration (namely the k-th iteration - see [1] and [2]). Then, our novel approach proposes to introduce the numeral *grossone*, as in [5–8], and follow some guidelines from [4], so that we can compute the lower block triangular matrix

$$
L_n =
\left(
\begin{array}{ccc|ccc|ccc}
1 & & & & & & & & \\
-\sqrt{\beta_1} & \ddots & & & & & & & \\
& \ddots & 1 & & & & & & \\
\hline
& & -\sqrt{\beta_{k-1}} & & & & & & \\
& & 0 & & V_k C_k^{-1} & & & & \\
& & & \left(-\dfrac{\beta_k \sqrt{\beta_{k+1}}}{\sqrt{\beta_k + \lambda_k^2}} \quad -\dfrac{\beta_k \sqrt{\beta_{k+1}}}{\sqrt{\beta_k + \lambda_{k+1}^2}}\right) C_k^{-1} & 1 & & & \\
\hline
& 0 & & & & -\sqrt{\beta_{k+2}} & \ddots & & \\
& & & & & & \ddots & 1 & \\
& & & & & & & -\sqrt{\beta_{n-1}} & 1
\end{array}
\right)
$$

and the diagonal matrix

$$\bar{B}_n = \begin{pmatrix} 1/a_1 & & & & & & \\ & \ddots & & & & 0 & \\ & & 1/a_{k-1} & & & & \\ & & & \frac{1}{\alpha_k s①} & & & \\ & 0 & & & \frac{s①}{\alpha_{k+1}} & & \\ & & & & & \ddots & \\ & & & & & & 1/a_n \end{pmatrix},$$

such that $L_n B_n L_n^T = \bar{L}_n \bar{B}_n \bar{L}_n^T$ and the symbol ① indicates *grossone*. Moreover,

$$C_k = \begin{pmatrix} 1/\sqrt{\lambda_k \alpha_k s①} & 0 \\ 0 & \sqrt{s①/(\lambda_{k+1}\alpha_{k+1})} \end{pmatrix} \in \mathbb{R}^{2\times 2},$$

being λ_k, λ_{k+1} the two eigenvalues (with $\lambda_k \lambda_{k+1} < 0$) of the 2×2 matrix

$$\begin{pmatrix} 0 & \sqrt{\beta_k} \\ \sqrt{\beta_k} & e_{k+1} \end{pmatrix}, \tag{2}$$

and the columns of the orthogonal matrix $V_k \in \mathbb{R}^{2\times 2}$ correspond to the normalized eigenvectors of the matrix in (2). It is shown in [10] that the latter arrangement can easily allow the computation of the sequence $\{d_j\}$ of negative curvature directions complying with Assumption 1 and Lemma 1.

References

1. Fasano, G.: Conjugate Gradient (CG)-type method for the solution of Newton's equation within optimization frameworks. Optim. Methods Softw. **19**(3–4), 267–290 (2004)
2. Fasano, G.: Planar-conjugate gradient algorithm for large scale unconstrained optimization, part 1: theory. J. Optim. Theory Appl. **125**(3), 523–541 (2005)
3. Fasano, G., Roma, M.: Iterative computation of negative curvature directions in large scale optimization. Comput. Optim. Appl. **38**(1), 81–104 (2007)
4. De Leone, R., Fasano, G., Sergeyev, Y.D.: Planar methods and grossone for the Conjugate Gradient breakdown in nonlinear programming. Comput. Optim. Appl. **71**(1), 73–93 (2018)
5. Sergeyev, Y.D.: Numerical infinities and infinitesimals: methodology, applications, and repercussions on two Hilbert problems. EMS Surv. Math. Sci. **4**, 219–320 (2017)
6. De Leone, R.: Nonlinear programming and grossone: quadratic programming and the role of constraint qualifications. Appl. Math. Comput. **318**, 290–297 (2018)
7. Gaudioso, M., Giallombardo, G., Mukhametzhanov, M.: Numerical infinitesimals in a variable metric method for convex nonsmooth optimization. Appl. Math. Comput. **318**, 312–320 (2018)
8. Sergeyev, Y.D., Kvasov, D.E., Mukhametzhanov, M.: On strong homogeneity of a class of global optimization algorithms working with infinite and infinitesimal scales. Commun. Nonlinear Sci. Numer. Simul. **59**, 319–330 (2018)
9. Moré, J., Sorensen, D.: On the use of directions of negative curvature in a modified Newton method. Math. Program. **16**, 1–20 (1979)
10. De Leone, R., Fasano, G., Roma, M., Sergeyev, Y.D.: Iterative ①-based computation of negative curvature directions in large scale optimization (submitted)

Solving Scalarized Subproblems within Evolutionary Algorithms for Multi-criteria Shortest Path Problems

Jakob Bossek$^{(\boxtimes)}$ and Christian Grimme

Information Systems and Statistics, University of Münster, Münster, Germany
{bossek,christian.grimme}@uni-muenster.de

Abstract. The $\mathcal{NP}$-hard multi-criteria shortest path problem (mcSPP) is of utmost practical relevance, e. g., in navigation system design and logistics. We address the problem of approximating the Pareto-front of the mcSPP with sum objectives. We do so by proposing a new mutation operator for multi-objective evolutionary algorithms that solves single-objective versions of the shortest path problem on subgraphs. A rigorous empirical benchmark on a diverse set of problem instances shows the effectiveness of the approach in comparison to a well-known mutation operator in terms of convergence speed and approximation quality. In addition, we glance at the neighbourhood structure and similarity of obtained Pareto-optimal solutions and derive promising directions for future work.

1 Introduction

Solving the shortest path problem plays a major role in tackling many network (flow) problems: When optimizing logistic networks, the routing of vehicles should ideally follow the shortest path from a starting location to a goal location. In computer networks, a shortest path may be the route with minimum latency. Luckily, with approaches from Dijkstra [9] to Floyd [13] there exist efficient approaches for finding optimal solutions - shortest paths from a starting location to all other or even between all location.

Often, however, problems are not restricted to finding a shortest path regarding a single objective. In logistic applications, costs for shortest paths may be computed based on distances, delivery time or fuel consumption (e. g., when considering the topology of a landscape and not only distances). Clearly, these objectives viewed separately may lead to different and contradicting solutions. Multi-objective optimization allows to consider contradicting objectives together and strives for a set of optimal compromises or Pareto-optimal solutions. These solutions cannot improve regarding one objective without deteriorating for another objective.

© Springer Nature Switzerland AG 2019
R. Battiti et al. (Eds.): LION 12 2018, LNCS 11353, pp. 184–198, 2019.
https://doi.org/10.1007/978-3-030-05348-2_17

The multi-criteria shortest path problem (mcSPP) is generally defined on a graph structure $G = (V, E, c)$, where V is the set of vertices, E is the set of edges, and $c : E \to \mathbb{R}_{>0}^m$ is a mapping of edges to an m-dimensional cost vector. Objective values for a path p through graph G result in an m-dimensional cost vector $F(p) = (f_1(p), \ldots, f_i(p), \ldots, f_m(p))^T \in \mathbb{R}_{\geq 0}^m$ with $f_i(p) = \sum_{e \in p} c_i(e)$, when so-called sum objectives are considered.

The mcSPP is $\mathcal{NP}$-hard, as it exposes exponentially many Pareto-optimal solutions and is thus intractable, in general [10]. Solution strategies for this problem class vary from enumerative methods [14,15,20] through FPTAS [21,22] up to interactive methods [7] and randomized approaches like evolutionary algorithms [5,17,19]. The very popular enumerative approaches determine possible solutions by labelling or ranking. For two objectives, labelling methods work similar to the approaches in the single-objective case: each node can hold mutually non-dominating labels. After the labelling approach terminated, the labels at the target node represent the efficient set of solutions. Contrary, for two-objective problems, ranking approaches solve the single-objective k-shortest path problem by starting at the shortest path for one objective and extending k until all efficient solutions are found. Although, from the practical point of view, many mcSPP instances do not have an exponential number of solutions [18], enumerative methods strongly depend on the amount of solutions on the Pareto-front. In case of exponentially many solutions, these methods are not applicable any more.

Over the last years several authors applied evolutionary algorithms as general heuristic approaches for solving mcSPP. These randomized approximation methods follow the principle of Darwinian evolution by implementing a population-based evolutionary optimization loop, which comprises variation as explorative and selection as goal-directed promoting operator. However, for mcSPP, the application of multi-objective evolutionary algorithms can currently be considered as case-studies in which existing, general algorithmic concepts have merely been applied. Existing knowledge on the problem class has not been considered. Additionally, the existing approaches were only evaluated on few or even single instances.

In this work, we approach the mcSPP with multi-objective evolutionary algorithms using a diverse set of test instances to get empirically broader insight into approximation results. Following the principles in [3,4], we systematically review and improve mutation operators as algorithmic components for including beneficial problem knowledge in order to solve the shortest path problem under multiple criteria.

From here, the paper continues with a description of notation, methods, existing operators. Thereafter, we detail the proposed operators. As preparation for evaluation and discussion of results, we detail the generation process for used test instances. Finally, and after a careful discussion of results we conclude the paper.

2   Problem Formulation

Let $G = (V, E, c)$ be an undirected simple graph with node set $V = \{1, \ldots, n\}$, edge set $E = \{\{v, w\} \mid v, w \in V\}$ and vector-valued cost function $c : E \to \mathbb{R}^m_{>0}$, which associates each edge $e \in E$ with $m \geq 2$ positive weights, e. g., distance and travel costs in case of G representing a street network. A sequence $p_{v_i, v_j} = (v_i, v_{i+1}, \ldots, v_{j-1}, v_j)$ is called a $v_i - v_j$-path in G if $\{v_k, v_{k+1}\} \in E$ for $k = i, \ldots, j - 1$. We define the i-th cost of a path p as the sum of i-th components of all edges lying on the path, i. e., $f_i(p) := \sum_{e \in p} c_i(e), i = 1, \ldots, m$. Wrapping up the path costs for all weights results in an m-dimensional cost vector $F(p) = (f_1(p), \ldots, f_m(p))^T \in \mathbb{R}^m_{>0}$. If $\mathcal{P}$ is the set of all feasible $s - d$-paths between a start node s and a destination node d the formulation

$$F(p) = (f_1(p), \ldots, f_m(p))^T \to \min!$$
$$\text{s. t. } p \in \mathcal{P}$$

defines the *multi-criteria shortest path problem* (mcSPP) with sum objectives[1]. In contrast to the single-objective case, where always an optimal (maybe not unique) solution exists, usually no single best solution can be found in the multi-criteria case. Instead, there exists a set of so-called non-dominated solutions $S = \{p \in \mathcal{P} \mid \nexists p' \in \mathcal{P} \text{ with } F(p') \preceq F(p)\}$. Here, the binary relation $\preceq$ is termed the *dominance relation* and is defined as follows: $p \preceq p'$, i. e., path p *dominates* path p', if $f_i(p) \leq f_i(p'), i = 1, \ldots, m$ and $\exists j \in \{1, \ldots, m\}, f_i(p) < f_i(p')$. In words: path p dominates another path p', if the cost vector of p is strictly better in at least one objective and not worse in the remaining objectives than the cost vector of p'. S is frequently termed *Pareto-set* and its image $F(S) \subset \mathbb{R}^m_{>0}$ *Pareto-front*. The goal is to approximate the Pareto-set and -front respectively. There exist classes of problems where $|S|$ is exponential in the number $n = |V|$ of nodes.

3   Considered Mutation Operators

Mutation operators have not received much attention in evolutionary multi-objective optimization for the mcSPP. A frequently adopted operator – termed random walk operator (RW) in the following – works as follows: Given a feasible $s - d$-path $p_{s,d} = (s = v_0, v_1, \ldots, v_l = d)$, select a random position $i \in \{0, \ldots, l - 1\}$ and replace the subpath $p_{v_i, d} = (v_i, \ldots, v_l = d)$ with a random $v_i - d$-path (see Fig. 1 for an example). Since this operator does not use any information about the edge weights usually the probability to come up with a dominated path is quite high.

---

[1] Other objective types are possible, e. g., bottleneck objectives, but not considered in this work.

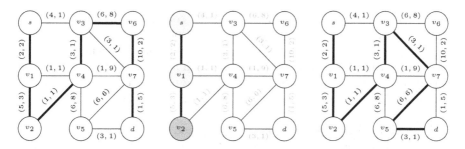

Fig. 1. Exemplary application of RW mutation. Left: initial path $p_{s,d}$ (bold edges). Center: select random position $v_i = v_2$, neglect all nodes located before v_2 in $p_{s,d}$ (gray nodes) and costs completely. Right: append random subpath $p_{v_2,d} = (v_2, v_4, v_3, v_7, v_5, d)$ to p_{s,v_2} (bold edges).

We propose the following alternatives which differ from RW by the method used to search for a new subpath. Here, we present the ideas for $m = 2$ objectives for sake of simplicity. However, adaptation to more than two objectives is straight-forward. Starting point is again a feasible path $p_{s,d} = (s = v_0, v_1, \ldots, v_l = d)$ and a randomly selected position $i \in \{0, \ldots, l-1\}$. Let p_{s,v_i} be the subpath from s to v_i. Instead of performing a random search, the first proposed *subgraph* operator (SG) selects a random cost component $o \in \{1, \ldots, m\}$ with equal probability, ignores the other cost components $\{1, \ldots, m\} \setminus \{o\}$ and searches for the shortest path from v_i to d with Dijkstra's algorithm. All nodes which are located prior to v_i on the input path $p_{s,d}$ are marked as visited to avoid loops. The resulting path $p_{v_i,d}$ is a minimal $v_i - d$-path regarding f_o. It is appended to p_{s,v_i} resulting in another feasible $s - d$-path $\tilde{p}_{s,d} = p_{s,v_i} \circ p_{v_i,d}$ where $\circ$ is the path concatenation. See Fig. 2 for an illustration of the working principle by example. Our second proposal is the *scalarized subgraph* operator (SGS) which is a generalization of SG. Instead of focusing on one of the objectives exclu-

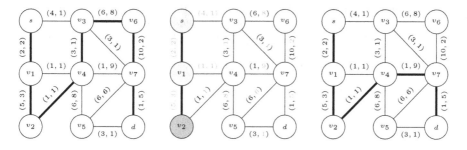

Fig. 2. Examplary application of subgraph mutation SG. Left: initial path $p_{s,d}$ (bold edges). Center: selected position is $v_i = v_2$. We sample the first objective, i. e., $o = 1$, ignore nodes on subpath p_{s,v_2} (gray nodes) and edge costs function $\neq o$. Right: find shortest $v_2 - d$-path $p_{v_2,d} = (v_2, v_4, v_7, d)$ with Dijkstra algorithm and append $p_{v_2,d}$ to p_{s,v_2} (bold edges).

sively, SGS samples a random weight $\lambda \in [0, 1]$, constructs the single-objective weighted sum problem $f_\lambda = \lambda f_1 + (1 - \lambda)f_2$ and solves this scalarized surrogate with a single-objective shortest path algorithm. As before, nodes already visited in p_{s,v_i} are marked visited. Again we end up with a $v_i - d$-path $p_{v_i,d}$, which can be appended to p_{s,v_i} resulting in a mutated $s - d$-path. An example is depicted in Fig. 3. Clearly, SG is a special case of SGS where $\lambda = 0$ or $\lambda = 1$ with equal probability.

The SGS mutation operator applies a weighted sum approach to subgraphs of G. Note, that weighted sum applied to G is only capable of finding so-called supported efficient solutions (see, e. g., [11]). These solutions are located on the convex hull of the Pareto-front. Solutions in concave regions cannot be detected with this approach. However, we expect, that applying weighted sum scalarization on subgraphs (ignoring already visited nodes) is able to (1) push solutions towards the Pareto-front rapidly and (2) identify solutions which are unreachable to the classic weighted sum approach. The first point is supported by the fact, that given a path p the path p' resulting from application of SG or SGS either dominates p or boths paths are incomparable. This is evident, since the appended subpath is a supported efficient solution of the shortest path problem on the subgraph and thus is non-dominated by any other possible subpath. We term this desirable behaviour *Pareto-beneficial* (see also [3]).

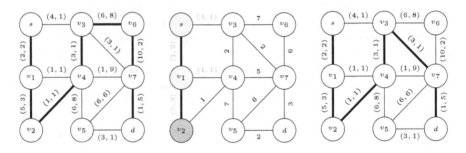

Fig. 3. Examplary application of SGS mutation. Left: Initial path $p_{s,d}$ (bold edges). Center: selected position is $v_i = v_2$. With the sampled weight (here $\lambda = 0.5$) we compute $f_\lambda(e) = \lambda f_1(e) + (1 - \lambda)f_2(e), \forall e \in E$ and ignore nodes located before v_i on the initial path (gray nodes). Right: find shortest $v_2 - d$-path regarding f_λ resulting in $p_{v_2,d} = (v_2, v_4, v_3, v_7, d)$ and append $p_{v_2,d}$ to p_{s,v_2} (bold edges).

4   Results

For an assessment of the proposed operators and their extensions, we conduct a series of experiments based on diverse graph topologies containing different amounts of nodes. Thus, we first detail the experimental setup and subsequently discuss the observations and results regarding this aspect. Afterwards, we briefly investigate solution properties based on the found solutions.

4.1 Graph Generation

In order to investigate the performance of the considered mutation operators we generated 150 random graphs in total: each 5 instances of 6 different topologies considering sizes $n \in \{50, 100, 250, 500, 1000\}$. The network topologies mimick real-world network structures and differ in network density, interconnection of nodes and (non)existence of clusters. The graph generation process is implemented in the R package `grapherator`[2] [2]. It follows a flexible three-step approach:

1. Place nodes in the Euclidean plane $[0, 100]^2$. Here we considered nodes placed uniformly at random within the bounding box or clustered nodes.
2. Establish links between nodes following different edge generation methods (complete graphs, edges based on Delauney triangulation or Waxmans probablistic model following [5]).
3. Associate each edge $e \in E$ with two additive uniform random weights $c_1(e) \in [20, 3000], c_2(e) \in [200, 5000]$. This last step is identical among all generated instances.

Figure 4 shows several generated networks by way of example.

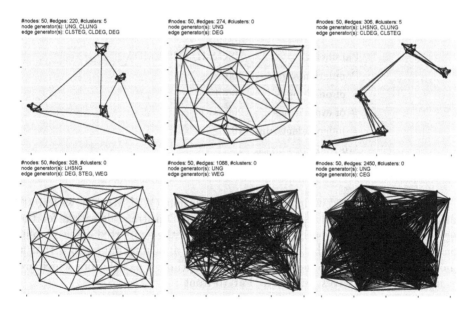

Fig. 4. Examplary graph topologies generated for our study.

---

[2] https://github.com/jakobbossek/grapherator.

4.2 Experimental Setup

We consider the three different mutation operators introduced in Sect. 2: random walk mutation (RW), subgraph mutation (SG) and scalarized subgraph mutation (SGS) as the generalization of SG. Two state-of-the-art evolutionary multi-objective algorithms are adopted as encapsulating meta-heuristics for each mutation operator. NSGA-II (non-dominated sorting genetic algorithm) [8] is a $(\mu + \lambda)$-strategy. It basically relies on non-dominated sorting as primary and crowding distance as secondary selection criterion. In contrast SMS-EMOA (S-metric selection) [12] – in its classical version – follows a $(\mu + 1)$ indicator based strategy. Here, the hypervolume [24] contribution of each individual is used directly for selection. Since our aim is the empirical investigation of mutation operators, recombination/crossover is not applied. However, we stress that the integration of recombination might be fruitful as well. All other parameters are wrapped up in Table 1. Each EMOA was executed 10 times for statistical soundness of subsequent performance assessment. We used the implementations from the R package `ecr` [1] as well as the packages' methods for performance assessment.

Table 1. Parameter settings for all configurations of the applied meta-heuristics NSGA-II and SMS-EMOA.

	Setting
Parameter	NSGA-II/SMS-EMOA
Population size μ	100
# of offspring λ	100/1
# of evaluations	$200 \cdot |V|$
# independent runs	10
Mutation prob. p_{mut}	1

In addition we used weighted sum scalarization (WSUM) [16] with 1000 equidistant weights $\lambda_k = \frac{k}{999}, k = 0, \ldots, 999$ and Dijkstra's algorithm for the single-objective shortest path problem to compute the supported efficient solutions as a baseline. Note that the drawback of this appealing approach is that it is not capable of finding non-supported Pareto-optima, i. e., solutions which are not located on the convex hull of the Pareto-front.

4.3 Performance of Mutation Operators

We start our observation of experimental data with considering different aggregated approximations of the Pareto-fronts in Fig. 5 for six graph instances with different topology, and sizes $n \in \{500, 1000\}$, respectively. The non-dominated solutions of all runs are aggregated for the application of the random walk (RW), subgraph (SG), and scalarized subgraph (SGS) mutation operators. In Fig. 5, we already qualitatively find that the SGS operator performs seemingly best in almost all cases. In this plot we also find an implicit further ranking of the other indicators: in many cases the SG operator is slightly superior to the random walk operator. Still, sometimes random walk mutation also produces good solution candidates in the final solution set.

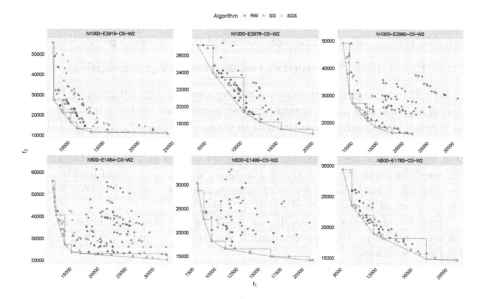

Fig. 5. Union of Pareto-front approximations of each 10 runs of all algorithms on some instances with $n = 500$ and $n = 1000$ nodes respectively. The instance names are encoded as follows: N<#nodes>-E<#edges>-C<#clusters>-W<#weights> where 0 clusters indicates unclustered, i. e., random, instances.

In Fig. 6 and Table 3, we analyze the observations systematically. Here, we restrict the analysis to NSGA-II results and the overall set of instances[3]. Note, that results for SMS-EMOA are congruent for the following observations. For two instances of each investigated number of vertices $n \in \{50, 100, 250, 500, 1000\}$ the distribution of hypervolume (HV) [24] values is given (top row of Fig. 6). This indicator measures convergence and diversity of solution sets by determining the enclosed volume of solutions and a reference point in objective space. Additionally, and considering the same instances, statistics for the cardinality of the final solution set (ONVG) [6] are given (bottom row of Fig. 6). For both indicators, the dashed line denotes the indicator values found by systematically iterated single-objective search using weighting of objectives (WSUM). This baseline is always

---

[3] Note, that we do not consider the topologies of generated instances due to space limitations. A detailed analysis of the topology's influence cannot be thought without considering the distribution of weights and the locations of start and destination notes. Therefore, this aspect is left for rigorous analysis in future work.

outperformed by the SGS operator, while the remaining operators sometimes stay below this baseline. Note, that the ONVG indicator is only meaningful, if it is considered secondary to the HV indicator. As a strongly dominated solution set can contain many (poor) solutions – often more, than a very good solution set – it is not sufficient as primary quality indicator. However, under the pre-condition of HV comparison, it enables us to discriminate solution quality in more detail. In order to get statistically sound quantitative results, we performed pairwise comparisons between the algorithms with the nonparametric Wilcoxon rank sum test on each test instance. We tested the hypothesis pair

$$H_0 : \mathrm{med}(\mathrm{HV}_A) \leq \mathrm{med}(\mathrm{HV}_B) \text{ vs. } H_1 : \mathrm{med}(\mathrm{HV}_A) > \mathrm{med}(\mathrm{HV}_B)$$

to check if the location shift between the hypervolume distributions HV_A and HV_B of algorithms A and B is significant at significance level $\alpha = 0.05$ adjusting the p-value with Bonferroni correction to avoid multiple-testing issues. Table 2 shows the aggregated test results. It turns out, that in the most interesting case (SGS vs. SG), the zero hypothesis is rejected in 109 of 150 cases ($\approx$ 73% of the cases). In particular with growing instance size the number of rejected tests increases: H_0 is not rejected for 34 of 90 instances of size $n \in \{50, 100, 250\}$, but only for 7 out of 60 instances of size $n \in \{500, 1000\}$.

Thus, for the presented result, we can conclude that SGS does not only dominate with respect to hypervolume. It is also able to find the largest number of optimal solutions.

To complement the discussion of indicator values for the three different operators, we provide exemplary trajectories for two 500-nodes-instances on the HV development aggregated over all runs in Fig. 7. We find that both new operators SG and SGS converge very fast. After only few generations, the operators almost

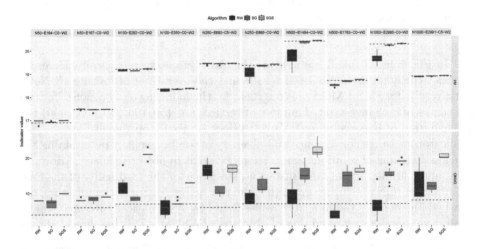

Fig. 6. Distributions of performance indicators HV (hypervolume on log-scale) and ONVG for each 2 exemplary instances of instance size $n \in \{50, 100, 250, 500, 1000\}$ respectively.

Table 2. Results of pairwise one-sided Wilcoxon rank sum test for location shift.

	H_0 not declined	H_0 declined
SG versus RW	96	54
SGS versus RW	47	103
SGS versus SG	41	109

reach their final HV value and thus the best approximation set. This speed-up is certainly related to the integration of local optimization on partial path. However, this speedup is also important in practise as the mutation operators SG and SGS are computationally more complex than the RW operator.

A final comparison of algorithm performance under different mutation operators is presented in Fig. 8. The shown heat-maps summarize the results of pairwise ε-indicator [24] evaluation. This indicator measures the degree of dominance between results of two algorithms A and B. It essentially states how far the result set of B has to be shifted towards the utopian point such that it is not dominated by A any more. Note, that this indicator is not symmetric. In other word, the larger the indicator for algorithm B compared to A and the smaller the indicator is for A compared to B, the better A performs compared to B. The results shown in Fig. 8 confirm the superiority of SGS over all other operators, the medium performance of SG, and the general low performance of the RW operator. For comparison reasons the WSUM approach for determining reference solutions is also included.

4.4   Properties of Solutions

To analyze the characteristics of solutions, we focus here on an exemplary instance comprising 500 nodes. We first create a scatter plot of the edge weights for this instance. Then we insert a frequency-based coloring of edges contained in solutions. As solutions we use the non-dominated set of solutions generated over all experimental runs for a given instance. Figure 9 shows the occurrence of edges in the solution set.

We can observe, that all (possibly) optimal solutions can be constructed from a subset of edges. This also means, that many edges are never part of (possibly) optimal solutions. At the same time, edges used in optimal solutions tend to gather in the lower left corner of the figure. Thus, edges with small costs in at least one component of the cost vector are more frequently used in (possibly) optimal solutions. The here exemplary presented observation is valid for all instances.

Table 3. Statistics of performance indicators for some exemplary instances.

Problem	Algorithm	avg.HV	sd.HV	avg.ONVG	sd.ONVG
N50-E164-C0-W21	RW	16.967	0.071	8.000	0.000
	SG	16.935	0.021	7.300	1.059
	SGS	**17.001**	0.000	**10.000**	0.000
N50-E167-C0-W24	RW	**17.493**	0.014	8.200	0.422
	SG	17.462	0.052	8.400	0.699
	SGS	17.486	0.004	**9.100**	0.316
N100-E282-C0-W24	RW	19.162	0.035	12.500	2.461
	SG	19.148	0.016	8.500	0.527
	SGS	**19.226**	0.000	**20.800**	1.135
N100-E350-C0-W25	RW	18.319	0.056	6.200	2.201
	SG	18.345	0.005	7.300	0.675
	SGS	**18.387**	0.000	**13.000**	0.000
N250-E692-C5-W21	RW	19.448	0.032	**16.800**	1.932
	SG	19.437	0.033	10.600	1.075
	SGS	**19.461**	0.003	16.600	1.713
N250-E865-C0-W25	RW	19.067	0.233	8.500	1.434
	SG	19.354	0.031	12.200	2.044
	SGS	**19.422**	0.002	**16.900**	0.316
N500-E1484-C0-W22	RW	19.724	0.361	9.000	3.399
	SG	20.391	0.045	15.600	2.366
	SGS	**20.453**	0.002	**22.200**	1.814
N500-E1763-C0-W25	RW	18.509	0.059	4.400	1.578
	SG	18.690	0.030	14.300	2.163
	SGS	**18.763**	0.001	**16.300**	1.059
N1000-E2980-C0-W23	RW	19.595	0.355	6.700	3.302
	SG	20.244	0.006	15.100	1.729
	SGS	**20.313**	0.001	**18.900**	0.568
N1000-E2991-C5-W23	RW	18.873	0.023	12.500	4.249
	SG	18.889	0.007	11.900	1.101
	SGS	**18.914**	0.001	**20.300**	0.483

Additionally, we use two visualizations based on the number of common edges (NCE) and the largest common sub-path (LCS) between two solutions. Per row/column, we visualize the comparison of a non-dominated solution with any other non-dominated solution of the solution set. The solutions are ordered according to objective value f_1. Thus, distant solutions in a row (column) are also distant in the non-dominated front; neighbouring solutions in a row (column) are also neighbours in the non-dominated front. The left-hand plot in Fig. 10

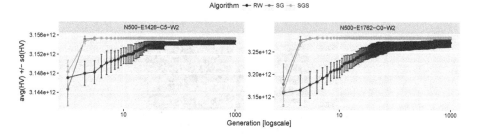

Fig. 7. Hypervolume trajectories (average of all runs ± standard devition for each EMOA for each generation) on two examplary graphs with 500 nodes.

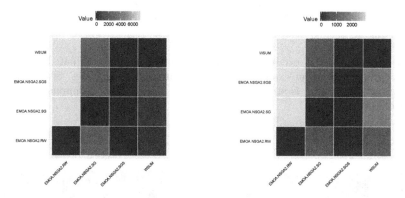

Fig. 8. Representative heat-maps of pairwise ε-indicator values for a problem with 500 nodes (left) and 1000 nodes (right).

shows the similarity of solutions regarding the NCE, while the right-hand plot shows the length of the largest common sub-path of two solutions, respectively.

In both visualizations we find, that there are only local neighbourhood relations between neighbouring solutions. Combined with the largest common sub-path behaviour (where the same effect can be observed) this means, that solutions can locally be transformed to neighbouring solutions by changing one or few edges. Interestingly, a little bit more distant solutions in objective space have almost nothing in common. As such, paths are completely disjunct for these solutions. That in turn suggests, that solutions cannot be constructed from each other by simply combining "optimal" building blocks. In fact, constructing new path (as it is done by the SGS operator, guided by a combined objective) can contribute to finding alternative non-dominated routes.

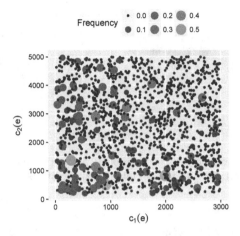

Fig. 9. Analysis of edge occurrence frequency for overall non-dominated $s - t$-paths in the solution sets of an examplary instance with $n = 500$ nodes. Scatter plot of edge weights. Each edge is sized/coloured by the fraction of non-dominated paths it is part of.

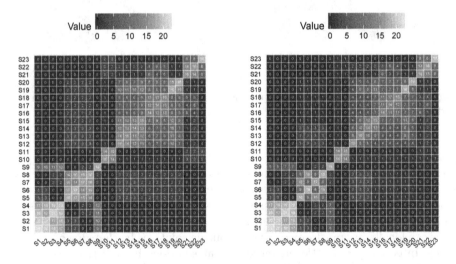

Fig. 10. Analysis of neighbourhood structure for overall non-dominated $s - t$-paths in the solution sets of an examplary instance with $n = 500$ nodes. Left: heat-map of pairwise NCE (number of common edges). Right: heat-map of pairwise size of largest common sub-path values.

5   Conclusions

This work deals with mutation in evolutionary algorithms for the multi-criteria shortest path problem. We present a mutation approach which follows the idea of replacing a subpath of an existing solution with another locally efficient subpath. Hence, the operator mimicks the well-known weighted sum scalarization approach in a smaller world and furthermore generates an offspring individual which is not dominated by its parent. A systematic comparison with simple random-path substitution and weighted sum scalarization reveals promising results. The new mutation operator does not only produce good approximations regarding dominated Hypervolume (statistically significant in most cases). It furthermore finds solutions in concave regions of the Pareto-front (which the global weighted sum approach is not capable of) and shows a rapid convergence behaviour.

Future work directions are manifold. We aim to test the proposed operator on specifically generated instances with a large number of non-dominated solutions. Furthermore, selection of the cutpoint in parent solutions is done uniformly at random at the moment. Here, we are confident, that smarter selection strategies may be beneficial in terms of even faster convergence. Additionally, multi-objective heuristics specifically designed for problem decomposition, such as MOEA/D [23], can be evaluated in conjunction with the new operators. Last but not least, theoretical analysis of the proposed mutation operator is desirable in order to understand its functionality in a more[4] rigorous way.

Acknowledgments. The authors acknowledge support from the *European Research Center for Information Systems (ERCIS)*.

References

1. Bossek, J.: ecr 2.0: a modular framework for evolutionary computation in R. In: Proceedings of the Genetic and Evolutionary Computation Conference Companion, GECCO 2017, pp. 1187–1193 (2017). https://doi.org/10.1145/3067695.3082470
2. Bossek, J.: grapherator: a modular multi-step graph generator. J. Open Source Softw. **3**(22), 528 (2018). https://doi.org/10.21105/joss.00528
3. Bossek, J., Grimme, C.: A pareto-beneficial sub-tree mutation for the multi-criteria minimum spanning tree problem. In: 2017 IEEE Symposium Series on Computational Intelligence (SSCI), pp. 3280–3287. IEEE, Honolulu (2017). https://doi.org/10.1109/SSCI.2017.8285183
4. Bossek, J., Grimme, C.: An extended mutation-based priority-rule integration concept for multi-objective machine scheduling. In: 2017 IEEE Symposium Series on Computational Intelligence (SSCI), pp. 3288–3295. IEEE, Honolulu (2017). https://doi.org/10.1109/SSCI.2017.8285224
5. Chitra, C., Subbaraj, P.: Multiobjective optimization solution for shortest path routing problem. Int. Sch. Sci. Res. Innov. **4**(1) (2010)
6. Coello Coello, C.A., Lamont, G.B., Van Veldhuizen, D.A.: Evolutionary Algorithms for Solving Multi-Objective Problems. Genetic and Evolutionary Computation, 2nd edn. Springer, New York (2007). https://doi.org/10.1007/978-0-387-36797-2
7. Coutinho-Rodrigues, J., Clímaco, J., Current, J.: An interactive bi-objective shortest path approach: searching for unsupported nondominated solutions. Comput. Oper. Res. **26**(8), 789–798 (1999). https://doi.org/10.1016/S0305-0548(98)00094-X

---

[4] https://www.ercis.org/.

8. Deb, K., Pratap, A., Agarwal, S., Meyarivan, T.: A fast and elitist multiobjective genetic algorithm: NSGA-II. IEEE Trans. Evolut. Comput. **6**(2), 182–197 (2002)
9. Dijkstra, E.W.: A note on two problems in connexion with graphs. Numerische Mathematik **1**, 269–271 (1959)
10. Ehrgott, M., Gandibleux, X.: A survey and annotated bibliography of multiobjective combinatorial optimization. OR-Spektrum **22**(4), 425–460 (2000). https://doi.org/10.1007/s002910000046
11. Ehrgott, M., Gandibleux, X.: A survey and annotated bibliography of multiobjective combinatorial optimization. OR-Spektrum **22**(4), 425–460 (2000). https://doi.org/10.1007/s002910000046
12. Emmerich, Michael, Beume, Nicola, Naujoks, Boris: An EMO algorithm using the hypervolume measure as selection criterion. In: Coello Coello, Carlos A., Hernández Aguirre, Arturo, Zitzler, Eckart (eds.) EMO 2005. LNCS, vol. 3410, pp. 62–76. Springer, Heidelberg (2005). https://doi.org/10.1007/978-3-540-31880-4_5
13. Floyd, R.W.: Algorithm 97: shortest path. Commun. ACM **5**(6), 345 (1962). https://doi.org/10.1145/367766.368168
14. Gandibleux, X., Beugnies, F., Randriamasy, S.: Martins' algorithm revisited for multi-objective shortest path problems with a maxmin cost function. 4OR **4**(1), 47–59 (2006). https://doi.org/10.1007/s10288-005-0074-x
15. Martins, E.Q.V.: On a multicriteria shortest path problem. Eur. J. Oper. Res. **16**(2), 236–245 (1984)
16. Miettinen, K.: Nonlinear Multiobjective Optimization. International Series in Operations Research & Management Science, vol. 12. Springer, New York (1998)
17. Mohamed, C., Bassem, J., Taicir, L.: A genetic algorithms to solve the bicriteria shortest path problem. Electron. Notes Discret. Math. **36**, 851–858 (2010). https://doi.org/10.1016/j.endm.2010.05.108. ISCO 2010 - International Symposium on Combinatorial Optimization
18. Müller-Hannemann, M., Weihe, K.: Pareto Shortest Paths is Often Feasible in Practice, pp. 185–197. Springer, Berlin (2001). https://doi.org/10.1007/3-540-44688-5_15
19. Pangilinan, J.M.A., Janssens, G.K.: Evolutionary algorithms for the multiobjective shortest path planning problem. In: International Journal of Computer and Information Science and Engineering, pp. 54–59 (2007)
20. Sanders, P., Mandow, L.: Parallel label-setting multi-objective shortest path search. In: 2013 IEEE 27th International Symposium on Parallel and Distributed Processing, pp. 215–224 (2013). https://doi.org/10.1109/IPDPS.2013.89
21. Tsaggouris, G., Zaroliagis, C.: Multiobjective Optimization: Improved FPTAS for Shortest Paths and Non-linear Objectives with Applications, pp. 389–398. Springer, Berlin (2006). https://doi.org/10.1007/11940128_40
22. Warburton, A.: Approximation of pareto optima in multiple-objective, shortest-path problems. Oper. Res. **35**(1), 70–79 (1987). https://doi.org/10.1287/opre.35.1.70
23. Zhang, Q., Li, H.: MOEA/D: a multiobjective evolutionary algorithm based on decomposition. IEEE Trans. Evolu. Comput. **11**(6), 712–731 (2007)
24. Zitzler, E., Thiele, L., Laumanns, M., Fonseca, C.M., Da Fonseca, V.G.: Performance assessment of multiobjective optimizers: an analysis and review. IEEE Trans. Evolu. Comput. **7**(2), 117–132 (2003)

Exact and Heuristic Approaches for the Longest Common Palindromic Subsequence Problem

Marko Djukanovic[1(✉)], Günther R. Raidl[1], and Christian Blum[2]

[1] Institute of Logic and Computation, TU Wien, Vienna, Austria
{djukanovic,raidl}@ac.tuwien.ac.at
[2] Artificial Intelligence Research Institute (IIIA-CSIC),
Campus UAB, Bellaterra, Spain
christian.blum@iiia.csic.es

Abstract. The longest common palindromic subsequence (LCPS) problem requires to find a longest palindromic string that appears as subsequence in each string from a given set of input strings. The algorithms that can be found in the related literature are specific for LCPS problems with only two input strings. In contrast, in this work we consider the general case with an arbitrary number of input strings, which is NP-hard. To solve this problem we propose a fast greedy heuristic, a beam search, and an exact A* algorithm. Moreover, A* is extended by a simple diving mechanism as well as a combination with beam search in order to find good quality solutions already early in the search process. The most important findings that result from the experimental evaluation include that (1) A* is able to efficiently find proven optimal solutions for smaller problem instances, (2) the anytime behavior of A* can be significantly improved by incorporating diving or beam search, and (3) beam search is best from a purely heuristic perspective.

Keywords: Longest common palindromic subsequence problem
A* search · Beam search · Hybrid optimization techniques

1 Introduction

In computer science, a *string* s is defined as a finite sequence of characters from a (generally finite) alphabet Σ. An important characteristic of a string s is its length, denoted by $|s|$. A string is generally used as a data type for representing and storing sequence information. Words, and even whole texts, may be stored by means of strings. Strings arise, in particular, in the field of bioinformatics, because most of the genetic instructions involved in the growth, development, functioning and reproduction of living organisms are stored in *Deoxyribonucleic acid* (DNA) molecules which can be represented by strings over the alphabet $\Sigma = \{A, C, T, G\}$. A string s is called a *palindrome* if $s = s^{\text{rev}}$, where s^{rev} is the reverse string of s; for example, madam is a palindrome.

© Springer Nature Switzerland AG 2019
R. Battiti et al. (Eds.): LION 12 2018, LNCS 11353, pp. 199–214, 2019.
https://doi.org/10.1007/978-3-030-05348-2_18

Note that, given a string s, any string t that can be obtained from s by deleting zero or more characters is called a *subsequence* of s. Palindromic subsequences are especially interesting in the biological context. In many genetic instructions, such as for example DNA sequences, palindromic motifs are found. In the context of a research project on genome sequencing it was discovered that many of the bases on the Y-chromosome are arranged as palindromes [17]. A palindrome structure allows the Y-chromosome to repair itself by bending over at the middle if one side is damaged. Moreover, it is believed that palindromes are also frequently found in proteins [10], but their role in the protein function is less understood. Biologists believe that identifying palindromic subsequences of DNA sequences may help to understand genomic instability [7,23]. Palindromic subsequences seem to be important for the regulation, for example, of gene activity, because they are often found close to promoters, introns and untranslated regions.

An important way for the comparison of two or more strings is to find long *common subsequences*. More specifically, given a set of m non-empty strings $S = \{s_1, \ldots, s_m\}$, a common subsequence of the strings in S is a subsequence that all strings in S have in common. Moreover, a *longest common subsequence* of the strings in S is a common subsequence of maximal length. The so-called *Longest Common Subsequence (LCS)* problem [18] is a classical optimization problem that aims at finding such a longest common subsequence of the strings in S. Apart from applications in computational biology [16], this problem appears, for example, in data compression [22] and the production of circuits in field programmable gate arrays [6]. Finally, a *common palindromic subsequence* of a set of strings S is a common subsequence of all strings in S which, at the same time, is a palindrome. For biologists it is not only of interest to identify the palindromic subsequences of an individual DNA string, for example, but it is also important to find longest common palindromic subsequences of multiple input strings in order to identify relationships among them.

1.1   Related Work

The LCS problem is known to be NP-hard for an arbitrary number (m) of input strings [18]. Note, however, that for any fixed m the problem is polynomially solvable by dynamic programming [11]. Standard dynamic programming approaches require $O(n^m)$ time and space, where n is the length of the longest input string. Even though this complexity can be reduced to $O(n^{m-1})$, see Bergoth et al. [1], dynamic programming becomes quickly impractical when m grows. Concerning simple approximate methods, the expansion algorithm in [5] and the Best-Next heuristic [9,14] are probably the best-known techniques. A break-through both in terms of computation time and solution quality was achieved with the beam search (BS) approach described by Blum et al. [2]. Beam search is an incomplete tree search algorithm which relies on a solution construction mechanism and bounding information. More specifically, the above BS uses the construction mechanism of the Best-Next heuristic and just a simple upper bound function. Nevertheless, this algorithm was able to outperform all existing algorithms at that time. Later, Mousavi and Tabataba [20] proposed a variant of this BS with a different heuristic function and a different pruning mechanism.

The specific problem tackled in this work—that is, the longest common palindromic subsequence (LCPS) problem—has so far only been studied for the case of $m = 2$ input strings (2-LCPS). Chowdhury et al. [8] propose two different algorithms: a conventional dynamic programming with time and space complexity $O(n^4)$, and a sparse dynamic programming algorithm with time complexity $O(R^2 \log^2 n \log \log n + n)$ and space complexity $O(R^2)$, where R is the number of matching position pairs between the two input strings. Furthermore, Hasan et al. [13] solved the 2-LCPS by making use of a so-called palindromic subsequence automaton (PSA). This algorithm has a time complexity of $O(n + R_1|\Sigma| + R_2|\Sigma| + n + R_1R_2|\Sigma|)$, where R_1 and R_2 denote the numbers of states of the two automata constructed for the two input strings and are bounded by $O(n^2)$. Finally, Inenaga and Hyyrö [15] present an algorithm that runs in $O(\sigma R^2 + n)$ time and uses $O(R^2 + n)$ space, where σ denotes the number of distinct characters occurring in both of the input strings.

By reducing the general LCS problem to the LCPS problem in polynomial time, it can be shown that the LCPS problem with an arbitrary number of input strings is NP-hard. To the best of our knowledge, no algorithm has been published yet for solving this general m-LCPS problem, which is henceforth simply called LCPS problem. An instance of the LCPS problem is denoted by (S, Σ), where S is the set of input strings over alphabet Σ.

1.2   Organization of the Paper

The rest of the paper is organized as follows. In Sect. 2, a fast greedy heuristic is presented. Moreover, two upper bound functions are proposed. In Sect. 3 we present two search algorithms that operate on the same search tree: an exact A^* search and a heuristic beam search. We further consider a variant of A^* that has a simple diving mechanism as well as beam search embedded in order to obtain promising complete solutions already early during the search process. In this way, both A^* and beam search can be used as heuristics to approach large instances. Experimental results are described in Sect. 4. Conclusions as well as an outlook to future work are finally provided in Sect. 5.

2   A Greedy Heuristic for the LCPS Problem

We first introduce some additional notations. As already mentioned before, let $n = \max_{s_i \in S} |s_i|$ be the maximum input string length. The j-th letter of a string s is stated by $s[j]$, with $j = 1, \ldots, |s|$. We further denote the concatenation of two strings by operator "$\cdot$", i.e., $s_1 \cdot s_2$ is the string obtained by appending string s_2 to string s_1. Notation $s[j, j']$, $j \leq j'$, refers to the substring of s starting at the j-th position and ending at position j'. The same notation refers to the empty string ε if $j > j'$. Finally, let $|s|_a$ be the number of occurrences of letter $a \in \Sigma$ in string s, and let $|s|_A = \sum_{a \in A} |s|_a$ be the total number of occurrences of letters from set $A \subseteq \Sigma$ in string s.

Inspired by the well-known Best-Next heuristic for the LCS problem [9], we present in the following a constructive greedy heuristic for the LCPS problem. Henceforth, a string s is called a *valid partial solution* concerning input strings $S = \{s_1, \ldots, s_m\}$, if $s \cdot s^{\text{rev}}$ or $s \cdot s[1, |s| - 1]^{\text{rev}}$ is a common palindromic subsequence of the strings in S. The greedy heuristic starts with an empty partial solution $s = \varepsilon$ and extends, at each construction step, the current partial solution by appending exactly one letter (if possible). During the whole process, the algorithm makes use of pointers $p_i^{\text{L}} \leq p_i^{\text{R}}$ that indicate for each input string s_i, $i = 1, \ldots, m$, the still *relevant substring* $s_i[p_i^{\text{L}}, p_i^{\text{R}}]$ from which the letter for extending s can be chosen. The choice of a letter with respect to a greedy criterion is explained below. At the start of the heuristic, i.e., when $s = \varepsilon$, the pointers are initialized to $p_i^{\text{L}} := 1$ and $p_i^{\text{R}} := |s_i|$, referring to the first, respectively, last letter of each string s_i, $i = 1, \ldots, m$. In other words, at each iteration the set of relevant substrings denoted by $S[p^{\text{L}}, p^{\text{R}}] = \{s_i[p_i^{\text{L}}, p_i^{\text{R}}] \mid i = 1, \ldots, m\}$ forms an LCPS subproblem, and the current partial solution s is ultimately extended by appending the solution to this subproblem.

One of the questions that remain is how to determine the subset of letters from Σ that can be used to extend a current partial solution s. For this purpose, let $c_a := \min_{i=1,\ldots,m} |s_i[p_i^{\text{L}}, p_i^{\text{R}}]|_a$ be the minimum number of occurrences of letter $a \in \Sigma$ in the relevant substrings with respect to s, and let $\Sigma_{(p^{\text{L}}, p^{\text{R}})} := \{a \in \Sigma \mid c_a \geq 1\}$ be the set of letters appearing at least once in each relevant substring. In principle, any letter from $\Sigma_{(p^{\text{L}}, p^{\text{R}})}$ might be used to extend s. However, there might be dominated letters in this set. In order to introduce the domination relation between two letters, we use the first and last positions at which each letter $a \in \Sigma_{(p^{\text{L}}, p^{\text{R}})}$ appears in each relevant substring $s_i[p_i^{\text{L}}, p_i^{\text{R}}]$:

$$q_{i,a}^{\text{L}} := \min \{j = p_i^{\text{L}}, \ldots, p_i^{\text{R}} \mid s_i[j] = a\}$$
$$q_{i,a}^{\text{R}} := \max \{j = p_i^{\text{L}}, \ldots, p_i^{\text{R}} \mid s_i[j] = a\}$$

A letter $a \in \Sigma_{(p^{\text{L}}, p^{\text{R}})}$ is called *dominated* if there exists a letter $b \in \Sigma_{(p^{\text{L}}, p^{\text{R}})}$, $b \neq a$, such that $q_{i,b}^{\text{L}} < q_{i,a}^{\text{L}} \wedge q_{i,b}^{\text{R}} > q_{i,a}^{\text{R}}$ for $i = 1, \ldots, m$. Clearly, it is better to delay the consideration of dominated letters and select a non-dominated letter for the extension of s. Furthermore, letters $a \in \Sigma_{(p^{\text{L}}, p^{\text{R}})}$ with $c_a = 1$, called *singletons*, should only be considered when no other letters remain in $\Sigma_{(p^{\text{L}}, p^{\text{R}})}$, since only one such letter can be chosen as single middle letter in the final solution. Accordingly, let the set of all non-dominated non-singleton letters from $\Sigma_{(p^{\text{L}}, p^{\text{R}})}$ with respect to s be denoted by $\Sigma_{(p^{\text{L}}, p^{\text{R}})}^{\text{nd}}$. Given a partial solution s, the selection of the letter to be appended to s and the adaption of the pointers work as follows:

1. If $\Sigma_{(p^{\text{L}}, p^{\text{R}})}$ is empty, the algorithm terminates with $s \cdot s^{\text{rev}}$ as resulting common palindromic subsequence, since no further extension is possible.
2. Otherwise, if $\Sigma_{(p^{\text{L}}, p^{\text{R}})}^{\text{nd}}$ is empty, only singletons remain in $\Sigma_{(p^{\text{L}}, p^{\text{R}})}$. The algorithm terminates with the common palindromic subsequence $s \cdot a \cdot s^{\text{rev}}$, where a is the first singleton from $\Sigma_{(p^{\text{L}}, p^{\text{R}})}$ in alphabetical order.

3. Otherwise, select a letter $a \in \Sigma^{\text{nd}}_{(p^L, p^R)}$ that minimizes the *greedy function* $g(a, p^L, p^R)$, which will be discussed in Sect. 2.1. Ties are broken randomly. Extend the current partial solution s and adapt the pointers as follows:

$$s := s \cdot a \tag{1}$$

$$p_i^L := q_{i,a}^L + 1 \qquad\qquad i = 1, \ldots, m \tag{2}$$

$$p_i^R := q_{i,a}^R - 1 \qquad\qquad i = 1, \ldots, m \tag{3}$$

2.1 Greedy Function

The greedy function that is used to evaluate any possible extension $a \in \Sigma^{\text{nd}}_{(p^L, p^R)}$ for a given partial solution extends the one used in [3] in a straight-forward manner. It calculates the sum of those fractions of the relevant substrings $s_i[p_i^L, p_i^R]$ that will be discarded from further consideration when appending a as next letter to the partial solution:

$$g(a, p^L, p^R) := \sum_{i=1}^{m} \frac{q_{i,a}^L - p_i^L + p_i^R - q_{i,a}^R}{p_i^R - p_i^L + 1}. \tag{4}$$

The major advantage of this function is its simplicity, as it can be calculated in time $O(m)$. Obviously, this function also has some weaknesses: (1) it does not take into account that, when choosing a specific letter, as a result, more or less letters might be excluded from further consideration, even in cases in which the chosen letter has a good (low) greedy function value; (2) it does not take into account that of all singletons at most one can finally be selected. Instead of improving the above greedy function along these lines, and thereby increasing its time complexity, we consider it more promising—especially with the type of algorithm in mind that will be presented in the next section—to develop upper bound functions for estimating the length of an LCPS. As we will see, these bounds can also be used as alternative greedy functions to evaluate possible extensions of partial solutions.

2.2 Upper Bounds for the Length of an LCPS

A first upper bound for the length of any palindromic subsequences obtainable for a set of strings S can be calculated by

$$\text{UB}_1(S) = \left(2 \sum_{a \in \Sigma_{(p^L, p^R)}} \left\lfloor \frac{c_a}{2} \right\rfloor \right) + \mathbb{1}_{\exists a \in \Sigma_{(p^L, p^R)} | c_a \bmod 2 = 1}. \tag{5}$$

The last term considers the fact that at most one singleton letter can be added at the end of a solution construction, with $\mathbb{1}$ denoting the unit step function that yields one iff the condition in the subscript is fulfilled, i.e., there exists a letter in $\Sigma_{(p^L, p^R)}$ with an odd value of c_a. Note that $\text{UB}_1(S)$ can be calculated in $O(mn)$ time, considering the required re-calculation of the counters c_a.

A second upper bound can be derived as follows. First, each relevant substring $s_i[p_i^L, p_i^R]$ is reduced by deleting all letters that are not in $\Sigma_{(p^L, p^R)}$. The resulting strings are denoted by $s_i[p_i^L, p_i^R]^{red}$, $i = 1, \ldots, m$. Then, a longest palindromic subsequence, denoted by $LPS(s_i[p_i^L, p_i^R]^{red})$, is calculated for each of these strings individually. As a longest common palindromic subsequence of all strings $S[p^L, p^R]^{red}$ cannot be longer than any individual longest palindromic subsequence, we obtain the upper bound

$$UB_2(S) = \min_{i=1,\ldots,m} |LPS(s_i[p_i^L, p_i^R]^{red})|. \tag{6}$$

A longest palindromic subsequence of a single string can be calculated by solving the LCS (longest common subsequence) problem for the problem instance that has as input strings the string itself and its reversal. This can be done by dynamic programming in $O(n^2)$ time. Masek and Paterson [19] presented a more specific and slightly faster algorithm that runs in $O(n^2/\log n)$ time.

Note that none of the two upper bounds dominates the other one. For example, for $S = \{abba, abab\}$ we get $UB_1(S) = 4$ and $UB_2(S) = 3$, whereas for $S = \{aba, bab\}$ we get $UB_1(S) = 1$ and $UB_2(S) = 3$. Therefore, it might be beneficial to consider the minimum of both bounds $UB_3(S) = \min(UB_1(S), UB_2(S))$.

Finally, observe that our upper bound functions can also directly be applied to evaluate any partial solution s with its still relevant substrings $S[p^L, p^R]$: While $UB_x(S[p^L, p^R])$, for $x \in \{1, 2, 3\}$, is an upper bound for the length by which s may still be extended, $|s| + UB_x(S[p^L, p^R])$ is an upper bound for the overall length of still achievable solutions. In the greedy heuristic, our upper bound functions can therefore be used instead of $g(a, p^L, p^R)$ by temporarily determining the updated p^L and p^R when one would accept letter a and calculating $UB_x(S[p^L, p^R])$.

3   Search Algorithms for the LCPS

In the following we first describe the state graph on which both A* and Beam Search will operate. This state graph is a directed acyclic multi-graph $G = (V, A)$ in which each node (state) $v = (p^{L,v}, p^{R,v}) \in V$ corresponds to a unique LCPS subproblem, i.e., a set of still relevant substrings indicated by the respective pointer vectors $p^{L,v}$ and $p^{R,v}$. Note that one such node will in general represent multiple different partial solutions in an efficient way. As an example consider $S = (abccdccba, baccdccab)$, and partial solutions $s = ac$ and $s' = bc$. It holds that $p^L = (4, 4)$ and $p^R = (6, 6)$ in both cases, and thus, both partial solutions will be represented by a common node. Here, s and s' have the same length, but this need not be the case in general. Our state graph has the root node $r = ((1, \ldots, 1), (|s_1|, \ldots |s_m|))$ corresponding to the original set of input strings S and representing the empty partial solution. Each node $v \in V$ stores as additional information the length l^v of a so far best—i.e., longest—partial solution represented by v. Furthermore, each node $v \in V$ has an outgoing arc $(v, v', a) \in A$ for each valid extension of the represented partial solutions by a non-dominated non-singleton letter $a \in \Sigma_v^{nd} = \Sigma_{(p^{L,v}, p^{R,v})}^{nd}$.

We emphasize that it is not necessary to store actual partial solutions s in the nodes. As pointed out already in Sect. 2, this is neither necessary for the greedy function evaluation, nor for the upper bound calculation. For any node in the graph a corresponding solution string can finally be efficiently derived in a backward manner by iteratively identifying predecessors in which the l^v-values always decrease by one.

3.1   A* Search for the LCPS Problem

A* is a widely used algorithm belonging to the class of informed search methods for finding shortest or longest paths [12]. It maintains two sets of nodes: N stores all so far reached nodes, while Q, the set of *open nodes*, is the subset of nodes in N that have not yet been *expanded*, i.e., whose outgoing arcs and respective neighbors have not yet been considered. We realize node set N by means of a hash map in order to be able to efficiently find an already existing node for a state (p^L, p^R), or to determine that no respective node exists yet. Moreover, Q is a priority queue in which nodes are sorted according to decreasing *priority values* $\pi(v) = l^v + \mathrm{UB}_x(S[p^{L,v}, p^{R,v}])$, where x specifies the used upper bound.

The pseudo-code of our A* search is shown in Algorithm 1. It starts with the root node as unique node in N and Q. At each step, the first node v from Q—that is, the highest priority node—is chosen and removed from Q. If this node is non-extensible, it is first checked if a singleton letter can be added, and afterwards the algorithm stops. Since our priority function is *admissible*, cf. [12], we can be sure that an optimal solution has been reached. Otherwise, node v is extended by considering all possible extensions from Σ_v^{nd}. For each obtained new state it is checked if a respective node exists already in N. If this is the case, the existing node's length-value is updated in case the new path to this node represents a new longest partial solution. Otherwise, a corresponding new node is created and added to N and Q.

Finally, we remark that both upper bound functions presented in Sect. 2.2, i.e., UB_1 and UB_2, as well as their combination UB_3, are *monotonic* (also called *consistent*) because the upper bound value of an extension of a node is always at most as high as the upper bound value of the originating node. Due to this property we can be sure that no re-expansions of already expanded nodes will be necessary, see again [12].

Diving in A*. One of the main advantages of A* is the fact that the search performs in an asymptotic optimal way with respect to the applied upper bound function, requiring the least possible number of node expansions in order to find a proven optimal solution. On the downside, good approximate solutions are typically only obtained, if at all, very late in the search. To improve this situation and turn our A* into an *anytime* algorithm, which can be terminated almost arbitrarily and still yields a reasonable solution, we augment it by switching in regular intervals to a temporary greedy depth-first search until no further extensible solution is obtained. We call this extension *diving*. More specifically, diving is initiated at the very beginning and after each δ regular A* iterations,

Algorithm 1 A* Search for the LCPS problem

1: **Input:** an instance (S, Σ)
2: **Output:** s_{bsf}, an optimal LCPS solution
3: $s_{\mathrm{bsf}} \leftarrow \varepsilon$
4: Create root node $r = ((1, \ldots, 1), (|s_1|, \ldots |s_m|))$ with $l^r = 0$
5: Add r to the initially empty node set N and priority queue Q
6: *optimal* $\leftarrow$ FALSE
7: **while** $Q \neq \emptyset$ **and** not *optimal* **do**
8:     Take the first node v from priority queue Q
9:     Determine Σ_v^{nd} from $p^{\mathrm{L},v}$ and $p^{\mathrm{R},v}$
10:    **if** $\Sigma_v^{\mathrm{nd}} = \emptyset$ **then**
11:        Derive a partial solution s represented by v
12:        **if** $\Sigma_v \neq \emptyset$ **then**
13:            Choose a singleton $a \in \Sigma_v$
14:            $s \leftarrow s \cdot a \cdot s^{\mathrm{rev}}$
15:        **else**
16:            $s \leftarrow s \cdot s^{\mathrm{rev}}$
17:        **end if**
18:        $s_{\mathrm{bsf}} \leftarrow s$
19:        *optimal* $\leftarrow$ TRUE
20:    **else**
21:        **for** $a \in \Sigma_v^{\mathrm{nd}}$ **do**
22:            Compute state v' that results from appending a at state v
23:            **if** $v' \in N$ **then**
24:                **if** $l^v + 1 > l^{v'}$ **then**
25:                    $l^{v'} \leftarrow l^v + 1$
26:                    Update entry for v' in Q with new priority value $\pi(v')$
27:                **end if**
28:            **else**
29:                Add new node v' with $l^{v'} = l^v + 1$ to N and Q
30:            **end if**
31:        **end for**
32:    **end if**
33:    Remove v from Q
34: **end while**

---

where δ is an external parameter. Starting from the first node taken from Q, i.e., the highest priority node, we always expand as next node a newly generated immediate successor with highest priority value. This depth-first search is performed until no further newly generated successor exists (i.e., we do not further follow any already previously created nodes). If a new best solution is obtained in this way, it is stored in s_{bsf}, which is returned in case of an early termination due to an imposed time limit. Note that, when extending a node during diving, the same steps regarding the update of the nodes in N and Q are performed as in A*. An important difference is, however, that nodes expanded during diving may now require a re-expansion at a later time when a longer partial solution is found for the respective state.

3.2 Beam Search for the LCPS Problem

With Beam Search (BS), we further consider an alternative, purely heuristic way of searching the state graph defined at the beginning of this section. BS [21] is a breadth-first search algorithm that explicitly limits the nodes examined at each level, for example, with an explicit upper bound of their number $\beta > 0$ called the *beam width*. Before presenting our specific BS for the LCPS problem, we define a dominance relation for nodes in the state graph considered at the same level of BS: Given nodes $u, v \in V$ we say u *dominates* v iff $u \neq v$ and $p_i^{\mathrm{L},u} < p_i^{\mathrm{L},v} \wedge p_i^{\mathrm{R},u} \geq p_i^{\mathrm{R},v}$ for all $i = 1, \dots, m$.

The pseudo-code of our BS is provided in Algorithm 2. The *beam B*— that is, the set of nodes considered at each step of the algorithm—is initialized with the root node r. Then, at each step, the nodes of the current beam are extended in all possible ways, dominated nodes are filtered in function RemoveDominatedEntries(V_{ext}), and the best β nodes with respect to their priority values are selected in function Reduce(V_{ext}, β) to obtain the beam for the next iteration.

Algorithm 2 Beam Search (BS) for the LCPS problem

1: **Input:** an instance (S, Σ)
2: **Output:** s_{bsf}, the best solution found
3: $s_{\mathrm{bsf}} \leftarrow \varepsilon$
4: Create root node $r = ((1, \dots, 1), (|s_1|, \dots |s_m|))$
5: Beam $B \leftarrow \{r\}$
6: **while** B is not empty **do**
7:     $V_{\mathrm{ext}} \leftarrow \emptyset$
8:     **for** each $v \in B$ **do**
9:         Determine Σ_v^{nd} from $p^{\mathrm{L},v}$ and $p^{\mathrm{R},v}$
10:        **if** $\Sigma_v^{\mathrm{nd}} = \emptyset$ **then**
11:            Derive a partial solution s represented by v
12:            **if** $\Sigma_v \neq \emptyset$ **then**
13:                Choose a singleton $a^* \in \Sigma_v$
14:                $s \leftarrow s \cdot a^* \cdot s^{\mathrm{rev}}$
15:            **else**
16:                $s \leftarrow s \cdot s^{\mathrm{rev}}$
17:            **end if**
18:            **if** $|s| > |s_{\mathrm{bsf}}|$ **then** $s_{\mathrm{bsf}} \leftarrow s$ **end if**
19:        **else**
20:            **for** $a \in \Sigma_v^{\mathrm{nd}}$ **do**
21:                Compute state v' that results from appending a at state v
22:                $V_{\mathrm{ext}} \leftarrow V_{\mathrm{ext}} \cup \{v'\}$
23:            **end for**
24:        **end if**
25:     **end for**
26:     $V_{\mathrm{ext}} \leftarrow$ RemoveDominatedEntries(V_{ext})
27:     $B \leftarrow$ Reduce(V_{ext}, β)
28: **end while**

3.3 Embedding Beam Search in A*

Instead of the simple diving described for A* in Sect. 3.1, we may also apply above BS embedded within A* at regular intervals, always starting with the first entry of Q as the initial node in beam B. As in simple diving, BS skips any already earlier encountered nodes (i.e., nodes that are already in N are not added to V_{ext}) in order to avoid ineffective re-considerations of parts of the state graph. Therefore, it might happen—just like in the case of simple diving—that the embedded BS ends without delivering any complete solution. Moreover, as in simple diving, for all considered extensions of nodes, the same steps regarding the update of the nodes in N and Q are performed as in A*, cf. Algorithm 1. Finally, note that with beam width $\beta = 1$ the embedded BS corresponds to simple diving.

3.4 Tie Breaking

While executing preliminary experiments for A*, we realized that many ties occur when ordering the nodes in the priority queue Q with respect to their priorities in $\pi(v)$. To guide the search in better ways, we decided to use the length of a represented longest partial solution as a secondary decision criterion in such cases. This improved the performance significantly, but still suffered from a significant number of ties. In order to also break these, it turned out to be beneficial to additionally consider the p-norm, which is for a node v defined as

$$||v||_p = \left(\sum_{i=1}^{m} \left| p_i^{R,v} - p_i^{L,v} \right|^p \right)^{1/p}. \tag{7}$$

Given two nodes $u \neq v$ with the same priority value and the same maximum length concerning the represented partial solutions, a node with a lower p-norm is finally preferred. The inspiration for making use of this norm is that the smallest still relevant substrings potentially have a higher impact on the final length of complete solutions than the larger ones. However, considering only the shortest one of the still relevant substrings—that is, applying the min norm—could be highly misleading. Therefore, a p value from $(0, 1)$ appears meaningful. Following further preliminary experiments, we finally chose $p = 0.5$ for all experiments discussed in the next section.

4 Experimental Results

The proposed algorithms were implemented in C++ using GCC 4.7.3 and all experiments were performed as single threads on Intel Xeon E5649 CPUs with 2.53 GHz and a memory limit of 15 GB.

The benchmark instances used in this work were initially introduced in [4] in the context of the LCS problem and are provided at https://www.ac.tuwien.ac.at/wp/wp-content/uploads/LCPS_instances.zip. This set consists for each combination of the number of input strings $m \in \{10, 50, 100, 150, 200\}$, the length of the input strings $n \in \{100, 500, 1000\}$ and the alphabet size $|\Sigma| \in \{4, 12, 20\}$ of 10 randomly generated instances, yielding a total of 450 problem instances.

Table 1. Comparison of BS with UB_1 to BS with UB_3 on the 150 problem instances with $|\Sigma| = 4$

m	n	BS with UB_1		BS with UB_3		UB_1 versus UB_2			
		$\overline{|s|}$	$\overline{t}[s]$	$\overline{|s|}$	$\overline{t}[s]$	$> (\%)$	$< (\%)$	$= (\%)$	$- (avg)$
	100	28.1	<0.0	**28.5**	0.3	74.7	10.0	15.3	5.5
10	500	150.7	0.1	**151.5**	76.8	95.6	2.0	2.4	54.3
	1000	**304.7**	0.4	304.3	656.1	97.9	1.0	1.2	122.5
	100	21.2	0.0	**21.4**	1.6	53.3	25.3	21.4	1.8
50	500	125.1	0.4	**125.4**	368.1	93.3	3.4	3.3	42.6
	1000	**256.5**	1.3	–	900.0	100.0	0.0	0.0	214.7
	100	19.5	0.1	**19.9**	3.1	48.8	27.0	24.2	1.3
100	500	118.3	0.7	**119.4**	174.5	93.4	3.4	3.2	42.6
	1000	**245.1**	2.3	–	900.0	100.0	0.0	0.0	209.4
	100	18.5	0.1	**18.6**	2.9	39.4	35.8	24.8	0.4
150	500	115.7	1.2	**116.5**	887.2	93.1	3.5	3.4	39.2
	1000	**240.9**	3.1	–	900.0	100.0	0.0	0.0	211.5
	100	17.9	0.1	**18.1**	3.6	39.5	35.6	24.8	0.4
200	500	**114.2**	1.4	–	900.0	92.0	4.3	3.7	36.5
	1000	**237.7**	4.0	–	900.0	100.0	0.0	0.0	212.4

4.1 Comparison of Upper Bound Functions

In order to study the differences and mutual benefits of the two upper bound functions from Sect. 2.2, BS with $\beta = 10$ was applied both using only UB_1 and using UB_3, that is, the minimum of UB_1 and UB_2. The outcome is presented in Table 1. Each row shows average results over the 10 problem instances for each combination of m and n. The results of BS with UB_1 are presented in terms of the obtained average solution quality ($\overline{|s|}$) and the average required computation time ($\overline{t}[s]$) in the third and fourth table column. The corresponding results of BS with UB_3 are listed in the fifth and sixth table column. The best result per table row is printed bold. The "–" symbol indicates that no complete solution of length greater than zero was derived within a CPU time limit of 900 s since the bound calculation took already too much time.

The following can be observed. First, when it is not too costly to calculate UB_2, as it is always the case for the instances with $n = 100$ and mostly when $n = 500$, BS using UB_3 is able to outperform BS using only UB_1. However, the high time complexity for calculating UB_2—that is, $O(mn^2)$—is a major obstacle in the context of larger problem instances. Because of these limitations, we perform all further experiments for BS, A^*, and the hybrid using only UB_1.

Nevertheless, the additional four columns in Table 1 clearly indicate that the usage of UB_2 can be promising also for larger instances. These columns show the percentages of nodes for which UB_2 dominates UB_1 ($> (\%)$), the percentages of nodes for which UB_1 dominates UB_2 ($< (\%)$), the percentage of nodes where both bounds are the same ($= (\%)$), and the average absolute values of subtracting UB_2 from UB_1 ($- (avg)$). Results show that UB_2 dominates UB_1 especially for long input strings. A promising idea seems to be to either limit the time for calculating UB$_2$ or to calculate this bound only for a suitably chosen subset of all nodes. However, these studies are left for future work.

4.2   Main Results

We now compare the performance of our four solution approaches: (1) the greedy algorithm from Sect. 2, henceforth referred to as Greedy; (2) BS; (3) A* with simple diving, henceforth referred to as A*+Dive; and (4) A* with embedded BS, henceforth referred to as A*+BS.

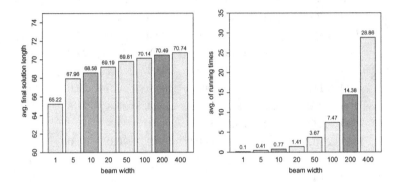

Fig. 1. Average final solution lengths and runtimes of BS with different beam widths β

Table 2. Results for $|\Sigma|=4$

m	n	Greedy		BS		A*+Dive			A*+BS		
		$|s|$	$\bar{t}[s]$	$|s|$	$\bar{t}[s]$	$|s|$	$\bar{t}[s]$	$\bar{t}_{best}[s]$	$|s|$	$\bar{t}[s]$	$\bar{t}_{best}[s]$
10	100	25.6	<0.1	*28.9	0.3	*28.9	13.2	5.6	*28.9	14.1	0.7
	500	143.6	<0.1	**157.4**	2.7	147.5	900.0	310.8	156.5	900.0	222.1
	1000	292.6	<0.1	**316.4**	7.1	291.8	900.0	376.3	313.3	900.0	413.8
50	100	19.4	<0.1	21.7	0.6	*21.8	6.6	1.9	*21.8	7.4	2.2
	500	117.6	<0.1	**128.0**	7.8	123.8	900.0	148.1	127.8	900.0	137.6
	1000	251.0	<0.1	**262.8**	22.5	252.4	900.0	227.8	260.9	900.0	176.5
100	100	18.1	<0.1	20.0	0.9	*20.1	8.8	2.0	*20.1	9.8	1.0
	500	112.2	<0.1	**121.4**	13.9	118.4	900.0	219.5	121.3	900.0	203.7
	1000	240.3	<0.1	**250.9**	41.9	242.8	900.0	130.1	249.7	900.0	281.2
150	100	16.3	<0.1	*19.0	1.2	*19.0	6.6	0.5	*19.0	7.6	0.2
	500	108.4	<0.1	118.1	20.0	115.6	900.0	359.7	**118.4**	900.0	324.8
	1000	234.0	0.1	**244.9**	58.8	238.5	900.0	176.1	244.4	900.0	251.8
200	100	16.4	<0.1	18.4	1.5	*18.5	8.7	2.1	*18.5	9.8	1.1
	500	107.1	<0.1	115.9	25.6	113.8	900.0	238.0	**116.5**	900.0	336.2
	1000	227.2	0.2	**241.4**	77.1	235.9	900.0	363.2	241.1	900.0	171.8

For deciding how to choose the beam width β in the stand-alone BS as well as in A*+BS, we applied BS to each of the 450 problem instances. Average final solution lengths and runtimes are shown in Fig. 1. As expected, with increasing beam width β also the solution quality increases. However, this comes at the cost of an approximately linear increase of the runtime. Since the solution quality for $\beta = 400$ is only slightly better than that with $\beta = 200$, but the required times are about twice as large, we chose $\beta = 200$ for the standalone BS. For the embedded BS, we decided to use $\beta = 10$ due to the still relatively good results and small average runtime of only 0.77 s per instance.

The two variants of A* further require a setting for δ, the number of regular A* iterations between diving/BS. We considered 5, 10, 50, 100, 500, and 1000 iterations and conducted preliminary experiments in a similar way as for β. Results (not shown) indicated that for $\delta = 10$, A* performs on average slightly but significantly better than with the other values. Therefore, we adopt this setting in our further tests for A*+Dive and A*+BS.

Results from the comparison of the four solution approaches are presented separately for instances of different alphabet sizes in Tables 2, 3 and 4. Again, shown values are averages over the 10 instances of the same type, and best results from each row are printed bold. Optimal solution values (as determined by A*+Dive and/or A*+BS) are marked with an asterisk. For each algorithm, the table shows final average solution lengths, average runtimes, and additionally, for the algorithms A*+Dive and A*+BS the column $\bar{t}_{best}[s]$ which shows the average computation times at which the best found solutions were obtained. A limit of 900 s was imposed per run. The following observations can be made:

Table 3. Results for $|\Sigma|=12$

m	n	Greedy		BS		A*+Dive			A*+BS		
		$|s|$	$\bar{t}[s]$	$|s|$	$\bar{t}[s]$	$|s|$	$\bar{t}[s]$	$\bar{t}_{best}[s]$	$|s|$	$\bar{t}[s]$	$\bar{t}_{best}[s]$
	100	8.9	<0.1	***9.6**	<0.1	***9.6**	<0.1	<0.1	***9.6**	<0.1	<0.1
10	500	54.1	<0.1	**60.6**	2.7	57.6	900.0	273.3	60.3	900.0	17.8
	1000	113.8	<0.1	**125.7**	6.6	115.1	900.0	396.7	124.0	900.0	270.4
	100	4.7	<0.1	***5.6**	<0.1	***5.6**	<0.1	<0.1	***5.6**	<0.1	<0.1
50	500	38.6	<0.1	**43.1**	6.1	41.8	900.0	124.0	**43.1**	900.0	40.1
	1000	83.9	<0.1	**90.4**	16.1	86.2	900.0	473.1	89.8	900.0	274.0
	100	3.9	<0.1	***4.6**	<0.1	***4.6**	<0.1	<0.1	***4.6**	<0.1	<0.1
100	500	35.0	<0.1	38.7	9.8	37.5	900.0	102.1	**39.0**	900.0	141.7
	1000	77.8	<0.1	**82.9**	27.4	79.9	900.0	130.0	82.7	900.0	117.4
	100	3.5	<0.1	***3.8**	<0.1	***3.8**	<0.1	<0.1	***3.8**	<0.1	<0.1
150	500	33.2	<0.1	**37.0**	13.7	36.0	900.0	61.4	**37.0**	900.0	105.7
	1000	72.7	0.1	79.2	37.2	77.2	900.0	211.8	**79.4**	900.0	94.0
	100	3.1	<0.1	***3.3**	<0.1	***3.3**	<0.1	<0.1	***3.3**	<0.1	<0.1
200	500	31.3	<0.1	**35.4**	17.5	35.0	900.0	202.0	35.3	900.0	57.9
	1000	71.1	0.2	**77.7**	50.5	75.3	900.0	208.6	77.3	900.0	159.5

Table 4. Results for $|\Sigma|=20$

m	n	Greedy		BS		A*+Dive			A*+BS		
		$\lvert s\rvert$	$\bar{t}[s]$	$\lvert s\rvert$	$\bar{t}[s]$	$\lvert s\rvert$	$\bar{t}[s]$	$\bar{t}_{best}[s]$	$\lvert s\rvert$	$\bar{t}[s]$	$\bar{t}_{best}[s]$
	100	5.0	<0.1	*5.4	<0.1	*5.4	<0.1	<0.1	*5.4	<0.1	<0.1
10	500	32.7	<0.1	38.3	2.8	36.6	900.0	107.0	**38.4**	900.0	166.6
	1000	70.4	<0.1	**79.3**	6.6	73.7	900.0	213.7	78.5	900.0	163.5
	100	2.3	<0.1	*2.5	<0.1	*2.5	<0.1	<0.1	*2.5	<0.1	<0.1
50	500	21.7	<0.1	**24.9**	5.5	24.5	900.0	70.4	**24.9**	900.0	3.8
	1000	48.7	<0.1	**54.3**	15.8	51.6	900.0	159.1	53.7	900.0	115.8
	100	*1.3	<0.1	*1.3	<0.1	*1.3	<0.1	<0.1	*1.3	<0.1	<0.1
100	500	18.5	<0.1	**21.9**	9.1	21.0	900.0	1.5	21.8	900.0	62.0
	1000	44.0	0.1	**48.7**	26.7	47.0	900.0	46.5	48.4	900.0	40.1
	100	*1.1	<0.1	*1.1	<0.1	*1.1	<0.1	<0.1	*1.1	<0.1	<0.1
150	500	17.8	<0.1	20.5	11.6	20.1	900.0	46.9	**20.6**	900.0	165.5
	1000	40.1	0.1	**46.0**	37.6	44.9	900.0	201.4	45.8	900.0	81.5
	100	*1.1	<0.1	*1.1	<0.1	*1.1	<0.1	<0.1	*1.1	<0.1	<0.1
200	500	16.9	<0.1	19.1	14.9	19.0	900.0	6.3	**19.4**	900.0	79.1
	1000	39.8	0.2	**44.7**	46.8	43.1	900.0	60.3	44.6	900.0	213.6

- By far the fastest algorithm is Greedy. However, Greedy also produces the weakest results in the comparison. Runtimes of the A* variants are generally higher, but of course these partly include proofs of optimality.
- Both A*+Dive and A*+BS are able to find optimal solutions for all instances with input string length $n = 100$. This corresponds to 15 out of 45 cases (table rows).
- In most of those cases in which the A* variants cannot find optimal solutions, A*+BS outperforms A*+Dive. This shows the benefit of using BS as embedded heuristic as opposed to simple diving.
- In those cases where the A* versions are able to find optimal solutions and prove their optimality, BS is most of the time also able to find solutions of equal quality. However, this seems to become more difficult for BS when the alphabet size decreases. In particular, BS failed to find all optimal solutions in three out of five cases with $|\Sigma| = 4$.
- From a pure heuristic point of view, BS outperforms A*+BS more and more when the length of the input strings increases. More specifically, while the results obtained by BS and the A*+BS are comparable for instances with $n = 500$, BS generally outperforms A*+BS for instances with $n = 1000$.

5   Conclusions and Future Work

We proposed different algorithms for solving the LCPS problem with an arbitrary number of strings heuristically as well as exactly. A general state graph was defined that can be searched by different strategies. With BS we provided a pure heuristic search that scales well to also large instances. With A^* we provided an efficient method for solving instances with up to 200 strings of lengths up to 100 to proven optimality. Since for instances with even larger strings, A^* search cannot find a complete solution in a reasonable time, it is upgraded to an anytime algorithm by embedding either the simple diving or the more advanced BS. For the instances where our hybrid algorithms do not find optimal solution, the optimality gaps between final (heuristic) solutions and the corresponding upper bounds produced by A^* are not so tight. The reason for this is that UB_1 partly provides only rather weak bounds. Using UB_1 in combination with UB_2, i.e., UB_3, would clearly be beneficial from the quality point-of-view, but the larger time complexity of UB_2 makes this approach prohibitive for larger instances. In future work the strengthening of the upper bounds seems to be most promising. We believe that this can be achieved by applying UB_2 only for subproblems up to a certain size or by finding an approximation of UB_2 that can be calculated in a faster way. Testing the algorithms with real world instances, e.g., coming from protein, DNA and virus structure sequences, would also be interesting, since such instances may have special structures on which the algorithms might perform differently or which might be further exploited.

Acknowledgments. We gratefully acknowledge the financial support of this project by the Doctoral Program "Vienna Graduate School on Computational Optimization" funded by the Austrian Science Foundation (FWF) under contract no. W1260-N35.

References

1. Bergroth, L., Hakonen, H., Raita, T.: A survey of longest common subsequence algorithms. In: Proceedings of SPIRE 2000–7th International Symposium on String Processing and Information Retrieval, pp. 39–48. IEEE press (2000). https://doi.org/10.1109/SPIRE.2000.878178

2. Blum, C., Blesa, M.J., López-Ibáñez, M.: Beam search for the longest common subsequence problem. Comput. Oper. Res. **36**(12), 3178–3186 (2009)

3. Blum, C., Festa, P.: Longest common subsequence problems. In: Metaheuristics for String Problems in Bioinformatics, Chap. 3, pp. 45–60. Wiley (2016)

4. Blum, C., Raidl, G.R.: Hybrid Metaheuristics: Powerful Tools for Optimization. Springer (2016)

5. Bonizzoni, P., Della Vedova, G., Mauri, G.: Experimenting an approximation algorithm for the LCS. Discret. Appl. Math. **110**(1), 13–24 (2001)

6. Brisk, P., Kaplan, A., Sarrafzadeh, M.: Area-efficient instruction set synthesis for reconfigurable system-on-chip design, pp. 395–400. IEEE Press (2004)

7. Choi, C.Q.: DNA palindromes found in cancer. Genome Biol. **6**, 1–3 (2005). https://doi.org/10.1186/gb-spotlight-20050216-01

8. Chowdhury, S.R., Hasan, M.M., Iqbal, S., Rahman, M.S.: Computing a longest common palindromic subsequence. Fund. Inf. **129**(4), 329–340 (2014). https://doi.org/10.3233/FI-2014-974

9. Fraser, C.B.: Subsequences and Supersequences of Strings. Ph.D. thesis, University of Glasgow, Glasgow, UK (1995)

10. Giel-Pietraszuk, M., Hoffmann, M., Dolecka, S., Rychlewski, J., Barciszewski, J.: Palindromes in proteins. J. Protein Chem. **22**(2), 109–113 (2003)

11. Gusfield, D.: Algorithms on Strings, Trees, and Sequences. Computer Science and Computational Biology. Cambridge University Press, Cambridge (1997)

12. Hart, P., Nilsson, N., Raphael, B.: A formal basis for the heuristic determination of minimum cost paths. IEEE Trans. Syst. Sci. Cybern. **4**(2), 100–107 (1968)

13. Hasan, M.M., Sohidull Islam, A.S.M., Sohel Rahman, M., Sen, A.: Palindromic subsequence automata and longest common palindromic subsequence. Math. Comput. Sci. **11**, 219–232 (2017). http://link.springer.com/10.1007/s11786-016-0288-7

14. Huang, K., Yang, C., Tseng, K.: Fast algorithms for finding the common subsequences of multiple sequences. In: Proceedings of the IEEE International Computer Symposium, pp. 1006–1011. IEEE Press (2004)

15. Inenaga, S., Hyyrö, H.: A hardness result and new algorithm for the longest common palindromic subsequence problem. Inf. Process. Lett. **129**(supplement C), 11–15 (2018)

16. Jiang, T., Lin, G., Ma, B., Zhang, K.: A general edit distance between RNA structures. J. Comput. Biol. **9**(2), 371–388 (2002)

17. Larionov, S., Loskutov, A., Ryadchenko, E.: Chromosome evolution with naked eye: palindromic context of the life origin. Chaos: Interdiscip. J. Nonlinear Sci. **18**(1) (2008)

18. Maier, D.: The complexity of some problems on subsequences and supersequences. J. ACM **25**(2), 322–336 (1978)

19. Masek, W.J., Paterson, M.S.: A faster computing string edit distances. Theor. Comput. Syst. Sci. **20**, 18–31 (1980)

20. Mousavi, S.R., Tabataba, F.: An improved algorithm for the longest common subsequence problem. Comput. Oper. Res. **39**(3), 512–520 (2012)

21. Ow, P.S., Morton, T.E.: Filtered beam search in scheduling. Int. J. Prod. Res. **26**, 297–307 (1988)

22. Storer, J.: Data Compression: Methods and Theory. Computer Science Press (1988)

23. Tanaka, H., Bergstrom, D.A., Yao, M.C., Tapscott, S.J.: Large DNA palindromes as a common form of structural chromosome aberrations in human cancers. Human Cell **19**(1), 17–23 (2006)

Multi-objective Performance Measurement: Alternatives to PAR10 and Expected Running Time

Jakob Bossek[(✉)] and Heike Trautmann

Information Systems and Statistics, University of Münster, Münster, Germany
{bossek,trautmann}@uni-muenster.de

Abstract. A multiobjective perspective onto common performance measures such as the PAR10 score or the expected runtime of single-objective stochastic solvers is presented by directly investigating the tradeoff between the fraction of failed runs and the average runtime. Multi-objective indicators operating in the bi-objective space allow for an overall performance comparison on a set of instances paving the way for instance-based automated algorithm selection techniques.

Keywords: Algorithm selection · Performance measurement

1 Introduction

Benchmarking and comparisons of (single-objective) optimization algorithms strongly rely on adequate performance measurement of the respective solvers. However, assessing solver performance in general is not straightforward at all as usually multiple views and requirements have to be considered simultaneously such as minimizing runtime, minimizing function evaluations, maximizing quality, etc.. For stochastic solvers specifically, minimizing variability across runs or maximizing the number of successful runs might be of interest. Quite often though, only a single indicator or a single-objective combination of indicators is focussed in practice.

Common measures like PAR10 (e.g. [1], mostly in combinatorial optimization) and Expected Running Time (ERT [4], mostly in continuous optimization) for example try to combine several aspects into a single performance indicator while the core concept is the distinction between successful and unsuccessful runs as well as possible penalization of the latter. Usually, a run is denoted as successful if it solves an instance to optimality within a given time limit, e. g., for the Traveling Salesperson Problem (TSP).

In our approach we propose to address the tradeoff between minimizing the fraction of unsuccessful runs and the minimization of the average running time directly by treating it as a multi-objective optimization problem for which specific multi-objective techniques and indicators can be used to measure overall algorithm performance across an instance/benchmark set. This provides the

© Springer Nature Switzerland AG 2019
R. Battiti et al. (Eds.): LION 12 2018, LNCS 11353, pp. 215–219, 2019.
https://doi.org/10.1007/978-3-030-05348-2_19

basis for automated instance-based algorithm selection techniques based on the suggested performance measure generated by multi-objective techniques related to the idea of automated multi-objective configuration presented in [2]. Moreover, the underlying concept generalizes to other kinds of performance indicator combinations (also of higher degree) as well.

2   Performance Measurement

Common Approaches. In combinatorial optimization the so-called penalised average runtime (**PAR10**, e.g. [1]) score is a widely used hybrid performance measure. It is defined as the average of the runtimes with unsolved instances penalized with $10 \cdot T$ where T is the cutoff time. It is thus a combined measure of number of successful runs and average running time.

The Expected Running Time (**ERT**, [4]) basically measures the average number of function evaluations (across multiple runs of a solver on an instance) that are needed to solve it. Usually applied in single-objective continuous black-box optimization success here means that the resulting solution differs at most by a predefined precision value from the global optimum. The ERT is a weighted sum of the expected average running time of the succesful runs and the cutoff time T, while it is weighted by the fraction of failure and success probability.

Multi-Objective Perspective. Both widely used performance measures PAR10 and ERT implicitly address the two goals of maximizing probability of success and minimizing the expected running time of a solver. However, the actual tradeoff between those objectives is concealed by solely focussing on the aggregated performance measure. Moreover, often, high penalty values bias the performance analysis while the extent of the used penalty is more or less arbitrary chosen. We therefore propose an alternative bi-criteria performance measure PF^{MO}. Let T_A^s be the random variable that describes the running time of successful runs of algorithm A and $p_f = 1 - p_s$ the probability of failure, then

$$\mathrm{PF}^{MO} := (r_s = E(T_A^s), p_f).$$

Obviously, we aim to minimize both criteria, i. e., a "good" algorithm will both minimize the number of unsuccessful runs and simultaneously minimize the expected running time of successful runs. Note, that $\mathrm{PF}^{MO} \in [0, T] \times [0, 1]$ and $(0, 0)$ is the desirable ideal or utopia point. The measure may be depicted in a 2D scatterplot either for each instance and algorithm combination or aggregated over all k instances of the respective instance set (see Fig. 2 for an example).

Multi-Objective Assessment. Within the resulting two-dimensional space, we may adopt the concept of *Pareto-dominance* [3] in order to compare solver performances on and across instances. Assuming point labels reflecting algorithms and both components of PF^{MO} as axis labels in Fig. 1 we see that while algorithms A and B are non-dominated, i. e.,

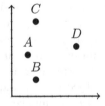

Fig. 1. Concept illustration.

A shows better average runtime, but worse failure rate than algorithm B and vice versa, e. g., A and B dominate C and D. Thus, we prefer algorithms that are located on low non-domination levels, ideally have a non-domination rank of one. By this means algorithms of the same rank become incomparable.

Moreover, the dominated Hypervolume (HV, [3,6]) of a point (i. e., its HV contribution) can be used to reflect a performance ranking of algorithms. It measures the size of the dominated space bounded by an anti-optimal reference point (which is implicitly given by $(T, 1)$). It is compliant with the Pareto dominance relation in that a lower nondomination rank leads to a higher HV value. HV contributions per algorithm can be aggregated over an instance set inducing a total order of algorithms.

3 Exemplary Illustration

This section is based on a benchmark study including feature-based automated algorithm selection of state-of-the-art inexact TSP solvers such as LKH, EAX as well as their restart variants and MAOS performed in [5]. Performances were measured in terms of PAR10 (using more robust median instead of mean for aggregation) on different representative kinds of instance sets (rue, tsplib, vlsi, netgen, morphed) and EAX+restart turned out to be the single best solver across all instance sets followed by LKH+restart. Figure 2 adopts the presented multi-objective view by showing the tradeoffs between both failure rate and average running time of successful runs.

In the aggregated version, i. e. averaged across the respective instance set per dimension, we see that EAX+restart clearly dominates the remaining solvers. However, differences to LKH and LKH+restart clearly are due to differences in runtime while both EAX and MAOS show substantially higher failure rates on average.

Table 1 shows summary statistics of the individual non-domination ranks and HV values of all algorithm performance results, i.e. of all points in the upper left subfigure of Fig. 2 allowing for an overall solver ranking across all instances by using averages across the instance set. Interestingly, the HV based ranking (see also bottom part of Fig. 2) is very much in line with the PAR10 based results in [5] while the non-domination ranks favor LKH+restart. However, differences in many cases are not statistically significant across the whole instance set but on several subsets such as e.g. netgen.

4 Conclusions

The bi-objective performance measure PF^{MO} as an alternative to PAR10 and ERT is presented by directly investigating the tradeoff between failure rate and average running time of successful runs. Thereby, the concept of Pareto dominance and multiobjective performance assessment in terms of dominated Hypervolume and non-dominated sorting offers very promising perspectives and new

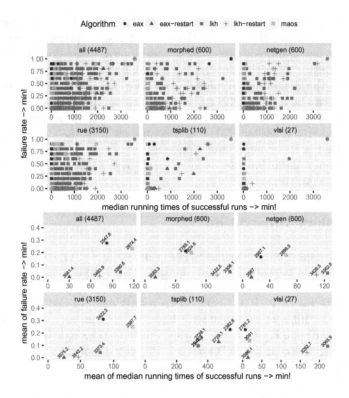

Fig. 2. Scatterplots of raw (all instances, top) and aggregated (r_S, p_f) vectors per instance set. Aggregation is performed componentwise, i. e., mean of failure rates and mean of running times of successful runs. The numbers indicate the unnormalized average dominated Hypervolume.

Table 1. Summary statistics of non-domination ranks and HV aggregated across all instances.

Algorithm	Avg. rank	SDev. rank	Avg. HV	SDev. HV
eax	2.25	1.05	2599.71	1007.71
eax-restart	2.28	1.03	**3569.18**	**178.45**
lkh	2.18	1.08	3382.47	603.23
lkh-restart	**2.03**	**0.97**	3510.41	360.99
maos	3.50	1.16	2749.20	1004.75

insights into algorithm behaviour. However, the core concept generalizes to other kinds of indicator combinations as well.

First conceptual studies of PF^{MO} on a TSP benchmark of inexact solvers hint at interesting aspects of solver behaviour which will be further analysed in future studies together with thoroughly comparing properties of PAR10, ERT and PF^{MO}.

Acknowledgements. The authors acknowledge support from the European Research Center for Information Systems (ERCIS) and the DAAD PPP project No. 57314626.

References

1. Bischl, B. et al.: ASlib: a benchmark library for algorithm selection. Artif. Intell. J. **237**, 41–58 (2016). https://doi.org/10.1016/j.artint.2016.04.003
2. Blot, A., Hoos, H., Jourdan, L., Marmion, M., Trautmann, H.: In: Joaquin, V. et al. (ed.) MO-ParamILS: A multi-objective automatic algorithm configuration framework, pp. 32–47. Springer International Publishing, Ischia (2016)
3. Coello Coello, C., Lamont, G.B., van Veldhuizen, D.: Evolutionary Algorithms for Solving Multi-objective Problems. Springer, Berlin (2007)
4. Hansen, N., Auger, A., Finck, S., Ros, R.: Real-Parameter Black-Box Optimization Benchmarking 2009: Experimental Setup. Technical Report RR-6828, INRIA (2009). https://hal.inria.fr/inria-00362649v3/document
5. Kerschke, P., Kotthoff, L., Bossek, J., Hoos, H.H., Trautmann, H.: Leveraging TSP solver complementarity through machine learning. Evol. Comput. **0**(0), 1–24 (2017). https://doi.org/10.1162/evco_a_00215, pMID: 28836836
6. Zitzler, E., Thiele, L.: Multiobjective evolutionary algorithms: a comparative case study and the strength Pareto approach. IEEE Trans. Evol. Comput. **3**(4), 257–271 (1999)

Algorithm Configuration: Learning Policies for the Quick Termination of Poor Performers

Daniel Karapetyan[1(✉)], Andrew J. Parkes[2], and Thomas Stützle[3]

[1] Institute for Analytics and Data Science, University of Essex, Colchester, UK
daniel.karapetyan@gmail.com
[2] ASAP Research Group, School of Computer Science, University of Nottingham, Nottingham, UK
[3] IRIDIA, CoDE, Universit Libre de Bruxelles (ULB), Brussels, Belgium

Abstract. One way to speed up the algorithm configuration task is to use short runs instead of long runs as much as possible, but without discarding the configurations that eventually do well on the long runs. We consider the problem of selecting the top performing configurations of Conditional Markov Chain Search (CMCS), a general algorithm schema that includes, for example, VNS. We investigate how the structure of performance on short tests links with those on long tests, showing that significant differences arise between test domains. We propose a "performance envelope" method to exploit the links; that learns when runs should be terminated, but that automatically adapts to the domain.

1 Introduction

Careful configuration of algorithms can lead to a significant improvement in performance [1, and many others]. This is usually done by searching in the space of configurations and evaluating each configuration on a set of target instances. However, such instances are often large and will require long runs, so direct and complete usage of such intended instances problems will be overly time-consuming. A natural desire is that, in a justified fashion, we should be able to reduce the run times by exploiting the results of "short runs" in order to configure for "long runs": There is a need to learn how to extrapolate from "short" to "long". This suggests that machine learning methods should be applied to collections of such "short-run data", to analyse patterns, and so produce predictions for the performance in the long runs. This view suggests that for algorithm configuration at least 3 different 'generic' spaces are relevant:

- C "Configuration space" – the direct parameters of the algorithms.
- S "Short-run space" – the space of (detailed) results using short runs.
- L "Long-run space" – results from long runs.

© Springer Nature Switzerland AG 2019
R. Battiti et al. (Eds.): LION 12 2018, LNCS 11353, pp. 220–224, 2019.
https://doi.org/10.1007/978-3-030-05348-2_20

A common procedure would be to do a (local) search in C-space, using fitnesses obtained from L-space. In this context, the usage of machine learning might be to develop a mapping from the C-space to the L-space, e.g. see SMAC [1]. In this paper, we instead study the potential for learning the mappings from S-space to L-space – aiming to exploit how short runs are able to predict longer term behaviours. The goal is to use such information to optimise the policies for when a trial of a particular configuration should be terminated – because there is high confidence it will not lead to a good final solution.

2   Experimental Setup

We used CMCS, a recent framework that defines the behaviour of a multi-component optimisation algorithm with a set of numeric parameters [2,3]. We used three problem domains: the Simple Plant Location Problem (SPLP) [2], the Far From Most String Problem (FFMSP) (the details of our components, testbed, etc. are not yet published) and the Bipartite Boolean Quadratic Problem (BBQP) [3]. We generated all 'meaningful' 3-component configurations with deterministic control mechanism, thus ending up with a finite number of configurations. We do not include details of exactly what these mean (see [2,3]), as for the purposes of this paper, these can simply be regarded as a categorical set of potential options, and defining the "C-space". The goal of the work is to find configurations that are the best performers, with respect to the long-run L-space, but exploiting their properties with respect to the short-run S-space in order to reduce the overall time budget.

The full time budget for each 'long' run was selected as 1024ms. Ten random instances were generated for each domain, with the size chosen to make them hard enough for this time budget. We then generated performance data[1] for each domain, by solving each of the ten instances by each of the configurations, and recording the solution quality at 1ms, 2ms, 4ms, ..., 1024ms. Objective values were scaled to $[0, 1]$ for each instance, and then averaged over the data instances. Hence, quality 0 (resp. 1) means that, for each instance, the configuration yielded solution as good (resp. bad) as any other configuration within the full 1024ms time budget.

In results for SPLP in Fig. 1(a), the solid line shows the Performance Profile (PP) of the top configuration, i.e. how the solution quality of the best-in-L-space configuration improves over time. The other two lines, which we call *cutoff lines*, are from aggregating PPs of a set of configurations. To obtain an $x\%$ cutoff line, we select the top-$x\%$ of the configurations in the S-space and then produce the cutoff line as a combined worst-case PP for all those configurations. (All three lines are monotonic, as CMCS records the best-so-far solutions.) As hoped, there is a strong correlation between the short- and long-run performance of the top configurations; specifically, the cutoff lines drop relatively quickly, suggesting that there is a potential for a significant speed up by terminating configurations

---

[1] http://csee.essex.ac.uk/staff/dkarap/?page=publications&
key=KarapetyanParkesStuetzlenst2018.

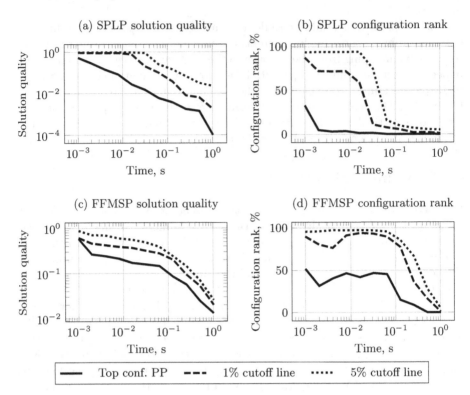

(a) SPLP solution quality   (b) SPLP configuration rank

(c) FFMSP solution quality   (d) FFMSP configuration rank

—— Top conf. PP   - - - 1% cutoff line   ····· 5% cutoff line

Fig. 1. Solution quality and configuration rank as they change throughout the run.

that perform poorly in the 'long' runs. To evaluate this, we plotted the ranks corresponding to the lines in Fig. 1(b). The lowness of the solid line indicates the top configuration was among the top performers throughout the run. The drop in the cutoff lines indicates that the runs longer than ∼20ms are likely to be useful in early prediction of the 'long' performance. E.g., at 32 ms, the 1% cutoff line potentially allows us to rule out 89% of all the configurations. Short-run performances do link to long-runs; hence, we can use this to quickly terminate configurations that are not likely to have good L-space performance. However, Fig. 1(c,d), showing the results for the FFMSP domain, demonstrate that in other domains one may need longer short-runs to predict 'long' performance.

Finding the exact cutoff lines in advance is impractical, as it requires long runs for all the configurations. The heuristic below obtains and exploits a reasonably reliable 1% cutoff line much quicker. Specific parameter values used are given, but are changeable.

1. Randomly select 1% out of all the configurations and place them into a pool P and run full-time tests for each $c \in P$. Generate the cutoff line as the combined worst case PP for all $c \in P$.
2. First pass: For each previously untested configuration c, start a full-time test and gradually build its PP. If at any of 1ms, 2ms, 4ms, ..., 512ms its PP

rises above the cutoff line by more than 20%, terminate the run. However, if c survives early termination, add c to P and remove the worst-performing (in the L-space) configuration from P; update the cutoff line.

3. Second pass: repeat Step 2, but scanning only through the configurations that were terminated early. This pass helps recovery from the potential bias at the beginning of the first pass if the initial cutoffs were too tight.

4. Return P as an approximation of the top 1% of all the configurations.

Table 1. Accuracy of the cutoff approximation algorithm.

Domain	#conf.	Speed up	Overlap (top conf.)	Overlap (1%)
SPLP	26,608	9.4	100%	99.3%
FFBSP	8,064	2.1	100%	99.9%
BBQP	9,860	9.2	100%	98.7%

Table 1 gives experimental results. (We did not report the results for BBQP in Fig. 1 due to lack of space). The 'Speed up' column tells how much quicker the heuristic is compared to evaluating all the configurations in S-space. The 'Overlap (top conf.)' tells how often the heuristic finds the best performing configuration (in 100 experiments), and the 'Overlap (1%)' column tells the average overlap between the true top 1% configurations and the ones found by our heuristic. Note the speed up factor significantly depends on the domain; the SPLP and BBQP domains gave more than 9x speed up, though in FFMSP, the gain is much more modest. In all the domains, our heuristic procedure successfully found the best performing configuration every time, and was 99% accurate in finding the top-1% of configurations.

3 Conclusions and Future Work

The PPs arising from CMCS were shown to have sufficient structure in their behaviours such that a "performance envelope/cut-off" could be constructed to effectively determine when terminating a test run was safe, in that most of the good configurations would be found. Behaviours of the PPs differed between domains, suggesting that dynamic adaptive methods are needed. We gave a heuristic method for this, that automatically strengthens the cut-offs as more PPs are collected, and so reduces overall runtime.

The work here only used a PP of a simple linear aggregate over a set of different test instances. However, future extensions should consider "Performance Trajectories"; using the entire, time-dependent vector of the performances over the set of test instances. Initial explorations have taken such performance vectors at a given "short time point" and then considered them as feature vectors with labels given by the ultimate aggregate quality with a long runtime. For SPLP, standard classification methods did identify the regions in the S-vector space that lead to longer term good performance. This suggests that information in performance trajectories is also available that can be extracted by machine learning in order to classify (or cluster) behaviours and so potentially be used to optimise policies for when tests on configurations can be terminated.

References

1. Hutter, Frank, Hoos, Holger H., Leyton-Brown, Kevin: Sequential model-based optimization for general algorithm configuration. In: Coello, Carlos A.Coello (ed.) LION 2011. LNCS, vol. 6683, pp. 507–523. Springer, Heidelberg (2011). https://doi.org/10.1007/978-3-642-25566-3_40
2. Karapetyan, D., Goldengorin, B.: Conditional Markov Chain Search for the simple plant location problem improves upper bounds on twelve Korkel-Ghosh instances. In: Goldengorin B. (ed.) Optimization Problems in Graph Theory, pp. 123–147. Springer (2018). https://doi.org/10.1007/978-3-319-94830-0
3. Karapetyan, D., Punnen, A.P., Parkes, A.J.: Markov chain methods for the bipartite boolean quadratic programming problem. Eur. J. Oper. Res. **260**(2), 494–506 (2017). https://doi.org/10.1016/j.ejor.2017.01.001

Probability Estimation by an Adapted Genetic Algorithm in Web Insurance

Anne-Lise Bedenel[1,2,3(✉)], Laetitia Jourdan[2], and Christophe Biernacki[3]

[1] MeilleureAssurance, Lille, France
anne-lise.bedenel@meilleureassurance.com
[2] Université Lille 1 CRIStAL, UMR 9189, Villeneuve-d'Ascq, France
laetitia.jourdan@univ-lille1.fr
[3] Inria, Université Lille 1, Villeneuve-d'Ascq, France
christophe.biernacki@inria.fr,christophe.biernacki@math.univ-lille1.fr

Abstract. In the insurance comparison domain, data constantly evolve, implying some difficulties to directly exploit them. Indeed, most of the classical learning methods require data descriptors equal to both learning and test samples. To answer business expectations, online forms where data come from are regularly modified. This constant modification of features and data descriptors makes statistical analysis more complex. A first work with statistical methods has been realized. This method relies on likelihood and models selection with the Bayesian information criterion. Unfortunately, this method is very expensive in computation time. Moreover, with this method, all models should be exhaustively compared, what is materially unattainable, so the search space is limited to a specific models family. In this work, we propose to use a genetic algorithm (GA) specifically adapted to overcome the statistical method defaults and shows its performances on real datasets provided by the company MeilleureAssurance.com.

Keywords: Genetic Algorithms · BIC · Insurance
WEB

1 Introduction

The objective of online insurance comparators is to propose to web users the offer the most adapted to their expectations, according to their profiles. Most of the online insurance comparators compare with only one criterion: the price. To improve web users comparison, the company MeilleureAssurance.com wishes to create a model allowing predicting the best offer according to web user profiles. It is a classical objective in statistics but, with the functioning of an insurance comparator, standard methods cannot be used. To do an online comparison, a web user has to fill a form of questions. When the form is filled, data are sent to insurer partners with a web service. So, they can send the real price of the offer back to the company. An insurance comparator adapts and changes regularly its forms:

© Springer Nature Switzerland AG 2019
R. Battiti et al. (Eds.): LION 12 2018, LNCS 11353, pp. 225–240, 2019.
https://doi.org/10.1007/978-3-030-05348-2_21

- For insurers: Each insurer has his particularly pricing system, questions are not homogeneous between all insurers partner.
- For web users: For more clarity and simplicity, questions are regularly adapted.

This adaptability is a specificity of insurance comparators. Due to this specificity, building a supervised classification model with these features becomes complex.

A first work has been realized to solve this problem, using several statistical tools as the likelihood, to estimate the parameters. In this first work, the modeling realized shows many constraints and involves a problem of model selection [1]. The model selection is realized with an exhaustive search. Indeed, with the statistical process, the selection model is performed by comparing all available models with an information criterion [2]. This method shows good results for the estimation and the model selection. However, it is time-consuming and the number of models to compare has to be limited. To avoid the exhaustive search, an optimization approach is considered.

In the literature, many papers can be found where a genetic algorithm is used to do a model selection [8–11] and parameters estimation [12]. In this work, we propose a new genetic algorithm to overcome defaults of statistical methods and new operators to have the best metaheuristics. Section 2 describes the statistical work with the modeling problem. Section 3 presents the algorithm used and new operators proposed. Section 4 drives experiments and compares results between statistical method and genetic algorithm on simulated and real datasets. Section 5, gives some conclusions and perspectives for future works.

2   Modeling of the Problem

2.1   Probabilistic Modeling

To introduce the general modeling, a use case is studied where the question is: "How web users react when data descriptors of feature change?"

To answer to this question, the feature *Coverage levels* is studied.

In a first time, this feature had four data descriptors {Third-party (T), Third-party++ (T++), third-party, fire and theft (TPFT), comprehensive (C)}. This feature denoted by $X \in \{1, \ldots, I\}$, I designating the number of data descriptors ($I = 4$ in our use case). In a second time, it has been decomposed into seven data descriptors: {Third-party (T), Third-party+ (T+), Third-party++ (T++), third-party, fire and theft (TPFT), comprehensive (C), comprehensive+ (C+), comprehensive++ (C++)}. This feature is symmetrically denoted by $Y \in \{1, \ldots, J\}$, J designating the new number of data descriptors ($J = 7$ in our use case).

Another specificity for an insurance comparator is that there is no data history of web users. So, the available observations on X and Y are never matched. More precisely, this property can be written like this:

Period before the modification N^- Web users have filled the feature X, so there are *observed* realisations $\mathbf{X}^- = (X_1^-, \ldots, X_{N^-}^-)$. The feature Y has never been filled, so there are *unobserved* realisations $\mathbf{Y}^- = (Y_1^-, \ldots, Y_{N^-}^-)$.

Period after the modification Symmetrically, N^+ Web users have filled the feature Y, so there are *observed* realisations $\mathbf{Y}^+ = (Y_1^+, \ldots, X_{N+}^+)$. The feature X has never been filled, so there are *unobserved* realisations $\mathbf{X}^+ = (X_1^+, \ldots, X_{N+}^+)$.

2.2   Parameters Estimation and Models Selection

We assume each couple (X_n^*, Y_n^*) is an independent and identically distributed realization of the couple (X, Y) with $n = 1, \ldots, N^*$ and $* \in \{-, +\}$. The distribution of the couple (X, Y) can be written as:

$$P(X = i, Y = j) = p_{ij} p_i \tag{1}$$

where $p_{ij} = P(Y = j | X = i)$ and $p_i = P(X = i)$, with $i = 1, \ldots, I$ and $j = 1, \ldots, J$. The interest, here, is to show the transition probabilities p_{ij} between data descriptors X and Y. The objective is to estimate the whole of transition probabilities $\mathbf{p}_{..} = (p_{ij})$. It can be noted $\mathbf{p}_{.} = (p_i)$, which is also an unkown parameter.

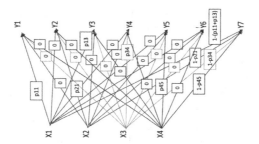

Fig. 1. Graph of possible matching between X- feature before modification and Y-feature after modification.

The set of transition (matching) probabilities $\mathbf{p}_{..}$ is introduced with the graph displayed on Fig. 1, where the oriented edges represent estimated parameters in the use case. It can be noted that the number of parameters $\mathbf{p}_{..}$ is larger than the number of parameters of Y distribution. So, the model is statistically over-parameterized and therefore has multiple solutions whose range can be found through repeated optimization from different starting points. More precisely, in the use case, there are 28 matching probabilities (so 24 free parameters). However, there are only 6 free parameters for Y distribution. So the model is said unidentifiable. To have an identifiable model, some constraints have to be imposed on $\mathbf{p}_{..}$ to limit the number of free parameters to 6 or less ($\dim(\mathbf{p}_{..}) \leq J - 1$). To respect the identifiability constraint, it has been proposed to fix some value of $\mathbf{p}_{..}$ to 0. This type of constraint will be a model noted $\mathbf{m}$. So, it leads to a set of models $\mathcal{M} = \{\mathbf{m}\}$ and they will be challenged. For one model $\mathbf{m}$, parameters $\mathbf{p}_{..}$ are estimated with log-likelihood maximization of observed data [3] defined by:

$$\ell_{\mathbf{m}}(\mathbf{p}_{..}, \mathbf{p}_{.}; \mathbf{X}^-, \mathbf{Y}^+) = \sum_{j=1}^{J} N_j^+ \ln \left(\sum_{i=1}^{I} p_{ij} p_i \right) + \sum_{i=1}^{I} N_i^- \ln p_i \qquad (2)$$

where $N_i^- = \#\{X_n^- = i, n = 1, \ldots, N^-\}$ and $N_j^+ = \#\{Y_n^+ = j, n = 1, \ldots, N^+\}$. With this maximization of log-likelihood, estimators of $\mathbf{p}_{.}$ and $\mathbf{p}_{..}$ (respectively $\hat{\mathbf{p}}_{.}$ and $\hat{\mathbf{p}}_{..}$) are obtained. To choice the best model $\mathbf{m}$ in the set $\mathcal{M}$, the conditional BIC criterion (Bayesian Information Criterion) [2], given by:

$$\mathrm{BIC}_{\mathbf{m}} = -2\ell_{\mathbf{m}}(\hat{\mathbf{p}}_{..}, \hat{\mathbf{p}}_{.}; \mathbf{Y}^+ | \mathbf{X}^-) + \nu_{\mathbf{m}} \ln N^+ \qquad (3)$$

where $\nu_{\mathbf{m}} = \dim(\mathbf{m})$ is the number of free parameters for the model $\mathbf{m}$, is used. The model $\hat{\mathbf{m}}$ having the lowest value of BIC criterion is selected. This method gives good results for the estimation and model selection [1]. However, it is an exhaustive method that is time-consuming. Indeed, the exhaustive method handles two problems. The first one is to find values of estimated parameters. The second one is to select the best model. So, in the exhaustive method, there is a continuous problem (estimation of parameters) for each model and a combinatorial problem (models selection). For example, in the use case, the feature X has 4 levels and Y has 7 levels. Therefore, there are $\binom{6*4}{6} = 134\,596$ possible models. For each model, probability values have to be found. So there are $134\,596$ continuous problems where probability values have to be found. All models have to be compared to select the best one. Among these possibilities, some of them are not allowed. Indeed, each X level and Y level has to be joined. Consequently, the whole of possibilities is reduced. However, it is not enough to do the exhaustive method. To do it, only a specific model family is compared which is defined by:

- The number of parameters is fixed and equal to the number of Y levels - 1.
- The parameters are probabilities, so for each X_i levels, the sum has to be equal to one. This constraint imposes, for each X_i levels, the last level of Y has to be equal to $(1 - \sum_{j=1}^{J} \mathbf{p}_{ij})$. Figure 1 illustrates this. So the last level of Y is fixed as no free parameters. It cannot be set to 0 and is not an estimated parameter.

Even with these constraints, the exhaustive method stays time-consuming. For example, in the use case, to do the estimation and the comparison of $4\,095$ models, the process needs 1h07. In a business context, it is very long. Moreover, the firm MeilleureAssurance.com could have features with more levels, so with more parameters to estimate and a larger combinatorial problem. The objective is to reduce the computation time to do the estimation and the comparison. To reach this objective, it can be interesting to perform a non-exhaustive search. The problem involves having a flexibility according to constraints and the objective is to have a high-quality solution. A stochastic method responds to these expectations. Moreover, according to the problem, a metaheuristic based on population (P-Metaheuristic), and more especially an evolutionary method, is adapted. To challenge the statistical method, we propose to use a genetic algorithm [16].

3   Algorithms

The use of a genetic algorithm will allow obtaining a good solution quickly. Moreover, a genetic algorithm is particularly adapted to real problems. To have a better speed of convergence, a steady-state algorithm (ssGA) [15] is also used.

3.1   Encoding and Evaluation

A genetic algorithm is defined by a potential solution, a population, an environment (search space) and a fitness function.

Definition of the Solution. To define the solution, the probabilistic modeling is used. A model **m** is a set of transition probabilities $p_{..}$ where some of them are estimated or fixed to 0. Work with probabilities involve: each estimated probability $p_{..} \in [0, 1]$ and is a real number. A solution corresponds to a model. Figure 2 shows an example of representation of model **m** and Fig. 3 his matrix form. In the solution, to each X level, the sum of probabilities has to be equal to 1. Figure 3 shows estimated probabilities of X for the level T which are 0.6 and 0.4.

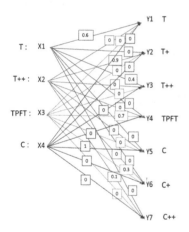

$X \backslash Y$	T	T+	T++	TPFT	C	C+	C++
T	0.6	0	0.4	0	0	0.	0
T++	0	0.9	0	0	0	0	0.1
TPFT	0	0	0	0.7	0	0.3	0
C	0	0	0	0	1	0	0

Fig. 2. Graph of possible matching between X- feature before modification and Y- feature after modification for a fixed model **m**.

Fig. 3. Matrix of transition corresponding to the model **m** between the levels X and Y.

Population. The population is composed of available models in the search space: the set of models $\mathcal{M}$. Contrary to the statistical method, the set of models $\mathcal{M}$ does not need to be reduced.

Fitness Function. To evaluate the solution (model), the BIC criterion defined by Eq. 3 is used as the fitness function which has to be minimized.

3.2   Operators

A genetic algorithm relies on three operators that are: selection, crossover and mutation. The choice of these operators depends on encoding parameters and for this problem a real encoding is used. The selection will be performed by a classical operator. To perform the crossover and mutation, the choice is more complex and these operators have to be adapted to the problem. Indeed, models have many 0 and the statistical process shows that the place of all 0 is a real information. The first objective will be to design operators adapted to manage the 0 information.

Selection Operator. The selection will be performed with a classical operator: the Binary Tournament Selection [6]. This operator selects randomly two solutions of population $\mathcal{M}$. In a second time, it evaluates solutions with BIC_m and selects the best solution $\hat{m}$ which is the solution with lowest BIC criterion. So, we have :

$$\hat{m} = \text{argminBIC}_m \qquad (4)$$

Crossover Operators. Two crossover operators relying on real encoding are compared to find the most efficient:

- Uniform crossover [5]: The uniform crossover operator is very simple, according to the crossover probability, each parameter can be crossed with a probability 0.5.
- SBX crossover (Simulated Binary Crossover) [4,13]: The SBX crossover is an operator which adapts to the evolution algorithm according to the fitness function of parents and their offspring. From two parents $p_1(i)$ and $p_2(i)$, this crossover generates two offspring $c_1(i)$ and $c_2(i)$ by the following relation:

$$\begin{cases} c_1(i) = 0.5[(1+\beta)p_1(i) + (1-\beta)p_2(i)] \\ c_2(i) = 0.5[(1-\beta)p_1(i) + (1+\beta)p_2(i)] \end{cases}$$

where β is a spread factor given by:

$$\beta = \begin{cases} (2u)^{\frac{1}{\eta+1}} & \text{if } u < 0.5 \\ (\frac{1}{2(1-u)})^{\frac{1}{\eta+1}} & \text{otherwise.} \end{cases}$$

where u is random number uniformly generated on the interval [0,1] and η a non-negative real parameter.

The constraints with parameters fixed to 0 in the probabilistic modeling involve that these operators have to be adapted. Indeed, in the problem, there are many 0. How handle them? Two solutions are proposed. In the first solution, the crossover is applied if and only if parameters of two parents are estimated else there is no crossover. In the second solution, the crossover is applied without constraints, so a 0 can become an estimated parameter and vice versa.

X\Y	T	T+	T++	TPFT	C	C+	C++
T	0	0	0	0	0.9	0.1	0
T++	0.19	0.07	0.73	0	0	0	0
TPFT	0	0.15	0	0.27	0	0	0.57
C	0	0	0.024	0.097	0	0	0

Parent 1

X\Y	T	T+	T++	TPFT	C	C+	C++
T	1	0	0	0	0	0	0
T++	0	0.25	0.4	0	0	0	0.35
TPFT	0	1	0	0	0	0	0
C	0	0.46	0.17	0.21	0.32	0.15	0

Parent 2

X\Y	T	T+	T++	TPFT	C	C+	C++
T	0	0	0	0	0.9	0.1	0
T++	0.19	0.07	0.73	0	0	0	0
TPFT	0	**0.97**	0	0.27	0	0	0.57
C	0	0	**0.018**	**0.16**	0	0	0

Offsprings generated after crossover without 0 changed

Fig. 4. Example of crossover keeping 0

The example Fig. 4, shows the process of the first solution. An SBX crossover is used and adapted to not to cross when the parameter is a 0. In the table showing the offspring, parameters (in bold) has been crossed. On the two first X levels (T and T++), there are no changes. On these two lines, only two parameters can be moved with a probability equals to 0.5 because these are estimated probabilities for both parents (1 and 2) at the same place. On the line of X level TPFT of the parent 2, the crossover operator can be applied only on the parameter equals to 1. In the last line (the X level C), two parameters estimated have been crossed because they are in common. This example shows when one of two parents has a 0, there is no crossover.

Table 1. Offsprings generated after crossover and correction operator

X\Y	T	T+	T++	TPFT	C	C+	C++
T	0	0	0	0	0.9	0.1	0
T++	0.19	0.07	0.73	0	0	0	0
TPFT	0	**0.54**	0	**0.15**	0	0	**0.31**
C	0	0	**0.1**	**0.9**	0	0	0

For each level of X, the sum of probabilities has to be equal to 1. As shown on Table 4, the crossover does not respect this constraint. Indeed, for the two parents, if the sum on each X level is performed, it will be equal to 1. But, for the offspring, the sum of probabilities of level TFPT or the level C is not equal to 1. For the level FPFT, it is equal to $0.97 + 0.27 + 0.57 = 1.8$. So to keep the sum of probabilities equals to 1 for each level of X, a correction operator (Algorithm 1) has been created.

Algorithm 1: Correction algorithm

Data: integer sup, i, j;
float difference, sum;
float value;
for *Each X levels* **do**
  | sum=0;
  | for *Each Y levels* **do**
  | | sum += Value of solution to the index(i+j*Y);
  | end
  | if *(sum ≥ 1)* **then**
  | | sup=1;
  | else
  | | sup=0;
  | end
  | difference = Math.abs(sum-1);
end
for *Each X levels* **do**
  | value = Value of solution to the index(i+j*Y);
  | if *(sup == 1)* **then**
  | | set the value by (value-difference*value/sum);
  | else
  | | set the value by (value+difference*value/sum);
  | end
end

For each X level, this operator sums probabilities. If the sum is equal to 1, it does nothing, else it adjusts the estimated parameters to have a sum equal to 1. Table 1 shows the offspring after the application of the correction operator. Parameters in bold are the parameters where a correction has been applied, so for the levels TPFT and C. After the correction, the sum of the parameters for the level TPFT is equal to $0.54 + 0.15 + 0.31 = 1$. It is the same for the level C. It can be noticed that each parameter is adjusted proportionally to his initial value. Two solutions have been compared to find the most relevant in the treatment of 0. The first one was to not cross the parameters equal to 0. On the contrary, the second one was to let the possible crossing on all the parameters.

Figure 5 shows the process of the second solution. As for the first solution, an SBX crossover is used but for this solution, it is allowed to apply on all parameters. The bold parameters in the table representing the offspring obtained correspond to the parameters where the crossover operator has been applied. The first parameter of level T of the feature X shows the difference with the first solution. Indeed, whereas the parameter is a 0 for the parent 2, the crossover operator has been applied and so this parameter becomes a 0 for the offspring. It is the same for the other parameters. In this solution, the information on the place of the parameters equal to 0 is not kept. As for the first example, a

$X\backslash Y$	T	T+	T++	TPFT	C	C+	C++
T	0.011	0	0.97	0	0.002	0.016	0
T++	0	0	0	0.62	0.14	0	0.23
TPFT	0	0.67	0	0.33	0	0	0
C	0	0	0	1	0	0	0

Parent 1

$X\backslash Y$	T	T+	T++	TPFT	C	C+	C++
T	0	0.29	0.48	0.21	0	0.02	0
T++	0	0	1	0	0	0	0
TPFT	1	0	0	0	0	0	0
C	0	0.002	0	0.03	0.87	0	0.093

Parent 2

$X\backslash Y$	T	T+	T++	TPFT	C	C+	C++
T	0	0.29	0.95	0.013	0.002	0.016	0
T++	0	0	0	0.021	0.15	0	0.23
TPFT	0	0.01	0	0.016	0	0	0
C	0	0	0	1	0.07	0	0.1

Offsprings generated after crossover with 0 changed

Fig. 5. Example of crossover with chaging 0 allowed

correction operator is applied to have the sum of probabilities of each X level equals to 1.

Mutation Operator. For the mutation operator, a polynomial mutation [13, 14] is used. In this operator, a polynomial probability distribution is used to perturb a solution in a parent's vicinity. The probability distribution in both left and right of a variable value is adjusted. So that no value outside the specified range [a, b] is created by the mutation operator. In the polynomial mutation the offspring is generated as follows:

$$x'_i = x_i + (x^u_i - x^L_i)\delta_i \tag{5}$$

where x^u_i (resp. x^L_i) represents the upper bound (resp. lower bound) for x_i. The parameter δ_i is computed from the polynomial probability distribution: $p(\delta) = 0.5(\eta_m + 1)(1-|\delta|^{\eta_m})$

$$\delta_i = \begin{cases} (2r_i)^{\frac{1}{\eta_m+1}} & \text{if } r_i < 0.5 \\ 1 - (2(1 - r_i))^{\frac{1}{\eta_m+1}} & \text{otherwise.} \end{cases}$$

where η_m is the distribution index and r_i is a random number in [0,1].

4 Experiments

4.1 Experimental Protocol

To compare 8 genetic algorithms obtained with different crossover and mutation operators, each meta-heuristic is run on the same simulated dataset. The simulated dataset is generated with the programming language R and the function Rmultinom(see Sect. 4.1). Each metaheuristic is implemented in JAVA with

JMETAL platform [7] on a Linux machine with processor Intel Core i5-4590 CPU 3.3GHz*4 and 4 GO of RAM. Each genetic algorithm is stopped after the same maximum number of iteration and is run 25 times. For each execution of the genetic algorithm, the best solution found is saved and stored. The average of 25 best solutions stored is calculated for each meta-heuristic. A statistical test of average comparison (Kruskal Wallis) is performed to compare the performance of these metaheuristics. This test has been chosen because the samples are small (25) and independent. If the test is significant, a left unilateral test of Mann Whitney will be performed to have the metaheuristics that have an average significantly lowest. A boxplot of each meta-heuristic is also realized. The meta-heuristics with the lowest average and the most efficiency is selected to be compared with the results of the statistical process.

To compare the performance of the best meta-heuristic selected to the statistical method, several simulated datasets have been generated. They are generated by the same process that for the comparison of the 8 metaheuristics. 2 real datasets are also used to compare the results. On each dataset, metaheuristic selected is run once and is stopped after a number of maximum iterations. The best solution is saved and the BIC value is compared to the BIC value found by the statistical process. Moreover, the place and the value of the estimated parameters is also studied to verify the simulated solution is found for both statistical method and genetic algorithm. In the simulated datasets, the place and value of the parameters are known, as the model to find is simulated. Contrary, to the real datasets where only the data in X and Y are known, and the model to find is unknown.

4.2   Description of Instances

Experiments are computed using 10 simulated datasets and 2 real datasets. Simulated datasets are generated with the software R and the function Rmultinom. This function allows generating datasets according to their multinomial distribution. Table 2 shows an example of data representation for simulated dataset. In this example, there are 10 000 couples (X, Y). As it is a simulated model, transition probabilities and the place of 0 are known. Table 3 shows probabilities associated to Table 2.

Table 2. Matrix of simulated datasets.

$X \backslash Y$	1	2	3	4	5	6	7
1	2122	1260	0	0	0	0	0
2	0	0	961	0	0	0	0
3	0	0	0	1975	0	0	0
4	0	0	0	0	3529	143	10

Table 3. Matrix of probabilities of simulated datasets.

$X \backslash Y$	1	2	3	4	5	6	7
1	0.63	0.37	0	0	0	0	0
2	0	0	1	0	0	0	0
3	0	0	0	1	0	0	0
4	0	0	0	0	0.96	0.038	0.002

Table 4 details information about the different datasets used for experiments. The dataset DS3 is used for the comparison of 8 metaheuristics. In these experiments, there are two real datasets (CoverageLevels, DriverStatut) that come from the company MeilleureAssurance.com. The dataset **CoverageLevels** corresponds to the use case used for the modeling.

Table 4. Instances description. The size of the feature X $|X|$ and Y $|Y|$ (i.e., number of observations), the number of levels of X ($\#X_i$) and Y ($\#Y_j$), the number of parameters to estimate ($\#p_{..}$) and the number of models compared with the statistical process $\#\mathcal{M}_s$

Name	$|X|$	$|Y|$	$\#X_i$	$\#Y_j$	$\#p_{..}$	$\#\mathcal{M}_s$
DS1	15000	15000	4	7	6	4095
DS2	11035	10035	4	7	6	4095
DS3	10000	10000	4	7	6	4095
DS4	5000	5000	4	7	6	4095
DS5	15000	15000	3	5	4	81
DS6	10000	10000	3	5	4	81
DS7	5000	5000	3	5	4	81
DS8	15000	15000	3	4	3	26
DS9	7880	7880	3	4	3	26
DS10	5000	5000	3	4	3	26
CoverageLevels	11441	8668	4	7	6	4095
DriverStatut	7438	8238	3	5	4	81

4.3   Parameters

Different parameter settings were studied before deciding which one to use for the final experiments.

All parameters have been chosen experimentally. To choose the number of maximum iteration a study of the convergence of the algorithm has been performed. The maximum iteration number and the size of the population differ according to the size of the search space. Table 5 shows parameters involved in this study.

4.4   Sensitivity Analysis of the Operators

The 8 metaheuristics compared are described as follows:

– C1Ua: A uniform crossover, with the first solution where the crossover is allowed only on estimated parameters, the polynomial mutation is realized on all estimated parameters according to the mutation probability.

Table 5. Genetic algorithm parameters

Parameters	Value
Number of maximum iteration for the comparison of metaheuristics	50 000
Number of maximum iteration for datasets (DS1,DS2,DS3,DS4 and CoverageLevels)	100 000
Number of maximum iteration for the other datasets	10 000
Size of population for datasets (DS1,DS2,DS3,DS4 CoverageLevels)	15 000
Size of population for the other datasets	500
Crossover probability	0.8
Mutation probability	0.8
Crossover index distribution	20
Mutation index distribution	20

- C2Ua: A uniform crossover, with the second solution where the crossover is allowed on all parameters, the polynomial mutation is realized on all estimated parameters according to the mutation probability.
- CSBX1a: An SBX crossover, with the first solution where the crossover is allowed only on estimated parameters, the polynomial mutation is realized on all estimated parameters according to the mutation probability.
- CSBX_class_a: An SBX crossover, with the second solution where the crossover is allowed on all parameters, the polynomial mutation is realized on all estimated parameters according to the mutation probability.
- CSBX1: An SBX crossover, with the first solution where the crossover is allowed only on estimated parameters, a one-point polynomial mutation is used.
- CSBX_class: A uniform crossover, with the second solution where the crossover is allowed on all parameters, a one-point polynomial mutation is used.
- C1U: A uniform crossover, with the first solution where the crossover is allowed only on estimated parameters, a one-point polynomial mutation is used.
- C2U: A uniform crossover, with the second solution where the crossover is allowed on all parameters, a one-point polynomial mutation is used.

The boxplot Fig. 6 shows the 8 compared metaheuristics. The metaheuristics C2Ua, C2U, CSBX_class and CSBX_class_a correspond to the metaheuristics which use the proposition where the crossover is allowed on all parameters (0 and estimated). The boxplot shows their results seem worse than the results of the other metaheuristics. The metaheuristics which seem to have better results correspond to the metaheuristic where the crossover operator is applied only to estimated parameters. The first idea will be to use this crossover. Metaheuristics with a multi-point polynomial mutation (C1Ua, CSBX1a) seems slightly better than the results with a one-point mutation.

Many statistical tests have been performed to compare these metaheuristics and to know if the difference is significant. The first test applied is a Kruskal Wallis test. This test allows to compare the 8 metaheuristics and to know if

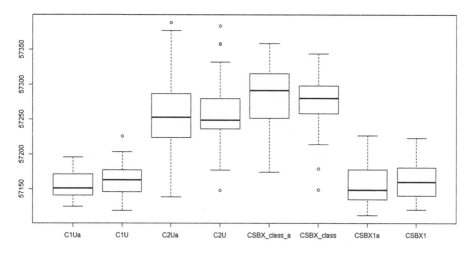

Fig. 6. Boxplot of 8 meta-heuristics compared.

there is really a significant difference between all these metaheuristics results. The Null hypothesis is the following:

$$\begin{cases} H0 : \text{All metaheuristic results are similar} \\ H1 : \text{There are differences between metaheuristics results.} \end{cases}$$

With a $p.value < 2.2e-16$ the null hypothesis is rejected. So, there is a significant difference between the metaheuristics results.

To know which metaheuristics have significantly different results, a Mann Withney test is performed. It is a test where the metaheuristics results are compared (in pairs). In a first time, a bilateral test has been applied. The assumptions are:

$$\begin{cases} H0 : \text{Metaheuristic results } (1) = \text{metaheuristic results } (2) \\ H1 : \text{Metaheuristic results } (1) \neq \text{metaheuristic results } (2). \end{cases}$$

This test shows, metaheuristics can be grouped in 3 subsets, which are: {C1Ua, C1U, CSBX1a, CSBX1}, {C2Ua, C2U} and {CSBX_class_a, CSBX_class}. The first subset group metaheuristics that have the smallest results, left unilateral tests of Mann Withney are performed on this subset to know if one of the metaheuristics is better. However, results show that there is not a significant difference. So, this comparison shows the choice of crossover is important and we have to use the metaheuristics with the crossover applied only to the estimated parameters. For the polynomial mutation, use a one-point mutation or a multi-point mutation does not have a significant impact on the results.

4.5   Comparing Statistical Results and Genetic Algorithm Results

We choose to use the metaheuristic CSBX1 to compare the genetic algorithm results to the statistical results. Table 6 shows the results of BIC criterion for

Table 6. Result of CSBX1 on simulated and real datasets.

Dataset	Best BIC GA	Best BIC SM	Time GA	Time SM	Average time ratio GA-SM
DS1	87 018.41	87 008.41	2.46 min	35 min	14.17
DS2	62 640.8	62 637.1	2.74 min	1 h17	**28.1**
DS3	57 105.8	57 101.31	2.5 min	1 h13	**29.2**
DS4	30 770.08	30 766.48	2.47 min	26 min	10.52
DS5	**64 084.99**	64 084.99	1.95 s	10 s	5.13
DS6	**49 875.41**	49 875.41	2.12 s	30 s	**14.15**
DS7	**23 604.8**	23 612.49	1.95 s	11 s	5.64
DS8	**67 325.2**	67 325.20	1.9 s	4 s	2.1
DS9	**32 836.38**	32 836.38	1.8 s	8 s	4.44
DS10	**21 358.85**	21 358.85	2.09 s	3 s	1.43
Real dataset 1	**58 287.7**	58 306.6	2.10 min	33 min	**15.71**
Real dataset 2	**25 558.9**	25 560.9	1.9 s	30 s	**15.78**

the statistical process and the GA. The execution time of each process is also shown in Table 6. On the different simulated datasets, the GA finds a result very close to the result of the statistical method. A test of Mann Whitney has been realized and gives a p-value = 0.9396. It means, the assumption H0 is accepted and there is no significant difference the statistical results and results of the genetic algorithm. Moreover, with the real dataset and the dataset DS7, the result of the BIC criterion found by the GA is better than the result found by the statistical method. The GA compares more models, so it can do better than the statistical method. For example, Fig. 7 shows the results of parameters estimation for DS7. Because of the constraint to reduce the set of models $\mathcal{M}$ in the statistical method, the statistical method can not find the true estimation of parameters. As the GA can cover the whole set of model $\mathcal{M}$, it can find the true estimation of parameters. Concerning the execution time, for each dataset, the GA is much smaller than the MS. Indeed, according to the dataset used, the GA is 30 times faster than the MS. Moreover, for the two real datasets, the GA is 15 times faster than the MS.

$X \backslash Y$	1	2	3	4	5
1	0.1	0	0	0	0.9
2	0	0.3	0.7	0	0
3	0.2	0	0	0.8	0

Simulated datatset

$X \backslash Y$	1	2	3	4	5
1	0	0	0	0	1
2	0	0.3	0.7	0	0
3	0.22	0	0	0.78	0

Results of SM

$X \backslash Y$	1	2	3	4	5
1	0.11	0	0	0	0.89
2	0	0.29	0.71	0	0
3	0.199	0	0	0.801	0

Results of GA

Fig. 7. Estimation parameter results for DS7.

With the real dataset, the best model is unknown, so the set of models of the statistical method can be too small and the genetic algorithm can find more quickly a better model which is not in the set of models of the statistical method.

5   Conclusion and Perspectives

In the insurance comparison domain, data constantly evolves. Indeed, to answer business expectations, online forms where data comes from are regularly modified. This constant modification of features and data descriptors makes analysis more complex. A first work has been realized to solve this problem, using several statistical tools as the likelihood, to estimate the parameters. In this first work, the modeling realized shows many constraints and involves a problem of model selection. In the statistical process, the model selection is an exhaustive search realized by comparing all models according to the BIC criterion. According to the feature studied, the combinatorial problem can quickly become huge and the method time-consuming. In this work, we propose to use a stochastic optimization method to overcome the statistical method defaults. According to the objective and the problem, a steady state genetic algorithm is used. In the genetic algorithm, a solution corresponds to a model which is composed of estimated parameters which are transitions probabilities and parameters fixed to 0. To handle parameters fixed to 0, a new crossover has been proposed to keep some information. It is compared with classical crossover applied on all parameters. Finally, to have the best results, 8 metaheuristics have been compared. As each solution is a probabilistic model, a new correction operator is applied after each crossover and mutation operator. This correction operator allows keeping a sum of probabilities equal to 1. The results of the comparison show that the GA with the new crossover is more efficient than a GA with the classical crossover applied to all parameters. The main objective of this work is to challenge the statistical method results. The results for BIC criterion with the GA and the SM are similar on the simulated datasets. So, with the simulated datasets, the simulated model and BIC criterion is found by both SM and GA. Moreover, as the GA compares more models, it can find a better model than the models found by the SM. It is the case with the real datasets. Concerning the execution time, the GA is much faster than the SM. On real datasets, GA is 15 times faster than SM. In this work, we show the GA can give good results for the estimations of parameters and very quickly. These results are very interesting and encouraging. Currently, operators chosen are basics so a perspective is to add more intelligence to these operators. For example, by using statistical knowledge to create new operators and to improve results.

References

1. Bedenel, A.-L., Biernacki, C., Jourdan, L.: Appariement de donnees evoluant en temps. Societe francaise de statistique (2016)
2. Burnham, K.P., Anderson, D.R.: Multimodel inference: understanding AIC and BIC in model selection. Sociol. Methods Res. **33**(2), 261–304 (2004)
3. McLachlan, G., Thriyambakam, K.: The EM Algorithm and Extensions, vol. 382. Wiley, New York (2007)
4. Agrawal, R.B., Deb, K.: Simulated binary crossover for continuous search space. Complex Syst. **9**, 115–148 (1995)
5. Reeves, C.: Genetic Algorithms. Handbook of Metaheuristics, pp. 55–82 (2003)
6. Miller, B.L., Goldberg, D.E.: Genetic algorithms, tournament selection, and the effects of noise. Complex Syst. **9**(3), 193–212 (1995)
7. Durillo, J.J., Nebro, A.J.: jMetal a java framework for multi-objective optimization. Adv. Eng. Softw. **42**, 760–771 (2011)
8. Balcombe, K.G.: Model selection using information criteria and genetic algorithms. Comput. Econ. **25**(3), 207–228 (2005)
9. Bies, R.R., Muldoon, M.F., Pollock, B.G., ManuckGwenn Smith, S., Sale, M.E.: A genetic algorithm-based, hybrid machine learning approach to model selection. J. Pharmacokinet. Pharmacodyn. **33**(2), 195–221 (2006)
10. Blauth, A., Pigeot, I.: Using Genetic Algorithms for Model Selection in Graphical Models Sonderforschungsbereich, vol. 386, Paper 278 (2002)
11. Paterlini, S., Minerva, T.: Regression model selection using genetic algorithms. In: 11th WSEAS, pp. 19–27 (2010)
12. Yao, L., Sethares, W.A.: Nonlinear parameter estimation via the genetic algorithm. IEEE Trans. Signal Process. **42**(4), 927–935 (1994)
13. Talbi, E.G.: Metaheuristics: from Design to Implementation, vol. 74. Wiley, New York (2009)
14. Deb, K., Deb, D.: Analysing mutation schemes for real-parameter genetic algorithms. Int. J. Artif. Intell. Soft Comput. **4**(1), 1–28 (2014)
15. Vavak, F., Fogarty, T.C.: A comparative study of steady state and generational genetic algorithms for use in nonstationary environments. In: Fogarty, Terence C. (ed.) AISB EC 1996. LNCS, vol. 1143, pp. 297–304. Springer, Heidelberg (1996). https://doi.org/10.1007/BFb0032791
16. Holland, J.H.: Adaptation in Natural and Artificial Systems: An Introductory Analysis with Applications to Biology, Control, and Artificial Intelligence. MIT Press, Cambridge (1992)

Adaptive Multi-objective Local Search Algorithms for the Permutation Flowshop Scheduling Problem

Aymeric Blot[1]([✉]) [iD], Marie-Éléonore Kessaci[1] [iD], Laetitia Jourdan[1] [iD], and Patrick De Causmaecker[2] [iD]

[1] Université de Lille, CNRS, UMR 9189 - CRIStAL, Lille, France
{aymeric.blot,me.kessaci,laetitia.jourdan}@univ-lille1.fr
[2] KU Leuven, CODeS and imec research groups, Kortrijk, Belgium
patrick.decausmaecker@kuleuven.be

Abstract. Automatic algorithm configuration (AAC) is an increasingly critical factor in the design of efficient metaheuristics. AAC was previously successfully applied to multi-objective local search (MOLS) algorithms using offline tools. However, offline approaches are usually very expensive, draw general recommendations regarding algorithm design for a given set of instances, and does generally not allow per-instance adaptation. Online techniques for automatic algorithm control are usually applied to single-objective evolutionary algorithms. In this work we investigate the impact of including control mechanisms to MOLS algorithms on a classical bi-objective permutation flowshop scheduling problem (PFSP), and demonstrate how even simple control mechanisms can complement traditional offline configuration techniques.

1 Introduction

Designing and tuning metaheuristics is a great challenge in the optimisation field. Offline configurators have been proposed to automatically configure optimisation algorithms according to a single indicator [20,24] and more recently, according to several indicators [3]. However, offline configuration requires a large amount of algorithm executions to test the different configurations and then, the output tuned algorithm is adapted to the training instances only. In this paper, we are interested in online configuration, also called parameter control, that adapts the algorithm during the execution [14,15,21].

Multi-objective local search (MOLS) algorithms are metaheuristics designed to solve multi-objective combinatorial optimisation problems. MOLS algorithms require the definition of several strategies such as the selection of the solution to explore, the exploration of the neighbourhood, the reference set to accept candidate neighbours, or the archive maintenance. They have successfully been used to tackle multi-objective permutation problems such as the travelling salesman problem, the quadratic assignment problem and the permutation flowshop

© Springer Nature Switzerland AG 2019
R. Battiti et al. (Eds.): LION 12 2018, LNCS 11353, pp. 241–256, 2019.
https://doi.org/10.1007/978-3-030-05348-2_22

scheduling problem [4,12,13,23]. Parameter control has mostly been applied to bio-inspired algorithms for single objective optimisation. In this paper, we propose to control multi-objective local search algorithms to solve the bi-objective permutation flowshop scheduling problem.

This paper is organised as follows. First, a literature review on adaptive metaheuristics is presented. Section 3 details the static and the adaptive versions of the multi-objective local search algorithms. Section 4 presents the experimental setup and Sect. 5 gives and discusses the results. Finally, Sect. 6 concludes the paper and draws some perspectives.

2   Adaptive Metaheuristics

Parameter control mechanisms are generally classified between deterministic, adaptive and self-adaptive approaches [14]. In the following, we focus on adaptive approaches that are parameter and algorithm-independent.

2.1   Adaptive Search

In their paper [30], Pisinger and Ropke describe a large neighbourhood search heuristic for a pick up and delivery problem which adapts during search through the selection probability of the neighbourhoods or "subheuristics" as the authors call them. The adaption strategy is steered by the history during the search and a reinforcement learning mechanism is employed to adjust the weight of each neighbourhood after a given number or a "segment" of iterations has been applied. The weight determines the probability of selecting a specific neighbourhood. Extra stochasticity is introduced through noise in some of the basic heuristics to create less deterministic behaviour and increase exploration.

An example of off-line learning can be found in [37] *e.g.*, a multi-mode scheduling problem is approached through an automaton model. Reinforcement learning agents learn the decision functions for these automatons off-line through observation of the history. The result is a fast constructive heuristic adapted to the real world situation from which problem instances originate. Off-line learning has taken on recently resulting in very strong results and frameworks. [3,19,20,24,26,32] and recently, tuning the tuner, in [8]. On-line as well as off-line learning are extensively used in hyperheuristics, see [6] for an overview.

2.2   Multi Armed Bandit

A good introduction on the subject of multi armed bandits (MAB), with the right balance between fundamental understanding and practical understanding, and without any unnecessary detail, can be found in [33]. MAB's, metaphorically referring to the infamous gambling machines, model a decision problem where the only (or main in some versions) source of information is history of previous selections. The only decision to be taken is which arm to pull next. The aim is to maximise the expected outcome of a finite series of decisions. This has been the subject of a fascinating piece of research leading to a large number of convergence results for various policies and their variations. In an adaptive local search setting, multi armed bandits can be used to model the decisions to be taken on which subalgorithm or neighbourhood to select in the next step given the history of the current search. MAB's have been applied in evolutionary algorithms, see e.g. [2] as well as in evolution based hyperheuristic settings [31]. An early example of an application to combinatorial optimisation in networks is [16]. The design of algorithms using MAB's was studied in [9,38].

2.3 ϵ-Greedy

MAB's, having only history to learn from, need to make decisions that both optimise the immediate result and optimise the lessons learned for better future results. Handling this exploitation/exploration dilemma is what makes a strategy for a MAB. A simple strategy is to pick the decision that has delivered the best result on average in the past. The MAB can then eventually start with an exploration phase where every handle is tried once or a predetermined number of times, after which the greedy strategy is used, adjusting the averages after every decision. In dynamical situations, the average may be weighted to introduce a bias towards recent history. These approaches are termed "greedy". In a slightly more explorative approach, a probability is introduced to allow for random selection of an arm, independently of its average success rate. This kind of approaches are called ϵ-greedy, see *e.g.* [22]. One possibility is to select at time $t+1$ the best performing arm with a probability of $(1-\varepsilon)$, leaving a probability of ε to uniformly select an arm at random (Eq. 2).

$$p_i(t+1) = \begin{cases} (1-\varepsilon) + \varepsilon/N, & \text{if } i = arg\ max_j\ \bar{r}_j(t) \\ \varepsilon/N, & \text{otherwise} \end{cases} \qquad (1)$$

2.4 Random

The most simple random control mechanism is based on uniform distribution of arms. However, other random approaches exist, such as *Softmax algorithms*, that use a specific distribution to select the next arm. Arms with better results on average in the past are assigned a higher probability to be selected. One possibility is to use a Boltzman distribution with the probability to select arm i at time $t+1$ is given in terms of the average rewards $\bar{r}_i(t)$ at time t as [22,33] (Eq. 2).

$$p_i(t+1) = \frac{e^{\bar{r}_i(t)/\tau}}{\sum\limits_{arms\ j} e^{\bar{r}_j(t)/\tau}}, \text{ for all arms } i \qquad (2)$$

where τ is a temperature parameter that can be taken constant or decreasing [1,7,22,36].

2.5   Adaptive Pursuit

Pursuit algorithms relate to classical techniques from machine learning in that the probability to select any arm is adjusted after each selection of a specific arm. Given a learning rate β, the probabilities at $t + 1$ are adapted after t as [22,29,33] (Eq. 3).

$$p_i(t + 1) = \begin{cases} p_i(t) + \beta(1 - p_i(t)), & \text{if } i = arg\ max_j\ \bar{r}_j(t) \\ p_i(t) + \beta(0 - p_i(t)), & \text{otherwise} \end{cases} \tag{3}$$

The probabilities at $t = 0$ are all equal. This works well for static systems, with convergence to one for the probability of selecting the best actions. It does not work well for dynamic systems exactly because of this convergence property. In order to alleviate this situation, *adaptive* pursuit was conceived [35]. In adaptive pursuit, a minimum and a maximum (P_{min} and P_{max}) for the probabilities p_i is introduced to leave room for exploration (Eq. 4).

$$p_i(t + 1) = \begin{cases} p_i(t) + \beta(P_{max} - p_i(t)), & \text{if } i = arg\ max_j\ \bar{r}_j(t) \\ p_i(t) + \beta(P_{min} - p_i(t)), & \text{otherwise} \end{cases} \tag{4}$$

Often, this is combined with a weighted average for the expected performances $\bar{r}_i$ coping better with dynamic situations.

2.6   Other Approaches

Other approaches are reinforcement comparison [22,33] and upper confidence bound [1]. These approaches offer more possibilities or a better behaviour in case of uncertainty at the cost of somewhat higher complexity.

In the following we will concentrate on the simpler methods mentioned above. These result in acceptable behaviour at a relatively low implementation cost which is important for local search algorithms.

3   Adaptive Multi-objective Local Search Algorithms

In this paper, we consider and solve multi-objective combinatorial optimisation (MOCO) problems using the Pareto dominance to compare two solutions s and s': s is said to dominate s' if and only if s is better or equal to s according to all criteria, and s is strictly better than s' at least for one criterion. If neither s dominates s' nor s' dominates s, both solutions are called incomparable. In order to handle the Pareto dominance, multi-objective metaheuristics, such as bio-inspired algorithms and multi-objective local search algorithms, can be applied to solve MOCO problems. The last ones have shown very good performance on multi-objective permutation problems such as the travelling salesman problem, the quadratic assignment problem and the permutation flowshop scheduling problem [4,12,13]. Therefore, we decided to focus on multi-objective local search algorithms in the rest of the paper. We present first a static version of multi-objective local search algorithms that will be used in the experiments, and then an adaptive version in which control mechanisms have been integrated into.

3.1   Static Multi-objective Local Search Algorithms

Multi-objective local search (MOLS) algorithms are originally adapted from stochastic local search algorithms [18] where the neighbourhood of a solution is explored to successively move to better and better solutions of the search space. Contrary to the single-objective case where a single current solution is considered only, an *archive* of solutions is maintained to store solutions found during the search. This archive is generally a set of Pareto solutions *i.e.*, all the solutions of the archive are incomparable. In the literature, many variants of multi-objective local search algorithms were proposed, such as PLS [28] and its numerous variants, iterated PLS [10], the stochastic PLS [11], and the anytime PLS [13]; or the DMLS [23].

The variants differ by several aspects: the selection strategy, the neighbourhood exploration strategy and the set of candidate neighbours to enter in the archive. Indeed, one, or several, or even all solutions of the archive can be explored according to the selection strategy. The neighbourhood of each solution selected in the previous phase can be fully or partially explored following the exploration strategy where a reference is used to compare the neighbours with the solution being explored. This reference is generally either the solution being explored or the entire archive. During the exploration, a policy can be used to accept neighbours as candidate solutions. This policy may be different from the termination of the neighbourhood exploration. For example, when the exploration is stopped as soon as a dominating neighbour has been found, both this neighbour and all the incomparable neighbours visited can be candidates.

Following the literature review presented in Sect. 2, the on-line adaptation of an algorithm is mainly done by changing one or two strategies only during the execution contrary to off-line adaptation where thousand combinations can be tested. In this paper, we limit our study to three different termination conditions in the neighbourhood exploration and fix the other strategies. Only one solution is randomly selected in the archive to be explored. The neighbourhood exploration of the selected solution stops when one neighbour verifies the condition where the reference is fixed being the entire archive. Three strategies will be considered in this study:

imp, stops when the visited neighbour dominates one of the solutions of the archive; this neighbour only is candidate,

imp_ndom, stops when the visited neighbour dominates one of the solutions of the archive; this neighbour and all the visited neighbours not dominated by the solutions of the archive are candidates,

ndom, stops when the visited neighbour is not dominated by the solutions of the archive; this neighbour only is candidate.

Algorithm 1: Basic Multi-Objective Local Search (`mols`)

Input: A set of solutions, an exploration strategy
Output: A set of solutions

`archive` ← initial set of solutions;
until *termination criterion is met* **do**
> /* Select a random solution */
> `selected` ← select_1_rand (`archive`);
> /* Apply the given exploration strategy */
> `reference` ← `archive`;
> `accepted` ← exploration_strat (`current, reference`);
> /* Update the archive with the accepted neighbours */
> `archive` ← bounded_pareto (`archive, accepted`);

return *archive*;

Finally, through the `combine()` function, all the candidate solutions are added to the archive, and only the incomparable solutions of the two sets are kept in the archive for the next iteration. Algorithm 1 outlines the basic multi-objective described above. The termination criterion of the algorithm can be an allocated time or number of evaluations for example.

The main drawback of the stochastic local search is they focus their search in a small part of the search space, and then mechanisms have to be integrated to enable them to visit several parts the search space. The iterated local search (ILS) algorithm [25] gives a way to do that for single-objective optimisation by adding a phase to perturb the current solution and so, diversify the search, and accept or not the new solution found. We give in Algorithm 2 a way to iterate a MOLS in this study, using a static exploration strategy and an initial archive. The final archive will contain the best Pareto solutions of the execution. The `kick_1()` function perturbs the execution of the `mols`: it randomly selects one solution from the final archive and it applies three random moves. Then, the algorithm has a chance to visit a new part of the search space. The `combine()` function merges the archive returned by the basic MOLS and the final archive of `s-mols` to keep the best Pareto solutions only. This MOLS is called *static* since all strategies are fixed before the execution.

3.2 Control in Multi-objective Local Search Algorithms

In this paper, we are interested in an adaptive version of the previous static multi-objective local search where mechanisms are added to modify the exploration strategies during the execution. These mechanisms aim at adapting the exploration strategies of the MOLS during the execution. From the static MOLS algorithm, we designed an adaptive MOLS where a `control_arm` method and an `update_rewards` method to compute the rewards associated to each strategy are added. First, the rewards of each strategy have to be initialized: each exploration strategy is tested. Then, at each iteration of the MOLS, the `control_arm`

Algorithm 2: Static MOLS (s-mols)

Output: A set of solutions

archive ← initialisation ();
until *termination criterion is met* **do**
> /* Perturb and apply the MOLS algorithm */
> current ← perturb (best_archive);
> current ← mols (current, exploration);
> /* Merge resulting archive */
> archive ← pareto (archive ∪ current);

return *archive*;

Algorithm 3: Adaptive MOLS (a-mols)

Output: A set of solutions

archive ← initialisation ();
until *termination criterion is met* **do**
> /* Select exploration strategy */
> exploration ← control_arm ();
> /* Perturb and apply the MOLS algorithm */
> current ← perturb (best_archive);
> current ← mols (current, exploration);
> /* Merge resulting archive and update rewards */
> tmp ← pareto (archive ∪ current);
> update_rewards (exploration, archive, tmp);
> archive ← tmp;

return *archive*;

---

method defines the exploration strategy for the next execution of the mols. The kick_1() and combine() functions are the same one as defined in s-mols. At the end of the iteration, the rewards of the exploration strategies are updated.

4   Experimental Setup

In this section, we present the bi-objective permutation optimisation problem tackled in this work, and detail the experimental protocol.

4.1   Permutation Flowshop Scheduling Problem

The Permutation Flowshop Scheduling Problem (PFSP) is a classical permutation problem which involves scheduling a set of N jobs $\{J_1, \ldots, J_N\}$ on a set of M machines $\{M_1, \ldots, M_M\}$. Each job J_i is processed sequentially on each of the M machines, with fixed processing times $\{p_{i,1}, \ldots, p_{i,M}\}$; machines can only process one job at a time. The sequencing of jobs is identical on every machine, so that a solution may be represented by a permutation of size N. In the following, we consider the bi-objective PFSP (bPFSP), minimising two widely studied objectives, the makespan C_{max} (Eq. 5), *i.e.*, the total completion time of the schedule, and the flowtime FT (Eq. 6), *i.e.*, the sum of the individual completion times C_i of the N jobs.

$$C_{\mathrm{max}} := \max_{i \in \{1,\ldots,N\}} C_i \tag{5}$$

$$FT := \sum_{i=1}^{N} C_i \tag{6}$$

Classical permutation neighbourhoods include the exchange neighbourhood, where the positions of two jobs are exchanged, and the insertion neighbourhood, where one job is reinserted at another position in the permutation. In this study, we consider the hybrid neighbourhood defined as the union of the exchange and insertion neighbourhoods, which is known for the bPFSP to lead to better performance than considering each neighbourhood independently [13].

We use the classical PFSP Taillard instances [34], widely used in the literature. They span numbers of jobs $N \in \{20, 50, 100, 200, 500\}$ and numbers of machines $M \in \{5, 10, 20\}$. There are 12 valid (N, M) combinations, with 10 available instances each, for a total of 120 instances.

In this study, we limit the initialisation of the MOLS algorithms to the use of a simple single-objective greedy algorithm on the two objectives independently. Indeed, using smarter initialisation procedures, the starting solutions would be too close to the optimal Pareto front, which is undoubtedly detrimental to the current study since we aim to emphasise the impact of the control mechanisms over the algorithm itself. To obtain two solutions of reasonable quality, we have chosen the NEH procedure [27] for the two objectives independently. It is often used to seed state-of-the-art bPFSP initialisation procedures (*e.g.*, the 2-phase local search algorithm [12]).

4.2   MOLS Algorithms and Control Mechanisms

Section 3 presented the static and the adaptive versions of a MOLS algorithm where only exploration strategies can be set. Indeed, all other strategies are fixed since earlier work [4,5] has shown that the exploration strategy is the most impactful MOLS component for the bPFSP.

In the experiments, we compare the three deterministic instantiations of Algorithm 2, each using a single exploration strategy (denoted simply by imp, imp_ndom, and ndom, respectively), to two adaptive algorithms (see Algorithm 3), using a basic random control mechanism or a ε-greedy control mechanism respectively. While many other more sophisticated mechanisms could have been compared (*e.g.*, see [17,35]), they would likely be similar because only three arms were considered, which considerably limits the number of dissimilar decision strategies.

In the random control mechanism decisions are uniformly taken at random, without any feedback from the search, in contrary to the ε-greedy control mechanism that uses feedback to take decisions. This feedback is computed every iteration using the hypervolume difference between the hypervolume of the new archive and the one of the previous iteration. It is then used to update the reward associated to the current strategy using a learning rate $\alpha = 0.8$ (Eq. 7, with r_i the reward of the arm i, and $f(t)$ the feedback obtained using the current arm).

$$r_i(t+1) = \begin{cases} \overline{r}_i(t) + \alpha(f(t) - \overline{r}_i(t)), & \text{if } i \text{ the current arm} \\ \overline{r}_i(t), & \text{otherwise} \end{cases} \tag{7}$$

In this study, we set $\varepsilon = 0.1$ so that the best performing strategy (*i.e.*, the arm $arg\ max_i\ \overline{r}_i(t)$) is chosen with 90% probability, either strategy being selected uniformly at random otherwise.

For both control strategies, we consider four different variants, that differ by the subset of exploration strategies that are available. First, the three exploration strategies are available for both adaptive algorithms (rand_3, greedy_3). Note that it is already known that the imp strategy leads to poorer results on the bPFSP. But, we still decide to make available this bad strategy in order to evaluate the control mechanism without any a priori knowledge. Secondly, we use this expertise and only make available the two strategies imp_ndom and ndom for both adaptive algorithms (rand_2, greedy_2). Finally, the last two variants introduce a *long-term learning* scheme, beginning with the three strategies but switching to only use the two best strategies during the search. Two possibilities are evaluated: either after half the total running time of both adaptive algorithms (rand_ltl_50, greedy_ltl_50), or after twenty percent of the total running time (rand_ltl_20, greedy_ltl_20).

The termination criterion of the three static algorithms and the eight variants of the adaptive algorithm is a total running time fixed to $n^2m/500$ s. The termination criterion of the inner MOLS (see Algorithm 1) is a combination of either n^2 solution evaluations or n iterations without improvement. This criterion is well adapted to the bPFSP since it enables a sufficient number of iterations of both the inner algorithm and the control mechanism. In the following experiments, about 1600 executions of the inner MOLS for instances with 20 jobs have been done, then about 750, 400, 250 and 100 iterations for instances with 50, 100, 200 and 500 jobs, respectively. This decrease of the number of executions when the number of jobs increases is explained by the exploration step that becomes more and more long and challenging as the size of the neighbourhood quickly grows.

4.3   Experimental Protocol

In this work, we propose to compare the performance of using traditional deterministic mechanisms (through three static algorithms) with different control mechanisms (through eight adaptive algorithms). Experiments are conducted

Table 1. Experimentations summary

Type	Approach	3 arms	2 arms	LTL
Deterministic	imp	✓		
Deterministic	imp_ndom	✓	✓	
Deterministic	ndom	✓	✓	
Random	rand_3	✓		✓
Random	rand_2		✓	✓
Random	rand_ltl_50			✓
Random	rand_ltl_20			✓
ε-greedy	greedy_3	✓		✓
ε-greedy	greedy_2		✓	✓
ε-greedy	greedy_ltl_50			✓
ε-greedy	greedy_ltl_20			✓

across all classical bPFSP Taillard instances, separated in twelve benchmarks of 10 instances sharing the same number of jobs and machines.

The experimental protocol is reduced to the exhaustive comparison of all approaches on all benchmark instances. Because of the stochasticity of both the algorithm and the control mechanisms, all approaches are run 20 times on each instance, using a given set of 20 random seeds.

In total, eleven approaches are compared, in 4 successive steps, as detailed in Table 1. First, we compare the three deterministic approaches with the two adaptive approaches that use all three explorations strategies. Then, we focus on the two best strategies, and compare the respective two deterministic approaches with the two adaptive approaches that use them only. Finally, we investigate the potential of a *long-term learning* scheme for the two control mechanisms independently, first by switching from three arms to two arms after half of the runtime has passed (rand_ltl_50, greedy_ltl_50), then after only twenty percent of the runtime (rand_ltl_20, greedy_ltl_20).

5   Experimental Results

Table 2 presents the rankings resulting of the four steps presented in the experimental protocol (see Sect. 4.3) for the 12 instance sizes together with the resulting average ranks. For each instance size, the ranking is computed using pairwise Wilcoxon signed rank tests and the Friedman test post hoc analysis checks the statistical equivalence between algorithms ranked 1, and their difference with the others.

First, we focus on the 3-arm adaptive approaches, rand_3 and greedy_3, comparing them to the three respective deterministic approaches imp, imp_ndom and ndom. As shown on Table 2, the imp and ndom approaches always perform

Table 2. Experimental ranking

Instance		3 arms					2 arms				LTL rand				LTL ε-greedy			
N	M	imp	imp_ndom	ndom	rand_3	greedy_3	imp_ndom	ndom	rand_2	greedy_2	rand_3	rand_2	rand_ltl_50	rand_ltl_20	greedy_3	greedy_2	greedy_ltl_50	greedy_ltl_20
20	5	5	4	1	1	1	4	1	1	1	4	1	3	1	1	1	1	1
20	10	5	4	1	1	1	4	1	1	1	4	1	1	1	1	1	1	1
20	20	5	3	3	1	1	3	3	1	1	2	1	2	2	1	1	1	1
50	5	5	4	1	1	1	4	1	1	1	4	1	1	1	1	1	1	1
50	10	5	4	1	1	1	4	1	1	1	4	1	1	1	4	1	1	3
50	20	5	4	1	1	1	4	1	1	1	4	1	1	1	4	1	1	1
100	5	5	4	1	1	1	4	1	1	1	4	1	3	1	4	1	3	1
100	10	5	1	1	1	1	4	1	1	1	4	1	3	1	4	1	3	1
100	20	5	2	1	2	2	4	1	2	2	4	1	3	1	4	1	3	1
200	10	5	1	1	3	3	4	1	1	1	4	1	2	2	4	1	3	1
200	20	5	2	1	3	3	4	1	1	1	4	1	3	2	4	1	2	2
500	20	5	1	1	3	3	1	1	1	1	3	1	3	2	3	1	3	2
average		5	2.8	**1.2**	1.6	1.6	3.7	**1.2**	**1.1**	**1.1**	3.8	1	2.2	1.3	2.9	1	1.9	1.3

very poorly and very well, respectively. Meanwhile, the imp_ndom approach performs rather poorly in small instances, but achieves very good results on the largest ones. Surprisingly, the two adaptive approaches (rand_3 and greedy_3) equivalently perform. More precisely, they perform very well on the first eight instances (rank 1), but their performance are more debatable on the four largest ones. Indeed, for instances with 100 jobs and 20 machines, they are outperformed by the deterministic approach ndom and equivalently perform with the imp_ndom approach. But, for instances with 200 and 500 jobs, they are also outperformed by this latter approach. In these cases, they are still better than the imp approach. This results show that the imp approach affects more the adaptive algorithms when the problem gets harder.

Then, in the second step, the imp approach is discarded and we focus on the 2-arm adaptive approaches (rand_2 and greedy_2) that have only the choice between the imp_ndom and ndom exploration strategies. Once again, the two adaptive approaches equivalently perform. However, they are rank 1 for all the instances except for the 100-jobs 20-instances (rank 2). Considering only the well performing exploration strategies largely improves the adaptive approaches for the largest instances.

Having validated that the imp arm should not be used on larger machines instances, we finally investigate a long-term learning scheme where arms can be discarded if they are worse than the others. Two approaches have been tested: the discard of the worst strategy is done after either fifty percent (rand_ltl_50,

Table 3. Complete ranking

N	M	imp	imp_ndom	ndom	rand_3	rand_2	rand_ltl_50	rand_ltl_20	greedy_3	greedy_2	greedy_ltl_50	greedy_ltl_20
20	5	11	10	1	9	1	8	1	1	1	1	1
20	10	11	10	1	9	1	1	1	1	1	1	1
20	20	11	9	9	6	1	6	6	1	1	1	1
50	5	11	10	1	9	1	1	1	1	1	1	1
50	10	11	10	1	9	1	5	5	8	1	1	5
50	20	11	10	1	7	1	1	1	9	1	7	1
100	5	11	9	1	10	1	7	5	8	1	5	1
100	10	11	7	1	10	1	7	1	9	1	6	1
100	20	11	6	1	9	2	6	2	9	2	8	2
200	10	11	4	1	9	1	6	6	9	1	6	4
200	20	11	4	1	9	1	7	5	10	1	7	6
500	20	11	1	1	9	1	7	5	9	1	7	5
average		11	7.5	**1.7**	8.8	**1.1**	5.2	3.2	6.2	**1.1**	4.2	**2.4**

greedy_ltl_50) or twenty percent (rand_ltl_20, greedy_ltl_20) of the total running time. In order to effectively analyse this long-term learning scheme, the two adaptive approaches are investigated and ranked separately. Unsurprisingly, the 2-arm versions of both adaptive approaches always statistically outperform their respective 3-arm versions and so for the versions using the long-term learning. Introducing long-term learning to only keep well-performing arms is efficient. The ranking between the two control mechanisms rand and greedy are not the same size by size, but the average ranks show that it is more efficient to discard an arm sooner since rand_ltl_20 and greedy_ltl_20 are better ranked than rand_ltl_50 and greedy_ltl_50 respectively. These results demonstrate how control mechanisms can effectively identify and evaluate the performance of strategies during the search.

Table 3 summarises all experiments and shows the overall ranking of the eleven approaches on each instance size and shows the final average ranks. Regarding the three deterministic approaches, the imp approach is always ranked last; the ndom approach is almost always ranked first, only beaten on the 20-jobs 20-machines instance where it is outperformed by all adaptive approaches. Regarding the adaptive approaches, both the 2-arm approaches rand_2 and greedy_2 are the best performing, then are ranked the ltl_20 ones and the ltl_50 ones. The approaches using the random control mechanism generally perform worse than the ones using the ε-greedy mechanism, especially for the

3-arm variants and the long-term learning variants. Interestingly, even the random adaptive approach performs really well when considering the two imp_ndom and ndom arms, potentially meaning that the adaptive algorithms will achieve very good results as long as there is no critically bad performing arm available. However, the long-term learning variants show that it is possible to identify, remove and recover in such event.

6   Conclusion

In this paper we presented an adaptive version of MOLS algorithms by introducing control mechanisms. If many control mechanisms can be found in the literature, they have mostly been applied to single objective bio-inspired algorithms; we investigated here their impact on a multi-objective iterative local search algorithm on the classical bi-objective bPFSP. We restrained the control on a single parameter, the exploration strategy, using two different types of control mechanisms: one uniform random and one feedback-based.

Our results show that on the studied problem, taken individually, the three considered explorations strategies performs very differently, and that adaptive approaches can achieve results statistically equivalent to the best strategy. We verified that identifying the best performing strategy before the search, using for example an offline automatic algorithm configurator, can lead to substantial improvement of the algorithm performance. Moreover, we show that very basic control mechanisms can achieve similar outcomes as long as the worst performing strategies are quickly discarded. Identification of the arm qualities continues to critically impact the performance of the overall algorithm, but the long-term learning approaches show that it is possible to introduce this knowledge during the search instead. It means that if a preliminary offline configuration of the algorithm results on multiple high performing mechanisms, it should be possible to simply postpone the decision and to adaptively decide during the search for each tackled instance.

In future work, it would be necessary to first investigate the addition of more possible arms to the MOLS structure, for example by differentiating the exploration reference, in order to compare more complex and potentially more efficient control mechanisms. Moreover, it would be helpful not to fix strategies a priori. Therefore, we would investigate the potential of control mechanisms to simultaneously manage different strategies. Finally, a single bi-objective problem was considered in this paper, in which the performance of each exploration strategy did not vary much. As MOLS algorithms have also been applied to other problems such as the travelling salesman problem or the quadratic assignment problem, for which the optimal MOLS configurations differs, it would be interesting to investigate adaptive MOLS on these problems.

References

1. Auer, P., Cesa-Bianchi, N., Fischer, P.: Finite-time analysis of the multiarmed bandit problem. Mach. Learn. **47**(2–3), 235–256 (2002). https://doi.org/10.1023/A:1013689704352
2. Belluz, J., Gaudesi, M., Squillero, G., Tonda, A.: Operator selection using improved dynamic multi-armed bandit. GECCO **2015**, 1311–1317 (2015). https://doi.org/10.1145/2739480.2754712
3. Blot, A., Hoos, H.H., Jourdan, L., Marmion, M.É., Trautmann, H.: MO-ParamILS: a multi-objective automatic algorithm configuration framework. LION **10**, 32–47 (2016)
4. Blot, A., Jourdan, L., Kessaci-Marmion, M.: Automatic design of multi-objective local search algorithms: case study on a bi-objective permutation flowshop scheduling problem. GECCO **2017**, 227–234 (2017)
5. Blot, A., Pernet, A., Jourdan, L., Kessaci-Marmion, M.É., Hoos, H.H.: Automatically configuring multi-objective local search using multi-objective optimisation. In: Trautmann, H., Rudolph, G., Klamroth, K., Schütze, O., Wiecek, M., Jin, Y., Grimme, C. (eds.) EMO 2017. LNCS, vol. 10173, pp. 61–76. Springer, Cham (2017). https://doi.org/10.1007/978-3-319-54157-0_5
6. Burke, E.K., et al..: Hyper-heuristics: a survey of the state of the art. J. Oper. Res. Soc. **64**(12), 1695–1724 (2013). https://doi.org/10.1057/jors.2013.71
7. Cesa-Bianchi, N., Fischer, P.: Finite-time regret bounds for the multiarmed bandit problem. ICML **1998**, 100–108 (1998)
8. Dang, N.T.T., Pérez Cáceres, L., Stützle, T., De Causmaecker, P.:Configuring irace using surrogate configuration benchmarks. In: GECCO (2017). https://lirias.kuleuven.be/handle/123456789/583393
9. Drugan, M.M., Nowé, A.: Designing multi-objective multi-armed bandits algorithms: A study. IJCNN **2013**, 1–8 (2013). https://doi.org/10.1109/IJCNN.2013.6707036
10. Drugan, M.M., Thierens, D.: Path-guided mutation for stochastic pareto local search algorithms. In: Schaefer, R., Cotta, C., Kołodziej, J., Rudolph, G. (eds.) PPSN 2010. LNCS, vol. 6238, pp. 485–495. Springer, Heidelberg (2010). https://doi.org/10.1007/978-3-642-15844-5_49
11. Drugan, M.M., Thierens, D.: Stochastic Pareto local search: pareto neighbourhood exploration and perturbation strategies. J. Heuristics **18**(5), 727–766 (2012)
12. Dubois-Lacoste, J., López-Ibáñez, M., Stützle, T.: A hybrid TP+PLS algorithm for bi-objective flow-shop scheduling problems. Comput. Oper. Res. **38**(8), 1219–1236 (2011)
13. Dubois-Lacoste, J., López-Ibáñez, M., Stützle, T.: Anytime Pareto local search. Eur. J. Oper. Res. **243**(2), 369–385 (2015)
14. Eiben, Á.E., Hinterding, R., Michalewicz, Z.: Parameter control in evolutionary algorithms. IEEE Trans. Evol. Comput. **3**(2), 124–141 (1999)
15. Eiben, A., Michalewicz, Z., Schoenauer, M., Smith, J.: Parameter Control in Evolutionary Algorithms. Parameter Setting in Evolutionary Algorithms, pp. 19–46 (2007)
16. Gai, Y., Krishnamachari, B., Jain, R.: Combinatorial network optimization with unknown variables: multi-armed bandits with linear rewards and individual observations. IEEE/ACM Trans. Netw. **20**(5), 1466–1478 (2012). https://doi.org/10.1109/TNET.2011.2181864

17. Gretsista, A., Burke, E.K.: An iterated local search framework with adaptive operator selection for nurse rostering. LION **11**, 93–108 (2017)
18. Hoos, H.H., Stützle, T.: Stochastic Local Search: Foundations & Applications. Elsevier, Morgan Kaufmann (2004)
19. Hutter, F., Hoos, H.H., Leyton-Brown, K.: Sequential model-based optimization for general algorithm configuration. LION **5**, 507–523 (2011)
20. Hutter, F., Hoos, H.H., Leyton-Brown, K., Stützle, T.: ParamILS: an automatic algorithm configuration framework. J. Artif. Intell. Res. **36**, 267–306 (2009)
21. Karafotias, G., Hoogendoorn, M., Eiben, Á.E.: Parameter control in evolutionary algorithms: trends and challenges. IEEE Trans. Evol. Comput. **19**(2), 167–187 (2015)
22. Kuleshov, V., Precup, D.: Algorithms for multi-armed bandit problems (2014). CoRR arXiv:abs/1402.6028
23. Liefooghe, A., Humeau, J., Mesmoudi, S., Jourdan, L., Talbi, E.: On dominance-based multiobjective local search: design, implementation and experimental analysis on scheduling and traveling salesman problems. J. Heuristics **18**(2), 317–352 (2012)
24. López-Ibáñez, M., Dubois-Lacoste, J., Cáceres, L.P., Birattari, M., Stützle, T.: The irace package: iterated racing for automatic algorithm configuration. Oper. Res. Perspect. **3**, 43–58 (2016)
25. Lourenço, H.R., Martin, O.C., Stützle, T.: Iterated local search. In: Handbook of Metaheuristics, pp. 320–353. Springer (2003)
26. Marmion, M.-E., Mascia, F., López-Ibáñez, M., Stützle, T.: Automatic design of hybrid stochastic local search algorithms. In: Blesa, M.J., Blum, C., Festa, P., Roli, A., Sampels, M. (eds.) HM 2013. LNCS, vol. 7919, pp. 144–158. Springer, Heidelberg (2013). https://doi.org/10.1007/978-3-642-38516-2_12
27. Nawaz, M., Enscore, E.E., Ham, I.: A heuristic algorithm for the m-machine, n-job flow-shop sequencing problem. Omega **11**(1), 91–95 (1983)
28. Paquete, L., Chiarandini, M., Stützle, T.: Pareto local optimum sets in the biobjective traveling salesman problem: an experimental study. In: Metaheuristics for Multiobjective Optimisation, pp. 177–199. Springer (2004)
29. Rajaraman, K., Sastry, P.S.: Finite time analysis of the pursuit algorithm for learning automata. IEEE Trans. Syst. Man Cybern. Part B **26**(4), 590–598 (1996). https://doi.org/10.1109/3477.517033
30. Ropke, S., Pisinger, D.: An adaptive large neighborhood search heuristic for the pickup and delivery problem with time windows. Transp. Sci. **40**(4), 455–472 (2006). https://doi.org/10.1287/trsc.1050.0135
31. Sabar, N.R., Ayob, M., Kendall, G., Qu, R.: A dynamic multiarmed bandit-geneexpression programming hyper-heuristic for combinatorial optimizationproblems. IEEE Trans. Cybern. **45**(2), 217–228 (2015). https://doi.org/10.1109/TCYB.2014.2323936
32. Smith-Miles, K.A.: Cross-disciplinary perspectives on meta-learning for algorithm selection. ACM Comput. Surv. **41**(1), 6:1–6:25 (2009). https://doi.org/10.1145/1456650.1456656
33. Sutton, R.S., Barto, A.G.: Reinforcement Learning: An Introduction. MIT Press, Cambridge (1998)
34. Taillard, E.: Benchmarks for basic scheduling problems. Eur. J. Oper. Res. **64**(2), 278–285 (1993)
35. Thierens, D.: An adaptive pursuit strategy for allocating operator probabilities. GECCO **2005**, 1539–1546 (2005)

36. Vermorel, J., Mohri, M.: Multi-armed bandit algorithms and empirical evaluation. In: Gama, J., Camacho, R., Brazdil, P.B., Jorge, A.M., Torgo, L. (eds.) ECML 2005. LNCS (LNAI), vol. 3720, pp. 437–448. Springer, Heidelberg (2005). https://doi.org/10.1007/11564096_42
37. Wauters, T., Verbeeck, K., Berghe, G.V., De Causmaecker, P.: Learning agents for the multi-mode project scheduling problem. J. Oper. Res. Soc. **62**(2), 281–290 (2011). https://doi.org/10.1057/jors.2010.101
38. Yahyaa, S.Q., Drugan, M.M., Manderick, B.: Annealing-pareto multi-objective multi-armed bandit algorithm. ADPRL **2014**, 1–8 (2014). https://doi.org/10.1109/ADPRL.2014.7010619

Portfolio Optimization via a Surrogate Risk Measure: Conditional Desirability Value at Risk (CDVaR)

İ. İlkay Boduroğlu$^{(\boxtimes)}$

Namık Kemal University, Çorlu Mühendislik Fakültesi,
Silahtarağa Mah, Çorlu, Tekirdağ 59860, Turkey
`ilkay.net@gmail.com`

Abstract. A *risk measure* that specifies minimum capital requirements is the amount of cash that must be added to a portfolio to make its risk acceptable to regulators. The 2008 financial crisis highlighted the demise of the most widely used risk measure, Value-at-Risk. Unlike the Conditional VaR model of Rockafellar & Uryasev, VaR ignores the possibility of abnormal returns and is not even a *coherent* risk measure as defined by Pflug. Both VaR and CVaR portfolio optimizers use asset-price return histories. Our novelty here is introducing an annual *Desirability Value* (DV) for a company and using the annual differences of DVs in CVaR optimization, instead of simply utilizing annual stock-price returns. The DV of a company is the perpendicular distance from the fundamental position of that company to the best separating hyperplane H_0 that separates profitable companies from losers during training. Thus, we introduce *both* a novel coherent *surrogate* risk measure, Conditional-Desirability-Value-at-Risk (CDVaR) *and* a direction along which to reduce (downside) surrogate risk, the perpendicular to H_0. Since it is a surrogate measure, CDVaR optimization does not produce a cash amount as the risk measure. However, the associated CVaR (or VaR) is trivially computable. Our machine-learning-fundamental-analysis-based CDVaR portfolio optimization results are comparable to those of mainstream price-returns-based CVaR optimizers.

Keywords: Portfolio optimization · Machine learning · Risk management · Downside risk · Conditional value at risk · Linear programming · Fundamental analysis · International financial reporting standards

1 Introduction

We employ a novel machine learning model to enhance an existing portfolio optimization tool (Rockafellar–Uryasev [23]) to come up with a well-balanced portfolio of stocks that will beat the market at the end of a preset portfolio duration. In our setting, a financial instrument, such as a stock or stock portfolio, is said to beat the market if and only if it returns more than a specific

© Springer Nature Switzerland AG 2019
R. Battiti et al. (Eds.): LION 12 2018, LNCS 11353, pp. 257–270, 2019.
https://doi.org/10.1007/978-3-030-05348-2_23

popular market index does, between Feb 16 and Dec 31. The reason for choosing a peculiar starting date for the comparison is as follows: We believe that it is the balance sheet quality that causes a stock to beat the market, and in Turkey, the year-end (third party audited) financial reports of companies are required to be made public before Feb 16 of the following year.

In portfolio optimization, beating the market is an issue mainly if no short positions are allowed in a portfolio. In this case, one is (a) forced to choose stocks that will outperform the market *and* (b) to allocate optimal amounts of capital to each such stock, which requires a portfolio optimization method.

If, on the other hand, short positions were allowed, we could use a market neutral portfolio selection method, an example of which is seen in Baronyan, Boduroğlu, and Sener [5]. Further portfolio optimization examples can be seen in [9] on pp. 150–152.

Here, we shall define what we mean by the "standard usage" of the Linear Programming (LP) model in [23]. If one employs stock-price-return histories to produce an optimal portfolio of financial instruments - optimal in the sense that it has minimum CVaR while having an expected return higher than a threshold - then one is using this model in the standard way. Moreover, no short positions are allowed in the standard usage.

A *risk measure* that specifies minimum capital requirements is the amount of cash that must be added to a portfolio to make its risk acceptable to regulators [16]. Pflug has a proof that shows CVaR is a *coherent* risk measure while VaR is not [21]. A simplified definition of a coherent risk measure can be found in [16]. Çobandağ-Güloğlu and Weber [7] describe a method to produce a Robust CVaR risk measure, RCVaR.

The Rockafellar–Uryasev CVaR portfolio optimizer is also important because, just as VaR does, it reduces only downside risk and does not touch "upside risk", a term that does not make much sense in portfolio optimization [28]. Our method goes one step further and reduces downside risk *selectively* using a surrogate risk measure. That is, we first define a direction along which to reduce a fundamental risk and then make an attempt to reduce it along that direction, using our surrogate risk measure. The concept of surrogate risk measures appear in Vos [27] and Johnson and Maxwell [20], among others. While Vos uses surrogate risk measures derived from balance sheets in an attempt to price small unlisted businesses for sale, Johnson and Maxwell describe how to cluster Australian companies into homogeneous risk groups using surrogate risk measures. Neither employs a mathematical portfolio optimization tool. To our knowledge, *ours is the first paper* that utilizes a surrogate risk measure in portfolio optimization. Note that a portfolio that is formed by optimizing our surrogate risk measure allows itself to assessment by classical monetary risk measures VaR and CVaR. So, there is nothing wrong with using a surrogate risk measure to form a well-balanced portfolio.

The book [8], whose editor is Abu-Mostafa, covers machine learning in portfolio optimization among other topics. Freitas et al. describe a prediction-based portfolio optimization model using Neural Networks [11]. Karaçor and Erkan compare the quantitative predictabilities of different financial instruments in the context of machine learning [17].

Organization of the paper: Sect. 2 describes the machine learning model that we introduce to come up with the best separating hyperplane H_0. Section 2.1 shows how to compute desirability values using H_0. Section 3 describes both the CVaR and the CDVaR portfolio optimization models. In Sect. 3.1, definitions of VaR and CVaR are listed. In Sect. 3.2, we discuss how to compute the VaR and CVaR of any portfolio, for the sake of completeness. In Sect. 3.3, the CVaR-optimizing LP model of [23] and the parameters that we employ are given. Moreover, we show how to tweak this LP model to compute the optimal money allocation to each stock when CDVaR is optimized instead. In Sect. 3.4, we define DVaR and CDVaR risk measures. We also explain how to compute these for *any* portfolio and also discuss how to read these off of an optimal solution of a CDVaR optimization problem. Section 4 provides detailed numerical results. Section 5 has the conclusions.

2 Machine Learning Model

Our goal, in this section, is to describe a novel Machine Learning model that will try to separate desirable companies from undesired ones via a separating hyperplane H_0. The perpendicular distance from a company's fundamental position to H_0 and the side on which the company stands w.r.t. H_0 determines the desirability value (DV)[1] of a company. The DV of a company is an issue when it comes to putting together a stock portfolio that will beat the market at the end of portfolio duration. So, we shall utilize the portfolio optimization technique of [23] in a way to make sure it takes into account a history of negative changes in DV as the *only* source of risk, while keeping the expected total desirability of the portfolio over a certain user defined threshold, μ.

By construction, the DV for a company depends upon its most recent two year-end balance sheets. By construction, a company with a higher DV will mean that it is more likely for this company to beat the market than a company with a lower DV. Every year, our portfolio duration is from Feb 16 to Dec 31 of that year.

Our training data set contains balance sheet information for all candidate companies for the years between 2006 to 2011, inclusive, and stock price information from Feb 16, 2007 to Dec 31, 2012. The testing data set contains balance sheet information from 2012 to 2016, and stock price information from Feb 16, 2013 to the last business day of 2017.

While the first stage of this paper is related to Machine Learning, the second stage is Optimization. The results from the Optimization stage will be provided for both the training and the testing data just like the results of the Machine Learning stage. However, during the Optimization stage, the testing data is a subset of the testing data in Machine Learning due to the looking-back behavior of the optimization.

---

[1] Our DV is not related to the desirability function of [14].

We picked the following 22 publicly traded companies in Borsa Istanbul (BIST) that are currently (or previously) listed in the BIST 50 index, with ticker symbols: AKSA, ARCLK, ASELS, AYGAZ, CCOLA, ECILC, ENKAI, EREGL, HURGZ, IZMDC,KRDMD, MGROS, SISE, SODA,TAVHL, TCELL, THYAO, TOASO, TRKCM, TTKOM, TUPRS, and ZOREN. Note that we did not include companies in the financial sector or holding companies in this list for the sake of simplicity. We employed 57 financial ratios from their year-end balance sheets, which we downloaded from [10]. These ratios are known to be some of the most important fundamental analysis factors [22]. The values of these ratios will be stored in $V(i, j, k)$ where i is the company index, $i \in \{1, \ldots, 22\}$, k is the fundamental ratio index, $k \in \{1, \ldots, 57\}$, and j is the balance-sheet year index, $j \in \{6, 7, 8, \ldots, 16\}$. For example, $j = t$ iff the year-end balance sheet of a company belonges to year $(2000 + t)$. Furthermore, if there is a portfolio that is based on this year-end balance sheet then the portfolio would be active in year $(2000 + t) + 1$, starting mid-February and ending at the end of December. Obviously, $\mathrm{ord}(i) = i$ and $\mathrm{ord}(j) = j - 5$. We shall also use an index m, where $m = \mathrm{ord}(i)\{\mathrm{ord}(j) - 1\}$ where $j > 6$. For the machine learning phase, $7 \le j \le 11$ denotes the balance-sheet years for the training period, and $12 \le j \le 16$ denotes the same for testing. Likewise, $14 \le j \le 16$ denotes the testing period for the optimization phase.

We also created discretized synthetic variables made out of $V(i, j, k)$ whose definitions will be given next. We discretized all real valued variables $V(i, j, k)$ using a 2-bin, equal-frequency form and assigned them values from the set $\{0, 1\}$. Thus, $Z(i, j, k)$ would be assigned the value 0 iff $V(i, j, k)$ was strictly below the *median* for variable k in year j. Note that $Z(i, j, k)$ were not explicitly entered into the analysis. Instead, we defined another synthetic variable $E(i, j, k)$, based on the 2-year-long path that $Z(i, j, k)$ has followed. Thus, the value of $E(i, j, k)$ would be 1 plus any one of the following 4 binary representations: $\{11, 10, 01, 00\}$. That is, $E(i, j, k) = 1 + (2a + b)$ iff $Z(i, j, k) = a$ and $Z(i, j - 1, k) = b$. This binary representation of the paths allows us to put more weight on more recent observations. Consequently, we treat $E(i, j, k)$ as ordinal variables because their values are ordered from 1 to 4, in increasing order of economic soundness.

The most important aspect in a supervised 2-class classification process is the description of the data objects used, along with how one assigns a tag to each data object in order for it to be classified. Our training data object $D(m)$ is defined to be a specific fundamental position of company i at the end of year j, for all i and j except for $j = 6$, along with a tag that shows if its stock beat the market the following year. In the training data set, $m \in \{1, 2, \ldots, 22 * 5 = 110\}$. So, each training data object $D(m)$ may be regarded as a row in a data matrix M that has 110 rows and $2 * 57 + 1 = 115$ columns[2] because the tag value is placed in column 115. The tag assigned to each data object is either 1, if the stock beats the popular market index BIST 100 the year after the balance sheet is announced or else, -1.

---

[2] We have both the $V(i, j, k)$ and the $E(i, j, k)$ columns.

The candidate basis vectors for the separating hyperplane H_0 are the first 114 columns of the data matrix M. Companies in Turkey report their balance sheets using International Financial Reporting Standards [19]. Historical split-adjusted (end-of-day) stock price data and the end-of-day BIST 100 index data were downloaded from [4]. All stock prices as well as the index values were converted to USD using the appropriate exchange rate USDTRY for the day.

2.1 How to Compute Desirability Values Using the Separating Hyperplane?

Using SPSS, we did a Machine Learning analysis via Logistic Discrimination (LD) [2]. The nonlinear objective function of LD is suitable to supervised training for classification. LD helped us eliminate 55 of the 57 balance sheet ratios that we started out with. Using only the training data, we discovered a separating hyperplane H_0 within a 2-D feature space. This space houses points that denote a company's fundamental position at a certain balance-sheet year. Since time is not a feature in this 2-D feature space, each company that we analyze may appear as a distinct point once every year. Actually, when the testing data is also allowed in this space, we observe 22 companies over 10 years. Thus, we shall have 220 not-necessarily distinct points in this space.

Here, we shall explain how we decided that H_0 was actually an acceptable separating hyperplane. We define $S_0(i, j)$ to be the balance sheet score of a company i in balance sheet year j. This score is calculated by taking the inner product of the two defining parameters of H_0 with the company's corresponding $E(m, k)$ and $V(m, k)$ values in Eq. (1). Recall that $m = \mathrm{ord}(i)\{\mathrm{ord}(j) - 1\}$. Since $j > 6$, we have $m > 0$. So, we define $S_1(m) = S_0(i, j)$. Thus,

$$S_1(m) = 0.247 * E(m, 27) - 0.011 * V(m, 32) \tag{1}$$

The above computation requires two distinct balance sheet ratios, the brief descriptions of which can be found in [22]. $E(m, 27)$ is the synthetic variable that is based on the 2-year-long route that the company's Earnings Per Share (EPS) has taken with regards to other competitors, as defined in Sect. 2. $V(m, 32)$ is the ratio of *Short-Term Liabilities* to Net Sales (%). The second term on the right hand side of Eq. (1) is pretty similar to the second term of the equation that defines the Turkish Economic Stability Index (TESI) introduced by Boduroğlu and Erenay [6]. The first terms of each equation are also similar in effect. Since there are no other terms except for a constant in TESI, these two analyses seem to support each other. That is, it seems as if what makes a company desirable for investors is very similar to what makes a country desirable for them. We suspect that this observation would also be seen in countries other than Turkey. This is because having a low short-term debt and high net sales in the corporate world and having a low short-term external debt and a high level of international reserves are fundamentally good for all corporations and countries, respectively. Likewise, having a high earnings per share for a company is similar to having a strong banking sector in a country.

As expected, H_0 is not a 100% separating hyperplane. We shall now try to give a statistical proof based only on the training data set to show that the separation of H_0 is good enough: First of all, the SPSS LD analysis showed that the Nagelkerke R^2 was 0.085. We shall not go into the discussion of whether, or not, this is a high enough R^2 because "low R^2 values in logistic regression are the norm" [15]. Table 1 shows that H_0 has an overall success rate of 58.2% for the training data. The overall success rate for the testing data, 56.4%, is not bad either. Also note that, in Fig. 1, both of the p-values are less than 10%. We also made sure that there was no multicollinearity between the two independent variables in Eq. (1). Based on all this evidence, we conclude that the separation of H_0 is good enough in the training set.

Table 1. Classification Statistics for H_0.

			Predicted	Predicted	Success
		Tag	-1	**1**	
TRAINING	Actual	-1	**19**	29	
	Actual	1	17	**45**	
	% Correct		%47	%61	**%58.2**
TESTING	Actual	-1	**23**	24	
	Actual	1	24	**39**	
	% Correct		%51	%62	**%56.4**

Variables in the Equation

		B	S.E.	Wald	df	Sig.	Exp(B)
Step 1[a]	V32	-.011	.006	2.931	1	.087	.989
	D27	.247	.096	6.544	1	.011	1.280

Fig. 1. The coefficients of the separating hyperplane H_0 and their p-values

Now that we have constructed H_0, it is easy to compute the perpendicular distance from H_0 to the point that denotes a certain company's fundamental position. Thus, we define the Desirability Value $DV(i, j)$ of a company i in balance sheet year j as

$$DV(i, j) = S_0(i, j) = S_1(m) \tag{2}$$

Note that

$$s_{DV}(i, j) = DV(i, j) - DV(i, j - 1) \tag{3}$$

While the standard usage of [23] requires the utilization of the relative change of the stock price with respect to its previous price, we use simple change from the previous DV of the company in the 2-D feature space.[3] We are comfortable with doing this because, unlike the stock price, the DV of a company has a meaning in and of itself.

To compare our method with the standard usage of [23], we will have three different experiments whose portfolio performances we shall compare with the performance of BIST 100:

- Using $s_{DV}(i, j)$, the annual differences of Desirability Values in the LP model (7). This is what we call the CDVaR-optimized portfolio.
- Using annual relative returns $s_{i,j}$ derived from prices in (7). This will be the CVaR-optimized portfolio, as in the standard usage of [23].
- An equally weighted portfolio (EWP) that accepts all 22 candidate stocks. This is the trivial solution since neither machine learning nor portfolio optimization is at work here.[4]

We found out that neither s_{DV} nor s were normally distributed in our experiments. However, since the CVaR-optimization model described in [23] does not specify the distribution of returns that can be used, we are fine.

3 CVaR and CDVaR Portfolio Optimization Models Based on LP

The general purpose in forming a CVaR portfolio optimization model is to minimize the coherent risk measure, CVaR, while keeping the portfolio's expected return greater than a threshold level [23]. The specific reason that we use this model in this paper is to minimize a coherent surrogate risk measure Conditional Value at Risk, CDVaR, while keeping our portfolio's expected surrogate return above a threshold. We do this by simply replacing $s_{i,j}$ with $s_{DV}(i, j)$, for all i and j.

3.1 Definitions of VaR and CVaR

While VaR asks the question "What is the *minimum* of how bad things can get (say, 95% of the time)?", CVaR asks "If things do get into the feared (5%) region, what is our expected loss?" [1,16]. (See Eqs. 5, 6, and Fig 2.) Obviously, the latter question is more meaningful because the former is an *optimistic* way of assessing risk, an optimism that is believed to have caused the 2008 financial meltdown, globally. The practice of risk management has been evolving quickly especially with the arrival of Basel III Standards [13]:

> For some time now global banks have attempted to capture market risk by means of VaR models. During the crisis, these models severely underestimated the tail events and the high loss correlations under systemic stress.

---

[3] Using relative returns of DV does produce very similar results.

[4] Even though the EWP consists of stocks from BIST 50, we make the comparison to the more popular sister index BIST 100.

In this section, the well-established mathematical definitions of Value at Risk (VaR) and Conditional Value at Risk (CVaR) will be given. We start by assuming that the predetermined portfolio weight vector $\mathbf{x}_j$ shall be kept constant from the inception of the portfolio at time t_0, till termination, $t_0 + T$,[5] where both endpoints are in year j. Note that this portfolio would be based on prices (of candidate companies) that are recorded as late as year $j - 1$. Of course, $x_{i,j}$ is the portfolio weight of asset i in year j with $\sum_i x_{i,j} = 1$ and $x_{i,j} \geq 0 \; \forall_{i,j}$ assuming an all-long portfolio. Let L_j be a deterministic non-negative liability payment that has to be met at $(t_0 + T)$. Let R_j be a scalar random variable that denotes the relative return of the portfolio wealth W at the end of year j. That is, $R_j = (W_{t_0+T} - W_{t_0})/W_{t_0}$. Let r_j be a theoretical realization of the random variable R_j in some portfolio cycle j.

Then, $r_{i,j}$ must be a *theoretical* realization of an individual asset return that belongs to asset i in cycle j such that $r_j = \sum_i x_i r_{i,j}$. Note that, we want to distinguish $r_{i,j}$ from $s_{i,j}$, which is a *recorded* realization of the individual asset return for asset i in cycle j. Also, let $\mathbf{\Theta}_j$ be a recorded vector of candidate asset returns for cycle j. Define a random variable Y_j, called shortfall, for year j, where

$$Y_j = L_j - R_j \tag{4}$$

We want to construct a portfolio whose $E[R_j] \geq L_j$, that is, we desire to get $E[Y_j] \leq 0$. Therefore, the undesired side of the distribution is on the right. See Fig. 2. Here, we consider two distinct risk measures, VaR and CVaR. By definition, $\text{VaR}_\beta(\mathbf{x})$ corresponds to the level of shortfall that will not be exceeded with probability β. See Fig. 2. Three typical values of β are 0.90, 0.95 and 0.99. So we have

$$\text{VaR}_\beta(\mathbf{x}) = \min\{\alpha : P(Y(\mathbf{x}) - \alpha \geq 0) \leq 1 - \beta\} \tag{5}$$

On the other hand, $\text{CVaR}_\beta(\mathbf{x})$ is the expected shortfall given that $Y(\mathbf{x}) \geq \text{VaR}_\beta(\mathbf{x})$. In other words,

$$\text{CVaR}_\beta(\mathbf{x}) = E\{Y(\mathbf{x})|Y(\mathbf{x}) \geq \text{VaR}_\beta(\mathbf{x})\} \tag{6}$$

3.2   How to Compute VaR and CVaR of Any Portfolio?

The VaR of *any* portfolio $\hat{\mathbf{x}}$ can easily be computed using the classical method: We would sort, in decreasing order, *all* available historical portfolio returns. We would then skip the first $100\beta\%$ elements of this list and pick the next one as VaR. On the other hand, CVaR is the average of all those entries that are worse than or equal to the VaR.

3.3   The LP Model for CVaR and How to Tweak it into CDVaR?

The following Linear Programming model that minimizes the coherent risk measure CVaR while keeping the expected portfolio return $E[R_{j+1}]$ above a certain

---

[5] T is referred to as the portfolio cycle length, which is 10.5 months in our case.

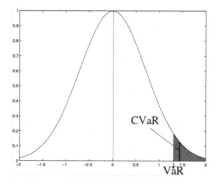

Fig. 2. The (unspecified) distribution of the shortfall, when the liability L_j is zero. By definition, VaR & CVaR levels are used for downside risk minimization only.

threshold comes from Rockafellar and Uryasev [23]. As before, let i be the index for candidate assets and j be the index for balance-sheet years. Each balance-sheet year j is followed by a fresh portfolio cycle $j + 1$. The input data contains:

β desired CVaR level close to1.

L_{j+1} liability$(L_{j+1} \geq 0)$

μ_j targeted expected return$(\mu_j \geq L_{j+1})$

I the set of candidate assets (stocks)

J the set of joint return samplesΘ_j

$q \leq |J|$ quota for the joint return samples in the optimization

$h_i = (\sum_j s_{i,j})/q$ the q-cycle moving average of the return of asset$i$

The CVaR-LP model is seen in the linear program (7). In this model, x_i are the decision variables that are the portfolio weights for assets i in portfolio cycle $j + 1$. Likewise, z_j is the nonnegative excess for sample Θ_j. Moreover, α_{j+1} is VaR$_\beta(\mathbf{x}_{j+1})$, as in Eq. (5), and γ_{j+1} is equal to the objective function value, CVaR$_\beta(\mathbf{x}_{j+1})$, as in Eq. (6).

$$\min_{\gamma_{j+1},\, \alpha_{j+1},\, \mathbf{x}_{j+1},\, \mathbf{z}_j} \qquad \gamma_{j+1}$$

$$\text{s.t.} \qquad \gamma_{j+1} = \alpha_{j+1} + \frac{1}{q(1-\beta)} \sum_j z_j$$

$$z_j \geq L_{j+1} - \sum_i x_i s_{i,j} - \alpha_{j+1} \qquad (7)$$

$$\sum_i x_i = 1$$
$$\sum_i x_i h_i \geq \mu_j$$
$$\mathbf{z_j} \geq 0$$
$$\mathbf{u}_{j+1} \geq \mathbf{x}_{j+1} \geq 0$$

This is an LP problem with $2 + |I| + q$ variables and $3 + q$ constraints, where $q + 1$ is what we call the look-back period of the optimization. That is, in order for us to run the CVaR optimization, we need $q + 1$ years of price data to look back on. Note that in LP (7), excesses z_j are forced to be nonnegative. Any positive excess will make the objective function more undesirable. Therefore, these excesses may be kept small by choosing an appropriate set of weights $\mathbf{x}_{j+1}$ and α_{j+1}. From this point on, we shall drop the subscript j.

There are a number of useful properties of $(\gamma^*, \alpha^*, \mathbf{x}^*, \mathbf{z}^*)$, the optimal solution of the LP (7). First, α^* gives $\text{VaR}_\beta(\mathbf{x}^*)$, which is VaR_β of the optimally resource-allocated portfolio, $\mathbf{x}^*$. Moreover, CVaR_β of the optimally allocated portfolio equals the optimal objective function value, γ^*.

We use the following values for the input parameters: $\beta = 0.95$, $L = 0$, $\mathbf{u} = 0.25$, $\mu = \min(0.01, \rho)$ where ρ is the mean of h_i among all i. Note that there were times when $\rho < 0$ even though, ideally, the targeted expected return $\mu \geq L \geq 0$. However, since the computation of the expected value is nowhere near being precise, we simply ignored this rare exceptional situation. Moreover, $q = 3$ for our CVaR optimization, and $q = 1$ for our CDVaR optimization. The q values for both optimizations were determined using the training data.

We assume zero transaction costs. Our three portfolios and BIST 100 are cash-only in USD, or flat, between Dec 31 and Feb 15 of each year awaiting the publication of annual balance sheets. We assume that no interest accrues during these 46 days.

The optimal solution of LP model (7) gives a CVaR-optimal portfolio when annual relative returns from price histories are substituted for $s_{i,j}$. On the other hand, it gives a CDVaR-optimal portfolio if s_{DV}, the annual difference in DV, is substituted into $s_{i,j}$. *This is our main contribution* in this paper.

All the three experimental portfolios listed in Sect. 2.1 were first composed on Feb 16, 2010. This is because the earliest $E(m)$ and consequently the earliest $S_1(m)$ were computed for $j = 7$. Thus, the earliest s_{DV} was computed for $j = 8$. If $(q+1)$, the look-back parameter for CDVaR, were as low as the lowest possible value of 2, then we could start the first CDVaR portfolio in mid-February 2009. However, since the best optimization look-back parameter for CVaR (discovered running the CVaR algorithm over the training stock-price data) was $q + 1 = 4$ years, we could start our very first CVaR portfolio in 2010 (after having received balance sheet results from the end of 2009, i.e. $j = 9$). Since the optimization look-back parameter of the CDVaR portfolio was $q + 1 = 2$ the first CDVaR portfolio was easily composed on Feb 16, 2010. For purposes of comparison, the first equally weighted portfolio, EWP, was also put together on Feb 16, 2010.

While a computer code for the CVaR-LP model (written in NuOpt, a high level programming language) can be found in [25], we did our own coding in GAMS [12,24]. Additional coding for I/O and bookkeeping was done in Python.

3.4   How to Define and Compute DVaR and CDVaR of Any Portfolio?

Our surrogate risk measure CDVaR also fits the description of coherence in [21] because we compute it in exactly the same manner as CVaR is computed using the LP model of [23]. The only difference in our case is the return distribution that we use as model input. However, since [23] does not specify the return distribution or how the returns are computed, we are on the safe side. Note that CDVaR is a coherent *surrogate* risk measure and not a coherent actual risk measure because the unit of CDVaR is not \$1 but an abstract distance in a 2-D feature space. However, this should not be a concern to the reader because once we compute our own optimal CDVaR portfolio, we can easily compute its actual (monetary) risk measure using CVaR or VaR. So, how we do the portfolio optimization is immaterial to the portfolio owner as long as the portfolio attains an acceptable risk level calculated by either CVaR or VaR, whichever is the choice of the portfolio owner, though we suggest the use of CVaR instead of VaR.

Here, we define DVaR and CDVaR. Let $Q_{i,j}$ be the random variable which takes on the values of $DV(i,j)$ seen in Eq. (2). Also let $U_j = \Sigma_i x_i Q_{i,j}$. Now, we substitute U_j instead of R_j into Eq. (4) and let $L_j = 0$ for the sake of simplicity. Thus, the definitions of surrogate risk measures Desirability Value at Risk (DVaR) and Conditional Desirability Value at Risk (CDVaR) are simply given by Eqs. (5) and (6), respectively.

Let us now describe how to find the *optimal* DVaR and CDVaR values of a CDVaR-optimal portfolio. If we have used the linear program (7) to put together a CDVaR-optimal portfolio, then both the DVaR and the CDVaR values can easily be read off of the optimal solution, remembering that $s_{i,j}$ were assigned the Desirability Value differences $s_{DV}(i,j)$ in the LP model, instead of returns derived from price.

However, if we needed to compute the DVaR or the CDVaR of *any* portfolio, $\hat{\mathbf{x}}$, then we would use the classical methods to compute the VaR and CVaR with *all* available historical differences based on DVs. Even though the basic unit of either of these surrogate risk measures is not related to \$1, it could be useful if one wants to compare the DVaR or the CDVaR of multiple portfolios.

4 Numerical Results

The optimal portfolio weights $\mathbf{x}^*$, described in Sect. 3 were never changed within a portfolio cycle, which starts in mid-February and ends at the end of the same year. First, we shall look at the portfolio optimization results with training and testing data combined because we have too few testing years. We compared the cumulative portfolio wealth of the portfolios that are (a) CDVaR-optimal (b) CVaR-optimal with that of the equally weighted portfolio (EWP) and with the cumulative return of the BIST 100 index, starting mid-February 2010 and ending at the end of 2017. Figure 3 displays the comparison results in USD and shows that the CDVaR-optimal portfolio beats the CVaR-optimal portfolio, the EWP, and BIST 100. CDVaR-optimal portfolio turned \$1 into \$2.89, CVaR-optimal portfolio turned \$1 into \$1.88, EWP turned turned \$1 into \$1.45, and BIST 100 turned \$1 into \$0.76. These results correspond to compounded average APRs (in USD) of 14.20, 8.22, 4.71, and -3.37, respectively. An average APR over 10% is generally accepted to be a good return for hedgefunds so CDVaR is the winner here. However, we also need to look at Sharpe Ratios. U.S. Federal Reserve Funds Rates [26] were used while determining these. The Sharpe Ratios were 0.63, 0.41, 0.29, and -0.01, respectively. CDVaR is the winner here too.

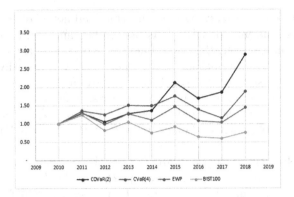

Fig. 3. Comparison of Portfolio Values in USD. Portfolio closing values in decreasing order: CDVaR(2), CVaR(4), EWP, BIST100. The arguments of CDVaR and CVaR denote the optimization-look-back period, $q+1$. EWP is the equally weighted portfolio. BIST100 is the stock market index.

Now, we shall look at portfolio optimization results with only the testing data. We made the same comparison between mid-February 2015 and the end of 2017. CDVaR-optimal portfolio turned \$1 into \$1.36, CVaR-optimal portfolio turned \$1 into \$1.07, EWP turned turned \$1 into \$0.94, and BIST 100 turned \$1 into \$0.78. These results correspond to compounded average APRs (in USD) of 10.89, 2.37, −2.14 and −7.76, respectively. Moreover, the Sharpe Ratios were 0.47, 0.20, 0.02, and −0.25, respectively. CDVaR is the clear winner in the testing data as well.

5   Conclusions

The CDVaR-optimized portfolio beats (in terms of both the total return and the Sharpe Ratio) the CVaR-optimized portfolio, the Equally Weighted Portfolio (formed using 22 stocks in BIST 50), and the market index BIST 100 when only the testing data was used. During testing, portfolio cycles started from mid-February 2015 and ended at the end of 2017. The reason EWP performed better than the BIST 100 may be due the fact that the former's components had been selected from BIST 50, the larger cap index than BIST 100. Moreover, no financial or holding companies were allowed in EWP, which might have also effected this outcome.

The same conclusion can also be made for the analysis over full data, training and testing, with portfolio cycles extending from mid-February 2010 to end of 2017. Since both the total return in USD terms and the Sharpe Ratio for CDVaR is higher than those of CVaR, EWP, and BIST 100, incorporating Machine Learning into portfolio optimization via CDVaR optimization seems to be a viable method even though we analyzed only a limited number of cases.

The fact that we have reduced the balance sheet reading problem into reading (or rather, computing) only two fundamental ratios is also important in and of itself. Also, the exactly two ratios selected for our CDVaR model are very similar in effect to the exactly two ratios that make up the Turkish Economic Stability Index [6]. So, it seems as if what makes a company desirable for investors is very similar to what makes a country desirable for them. We suspect that this observation would also be seen in countries other than Turkey.

Acknowledgements. Attila Odabaşı initialized the author's thoughts on using machine learning techniques in fundamental analysis. Ahmet Boyalı did the initial calculations in the machine learning problem. Murat G. Aktaş, CEO of Finnet Corp., provided us with guidance along with balance sheet data. Selahaddin Yıldırım wrote the Python code that handled the portfolio bookkeeping. The author is also grateful to the organizers of the LION 12 Conference at Kalamata. He also thanks the three anonymous referees, as well as Wolfgang Hörmann and Sevda Akyüz for reading the paper and providing him with ideas for a better presentation.

References

1. Acerbi, C., Tasche, D.: Expected shortfall: a natural coherent alternative to value at risk. Econ. Notes **31**(2), 379–388 (2002)
2. Alpaydin, E.: Introduction to Machine Learning. MIT Press, Cambridge (2004)
3. Artzner, P.: Coherent measures of risk. Math. Financ. **9**, 203–228 (1999)
4. http://bigpara.hurriyet.com.tr/borsa/gecmis-kapanislar (2018)
5. Baronyan, S.R., Boduroğlu, I.I., Sener, E.: Investigation of stochastic pairs trading strategies under different volatility regimes. Manch. Sch. **78**, 114–134 (2010). https://doi.org/10.1111/j.1467-9957.2010.02204.x
6. Boduroğlu, I.I., Erenay, Z.: A machine learning model for predicting a financial crisis in Turkey: Turkish economic stability index. Int. J. High Perform. Comput. Appl. **21**(1), 5–20 (2007)
7. Çobandağ-Güloğlu Z., Weber G.W. (2017) Risk Modeling in Optimization Problems via Value at Risk, Conditional Value at Risk, and Its Robustification. In: Pinto A., Zilberman D. (eds) Modeling, Dynamics, Optimization and Bioeconomics II. DGS 2014. Springer Proceedings in Mathematics & Statistics, vol 195. Springer, Cham
8. Computational Finance, Abu-Mostafa, Y. (Ed.): Computational Finance 1999, 2nd edn. MIT Press, Cambridge (2001)
9. Elton, E.J., Gruber, M.J., Brown, S.J., Goetzman, W.N.: Modern Portfolio Theory and Investment Analysis, 7th edn. Wiley, New York (2007)
10. http://www.finnet.gen.tr/urun/fma (2018)
11. Freitas, F.D., De Souza, A.F., de Almeida, A.R.: Prediction-based portfolio optimization model using neural networks. Neurocomputing **72**(10–12), 155–2170 (2009). ISSN 0925–2312, https://doi.org/10.1016/j.neucom.2008.08.019
12. GAMS Development Corp.: GAMS: The Solver Manuals, GAMS Development Corp. Washington (2017)
13. Hannoun, H.: The Basel III Capital Framework: a decisive breakthrough. BoJ-BIS High Level Seminar on Financial Regulatory Reform: Implications for Asia and the Pacific. Hong Kong SAR, 22 Nov 2010 (2010)
14. Harington, J.: The desirability function. Ind. Quality Control **21**, 494–498 (1965)
15. Hosmer, D.W., Lemeshow, S.: Applied Logistic Regression, 2nd edn. Wiley, New York (2000)

16. Hull, J.: Risk Management and Financial Institutions. Prentice Hall, Upper Saddle River (2006)
17. Karaçor, A.G., Erkan, T.E.: In: Çelebi N. (ed.) On the Comparison of Quantitative Predictabilities of Dierent Financial Instruments, Chapter in Intelligent Techniques for Data Analysis in Diverse Settings (Advances in Data Mining and Database Management), p. 282 (2016)
18. Ince, H., Trafalis, T.B.: Kernel methods for short-term portfolio management. Expert Syst. with Appl. **30**(3), 535–542 (2006). ISSN 0957-4174, 2006, https://doi.org/10.1016/j.eswa.2005.10.008
19. International Financial Reporting Standards: www.iasb.org/IFRS+Summaries (2008)
20. Johnson, T., Maxwell, P.A.R.: Homogeneous Risk Classifications for Industry Studies, Wiley Online Library (2007). https://doi.org/10.1111/j.1475-4932.1976.tb01570.x
21. Ch, G.: Pflug, Some remarks on the value-at-risk and conditional-value-at-risk. In: Uryasev, S. (ed.) Probabilistic Constrained Optimization, Methodology and Applications. Kluwer (2000)
22. Press, E.: Analyzing Financial Statements, Lebahar-Friedman (1999)
23. Rockafellar, R.T., Uryasev, S.: Optimization of conditional value-at-risk. J. Risk (2000)
24. Rosenthal, R.E.: GAMS: A User's Guide, GAMS Development Corp. Washington (2017)
25. Scherer, B., Martin, D.: Intro to Modern Portfolio Optimization with NuOPT and S^+ Bayes. Springer (2005)
26. Trading Economics Web Page (2018). https://tradingeconomics.com/united-states/interest-rate
27. Vos, E.: Risk, return, price: small unlisted businesses examined. J. SEAANZ **3**(1–2), 12–120 (1995)
28. Ziemba, W.T.: The symmetric downside-risk sharpe ratio. J. Portf. Manag. **32**(1), 108–122 (2005)

Rover Descent: Learning to Optimize by Learning to Navigate on Prototypical Loss Surfaces

Louis Faury[1,2(✉)] and Flavian Vasile[1]

[1] Criteo Research, Paris, France
f.vasile@criteo.com
[2] Ecole Polytechnique Federale de Lausanne, Lausanne, Switzerland
l.faury@criteo.com

Abstract. Learning to optimize - the idea that we can learn from data algorithms that optimize a numerical criterion - has recently been at the heart of a growing number of research efforts. One of the most challenging issues within this approach is to learn a policy that is able to optimize over classes of functions that are different from the classes that the policy was trained on. We propose a novel way of framing learning to optimize as a problem of learning a good navigation policy on a partially observable loss surface. To this end, we develop *Rover Descent*, a solution that allows us to learn a broad optimization policy from training only on a small set of *prototypical two-dimensional surfaces* that encompasses classically hard cases such as valleys, plateaus, cliffs and saddles and by using strictly zeroth-order information. We show that, without having access to gradient or curvature information, we achieve fast convergence on optimization problems not presented at training time, such as the Rosenbrock function and other two-dimensional hard functions. We extend our framework to optimize over high dimensional functions and show good preliminary results.

1 Introduction

Finding the minimizer θ^* of a function f over some domain Ω is a recurrent problem in a large variety of engineering and scientific tasks. Instances of this problem appear in machine learning, optimal control, inventory management, portfolio optimization, and many other applications. This great diversity of problems has led over the years to the development of a large body of optimization algorithms, from very general to problem-specific ones.

Recently, the advent of deep learning led to the creation of several methods targeting high-dimensional, non-convex problems (the most famous ones being momentum [26], Adadelta [34] and Adam [18]), now used as black-box algorithms by a majority of practitioners. Other attempts in this field use some additional problem-specific structure, like the work by [22] that leverages fast multiplication by the Hessian to yield computationally demanding, though better performing optimization policies. A common point to all these algorithms

© Springer Nature Switzerland AG 2019
R. Battiti et al. (Eds.): LION 12 2018, LNCS 11353, pp. 271–287, 2019.
https://doi.org/10.1007/978-3-030-05348-2_24

is that they leverage human-based understanding of loss surfaces, and usually require tuning hyper-parameters to achieve state-of-the-art performance. This tuning process can sometimes reveal mysterious behavior of the handled optimizers, making it reserved to human experts or the subject of a long and tedious search. Also, its result is a static optimizer excelling at a specific task, but likely to perform poorly on others.

If the limitations of hand-designed algorithms come from poor human understanding of high-dimensional loss landscapes, it is natural to ask what machine learning can do for the design of optimization algorithms. Recently, [2,19] both introduced two frameworks for learning optimization algorithms. While the former proposes to learn task-specific optimizers, the latter aims to produce task-independent optimization policies. While in the most general case this is bound to fail - as suggested by the *No Free Lunch* theorem for combinatorial optimization [33] - we also believe that data driven techniques can be robust on a great variety of problems.

Most optimization algorithms can be framed, for a given objective function f and a current iterate θ_i, as the problem of finding an appropriate update $\Delta\theta_i$. This update can for instance depend on past gradient information, rescaled gradient using curvature information or many other features. In a general manner, we can write $\Delta\theta_i = \phi(\theta_i, h(f, \theta_{i-1}, \ldots, \theta_0), \xi)$ where $h(\cdot)$ denotes the set of features accumulated during the optimization procedure, and ξ denotes the optimization hyper-parameters.

In our approach, we aim to bypass computing gradient and curvature information and learn the optimization features directly from data. This should allows us to obtain local state descriptors that can outperform classical features in terms of generalization on unseen loss functions and input data distributions. In this vein, we draw an analogy between learning an optimization algorithm and learning a navigation policy while having access to raw local observations of the landscape, which is also the inspiration for the name of our method, *Rover Descent*. Our algorithm contains three chained predictors that compute the angle of the move, the magnitude of the move (e.g. learning rate) and the resolution of the grid of the zeroth-order samples at the landing point. We train our navigation agent on hard *prototypical two-dimensional surfaces* in order to make sure we develop feature detectors and subsequent policies that will be able to lead to good decisions in difficult areas of the loss function. We pose both the learning rate and resolution predictor as reinforcement learning problems and introduce a reward-shaping formula that allows us to learn from functions with different magnitude and from multiple proto-families. In our experiments this was a crucial factor in being able to generalize on many different types of evaluation functions.

We show that this setup leads to very good convergence speeds both in two and higher dimensions, on evaluation functions that are not presented at training time. For a zeroth-order optimization algorithm, the convergence performance is surprisingly good, leading to results comparative to or better than the task-specific optimizer (*e.g.* the best one out of set of specifically tuned first and second order optimizers). In conclusion, we believe that our main contributions are the following: framing the learning to optimize problem as a navigation task, proposing a zeroth-order information-based learning architecture, coupled with a proper training procedure on prototypical two-dimensional surfaces and a reward shaping formula and showing experimentally that it can match/outperform first and second order techniques on meta-generalization tasks.

The rest of the paper is organized as follows. We first give a brief summary of past and recent related work in the field of learning to learn and learning to optimize, and position our approach with respect to existing work in Sect. 2. We then develop in Sect. 3.2 our approach in the two-dimensional case, before extending it to a higher dimensional setting in Sect. 3.3. We present experimental results in Sect. 4 that show the validity of our approach in a variety of setups. We finally develop potential ideas for future work in Sect. 5.

2   Related Work

2.1   Numerical Optimization

The field of optimization has been studied for many years and for a great diversity of problems. Providing a complete review of the subject would be out of the scope of this paper, and therefore we provide only a short reminder on different approaches in the domain.

Some simple settings (convexity, L-smoothness, ..) have been intensively exploited to devise a large number of optimizers, derive upper-bound convergence rates [25] and even some information theoretical complexity lower bounds for black-box optimizers [1]. More recently, motivated by the growing interest in deep learning, a lot of research efforts were also invested in devising smart, adaptative optimizers for complicated, very high dimensional objectives.

Part of the diversity of existing optimizers is explained by the different types of oracles (possibly noisy, second, first, zeroth order evaluation oracles or even comparison oracle) available for a given problem. The case of noisy first order oracles has been widely adopted in the machine learning community and led to many innovations (a detailed survey can be found in [5]). Noisy zeroth order oracles also received a lot of attention from the Bayesian optimization and the evolutionary optimization communities (one of the most successful method being Covariance Matrix Adaptation Evolutionary Strategy [13]), and have also seen a few heuristics approaches (the Nelder–Mead heuristic [24] being one of them).

2.2   Meta-learning

Learning to learn (or meta-learning) is not a recent idea. Reference [29] thought of a Recurrent Neural Network (RNN) able to modify its own weights, building a fully differentiable system allowing the training to be learned by gradient descent. Reference [17] proposed to discover optimizers by gradient descent, optimizing RNNs modeling the optimization sequence with a learning signal emerging from backpropagation on a first network.

Recently, some meta-learning tentatives have shown great progress in different optimization fields. Various attempts tried to dynamically adapt the hyper-parameters of hand-designed algorithms, like [7] or [14]. Using gradient statistics as an input for a recurrent neural net, [20] were able to reinforcement learn a policy effective for training deep neural networks. In [2], the authors show that when leveraging first-order information one could learn by gradient-descent optimizers that outperforms current state-of-the-art of existing problems - however only when the meta-train dataset is made of the same class of problem. When confronted to a different class of functions, the meta-learner is unable to infer efficient optimization moves. With the same idea of using gradient-descent for training the optimizer, [6] use zeroth order information in order to learn an optimizer for the Bayesian optimization setting.

One could think that by showing enough examples to a meta-learner (namely made up of instances where traditional optimizers reach their limits), and adapting its structure to cover a large number of classes of functions, it could adapt to unknown loss landscapes. This idea was exploited by [32], who manage to learn optimizers that generalizes to completely unseen data, while still being able to scale up to high-dimensional problems. Their process namely involves training by gradient descent hierarchical RNNs and showing it a great variety of examples. However, their optimizer's structure remains quite complicated, and doesn't provide human-level understanding of the features leveraged by the meta-learner. We believe that a more intelligible architecture could enable us to understand better what the network is learning, while still being effective on a large class of functions, when trained on a selected number of meta-examples.

2.3 Reinforcement Learning Preliminaries

Reinforcement learning is a framework in which an agent learns its actions from interaction with its environment. The environment generates scalar values called rewards, that the agent is seeking to maximize over time. The environment is modeled as a Partially Observable Markov Decision Process (POMDP), defined to be the tuple $(\mathcal{O}, \mathcal{S}, \mathcal{A}, p_0, p, q, r)$ where $\mathcal{O}$ is the set of observations, $\mathcal{S}$ the set of states and $\mathcal{A}$ the set of actions. $p_0(s)$ is the initial probability distribution over the states, $p(s'|s, a)$ the transition model, $p(o|s)$ the distribution of an observation conditionally to a state and $r : \mathcal{S} \to \mathbb{R}$ a function that assigns a reward to each state. The objective is to learn a policy $\pi(a|s) : \mathcal{S} \to \mathcal{A}$ providing the probability of choosing action a in state s. This policy should maximize the discounted expected return $\bar{R} = \mathbb{E}_\rho \left[\sum_{t=0}^{T} \gamma^t r(s_t) \right]$ where $\gamma \in (0, 1)$ is a discount factor allowing the agent to be more sensitive to rewards it will get in a close future, and the expectation being taken with respect to the state-action distribution ρ. A complete introduction to the reinforcement learning framework can be found in [31].

Policy search is a family of algorithms that directly search in the policy space for $\pi^* = \underset{\pi}{\mathrm{argmax}} \{\bar{R}\}$. To make this search tractable, π is usually tied to some parametrized family. A popular algorithm to perform that search is the Deterministic Policy Gradient [30] (DPG) where we learn a deterministic parametrized policy $\pi_\eta(a|s) = \mu(\eta, s)$ in a fully observable Markov decision process ($\mathcal{O} = \mathcal{S}$). The system is composed of two entities, an actor and a critic. The critic, parametrized by ω, has the role to evaluate the Q-values (expected return when taking an action a in state s) of the current policy induced by the actor (parametrized by η). As it is common in actor-critic approaches, the critic is updated by batch of logged experience to minimize the squared temporal difference error $(r_t + \gamma Q_\omega(s_{t+1}, a_{t+1}) - Q_\omega(s_t, a_t))^2$. The actor's parameters are updated in the direction that maximizes the Q-values for a batch of logged states: $\Delta\eta \propto \nabla_a Q_\omega(s, a)^T \nabla_\eta \mu(\eta, s)$. Reference [21] applied this algorithm to deep neural networks as function approximators, using techniques that were proven successful in deep Q-learning [23], like target networks and experience replay. Reference [15] also extended this approach for POMDP, where it is useful to use Recurrent Neural Networks as models for the policy.

3   Our Approach

3.1   Intuition

In this paper, we propose framing the problem of learning to optimize as a problem of navigation on the partially observable error surface. The error surface is defined by the values of the loss function taken over the range of its inputs. In this framework, the optimizer is an agent that starting from the initial point, attempts to reach the lowest point on the surface with the smallest number of actions (where an action is a move to an arbitrary point on the landscape), while observing only a set of points sampled from the loss surface. Our goal is to learn the navigation policy that maps the current state of the agent to a move on the surface.

To this end, we decide to divide the architecture of our agent in three sequential modules: the normalized update direction predictor Δ, that predicts the angle of the update, the learning rate predictor that predicts the magnitude of the update α, and the resolution predictor, that predicts the scale δ of the observation set at the landing point. This choice is motivated by our intuition that these steps can be approached in a hierarchical way (first choose a direction, then a step size accordingly for instance) and therefore might involve different training methods and procedures. Furthermore, each of the modules can act as a correction factor on the other two modules. For example, if the update angle is not correct, the learning rate module can compensate by making the move very small and the resolution module can zoom out/in to make the next angle prediction task easier. Figure 1 sums up the proposed architecture in the two-dimensional case.

3.2   The Two-Dimensional Case

In the following subsection, we consider the simple case $d = 2$ to develop our experimental set-up. A generalization for higher dimensions can be found in Sect. 3.3.

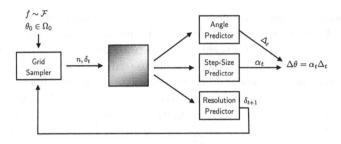

Fig. 1. Decomposition of the optimization step in three independent modules: angle prediction, step-size prediction and resolution prediction.

Prototypical Landscapes. Because our end goal is to be able to optimize complex loss landscapes, we are interested in selecting a small but sufficient set of prototypical landscapes as our meta-training set $\mathcal{F}_{train}$. More precisely, we decide to target surface degeneracies that are common when learning the weights of deep neural networks. These namely include *valleys*, *plateaus*, but also *cliffs* [3] and *saddles* [8]. We also decide to add *quadratic bowls* to that list, to provide simpler and saner landscapes. Figure 3 provides a visualisation of each of these landscapes as generated by our meta-training algorithm. Interestingly enough, all of these landscapes were listed in [28], which provides a collection of unit tests for optimization. In the line of this work, we estimate that learning an optimizer over such landscapes can result in a robust algorithm. Also, because it is frequent in real world applications to only have access to noisy samples of the function we wish to optimize, our framework provides noisy versions of the landscapes described above.

Input Design. Let us consider two classical optimizers: gradient-descent and Newton descent (see [27] for complete details). While gradient descend can escape non-strict saddle points but shows shattering behavior insides valleys (see Fig. 2), second-order methods can leverage curvature to make quick progress inside valleys. However, saddles are attraction points for such methods. To get the best of both worlds, we want our local descriptor to be able to represent both first and second order information. Finite difference provides an easy way to approximate them from zeroth order sample of a function. However, the precision of finite difference can be severely impacted by noisy oracles, although this can be alleviated by a pre-filtering of the function samples (like low-pass filtering for removing white noise).

Let f the loss landscape we are optimizing, mapping $\Omega_f \subset \mathbb{R}^d$ into $\mathbb{R}$, and $\theta \in \Omega_f$. A natural way to describe the surroundings of θ is to sample a grid centered on θ. Given a budget of n^2 samples and a resolution δ, we note $s_\delta^n(f, \theta)$ the resulting two-dimensional grid.

$$s_\delta^n(f, \theta) \triangleq \left(f\left(\begin{pmatrix} \theta_1 \\ \theta_2 \end{pmatrix} - \delta \cdot \begin{pmatrix} i - n/2 \\ j - n/2 \end{pmatrix} \right) \right)_{i,j \in \{1,\ldots,n\}^2} \tag{1}$$

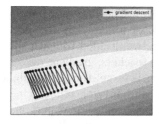
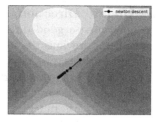

Fig. 2. *Left*: gradient descent has a shattering behavior in narrow valleys. *Right*: saddle points are attractors for Newton descent.

This state representation has three advantages; it allows us to have a human-understandable input to our model, represents the surroundings of the current iterate and can approximate the inputs taken by gradient descent and Newton descend via finite difference. Another advantage is that pre-filtering can be efficiently applied by convolutions. For this state representation to represent compactly various functions, independently of their magnitude, we linearly rescale it to take its values in $[0, 1]$.

It is important to note that such an input becomes extremely expensive to compute as the dimensionality of the problem grows, as the size of $s_\delta^n(\cdot)$ grows exponentially with d. Therefore, we will use this solution for $d = 2$, and discuss different ways of scaling to higher dimension in Sect. 3.3. In the rest of this paper, we will use a fixed value for $n = 15$.

One could argue that using noisy zeroth order oracle for optimization is uncompetitive compared to higher order methods. Indeed, the study of convergence rates and lower bounds for convex optimization problem shows the superiority of first-order oracles over single zeroth-order function evaluation [25]. However, it was proven in [9] that by using two function evaluations, the oracle complexity of the latter type of algorithms could compete with the former, up to a low-order polynomial of the dimension factor (in the convex case). Because we use many of such samples, we are confident that we will be competitive against higher order oracles.

Fig. 3. Instances of $\mathcal{F}_{train}$. In order: quadratic bowl, valley, (plateau+cliff), saddle. Best viewed in color (color figure online).

Learning the Update Direction/Angle. The first step of our three-step optimizer is to determine a good direction of update, given a grid of samples $s_\delta^n(f, \theta)$. In light of the previous discussion, we decided to learn this angle prediction by *imitation learning*, provided two teachers: gradient descent and Newton descent. The field of imitation learning is large, though dominated by two antagonist approaches: *behavioral cloning* and *inverse reinforcement learning*. The latter recovers the cost function that a teacher or expert is minimizing, while the former involves training a complex model (usually a deep neural network) in a supervised fashion so that it mimics a teacher. Thorough details on both these methods, as well as a complete survey of the field of imitation learning can be found in [4]. Behavioral cloning, while being straight forward and simple to implement, is known to require a large amount of data and to be prone to *compounding errors*, leading to divergence between the teacher's and the learner's paths. On the other side, inverse reinforcement learning allows the imitator to interact with the environment, and fit its behavior over whole trajectories (therefore is not affected by the compounding error issue). However, it often implies using reinforcement learning in a inner loop, making this technique rather costly to use. In our set-up, we decided to use a behavioral cloning approach for learning the angle predictor. We can indeed easily generate large amounts of training data, and are not trying to fit the entire teacher behavior but only a subpart - the direction, not the step-size.

We collect our training data by launching optimization runs, where we follow the best out of the teachers (here best means leading to the largest decrease of the objective function). At each step, we record the local grid sample with a predetermined resolution, as well as two opposite directions of update: the optimal one d_* (given by the best teacher) and a set of opposite randomly generated ones $\tilde{d}_*$ (sampled to lie in the half-space defined by $\{d \mid d^T d_* < 0\}$). The expert move d_* and its negative counterparts $\tilde{d}_*$ are normalized to create two actions $a_* = d_*/\|d_*\|_2$ and $\tilde{a}_*$ similarly. We then create two state-action pairs with respective label $t = 1$ and $t = 0$, corresponding to a positive and a negative sample. In practice, we sampled $5 \cdot 10^4$ functions from $\mathcal{F}_{train}$ and let the optimization procedure run for 10 steps on each functions, creating 10^6 (state, action, label) tuples to train on, stored in $\mathcal{D}_{train}$.

To fit the resulting (state-action) pairs, we design a simple neural network made of two convolutional layers followed by two fully connected layers. The idea of using convolutional layer is related to the problem of filtering we mentioned earlier, and to the idea that the learnt filters in the convolutional layer can act as identifiers for the different landscapes encountered during an optimization run. For a given grid of sample s, we denote $y(\omega, s)$ the output of this model, parametrized by the weights ω of the network. We use batch-normalization layers after the convolutional layers, and train the model to minimize the cross-entropy loss:

$$J(\mathcal{D}_{train}, \omega) = \sum_{(s,a,t) \in \mathcal{D}_{train}} \left[t \log \{\sigma(y(\omega, s), a)\} + (1-t) \log \{1 - \sigma(y(\omega, s), a)\} \right]$$

$$(2)$$

with $\sigma(y(s), a) = (1 + e^{-y(\omega, s)^T a})^{-1}$. The objective is to learn to correlate the output of the model when the action a has a positive label (it was sampled from the best teacher). The idea of storing negative versions of that optimal action can be understood as negative sampling, or noise contrasted estimation [11]. We found that this approach, over the other ones we tried, leads to better performances while greatly reducing overfitting. Because we only want to use this model as an angle predictor, we will use a normalized version of the output: $\Delta(s) = y(s)/\|y(s)\|$.

Learning the Step-Size and the Resolution. At this point, we have learnt a good angle predictor. We now want to learn two new behaviors: the step-size to apply to the update, as well as the next resolution of the sample grid. Learning the step-size is obviously crucial for the optimization step. Learning the resolution is also extremely important: far from an optimum, we'd like to zoom-out to get a better understanding of the landscape. Close to an optimum, we expect an efficient system to zoom in to refine its estimation of the localization of the optimal point. Those two behaviors can't be learnt efficiently from a teacher (line-search is an unfairly good teacher for the step-size, and we simply don't have available a hand-designed teacher for the resolution).

We consider the following environment for our problem. Let, for a given loss function f, the full state space $\mathcal{S}_f = \{\theta, \alpha, \delta\}$ and the observations $\mathcal{O}_f = (s_\delta^n(f, x))$. The agent hence only has access to the current grid of samples around θ with resolution δ, but not to the current iterate position θ, the current-step size α or the current resolution δ. The idea behind this is to be able to generalize to unseen landscapes, and be robust to transformations such as rescaling or translations. The only events that should impact the agent's behavior is a sharp change in the neighboring landscape around the current iterate. The action space is set to be $\mathcal{A} = \{\Delta\alpha, \Delta\delta\} \subseteq [-0.5, 1]^2$ which constitutes the update rate of the step-size and the resolution. We consider deterministic transitions:

$$\theta_{t+1} = \theta_t + \alpha_t \Delta(s_{\delta_t}^n(f, \theta_t)), \quad \alpha_{t+1} = \alpha_t(1 + \Delta\alpha_t), \quad \delta_{t+1} = \delta_t(1 + \Delta\delta_t) \quad (3)$$

where the current iterate θ_t is updated along the direction $\Delta(s_{\delta_t}^n(f, \theta_t))$ with step-size α_t.

We have several options for the reward function. One possibility is to consider a budgeted optimization scheme, with reward $r(s) = -f(\theta_t)\mathbb{1}_{t=T}$ (the reward is only given by the final value of the function at the last step). In this case, the reward is rather sparse, and leads the trajectory search in ambiguous ways. We can prefer another solution, where the whole trajectory of the agent over the landscapes is evaluated: $r_f(s_t) = -f(\theta_t)$. This leads the policy search to optimize for the return $R_f = -\sum_{t=0}^T \gamma^t f(\theta_t)$. Note that for $\gamma = 1$ this leads us to optimize over the same criterion that [2, 20] (in the former, the authors call this the meta-loss).

It is important to note that the previously described POMDP, that we will denote as $\mathcal{M}_f$, is parametrized by a function f sampled inside $\mathcal{F}_{train}$. This induces a distribution $\mathcal{M}_{train}$ over POMDPs. In the following experiments, we won't make that distinction and train a single parametrized policy on the resulting POMDP distribution - that implies that every new episode is generated with $\mathcal{M}_f \sim \mathcal{M}_{train}$. This induces a difficulty over the learning task: both the transitions (3) and the reward function change between every episode. To help the agent figure out optimal moves, we can change the reward so that it becomes insensitive to the magnitude of the sampled function f and the position of the initial iterate θ_0:

$$r_f(s_t) = -\frac{f(\theta_t) - f(\theta_f^*)}{f(\theta_0) - f(\theta_f^*)} \tag{4}$$

with $\theta_f^* = \arg\min_\theta f(\theta)$. To also help the agent optimize over long trajectories where the magnitude of $f(\theta_0)$ largely surpasses $f(\theta_f^*)$, we propose a second version of the reward function:

$$r_f(s_t) = -\frac{f(\theta_t) - f(\theta_f^*)}{\bar{f}_k - f(\theta_f^*)} \tag{5}$$

with $\bar{f}_k$ being the mean value of the objective function over the last k iterates (we found that in our set-up, $k = 5$ provides good results). The use of this reward function was a crucial element in the success of our reinforcement learning approach.

We model the agent policy by a recurrent neural network, made up of two convolutional layers, followed by a Long-Short Term-Memory cell (LSTM, introduced by [16]), followed itself by two fully-connected layers. The critic is modeled by a similar network, and both were trained using the DPG algorithm. During training, we sample $f \sim \mathcal{F}_{train}$ at the beginning of each episode. The initial iterate is randomly sampled in the landscape so that it is far away enough from the optimum of the loss function. The episode is ran for a fixed horizon $T = 30$ and we fix the discount factor γ to 1.

3.3 Architecture for $d > 2$

The idea of using grid samples $s_\delta^n(f, \theta)$ can't be exploited in high-dimensional problems as its size grows exponentially with d. To extend our framework for $d > 2$, we consider the following set-up: let $f : \mathbb{R}^d \to \mathbb{R}$ and θ the initial iterate. We note $s(f, \theta, i, j)$ the vector that contains the two-dimensional grid sampled at θ along the dimensions i and j. In other words, with δ_i the d-dimensional vector whose entries are all 0 but the ith one that is set to 1, and $E_{i,j} = (\delta_i, \delta_j)$ a $d \times 2$ matrix, we note: $s_\delta^n(f, \theta, i, j) = s_\delta^n(f, E_{i,j}^T \theta)$.

By considering all pairs of dimensions, we can compute $d(d-1)/2$ of such grids, leading to the prediction of as many angles $\Delta_{i,j}(f, \theta) = \Delta(s_\delta^n(f, \theta, i, j))$, step-size updates $\Delta\alpha_{i,j}$ and resolution updates $\Delta\delta_{i,j}$ - all predicted with the models trained in the two dimensional case. Therefore, if we keep record of all step-size and resolution for every pair of dimensions (i, j) we can compute $d(d-1)/2$ updates $\Delta\theta_{i,j} = \alpha_{i,j}\Delta_{i,j}(f, \theta)$. We can consider each of these outputs like the $d(d-1)/2$ two-dimensional projections of the true d-dimensional update

$\Delta\theta$ so that $\Delta\theta_{i,j} = E_{i,j}^T \Delta\theta$. We can therefore try to retrieve $\Delta\theta$ by a least-square approach, and find $\Delta\hat{\theta}$:

$$\Delta\hat{\theta} \triangleq \arg\min_{\delta\theta} \sum_{1 \leq i < j \leq d} \left(E_{i,j}^T \delta\theta - \Delta\theta_{i,j}\right)^2 \tag{6}$$

Solving this equation leads to the analytical expression:

$$\Delta\hat{\theta} = \frac{1}{d-1} \sum_{1 \leq i < j \leq d} \alpha_{i,j} E_{i,j} \Delta_{i,j}(f,\theta) \tag{7}$$

Each pair of dimension has a corresponding step-size $\alpha_{i,j}$ and resolution $\delta_{i,j}$, which are updated by running the associated two-dimensional grid through the system described in Sect. 3.2. This computation requires maintaining $d(d-1)/2$ learning rates and resolutions, and computing as many grid samples. Because of this quadratical growth with the dimension, this leads to clock-time and memory issues for large values of d. A simple way round this problem is to sample $k < d$ pairs of dimensions, compute $\Delta\hat{\theta}$ based only on these k pairs and update their corresponding learning rates and resolutions. If we note Θ_k the set of k pairs we sampled:

$$\Delta\hat{\theta}_k = \frac{1}{k-1} \sum_{(i,j) \in \Theta_k} \alpha_{i,j} E_{i,j} \Delta_{i,j}(f,\theta) \tag{8}$$

Different strategies can be employed to sample Θ_k, possibly leveraging some knowledge about the optimization problem's structure. Such strategies are experimentally investigated in Sect. 4.2.

4   Results

4.1   Two-Dimensional Experiments

We first evaluate the optimizer resulting from the model we introduced on $\mathcal{F}_{train}$ to analyze its behavior on known landscapes. We can already interpret this set-up as meta-testing on some hold-out since we only sample in $\mathcal{F}_{train}$, which contain an infinite number of functions. Therefore, we can assume that whenever we sample in $\mathcal{F}_{train}$, we obtain a function that the optimizer has not seen during training. We follow a simple procedure: we sample $f \in \mathcal{F}_{train}$, an initial point $\theta_0 \in \Omega_0$, and sample an initial step-size and an initial resolution inside the distribution used at meta-training time. We then add a small perturbation to the initial iterate and run an optimization trajectory with a fixed horizon. We repeat this procedure to evaluate the global sensitivity of our algorithm to the position of the first iterate. To compare its performance with a broad variety of optimizers, we decide to evaluate with the same procedure a collection of optimization algorithms that include: gradient descent, Nesterov accelerated gradient descent, Newton Descent, Covariance Matrix Adaptation Evolution Strategy (CMA-ES, used with a population size that equals the n^2 number of function evaluations our algorithm performs) and the Nelder–Mead method. The results are regrouped in Fig. 4, and are consistent in our experiments: we have learnt to compete with a wide variety of hand-designed algorithms. Their respective hyper-parameters are modified at each time to perform as well as possible on the whole modality of $\mathcal{F}_{train}$ we are testing on. This means that our learnt optimizer sometimes compete with unfairly good algorithms (like Newton descent on a quadratic loss). In some cases, the apparent lack of trajectory envelope is due to the fact that the perturbation on the initial point sometimes have to be reduced for visualization purposes.

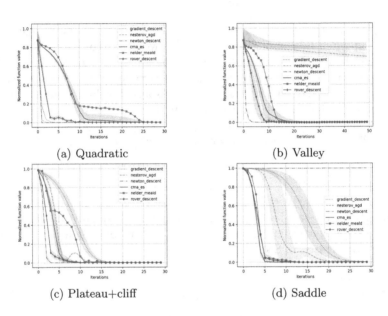

(a) Quadratic

(b) Valley

(c) Plateau+cliff

(d) Saddle

Fig. 4. Tests runs on instances of modalities of $\mathcal{F}_{train}$. The shades represents the envelope of the trajectories over an entire 20-fold. Best viewed in color (color figure online).

To evaluate the meta-generalization abilities of our learnt optimizer, we also evaluate it on a two-dimensional meta-testing dataset $\mathcal{F}_{test}$. We selected various two-dimensional optimization problems known to be challenging for general optimization methods. The complete list contains Rosenbrock, Ackley, Rastrigin, Maccornick, Beale and Styblinksi's function. It is important to note that none of these landscapes were seen by the optimizer during its training. For each of those functions, we select a starting point that constitute a challenge for all compared optimizers. We then followed the previously described procedure, and set the hand-designed optimizers hyper-parameters to show good behavior for every small perturbation of the initial iterate. The results are displayed in Fig. 5, and remain consistent when changing the starting point for each of the meta-test functions. Our learnt optimizer can generalize to new landscapes, even multimodal ones, and compete with a wide variety of optimizers. On these multimodal landscapes, CMA-ES and the Nelder–Mead method provide two strong baselines (only the mean trajectory appear for such functions for visualisation purposes). On Fig. 5e, they are the only two algorithms with our method that find the global optimum. However, in Fig. 5f, only our method finds for every perturbation of the initial iterate the global minimum.

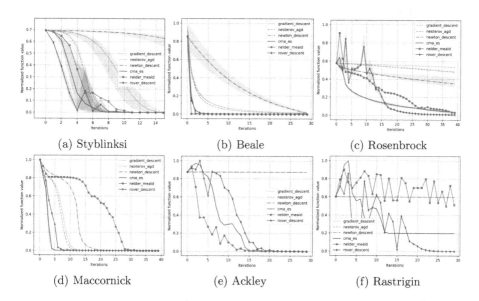

(a) Styblinksi   (b) Beale   (c) Rosenbrock

(d) Maccornick   (e) Ackley   (f) Rastrigin

Fig. 5. Tests runs on instances of modalities of $\mathcal{F}_{test}$. The shades represents the envelope of the trajectories over an entire 20-fold. Best viewed in color (color figure online).

4.2   High-Dimensional Experiments

We now test the procedure described in Sect. 3.3 for problem of dimensions $d > 2$. We propose to do this by considering a linear classifier for a binary classification task, and a small neural network classifier.

Linear Binary Classification. We generate random binary classification tasks in dimension $d > 2$, according to the framework described in [12]. We want to optimize over the cross-entropy loss induced by this dataset. We test against two fairly good optimizers for this task: tuned gradient descent and Newton descent, and randomly sample the initial iterate. We use a fixed budget $k = 10$ of dimensions we can sample at each iteration - as described in Sect. 3.3. The results for three different randomly generated datasets of different dimensions are presented in Fig. 6.

The results presented here are consistent in our experiments: our learnt procedure competes with tuned optimizers that use respectively first and second order information. However, one major downside of our optimizer is clock-time performances - one optimization run in this simple set-up can take up to a minute for $d > 50$, against a few seconds for gradient descent on the machine used for our experiments. Also, its performance is impacted by the under sampling that happens when the budget k is significantly smaller than the dimension d. Increasing the budget k improves the per-iteration performance but greatly impacts the algorithm's clock-time (as the number of operations grows quadratically with k). In the following experiments, we propose sampling strategies for the pairs of dimensions used at every update that take advantage of the problem's structure to cope with this limitation.

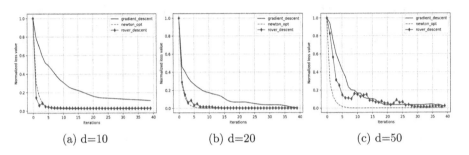

(a) d=10   (b) d=20   (c) d=50

Fig. 6. Test runs for cross entropy loss of randomly generated binary classifications tasks of dimension d, with budget $k = 10$ of pair samples per iterations.

Small Neural Network. We now want to use *Rover Descent* for a more complicated task. We consider using a small neural network in order to solve the Iris dataset [10], which consists of 150 instances of 4-dimensional inputs and their respectives labels. The model we use is a small neural network, with two hidden-layer of width 10 and a softmax output activation, alongside a cross-entropy loss. The dimension of its loss landscape is $d = 193$. We compare our results with tuned gradient descent and Adam.

Figure 7a shows the results we obtained using the same sampling strategy as presented earlier (sampling uniformly at random k dimensions for which we create all possible $k(k-1)/2$ slices). It now appears that simply increasing k is not enough to ensure good behavior. Also, increasing k implies that for every dimension we sample, we increase the number of two-dimensional moves our algorithm receives. Unlike for the convex case of linear binary classification, it appears that taking their mean value is a poor strategy.

To improve those results, we decide to use a different sampling strategy. For *every* dimension, we are going to create pairs with l other dimensions, sampled uniformly at random. The difference is that now every dimension will be used at least once at every update. The number of pairs we create is now $l \times d$. Figure 7b presents the result obtained for different values of l and proves the superiority of this approach over the previous one.

Finally, we decided to use the special structure of the neural network to improve our algorithm. We use the same sample strategy we just presented, except that now the l dimensions needed for every dimensions are sampled within a pre-defined subset. More precisely, we want to leverage a block-diagonal structure of the Hessian of the neural network: for every dimension (which corresponds to a weight or a bias of the neural network), we only create pairs with dimensions corresponding to a weight or bias belonging to the same layer. Figure 7c presents the results obtained, which again improves against the last one and compete with Adam on this task.

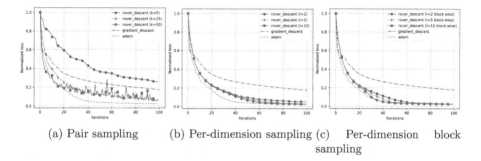

(a) Pair sampling   (b) Per-dimension sampling (c)   Per-dimension   block sampling

Fig. 7. Test runs on a small neural network for Iris dataset, with different sampling strategies for the two-dimensional slices. Best viewed in color (color figure online).

5   Conclusion

We introduced a new framework in order to achieve meta-generalization when learning to optimize. By combining tools from imitation learning and reinforcement learning, and defining the meta-dataset as a small set of prototypical functions that frequently appear in optimization problems, we were able to learn an optimization algorithm that generalizes well to unseen loss landscapes. Though this learnt optimizer is dependent on the setting of given hyper-parameters, we show that their tuning doesn't have a noticeable impact on performance, as the learnt optimizer is quite robust and can recover from bad initializations.

In our future work, we plan on running our framework on more complex, high-dimensional non-convex problems, which will require some adaptations for reducing the clock-time of our optimizer. Also, we can imagine having more control on the meta-train dataset in order to scan for complicated landscapes. For instance, following a leave-one-out procedure, we could estimate the prototypical landscapes that are needed by a learnt optimizer to train a deep neural network, and draw conclusions regarding its loss landscapes. We also plan on releasing an open-source version of our code shortly after the publication of this paper.

References

1. Agarwal, A., Wainwright, M.J., Bartlett, P.L., Ravikumar, P.K.: Information-theoretic lower bounds on the oracle complexity of convex optimization. In: Advances in Neural Information Processing Systems, pp. 1–9 (2009)
2. Andrychowicz, M., et al.: Learning to learn by gradient descent by gradient descent. In: Advances in Neural Information Processing Systems, pp. 3981–3989 (2016)
3. Bengio, Y., Simard, P., Frasconi, P.: Learning long-term dependencies with gradient descent is difficult. In: IEEE Transactions on Neural Networks (1994)
4. Billard, A.G., Calinon, S., Dillmann, R.: Learning from Humans, pp. 1995–2014. Springer International Publishing, Cham (2016). https://doi.org/10.1007/978-3-319-32552-1_74
5. Bottou, L., Curtis, F.E., Nocedal, J.: Optimization methods for large-scale machine learning (2016). arXiv:1606.04838
6. Chen, Y., et al.: Learning to learn without gradient descent by gradient descent. Learning (2017)
7. Daniel, C., Taylor, J., Nowozin, S.: Learning step size controllers for robust neural network training. In: AAAI, pp. 1519–1525 (2016)
8. Dauphin, Y.N., Pascanu, R., Gulcehre, C., Cho, K., Ganguli, S., Bengio, Y.: Identifying and attacking the saddle point problem in high-dimensional non-convex optimization. In: Advances in Neural Information Processing Systems (2014)
9. Duchi, J.C., Jordan, M.I., Wainwright, M.J., Wibisono, A.: Optimal rates for zero-order convex optimization: The power of two function evaluations. In: IEEE Transactions on Information Theory (2015)
10. Fisher, R.A.: The use of multiple measurements in taxonomic problems. Ann. Hum. Genet. 7(2), 179–188 (1936)
11. Gutmann, M., Hyvärinen, A.: Noise-contrastive estimation: a new estimation principle for unnormalized statistical models. In: Proceedings of the Thirteenth International Conference on Artificial Intelligence and Statistics, pp. 297–304 (2010)
12. Guyon, I.: Design of experiments for the NIPS 2003 variable selection benchmark (2003)
13. Hansen, N.: The CMA evolution strategy: a tutorial (2016). arXiv:1604.00772
14. Hansen, S.: Using deep Q-learning to control optimization hyperparameters (2016). arXiv:1602.04062
15. Heess, N., Hunt, J.J., Lillicrap, T.P., Silver, D.: Memory-based control with recurrent neural networks (2015). arXiv:1512.04455
16. Hochreiter, S., Schmidhuber, J.: Long short-term memory. In: Neural Computation (1997)
17. Hochreiter, S., Younger, A.S., Conwell, P.R.: Learning to learn using gradient descent. In: International Conference on Artificial Neural Networks (2001)
18. Kingma, D., Ba, J.: Adam: a method for stochastic optimization (2014). arXiv:1412.6980
19. Li, K., Malik, J.: Learning to optimize (2016). arXiv:1606.01885
20. Li, K., Malik, J.: Learning to optimize neural nets (2017). arXiv:1703.00441
21. Lillicrap, T.P., et al.: Continuous control with deep reinforcement learning (2015). arXiv:1509.02971
22. Martens, J.: Deep learning via Hessian-free optimization. In: ICML (2010)
23. Mnih, V., et al.: Human-level control through deep reinforcement learning. Nature 518(7540), 529–533 (2015)

24. Nelder, J.A., Mead, R.: A simplex method for function minimization. Comput. J. **7**(4), 308–313 (1965)
25. Nemirovskii, A., Yudin, D.B., Dawson, E.R.: Problem complexity and method efficiency in optimization (1983)
26. Nesterov, Y.: A method of solving a convex programming problem with convergence rate O(1/k2). Sov. Math. Dokl. **27**, 372–376 (1983)
27. Nocedal, J., Wright, S.: Numerical Optimization. Springer Science & Business Media, Berlin (2006)
28. Schaul, T., Antonoglou, I., Silver, D.: Unit tests for stochastic optimization (2013). arXiv:1312.6055
29. Schmidhuber, J.: Evolutionary Principles in Self-Referential Learning. On Learning now to Learn: The Meta-Meta-Meta...-Hook. Master's thesis (1987)
30. Silver, D., Lever, G., Heess, N., Degris, T., Wierstra, D., Riedmiller, M.: Deterministic policy gradient algorithms. In: Proceedings of the 31st International Conference on Machine Learning (ICML-14) (2014)
31. Sutton, R.S., Barto, A.G.: Reinforcement Learning: An Introduction. MIT Press, Cambridge (1998)
32. Wichrowska, O., et al.: Learned optimizers that scale and generalize. In: CoRR (2017)
33. Wolpert, D.H., Macready, W.G.: No free lunch theorems for optimization. In: IEEE Transactions on Evolutionary Computation (1997)
34. Zeiler, M.D.: Adadelta: an adaptive learning rate method (2012). arXiv:1212.5701

Analysis of Algorithm Components and Parameters: Some Case Studies

Nguyen Dang[1,2](✉) and Patrick De Causmaecker[1]

[1] Department of Computer Science, KU Leuven, CODeS & KULAK, Leuven, Belgium
patrick.decausmaecker@kuleuven.be
[2] School of Computer Science, University of St Andrews, St Andrews, UK
nttd@st-andrews.ac.uk

Abstract. Modern high-performing algorithms are usually highly parameterised, and can be configured either manually or by an automatic algorithm configurator. The algorithm performance dataset obtained after the configuration step can be used to gain insights into how different algorithm parameters influence algorithm performance. This can be done by a number of analysis methods that exploit the idea of learning prediction models from an algorithm performance dataset and then using them for the data analysis on the importance of variables. In this paper, we demonstrate the complementary usage of three methods along this line, namely forward selection, fANOVA and ablation analysis with surrogates on three case studies, each of which represents some special situations that the analyses can fall into. By these examples, we illustrate how to interpret analysis results and discuss the advantage of combining different analysis methods.

Keywords: Forward selection · fANOVA · Ablation analysis with surrogates · Parameter analysis

1 Introduction

Given a parameterised algorithm with different design choices and a distribution of problem instances to be solved, *algorithm configuration* is the task of choosing a good design and parameter setting (a *configuration*) of the algorithm to be used. This can be done either manually or by an automatic algorithm configurator such as *ParamILS* [11], *SMAC* [8] or *irace* [14]. Besides finding a good algorithm configuration, algorithm developers are usually also interested in understanding how different algorithm components and parameters influence algorithm performance. Various approaches for analysing the importance of algorithm parameters have been proposed, such as [1,3,6,9,10]. The insights provided by these methods may produce useful knowledge that can be transferred back into the algorithm development process in an iterative manner, leading to higher performing algorithms.

© Springer Nature Switzerland AG 2019
R. Battiti et al. (Eds.): LION 12 2018, LNCS 11353, pp. 288–303, 2019.
https://doi.org/10.1007/978-3-030-05348-2_25

Many analysis methods exploit the idea of learning prediction models from an algorithm performance dataset and then using them for the data analysis on the importance of variables. Among them, there is a group of methods that do not require a specific way of sampling algorithm performance data or particular experimental designs, are able to handle any types of algorithm parameters (including categorical ones), and have been shown to be applicable to cases with a large number of parameters. These include forward selection [9], fANOVA [10] and ablation analysis with surrogates [2]. Due to the general applicability of these methods, they can be used as a next step after the automatic algorithm configuration procedure to give more insights into the decisions of the configurator. More specifically, results of algorithm runs called by the configurator can be given to these analysis methods for building their prediction models.

Unlike the algorithm configuration problem, where the result obtained is a well-performing algorithm configuration, the analysis of algorithm components and parameters does not have a unique output. Different analysis methods can provide results on different perspectives. Forward selection identifies a key subset of algorithm parameters. fANOVA is based on functional analysis of variance [7], where the overall algorithm performance variance is decomposed into components associated with parameter subsets. Ablation analysis [6] addresses the local region between two algorithm configurations, and helps to recognise the contribution of each algorithm component or parameter on performance gain between two configurations under study.

The question of when to apply these methods is related to the kind of information we want to gain for a deeper understanding of how the analysed algorithm works. It might not always be straightforward for an algorithm developer to interpret analysis results and decide what to use for improving his/her algorithm. In the original papers of the three analysis methods [2,6,9,10], interesting example applications in domains of machine learning, propositional satisfiability, mixed-integer programming, answer set programming, and automated planning and scheduling have been discussed. The main aim of this paper is to add to this discussion by illustrating the applications of these analysis methods on a number of case studies, each of which represents a specific situation where the advantage of combining different analyses is illustrated.

The implementation of all three analysis methods is from *PIMP* (https:// github.com/automl/ParameterImportance), a Python-based package for analysis parameter importance provided by the *ML4AAD Group*.[1]

The paper is organised as follows. The three analysis methods [2,9,10] are briefly described in Sect. 2. The application and combination of the analysis methods on three case studies are then explained in Sect. 3. Finally, conclusions and future work are given in Sect. 4.

---

[1] http://www.ml4aad.org/.

2   Analysis Methods

2.1   fANOVA

The *fANOVA* method [10] is an approach for analysing the importance of algorithm parameters using a random forest prediction model and the *functional analysis of variance* [7]. Given an algorithm performance dataset, fANOVA first builds a random forest-based prediction model to predict the average performance of every algorithm configuration over the whole problem instance space. Afterwards, the functional analysis of variance is applied to the prediction model to decompose the overall algorithm performance variance into additive components, each of which corresponds to a subset of the algorithm parameters. The ratio between the variance associated with each component and the overall performance variance is then used as an indicator of the importance of the corresponding algorithm parameter subset. fANOVA also provides some insights on which regions are good and bad (with a degree of uncertainty) for each parameter inside the subset through marginal plots. Given a specific value for each algorithm parameter in the subset, the marginal prediction value is the average performance value of the algorithm over the whole configuration space associated with all parameters not belonging to the subset. A marginal plot shows the mean and the variance of the marginal prediction values given by the random forest's individual trees.

An implementation of fANOVA is provided as a Python package,[2] wrapped by the PIMP package. As a choice of implementation, the package only lists the contribution percentage and shows marginal plots of all single parameters and pairwise interactions. It is also possible to acquire the importance of a specific higher-order interaction given a fixed value for each parameter in the interaction.

2.2   Forward Selection

Forward selection is a popular method for selecting key variables in model construction. In [9], it was used to identify a subset of key algorithm parameters. Given a performance dataset of an algorithm, the method first splits this dataset into training and validation sets. Starting from an empty subset of parameters, the method iteratively adds one parameter at a time to the subset, in such a way that the regression model built on the training set using the resulting parameter subset yields the lowest root mean square error (RMSE) on the validation set. Random forest is used as the regression model, since it has been shown to be generally the best for predicting optimization algorithm performance in the literature [12]. Problem instance features can also be added into the analysis by treating them just as algorithm parameters. It should be noted that the resulting selection paths of different runs of this analysis could be different, due to a number of factors: (i) the prediction model's randomness (ii) the availability of correlated variables (iii) different splits of training and validation sets.

---

[2] https://github.com/automl/fanova.

2.3   Ablation Analysis with Surrogates

If an algorithm developer already has a default configuration for their algorithm in mind, and receives a better performing configuration from an automatic algorithm configurator, he/she might wonder which parameters in the default configuration should be modified in order to gain such performance improvement. The ablation analysis [6] answers this question by examining performance changes in the local path between the two configurations in the algorithm configuration space. Starting from the default configuration, the method iteratively modifies one parameter at a time from its default value to the value in the tuned configuration in such a way that the resulting configuration gains the largest amount of performance improvement. Important parameters in the local path can then be recognised by the percentage of their contribution to the total performance gains. The analysis can also be done in the reverse direction, in which parameters that yield the smallest performance loss are chosen first. Since results of the reverse path can be different from the original path, doing ablation in both directions was recommended.

The original ablation method [6] performs real algorithm runs during its search. Instead, in this paper, we take the surrogate-based version [2], where algorithm performance is provided by a prediction model. This allows re-using the algorithm performance dataset given by automatic algorithm configuration, hence reducing the computational cost of the ablation analysis significantly.

3   Case Studies

3.1   Case Study 1: Ant Colony Optimization Algorithms for the Travelling Salesman Problem

In this case study, we consider ACOTSP [15], a software package that implements various Ant Colony Optimization algorithms for the symmetric Travelling Salesman Problem. The algorithm has 11 parameters, including two categorical parameters (*algorithm, localsearch*), 4 integer parameters (*ants, nnls, rasrank, elitistants*), and 5 continuous parameters (*alpha, beta, rho, dlb, q0*). These are configured by the automatic algorithm configuration tool irace [14] with a budget of 5000 algorithm runs. ACOTSP's default configuration and the five best configurations returned by irace are listed in Table 1. The improvement over the default configuration is statistically significant (Wilcoxon signed rank test with a confidence level of 99%). We apply fANOVA and ablation analyses with surrogates on the performance dataset obtained after the configuration step. The analyses aim at explaining the choice of irace on selecting the best configurations.

fANOVA Analysis. There is a difference between fANOVA and irace (default version) in the way the algorithm performance is measured. The random forest in fANOVA evaluates performance of an algorithm configuration as the mean of performance values across all problem instances. The default setting of irace,

Table 1. The five best configurations returned by irace and the default configuration of ACOTSP

algorithm	localsearch	alpha	beta	rho	ants	nnls	q0	dlb	rasrank	elitistants
(Best configurations)										
acs	3	2.31	8.77	0.48	34	10	0.45	1		
acs	3	2.95	8.64	0.49	48	14	0.54	1		
acs	3	2.67	7.78	0.46	47	12	0.51	1		
acs	3	2.98	9.58	0.14	31	15	0.81	1		
acs	3	2.44	8.13	0.48	42	13	0.53	1		
(Default configuration)										
mmas	1	1.64	4.65	0.5	50	31		0		

on the other hand, uses the Friedman test as the statistical test for identifying bad configurations, which means that the performances of configurations are converted to ranks before being compared. Each strategy has its own advantage. When the ranges of performance values greatly vary among different instances (for example, due to different problem sizes), the Friedman test can avoid the dominance of instances with large performance values. However, there are cases where the magnitude of performance difference is important. For example, when a configuration is slightly better than another one on many instances, but performs dramatically worse on a smaller fraction of instances, the latter configuration may be preferable. In order to use fANOVA results to explain irace's decisions, performance values should be normalised so that they belong to the same range across different problem instances before they are given to fANOVA. For each problem instance, the normalisation value is calculated by Eq. 1:

$$normalised\_cost = \frac{cost - min\_cost}{max\_cost - min\_cost} \tag{1}$$

where min_cost and max_cost are the smallest and the largest cost values in the performance dataset for the corresponding problem instance.

Following is fANOVA's partial output on the normalised performance dataset:

```
All single-parameter effects: 87.63%
All pairwise interaction effects: 11.72%
localsearch: 77.85%
rho: 4.86%
... (remaining effects: < 4%)
```

Results indicate that single parameters and pairwise interactions can explain 99.35% of the total performance variance. The categorical parameter *localsearch*, which defines the choice of the local search used inside ACOTSP, is obviously the most important parameter. Its effect clearly dominates all of the others, as it explains 77.85% of the total algorithm performance variance. Its marginal plot shown in Fig. 1 points out that 3 is the best value for this parameter, which explains the consistent choice of *localsearch* = 3 in the best configurations returned by irace.

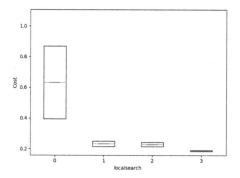

Fig. 1. Marginal plot of parameter *localsearch* given by fANOVA

Ablation Analysis. Next, we apply the ablation analysis with surrogates on the path from the default configuration to the best one using the same normalised performance dataset. Figure 2 shows the order of parameters chosen and their corresponding percentage of performance gains along the path. Again, the *localsearch* parameter shows a clear dominance over the others, which is in line with fANOVA findings. In this local region between the two configurations, the influence of this parameter is quite strong, as it provides 92% of the improvement gain when switching from the value of 2 (default) to 3.

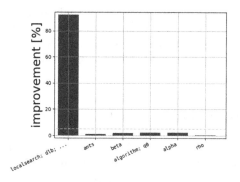

Fig. 2. Ablation analysis (with surrogate) on the path from ACOTSP default configuration to the best configuration returned by irace. The y-axis shows the percentage of performance gain when switching the value of each parameter.

It should be noted that the finding of fANOVA and ablation analysis on the most important parameters is not necessarily consistent in all cases. fANOVA analysis works on the whole configuration space, while ablation analysis focuses on the local path between two specific configurations. Therefore, the dominance of parameter *localsearch* given by both analyses emphasises the importance of this parameter on both the global and the local scale.

In the next case study, we will demonstrate an opposite situation where the analyses become more complicated and the impact of parameters on irace's choices is quite local.

3.2   Case Study 2: Tuning Irace on a Configuration Benchmark

irace is an automatic algorithm configurator, but it also has its own parameters. In [5], irace is used to configure irace on different configuration benchmarks in a procedure called *meta-tuning*. To avoid confusion, the higher-layer irace is named *meta-irace*. In an algorithm configuration setting, the tuned irace is simply considered as a parametrised optimisation algorithm, and each algorithm configuration benchmark plays the role of a problem instance. In this case study, we consider one of the meta-tuning experiments in [5], where irace is tuned on a configuration benchmark of the mixed-integer programming solver *CPLEX* [13], namely *CPLEX-REG*, with a budget of 5000 irace runs. The tuned irace has 9 parameters, including 3 categorical parameters (*elitist, testType, softRestart*), 5 integer parameters (*nbIterations, minNbSurvival, elitistInstances, mu, firstTest*) and one continuous parameter (*confidence*). The default configuration of irace and the five best configurations returned by meta-irace are shown in Table 2. The improvement given by meta-irace over the default configuration is statistically significant (Wilcoxon signed rank test with a confidence level of 99%).

The aim of the analyses in this case study is to understand the choice of the configurator (here, meta-irace) on the best irace's configurations. We will show that the interpretation of the analyses here is more complicated compared to the previous case study. In particular, the ablation analysis in the two directions can provide different results due to the complex interactions between irace's parameters. Moreover, a complementary usage of various fANOVA package's functionalities to gain more insights into the findings given by the ablation analysis is also presented.

Ablation Analysis. Ablation in both directions between two algorithm configurations was recommended [6]. The rationale for this is due to the possibility of interactions between parameters. This case study is a clear example of this situation. We will show that the reverse path can give interesting information that is not really obvious in the original path.

First, we apply the ablation from the default irace configuration to the best one for ten times. One thing we notice from the results is that the orders in which the parameters are chosen as well as their contribution on the performance gain vary quite a bit among different runs. This can be seen in the plots of three example paths shown in Fig. 3. Because of those fluctuations, it is difficult to draw any conclusions from the ablation analysis in this direction.

Table 2. irace's default configuration and the five best configurations given by meta-irace.

nbIterations	minNbSurvival	confidence	elitist	elitistInstances	testType	mu	firstTest	softRestart
(Elite configurations)								
35	2	0.52	false		t-test-holm	4	6	true
29	1	0.55	false		t-test-holm	4	5	true
37	1	0.51	false		t-test-holm	4	5	false
29	1	0.55	false		t-test-holm	3	5	true
12	1	0.77	false		t-test	7	5	true
(Default configuration)								
8	8	0.95	true	1	t-test	5	5	true

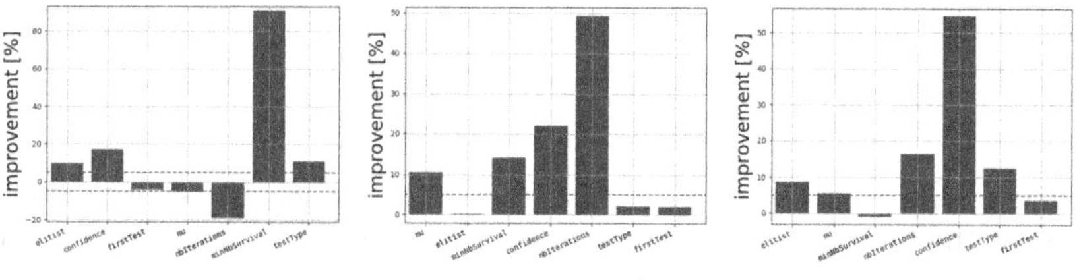

(a) Run 1   (b) Run 2   (c) Run 3

Fig. 3. Three ablation paths from the irace default configuration to the best one returned by meta-irace.

We apply another ten times the ablation, now on the reverse path from the best configuration to the default configuration. At each step, the parameter that introduces the least performance loss when changing its value from the best configuration to the default one is chosen. Figure 4 shows the reverse paths with the same random seeds as in the ones used in Fig. 3. We can see that the orders in which the parameters are chosen along the paths are more consistent among different runs. In particular, the four parameters *nbIterations, confidence, minNbSurvival* and *elitist* are always chosen at the end of the path, which means that changing their values in the best configuration will cause more performance loss than all other parameters. Indeed, most of the performance loss on the paths are due to the three parameters *nbIterations, confidence, minNbSurvival*. Parameter *elitist* is constantly the last one chosen in the ten paths although it is not explicitly associated with a big loss in performance. This indicates two things: (i) the importance of setting this parameter at the right value for achieving the good performance of the best configuration, as changing its value in the path between the best configuration to the default one will induce larger performance loss than any other parameters, (ii) the strong dependency of this parameter and the others, especially the three parameters *nbIterations, confidence, minNb-*

Survival; or, in other words, the possibly complicated interactions between these parameters in the local region between the two configurations under study.

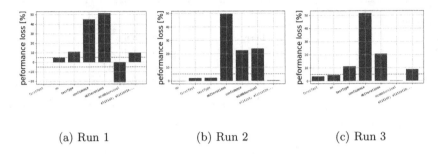

      (a) Run 1   (b) Run 2   (c) Run 3

Fig. 4. Three ablation paths from the best one returned by meta-irace the default irace configuration.

fANOVA Analysis. Next, we use fANOVA analysis to gain additional information on the findings given by the ablation. In particular, we want to see how parameter *elitist* interacts with the others, especially the three parameters *nbIterations, confidence, minNbSurvival*, and whether their impacts are global.

We do two fANOVA analyses, one on the global space, dubbed *global-fANOVA* and one on the local space where the algorithm performance is not worse than the default one, namely *capped-fANOVA*. Parts of them are shown in Table 3.

Table 3. Partial results of fANOVA analyses on irace's performance data

global-fANOVA	capped-fANOVA
All single-parameter effects: 38.01%	All single-parameter effects: 11.69%
All pairwise interaction effects: 34.31%	All pairwise interaction effects: 28.17%
nbIterations: 11.8%	firstTest x nbIterations: 3.89%
minNbSurvival: 6.22%	minNbSurvival: 3.28%
firstTest: 5.73%	firstTest x minNbSurvival: 2.82%
testType: 5.25%	firstTest: 2.55%
confidence: 5.21%	minNbSurvival x nbIterations: 2.26%
... (remaining effects: <4%)	... (remaining effects: <2%)

The differences between the two analyses are that in the local one, single parameter effects become less important, both single parameter and pairwise interaction effects have less contribution to the total performance variance, and their contribution is more widely spread. One possible explanation for these differences is that in the local space, the interactions of algorithm parameters are more complicated. Anyway, one clear thing is that *elitist* parameter is not among the top list effects in both analyses.

In capped-fANOVA results, the total contribution of all single parameters and pairwise interactions on the overall performance variance is less than 40%, leaving a potential for some important higher-order interaction. We can check if this is the case for any of the interactions between *elitist* and the three parameters *nbIterations*, *confidence*, *minNbSurvival* by giving those four parameters to fANOVA and asking for the percentage of variance explained by all relevant interactions. Results show that the total contribution of all (pairwise and higher-order) interactions between *elitist* and the other three parameters is only 5.18%, and it generally is spread evenly among them. It means that there are no important interactions as expected. Since capped-fANOVA works on the configuration space where performance value is not worse than the default configuration, the observations in this analysis indicate that the importance of *elitist* is even more local, i.e., it only works when not only the three parameters *nbIterations*, *confidence*, *minNbSurvival* are set properly, but also the other ones have to be in the appropriate ranges. In fact, if we look at the values of the other parameters in the default and the best configurations, we can see that they are already quite close to each other. For example, the parameter *firstTest* has the values of 5 and 6 in the default and the best configurations, respectively, while its domain is [4, 20]. We can further confirm our conjecture as follows. First, we ask fANOVA for the predicted average cost of the default and the best configurations (here we get 138.92 and 135.81). Next, we modify the value of parameter *firstTest* in the best configuration to a more distant one, say 12. Then we ask fANOVA for the predicted cost of this new configuration (here we get 138.78). This increasing cost implies the importance of having *firstTest* near the local region around 6.

In summary, the conclusion from this fANOVA analysis is that the importance of irace parameters and the choice of the meta-irace are very local, i.e., although the four parameters *elitist*, *nbIterations*, *confidence*, *minNbSurvival* play essential roles in improving the performance between the default and the best configurations, the other parameters also need to be set in the appropriate ranges that are not far from their values indicated in the best configuration. Therefore, if irace's developers want to explain what changes in irace's behaviours (according to its parameters) make the performance improved, they will need to associate those changes with all parameters instead of only a few of them.

3.3 Case Study 3: A Large Neighbourhood Search for a Vehicle Routing Problem with Time Windows

The example analyses so far are only on the algorithm configuration space. In this case study, we illustrate the integration of problem instance features into fANOVA analysis, so that we can study not only the impact of algorithm parameters, but also their interactions with instance features.

The algorithm and problem instances considered in this case study are provided in [4]. The algorithm is a Large Neighbourhood Search (LNS) metaheuristic algorithm for solving a typical Vehicle Routing Problem with Time Windows. The algorithm has 8 parameters, including 2 categorical parameters (*repair*, *destroy*), 2 integer parameters (*random_seed*, *deterministic_parameter*) and 4 continuous parameters (*noise_parameter*, *cooling_rate*, *control_parameter*, *start_temperature*). There are 200 problem instances generated randomly according to 5 problem features, including 3 integer features (*customer_numer*, *customer_demand*, *max_running_time*) and 2 continuous features (*average_service_time*, *average_time_windows*). For each problem instance, 20 random algorithm configurations are randomly generated and tested. In total, the performance dataset contains 4000 data points.

fANOVA Analysis. We can add instance features into fANOVA analysis by treating them as algorithm parameters, i.e., each combination of algorithm configuration and instance feature is given to fANOVA as an algorithm configuration. It must be noted that this integration can only be done when the instance features are independent, since this is one of the key assumptions fANOVA makes on its input variables. In this case study, the problem instances are generated based on random values drawn from all instance features in an independent way. A part of fANOVA's output is given in the left column of Table 4.

Table 4. Partial results of fANOVA analyses on LNS algorithm performance data

On original performance data	On normalised performance data
All single-parameter effects: 80.87%	All single-parameter effects: 60.74%
All pairwise interaction effects: 12.23%	All pairwise interaction effects: 15.38%
customer_number: 75.24%	repair: 52.68%
average_service_time x customer_number: 5.24%	destroy x repair: 4.42%
average_service_time: 3.78%	customer_number x repair: 4.19%
... (remaining effects: <2%)	customer_number: 4.0%
	destroy: 3.14%
	... (remaining effects: <3%)

An obvious observation from this result is that only instance features are listed as the most important variables. In particular, the feature *customer_number* has a clear dominant influence, as it explains 75.24% of the total performance variance. Looking at its marginal plot shown in Fig. 5, we can see that the average performance values increase according to this feature. This means that the range of the algorithm performance values greatly depends on problem instance size (here it is the number of customers).

In situations like this, depending on the aim of our analysis, normalisation can be necessary. Here we aim to study the algorithm performance without any

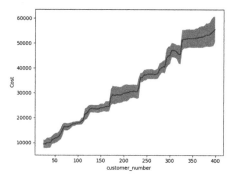

Fig. 5. Marginal plot of instance feature *customer_number* given by fANOVA

bias towards particular instances, which is similar to what irace's default setting does as described in Sect. 3.1. Therefore, we re-run the fANOVA analysis, now with normalised performance values. A part of the new results is listed in the right column of Table 4. The impact of *customer_number* is now significantly reduced, so that the influence of algorithm parameters becomes visible. The two categorical algorithm parameters *repair* and *destroy*, which represent the choices of the repair and the destroy operators inside the Large Neighbourhood Search, now involves in the top list of important parameters and pairwise interactions. In particular, the high contribution percentage of parameter *repair* (52.68%) on the total performance variance and its marginal plot (Fig. 6a) show that the repair operators Regret2 and Greedy work generally best and worst, respectively, across the whole problem instance space. In addition, we can also see how much influence the choice is according to different problem sizes by looking at the marginal plot of the pairwise interaction *repair* x *customer_number* (Fig. 6b): on instances with the number of customers less than 50, the bad performance of Greedy is not as obvious as on instances with larger sizes; for the other two choices, Regret2 and GreedyRegret2, their overall performance difference also gets clearer when the number of customers increases. Other problem instance features do not have important interactions with algorithm parameters.

Forward Selection. Although the fANOVA analysis provides interesting insights into interactions between algorithm parameters and instance features, such an analysis, in this case study, takes a large amount of time to run (16 hours on a single core Intel CPU 2.4Ghz)[3]. The reason is due to the heavy computation of all pairwise interactions. In this part, we demonstrate a cheaper alternative for this particular case using a combination of fANOVA and forward selection. First, we use fANOVA to calculate the importance of single variables. Next, we use forward selection to identify the key subset of variables, and then

---

[3] This amount of running time is reported on the Python-based fANOVA package linked by PIMP. The previous fANOVA version, which is Java-based, is faster, although it still needs several hours to finish this analysis.

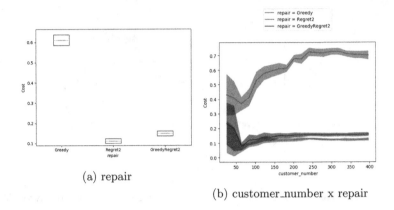

(a) repair

(b) customer_number x repair

Fig. 6. Marginal plots of the algorithm parameter *repair* and its pairwise interaction with the instance feature *customer_number* given by fANOVA analysis with performance normalisation.

use fANOVA to examine interactions of the variables in that set only. In this way, the whole analysis takes less than half an hour.

First, we apply fANOVA on single parameter effects only, which takes less than one minute to finish. Next, we apply forward selection to the normalised performance dataset. Since analysis results might vary among different runs, we repeat it ten times. It takes about 3 min in total. The visualisation of one run's result, which is a path of sequentially adding one variable to the model at a time so that the resulting root mean square error on the validation set is minimised, is shown in Fig. 7a. We can see that the first three parameters chosen are *repair*, *destroy* and *customer_number*. This happens to be the case for all the ten runs; these three parameters are always chosen first, while the fourth parameter onwards in the path can vary among different runs. This indicates the high relevance of these three parameters on the prediction model.

Since these experiments are computationally cheap, we also apply backward elimination, the reverse procedure of forward selection, where starting from the whole variable set, one variable resulting the least decrease in the prediction error is removed sequentially. This may help to take into account dependency between the parameters. These runs take about 5 min in total. The chosen path and the error values in one of the ten runs is shown in Fig. 7b. The three parameters *repair*, *destroy* and *customer_number* are the last one to be chosen, and this is again consistent among the ten runs. This implies the importance of these parameters' interactions.

Given those indications, we can now focus on analysing the interactions between these three variables using fANOVA. Results of the importance quantification and marginal plots given the three variables *repair*, *destroy* and *customer_number* are obviously the same as the original fANOVA, with an addition of the three-way interaction between those variables. However, since the importance of this interaction is rather small (2.3%), we can simply ignore it. This

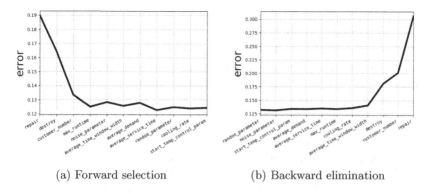

(a) Forward selection   (b) Backward elimination

Fig. 7. Prediction error of the model trained on a subset of variables chosen during one run of forward selection and backward elimination analyses.

particular fANOVA analysis takes about two minutes. So the whole analysis only needs 12 min.

4   Conclusions and Future Work

In this paper, we have demonstrated and discussed the complementary usage of three parameter analysis techniques, namely forward selection [9], fANOVA [10] and ablation [2,6] analyses. This is done using three case studies, each of which represents some special situations that the analyses can fall into. In the first and the second case study, where the aim is to use fANOVA and ablation analyses to understand the choice of the best configuration given by an automatic algorithm configuration tool (here irace), analysis conclusions diverge: in the first one, there is a dominant parameter in both the global configuration space and the local space between the default and best algorithm configurations, so that the algorithm developers can focus on this particular parameter for explaining the improvement over the default configuration; in the second one, the impact of parameters are very local and the algorithm performance improvement gained requires all parameters to be in their reasonable ranges of values, which means that the developers need to look at all parameter values as a whole instead of focusing on a few of them. In the third case study, the integration of problem instance features into fANOVA analysis is illustrated. This shows how the range of algorithm performance values changes according to problem instance sizes; hence, it emphasises the necessity of normalising performance data to avoid bias. The interaction between algorithm parameters and instance features also provides interesting insights into how the impact of important algorithm parameters can vary among different problem sizes. Moreover, a complementary usage of the forward selection analysis that can help to significantly reduce the computation cost of fANOVA analysis in this particular case study is also presented.

In future work, more case studies can be investigated, especially the ones for local search metaheuristics and evolutionary algorithms in solving combinatorial optimisation problems, where the applications of these analysis techniques are still scarce. Moreover, the integration of problem instance features into the combined analyses can be further investigated. As performance of optimisation algorithms might greatly vary according to different types of problem instances, insights acquired by studying both algorithm parameters and problem instance features would provide useful information for algorithm developers.

Acknowledgement. This work is funded by COMEX (Project P7/36), a BEL-SPO/IAP Programme. The computational resources and services were provided by the VSC (Flemish Supercomputer Center), funded by the Research Foundation - Flanders (FWO) and the Flemish Government - department EWI. The authors are grateful to Thomas Stützle and the anonymous reviewers for their valuable comments, which help to improve the quality of the paper.

References

1. Bartz-Beielstein, T., Lasarczyk, C., Preuss, M.: The sequential parameter optimization toolbox. In: Bartz-Beielstein, T., Chiarandini, M., Paquete, L., Preuss, M. (eds.) Experimental Methods for the Analysis of Optimization Algorithms, pp. 337–362. Springer, Heidelberg (2010). https://doi.org/10.1007/978-3-642-02538-9_14
2. Biedenkapp, A., Lindauer, M., Eggensperger, K., Hutter, F., Fawcett, C., Hoos, H.H.: Efficient parameter importance analysis via ablation with surrogates. In: Singh, S.P., Markovitch, A. (eds.) AAAI Conference on Artificial Intelligence. AAAI Press (2017)
3. Chiarandini, M., Goegebeur, Y.: Mixed models for the analysis of optimization algorithms. Exp. Methods Anal. Optim. Algorithms 1, 225 (2010)
4. Corstjens, J., Caris, A., Depaire, B., Sörensen, K.: A multilevel methodology for analysing metaheuristic algorithms for the VRPTW
5. Dang, N., Pérez Cáceres, L., De Causmaecker, P., Stützle, T.: Configuring irace using surrogate configuration benchmarks. In: Proceedings of the Genetic and Evolutionary Computation Conference, pp. 243–250. ACM (2017)
6. Fawcett, C., Hoos, H.H.: Analysing differences between algorithm configurations through ablation. J. Heuristics **22**(4), 431–458 (2016)
7. Hooker, G.: Generalized functional ANOVA diagnostics for high-dimensional functions of dependent variables. J. Comput. Graph. Stat **16**(3), 709–732 (2012)
8. Hutter, F., Hoos, H.H., Leyton-Brown, K.: Sequential model-based optimization for general algorithm configuration. In: Coello, C.A.C. (ed.) LION 2011. LNCS, vol. 6683, pp. 507–523. Springer, Heidelberg (2011). https://doi.org/10.1007/978-3-642-25566-3_40
9. Hutter, F., Hoos, H.H., Leyton-Brown, K.: Identifying key algorithm parameters and instance features using forward selection. In: Nicosia, G., Pardalos, P. (eds.) LION 2013. LNCS, vol. 7997, pp. 364–381. Springer, Heidelberg (2013). https://doi.org/10.1007/978-3-642-44973-4_40
10. Hutter, F., Hoos, H.H., Leyton-Brown, K.: An efficient approach for assessing hyperparameter importance. In: Proceedings of the 31th International Conference on Machine Learning, vol. 32, pp. 754–762 (2014)
11. Hutter, F., Hoos, H.H., Leyton-Brown, K., Stützle, T.: ParamILS: an automatic algorithm configuration framework. J. Artif. Intell. Res **36**, 267–306 (2009). Oct

12. Hutter, F., Xu, L., Hoos, H.H., Leyton-Brown, K.: Algorithm runtime prediction: methods and evaluation. Artif. Intell. **206**, 79–111 (2014)
13. IBM. ILOG CPLEX optimizer (2017). http://www.ibm.com/software/integration/optimization/cplex-optimizer/
14. López-Ibáñez, M., Dubois-Lacoste, J., Pérez Cáceres, L., Stützle, T., Birattari, M.: The irace package: Iterated racing for automatic algorithm configuration. Oper. Res. Perspect. **3**, 43–58 (2016)
15. Stützle, T.: ACOTSP: a software package of various ant colony optimization algorithms applied to the symmetric traveling salesman problem (2002)

Optimality of Multiple Decision Statistical Procedure for Gaussian Graphical Model Selection

Valery A. Kalyagin[1,2]([✉]), Alexander P. Koldanov[1,2], Petr A. Koldanov[1,2], and Panos M. Pardalos[1,2]

[1] Laboratory of Algorithms and Technologies for Network Analysis, National Research University Higher School of Economics, Bolshaya Pecherskaya 25/12, Nizhny Novgorod, Russia
vkalyagin@hse.ru, pardalos@ufl.edu
[2] Center for Applied Optimization, University of Florida, 401 Weil Hall, P.O. Box 116595, Gainesville, FL 32611-6595, USA

Abstract. Gaussian graphical model selection is a statistical problem that identifies the Gaussian graphical model from observations. Existing Gaussian graphical model selection methods focus on the error rate for incorrect edge inclusion. However, when comparing statistical procedures, it is also important to take into account the error rate for incorrect edge exclusion. To handle this issue we consider the graphical model selection problem in the framework of multiple decision theory. We show that the statistical procedure based on simultaneous inference with UMPU individual tests is optimal in the class of unbiased procedures.

Keywords: Gaussian graphical models · Multiple Decision · Optimal multiple decision statistical procedures · Unbiased multiple decision statistical procedures

1 Introduction

Gaussian graphical models are known to be a useful tool in different applied fields such as bioinformatics, error-control codes, speech and language recognition, and information retrieval [8]. Given a sample of observations, an important question is how to recover the structure of an undirected Gaussian graph. This problem is called the Gaussian graphical model selection problem (GGMS). A comprehensive survey of different approaches to this problem is given in [2].

Traditional measures of quality of statistical procedures for GGMS used in the literature include the FWER (family wise error rate), FDR (false discovery rate), and FDP (false discovery proportion) [2]. Most of these measures are connected with Type I error (false edge inclusions). However, when analyzing the quality of GGMS procedures it is important to take into account both types of errors, Type I and Type II errors. To deal with this issue, we consider the GGMS

© Springer Nature Switzerland AG 2019
R. Battiti et al. (Eds.): LION 12 2018, LNCS 11353, pp. 304–308, 2019.
https://doi.org/10.1007/978-3-030-05348-2_26

problem within the framework of multiple decision theory [5]. We measure the quality of statistical procedures using a risk function with additive losses. The main topic addressed in this paper is construct an optimal (in the sense of the value of risk function) statistical procedures for GGMS. To find an optimal statistical procedure, we combine Neyman structure tests for individual hypotheses with simultaneous inference. We show that obtained multiple decision statistical procedure is optimal in the class of unbiased multiple decision procedures.

2 Problem Statement

Let $X = (X_1, X_2, \ldots, X_p)$ be a random vector with the multivariate Gaussian distribution from $N(\mu, \Sigma)$, where $\mu = (\mu_1, \mu_2, \ldots, \mu_p)$ is the vector of means and $\Sigma = (\sigma_{i,j})$ is the covariance matrix, $\sigma_{i,j} = \text{cov}(X_i, X_j)$, $i, j = 1, 2, \ldots, p$. Let $x(t)$, $t = 1, 2, \ldots, n$ be a sample of size n from the distribution of X. We assume in this paper that $n > p$, and that the matrix Σ is non degenerate. The undirected Gaussian graphical model is an undirected graph with p nodes. The nodes of the graph are associated with the random variables $X_1, X_2, \ldots, X_p$, edge (i, j) is included in the graph if the random variables X_i, X_j are conditionally dependent [1,4]. Gaussian graphical model selection problem consists of the identification of a graphical model from observations.

It is known [1] that conditional independence of the random variables X_i, X_j is equivalent to the following relation $\rho^{i,j} = 0$, where $\rho^{i,j}$ is the partial correlation between X_i, X_j given all other components of X. Hence, the problem of pairwise conditional independence testing is equivalent to

$$h_{i,j} : \rho^{i,j} = 0 \text{ vs } k_{i,j} : \rho^{i,j} \neq 0, \quad i \neq j, i, j = 1, 2, \ldots, p \qquad (1)$$

3 Multiple Decision Approach

In this Section we consider the GGMS problem in the framework of decision theory [7]. According to this approach we specify the risk function. Let $X = (X_1, X_2, \ldots, X_p)$ be a random vector with multivariate Gaussian distribution from $N(\mu, \Sigma)$. Observations are modeled in the GGMS study as a sequence of random vectors $X(t)$, $t = 1, 2, \ldots, n$ where n is the sample size and vectors $X(t)$ are i.i.d. as X. Let $x = (x_i(t))$ be observations of the random variables $X_i(t)$. Consider the set $\mathcal{G}$ of all $p \times p$ symmetric matrices $G = (g_{i,j})$ with $g_{i,j} \in \{0, 1\}$, $i, j = 1, 2, \ldots, p$, $g_{i,i} = 0$, $i = 1, 2, \ldots, p$. Matrices $G \in \mathcal{G}$ represent adjacency matrices of all simple undirected graphs with p vertices. The total number of matrices in $\mathcal{G}$ is equal to $L = 2^M$ with $M = p(p-1)/2$. The *GGMS problem* can be formulated as a multiple decision problem of the choice of one from L hypotheses:

$$H_G : \rho^{i,j} = 0, if g_{i,j} = 0, \quad \rho^{i,j} \neq 0 \text{ if } g_{i,j} = 1; \quad i \neq j, \quad i, j = 1, 2, \ldots, p \qquad (2)$$

The multiple decision statistical procedure $\delta(x)$ is a map from the sample space $R^{p \times n}$ to the decision space $D = \{d_G, g \in \mathcal{G}\}$, where the decision d_G is the acceptance of hypothesis H_G, $G \in \mathcal{G}$. Let $\varphi_{i,j}(x)$ be tests for the individual hypothesis (1). More precisely, $\varphi_{i,j}(x) = 1$ means that edge (i,j) is included in the graphical model, and $\varphi_{i,j}(x) = 0$ means that edge (i,j) is not included in the graphical model. Let $\Phi(x)$ be the matrix

$$\Phi(x) = \begin{pmatrix} 0 & \varphi_{1,2}(x) & \cdots & \varphi_{1,p}(x) \\ \varphi_{2,1}(x) & 0 & \cdots & \varphi_{2,p}(x) \\ \cdots & \cdots & \cdots & \cdots \\ \varphi_{p,1}(x) & \varphi_{p,2}(x) & \cdots & 0 \end{pmatrix}. \tag{3}$$

Any multiple decision statistical procedure $\delta(x)$ based on the simultaneous inference of individual edge tests (1) can be written as $\delta(x) = d_G$ iff $\Phi(x) = G$.

Let $a_{i,j}$ be the loss from the false inclusion of edge (i,j) in the graphical model, and let $b_{i,j}$, be the loss from the false non inclusion of the edge (i,j) in the graphical model. We say that the loss function is additive, if the total loss from the misclassification of the decision d_S when d_Q is true is the sum of losses from the misclassification of individual edges:

$$w(S,Q) = \sum_{\{i,j:s_{i,j}=0;q_{i,j}=1\}} a_{i,j} + \sum_{\{i,j:s_{i,j}=1;q_{i,j}=0\}} b_{i,j} \tag{4}$$

4   Optimal Multiple Decision Procedures

According to Wald [7] the procedure δ^* is referred to as optimal in the class of statistical procedures $\mathcal{C}$ if

$$Risk(S,\theta;\delta^*) \leq Risk(S,\theta;\delta), \tag{5}$$

for any $S \in \mathcal{G}$, $\theta \in \Omega_S$, $\delta \in \mathcal{C}$.

In this paper we consider the class of w-unbiased statistical procedures. Statistical procedure $\delta(x)$ is referred to as w-unbiased if one has

$$Risk(S,\theta;\delta) = E_\theta w(S;\delta) \leq E_\theta w(S';\delta) = Risk(S',\theta;\delta), \tag{6}$$

for any $S, S' \in \mathcal{G}$, $\theta \in \Omega_S$. The unbiasedness of statistical procedure δ for a general risk function means that δ comes closer to the true decision than to any wrong [6], Ch. 1.

Optimal unbiased tests for individual hypothesis $h_{i,j}$ were constructed in our paper [3] using the tests of a Newman structure. It was shown that optimal unbiased test is equivalent to the classical partial correlation test

$$\varphi_{i,j}^{opt} = \begin{cases} 0, 2q_{i,j} - 1 < r^{i,j} < 1 - 2q_{i,j} \\ 1, \text{otherwise}, \end{cases} \tag{7}$$

where $r^{i,j}$ is the sample partial correlation, and $q_{i,j}$ is the $(\alpha_{i,j}/2)$-quantile of the beta distribution $Be(\frac{n-p}{2}, \frac{n-p}{2})$.

Define the multiple decision statistical procedure δ^{opt} by (3) with $\varphi_{i,j} = \varphi_{i,j}^{opt}$. The following theorem shows that δ^{opt} is optimal in the class of w-unbiased multiple decision procedures if significance levels of individual tests are related with individual losses.

Theorem 1. *Let the loss function w be defined by (4) and*

$$\alpha_{i,j} = \frac{b_{i,j}}{a_{i,j} + b_{i,j}}, \ i \neq j, \ i,j = 1,2,\ldots,p. \tag{8}$$

Then the procedure δ^{opt} is optimal multiple decision statistical procedure for Gaussian graphical model selection in the class of w-unbiased procedures.

Sketch of the proof. We use a basic approach by Lehman [5]. Let δ be a statistical procedure defined by (3). If the loss function is additive then the risk of statistical procedure δ is the sum of the risks of individual tests $\varphi_{i,j}$. Then we prove that δ^{opt} is a w-unbiased multiple decision procedure. Using this, we prove that δ^{opt} is optimal in the class of w-unbiased statistical procedures.

5   Discussion

The main goal of this paper is to develop the multiple decision approach by Lehmann for the GGMS problem. In our approach, we measure the quality of selected statistical procedures using a risk function. The GGMS problem results presented in [2] can be considered within the framework of this general approach as a specific choice of the risk function. Namely, the FWER (probability of at least one false edge inclusion) used in [2] is equal to the risk function where the losses are defined as $w(S,Q) = 1$ if there exists a pair (i,j) such that $s_{i,j} = 0$ but $q_{i,j} = 1$, and $w(S,Q) = 0$ otherwise. The new approach allows us to define and construct an optimal multiple decision statistical procedure for GGMS. Optimality of GGMS statistical procedures is an important question, which is not well investigated. We present here a first step in this direction.

Acknowledgments. The Sects. 1 and 2 of the article were prepared within the framework of the Basic Research Program at the National Research University Higher School of Economics (HSE). The Sect. 4 was prepared with a support of RSF grant 14-41-00039.

References

1. Anderson, T.W.: An Introduction to Multivariate Statistical Analysis, 3rd edn. Wiley-Interscience, New York (2003)
2. Drton, M., Perlman, M.: Multiple testing and error control in Gaussian graphical model selection. Stat. Sci. **22**(3), 430–449 (2007)
3. Koldanov, P., Koldanov, A., Kalyagin, V., Pardalos, P.: Uniformly most powerful unbiased test for conditional independence in Gaussian graphical model. Stat. Probab. Lett. **122**, 90–95 (2017)

4. Lauritzen, S.L.: Graphical Models. Oxford University Press, Oxford (1996)
5. Lehmann, E.L.: A theory of some multiple decision problems, I. Ann. Math. Stat. **28**, 1–25 (1957)
6. Lehmann, E.L., Romano, J.P.: Testing Statistical Hypotheses. Springer, New York (2005)
7. Wald, A.: Statistical Decision Functions. Wiley, New York (1950)
8. Wainwright, M.J., Jordan, M.I.: Graphical models, exponential families, and variational inference. Found. Trends Mach. Learn. **1**(1–2), 1–305 (2008)

Hyper-Reactive Tabu Search for MaxSAT

Carlos Ansótegui[1], Britta Heymann[3], Josep Pon[1], Meinolf Sellmann[4],
and Kevin Tierney[2(✉)]

[1] DIEI, Universitat de Lleida, Lleida, Spain
{carlos,jponfarreny}@diei.udl.ca
[2] Bielefeld University, Bielefeld, Germany
kevin.tierney@uni-bielefeld.de
[3] ORCONOMY GmbH and Paderborn University, Paderborn, Germany
britta.heymann@orconomy.de
[4] General Electric, Niskayuna, NY, USA
meinolf@ge.com

Abstract. Local search metaheuristics have been developed as a general tool for solving hard combinatorial search problems. However, in practice, metaheuristics very rarely work straight out of the box. An expert is frequently needed to experiment with an approach and tweak parameters, remodel the problem, and adjust search concepts to achieve a reasonably effective approach. Reactive search techniques aim to liberate the user from having to manually tweak all of the parameters of their approach. In this paper, we focus on one of the most well-known and widely used reactive techniques, reactive tabu search (RTS) [7], and propose a hyper-parameterized tabu search approach that dynamically adjusts key parameters of the search using a learned strategy. Experiments on MaxSAT show that this approach can lead to state-of-the-art performance without any expert user involvement, even when the metaheuristic knows nothing more about the underlying combinatorial problem than how to evaluate the objective function.

1 Introduction

Making a local search approach work effectively for a practical problem is often tedious work. Frequently, it requires an expert who can reformulate the problem and adjust search parameters in such a manner that real-world problem instances can be solved with sufficient performance. Historically, metaheuristics that did not require tuning parameters have been favored in the literature, as this should liberate the user from tediously searching for an instantiation of the metaheuristic that leads to good practical performance for the problem at hand.

To provide good performance even without many parameters, reactive search approaches (see [8]) have been developed that use built-in strategies to automatically and dynamically adjust key parameters during search. For example, the canonical reactive search algorithm, reactive tabu search (RTS) [7], adapts the central parameter of tabu search, namely the length of the tabu list. RTS increases the tabu list length when cycles that occur in the search are shorter

© Springer Nature Switzerland AG 2019
R. Battiti et al. (Eds.): LION 12 2018, LNCS 11353, pp. 309–325, 2019.
https://doi.org/10.1007/978-3-030-05348-2_27

than a fixed threshold. The list length is decreased when all potential moves are tabu as well as when the number of steps taken since the last list length adjustment grows beyond the moving average of recent cycle lengths.

Hyper-reactive search replaces the static parameter adjustment strategies of reactive search with dynamic strategies that are tuned through an offline learning process. For example, the dialectic search procedure [16] is extended in [5] so that its parameters are adjusted online according to a strategy learned offline. The resulting hyper-reactive dialectic search (HRDS) algorithm is shown in [5] to outperform state-of-the-art MaxSAT solvers.

Given the success of RTS in the literature, a key question is whether it can be transformed into a hyper-reactive approach. In this paper, we pursue three objectives.

- First, we devise an approach that self-adjusts even more tabu search parameters than RTS.
- Second, we open the way the approach adjusts its parameters to the outside so that the overall approach can be customized and tailored for the problem at hand using a standard parameter tuner.
- Finally, we use the resulting, generally applicable approach to tackle the MaxSAT problem and demonstrate high quality performance.

In the following, we first discuss related work and then devise the hyper-parameterized reactive tabu search approach that self-adjusts all key tabu search parameters based on dynamic search statistics. We then show how this new approach can be used to tackle the MaxSAT problem. Finally, we report on our experimental results on the MaxSAT problem.

2   Related Work

The goal of reducing the reliance on human experts when solving search problems has been the focus of a number of communities and works. We provide a brief overview of the literature in this area.

Hyper-heuristics [9,10] are "heuristics to choose heuristics". Given a combinatorial optimization problem to be solved, hyper-heuristics attempt to dynamically choose the correct heuristic to apply at any given time, potentially switching between multiple search paradigms (genetic algorithms, local search, etc.) during a single solution procedure. Reinforcement learning is used in [22] to adjust the search strategies and parameters for a hyper-heuristic based great deluge algorithm applied to a timetable examination problem. A further work in this area, [21], adjusts heuristic selections and adapts parameters dynamically with static, reactive strategies during search.

Doerr and Doerr show that the population size of an evolutionary algorithm can be optimally controlled for a generalized ONEMAX problem [11,12]. While this result it not for a "real-world" problem, it provides strong evidence that online adjustment strategies are much more effective than their static counterparts.

The idea of hyper-parameterizing local search was pioneered in [17] who invented a solver named SATenstein. SATenstein is an approach for tackling the SAT problem and can be instantiated to process like a number of successful local search approaches that have been developed for SAT earlier. That is, SATenstein's hyper-parameters determine what search strategy is used. Therefore, SATenstein can be trained for specific families of SAT instances using a parameter tuner. In fact, using instance-specific tuning, a solver like SATenstein can even be configured to set its own hyper-parameters based on characteristics of the concrete instance to be solved [15,27]. Essentially, different instantiations of the same solver then form an algorithm portfolio [19].

Building on this idea, a hyper-parameterization of dialectic search [16] was proposed in [5]. Here, hyper-parameters are used to determine how the standard parameters of dialectic search are adjusted *dynamically* based on runtime statistics that characterize how the search is progressing. In the following, we present how this idea can be realized for the Tabu Search metaheuristic.

3   Hyper-Parameterized Reactive Tabu Search

We now describe a hyper-parameterized tabu search that is based on RTS [7].

3.1   Tabu Search Parameters

We assume the reader is familiar with tabu search in general. For an introduction, we refer to [13,14]. The key parameters of the general tabu search metaheuristic are the following:

1. Length of the tabu list(s)
2. Escape rules
3. Neighborhood size
4. Aspiration criterion

The tabu search framework we develop here is calibrated for binary search problems, meaning unconstrained optimization problems where all variables take two values (0/1 or true/false). In this context, let us consider the above parameters in detail.

Length of Tabu Lists: We use two lists, one that prevents recent solutions from being revisited, and another that keeps track of recently flipped variables that are kept from flipping again too quickly (unless an aspiration criterion allows it anyway). The length of the lists is governed by four dynamically self-updating continuous parameters in $[0, 1]$; two for each list.

After every tenth local search step, for each list, we consider the difference between both parameters. If the difference is positive, the list size is considered to be increased, otherwise it is considered to be decreased, whereby the absolute value of the difference determines the probability that a change in list length occurs. If a change occurs, the list of previously visited solutions grows or decreases by a factor of 1.01. The list of recently flipped variables increases or decreases by one one thousandth of the total number of variables.

As an aside: For the list of previously visited solutions we actually only record the variables that have been flipped. Then, to determine which variables are not allowed to be flipped in the current step, we simply traverse backwards through the list and add or remove the variables to/from a set. Every time the set has cardinality one, the remaining variable in the set cannot be flipped as otherwise a previously visited solution recurs.

Consider this example: The variables most recently flipped (most recent first) had indices $[1, 2, 3, 4, 3, 2, 1, 5, 4]$. We first add variable 1 to the set: $\{1\}$. This set has cardinality 1, which means that the variable in the set may not be flipped (naturally, as that would directly undo the most recent move and thus return us directly to a recently visited solution). Next we add 2, 3, and 4. The set now contains $\{1, 2, 3, 4\}$. Next we consider variable 3, but since it is already in the set, we remove it: $\{1, 2, 4\}$. After also removing 2 and 1, we get: $\{4\}$. This set now has cardinality 1, which means that we cannot flip variable 4 either or we will return to the solution visited 7 steps earlier. We then add 5 and remove 4, leading to the final set of cardinality 1: $\{5\}$. Consequently, at the current step we may neither flip variables 1, 4 or 5 if we want to avoid revisiting a solution that already occurred within the most recent 9 steps.

Escape Rules: Four dynamically self-updating parameters govern the escape behavior of the tabu search. After every tabu step, the first parameter, continuous in $[0, 1]$, determines the probability of an escape move. If an escape move is initiated, the next two parameters describe the minimum and maximum number of variables to be flipped. The concrete number of variables to be randomly selected and flipped is chosen uniformly at random in the given interval.

The last parameter $p \in [0, 1]$ then determines whether the new point is accepted as a new starting point for a new series of regular tabu search steps. To determine acceptance, we maintain a hash table of all previously visited solutions. The new potential starting point is hashed using a universal hash function and we then consider how many previously visited solutions r hash to the same value.

- If this hash value has seen more than the average μ solutions compared to the other values, the new point is rejected and we repeat the process of building a new starting solution *starting from the original solution where the tabu search last stopped*.
- If the new solution falls into a bucket that has seen less than the average number of solutions μ per bucket, we check if it is even less than $(1 + p)$ standard deviations (σ) below the average. If so, the new point is accepted as our new starting point.
- If the point hashes to a value that has been hashed to q times before, and q falls somewhere in the interval $[\mu - (1 + p)\sigma, \mu]$, then we consider the ratio

$r \leftarrow \frac{\mu - q}{(1+p)\sigma} \in [0, 1]$. The new point is then be accepted as new starting point with probability r.

Neighborhood Size: Especially when tackling problems with many variables, we may not want to consider all variables for flipping to determine which move would result in the best neighboring solution. We therefore introduce another continuous, dynamically self-updating parameter in $[0, 1]$ that determines the percentage of variables that are considered for flipping. If none of these variables yields a non-tabu solution, we initiate an escape move. Otherwise, we choose the first non-tabu variable, in random order, that results in an objective function improvement, or the variable that results in the least performance decrease.

Aspiration Criterion: In a tabu search, when a tabu criteria is used that goes beyond recording which solutions have already been visited, we may accidentally prevent moves that would lead to improving solutions. In our case, we use a second tabu list that prevents recently flipped variables from being flipped again quickly.

After each escape movement, we keep a running average μ_a of the cost (assuming a minimization problem) of all solutions as well as the cost of the best solution b_a found so far in the current "regime" (as the search period between two escape moves is often called in the literature). Then, we override the tabu lists if the solution we would arrive at has cost of at most $b_a + (\mu_a - b_a) * s$, where s is a continuous, dynamically self-updating parameter in $[0, 1]$.

3.2 Dynamic Search Features

Note that all parameters listed above take continuous values in $[0, 1]$, but we did not explain how they self update. In this section, we explain how.

Over the course of the tabu search, we keep track of a number of dynamic search features (runtime statistics) that are meant to characterize the status of the search. In total, we track eleven search features:

PercentTimeElapsed: At runtime, the user must specify how much CPU time is available for the search. This feature keeps track what percentage of CPU time has already elapsed.

BestUpdates: This feature counts the number of times we improve the objective function during the course of the entire optimization.

MovesAfterLastBestUpdate: In this dynamic search feature, we count the number of tabu search moves that have taken place since the last overall best solution was found.

MovesAfterLastImprovementInCurrentRegime: Analogous to the previous feature, in this value we count the number of moves that took place since the last improvement was found within the current regime. That is, after an escape move, we keep track of the best solution quality found afterwards. In this feature, we record how many moves have occurred since this improvement was found.

MovesInCurrentRegime: We count the number of tabu search moves after the last escape move.

TabuSolutions: Here we maintain the current length of the first tabu list which forbids recently visited solutions to be revisited.

TabuVariables: Analogous to the previous dynamic search features, in this value we record the current length of the second tabu list, which prevents specific variables that were recently flipped, from being flipped again.

AverageQualityThisRegime: Starting with the quality of the initial solution after the last escape move, we keep track of the average solution quality within the current regime.

AverageBestQualityThisRegime: In this statistic, we maintain the average quality of the best solution found within the current regime. That is, for every tabu search move after the last escape action, we make an entry of the best quality that was found within the current regime until that move. This dynamic search feature is the average of this list of values.

BestQualityThisRegime: Here we record the best solution seen after the last escape move.

AverageBestQualityAtTheEndOfRegimes: Right before each escape move, we consider the best solution quality that was found within the regime that is now ending. This statistic maintains the average over these values for all completed regimes.

Note that the search features listed above may vary a lot from instance to instance, because of differences in instance sizes, different objective scales etc. Consequently, to allow offline training (see below), we need to normalize these features.

To normalize the number of overall best solution updates, we divide that number by an estimate of the total number of tabu search moves we will be able to conduct within the given time limit. This estimate is updated based on the actual observed time for the tabu search moves. This same estimate is also used to normalize the three following features which all count numbers of tabu search moves. The tabu list lengths values are divided by the number of binary variables in the problem instance to be solved.

Finally, to normalize all quality features we maintain the moving average of the quality of all accepted solutions as well as the quality of the current overall best quality found. The quality features are normalized by dividing the respective quality by the difference between average quality and best quality found so far.

3.3   Hyper-Parameterization

Given the set of dynamic search features defined in the previous section, the next core decision is how these features should determine the setting of the tabu search parameters. More formally, we need to define algorithmically how the search features as inputs ought to be used to compute the output: the current search parameters. Note that, historically, this was the task of the algorithm designer. Consider RTS from [7], for example. Here, the mechanism by which the length of the tabu list is increased or decreased is prescribed by the authors, whereby the decision to increase or decrease the length is dependent on search characteristics, such as the average cycle length observed during the most recent search.

The paradigm shift introduced in [5] is to let the machine *learn* this functional dependency between search features and search parameters. The decision that is left to the algorithm designer is merely what class of functions ought to be considered by the machine. Given the tremendous success of deep learning, it is tempting to propose a neural network structure that transforms features into parameters, in which the weights in the network would be learned during a training phase. The problem with this approach is the number of weights that would need to be considered as hyper-parameters. Note that, in the context of a local search metaheuristic, we do not have a fixed training set with supervised labels available to us. In fact, earlier search decisions affect the distribution of search decisions that need to be taken later (this problem was also identified when trying to devise search guidance for systematic search approaches, see for example [18]). Moreover, the quality of the decisions taken can only be judged by running the search multiple times and possibly on a whole set of problem instances.

That is why we use an algorithm configurator to "learn" the hyper-parameters that ultimately determine how search features are transformed into search parameters. However, local search performance is not differentiable, and therefore there is no simple gradient decent approach available to us that would provide high-quality parameters very quickly. Consequently, the search for superior hyper-parameters is very tedious (thankfully only for the machine and not the user) and the search becomes more difficult the more hyper-parameters need to be tuned.

To keep the search for hyper-parameters manageable, we therefore need to choose a functional dependency between search features and search parameters that requires few hyper-parameters. A simple concept class consists of logistic regression functions which were also used in [5]. For each search parameter p, using k search features $f_1, \ldots, f_k$, we define a function $p \leftarrow \frac{1}{1+e^{w_0^p + \sum_{i=1}^{k} w_i^p f_i}}$, in which $w_0^p, \ldots, w_k^p$ denote the hyper-parameters that govern the relationship between the k search features and parameter p. Note that, for each search parameter, this choice requires just one more hyper-parameter than there are features. Moreover, the resulting values naturally cover the continuous interval $[0, 1]$ that all our parameters fell into.

4   MaxSAT

The MaxSAT problem is the optimization version of the Satisfiability (SAT) problem.

Definition 1. *A* truth variable *is a variable that either takes true or false as its value. Such an assignment is called a* truth assignment. *A literal is a truth* *variable (positive literal) or its negation (negative literal). A positive literal is said to evaluate to true iff its variable is set to true. A negative literal is said to evaluate to true iff its variable is set to false.*

A clause *is a disjunction of literals. A clause is said to be* satisfied *under a truth assignment to its variables if at least one of the literals in the clause evaluates to true. Otherwise the clause is said to be* falsified. *A weighted clause is a pair (C, w), where C is a clause and w is a natural number or infinity, indicating the cost for falsifying clause C.*

A Weighted Partial MaxSAT formula *(WPMS) is a set of weighted clauses $\varphi = \{(C_1, w_1), ..., (C_m, w_m), (C_{m+1}, \infty), ..., (C_{m+m'}, \infty)\}$ where the first m clauses are soft and the last m' clauses are hard. The objective of the MaxSAT problem is to find a truth assignment that satisfies all hard clauses while minimizing the total cost of all soft clauses that are falsified.*

A Partial MaxSAT formula *(PMS) is a WPMS formula where the weights of soft clauses are all equal. A* (plain) MaxSAT formula *is a partial MaxSAT formula where all clauses are soft.*

MaxSAT is an important problem because several significant practical problems can be formulated naturally as MaxSAT problems. Examples range from scheduling [25] to FPGA routing [26] to circuit design and debugging [23], to name just a few.

From a research perspective, MaxSAT is an interesting problem as it requires the ability to reason simultaneously about both optimality and scarce feasibility. Depending on the particular problem instance being solved, it is more important to emphasize one aspect or the other.

Driven by the annual *MaxSAT Evaluation* (MSE) [1,6], MaxSAT solvers have seen significant performance improvements in recent years. Therefore, MaxSAT makes an excellent benchmark for a general-purpose metaheuristic.

To evaluate the hyper-reactive meta-heuristic framework that we developed, it is important for us not to change the general framework to accommodate the particular problem class at hand. Consequently, the only calibration we allow for the particular application is the provisioning of an efficient way how to evaluate the objective function of a new assignment and, incrementally, the objective value change when one variable is flipped.

For MaxSAT, to enable fast incremental objective updates, we pre-compute once in the beginning which variables occur in which clauses and we maintain, for each clause, the set of variables that currently support the clause, if any. Then, when we flip a variable, we can easily consider only those clauses that the variable appears in and determine whether the variable is now, or no longer is, supporting the clause. By looking at the cardinality of the set of supporting variables, we can then quickly determine for which clauses we must add or subtract its penalty value to or from the previous objective value due to the variable change.

5   Numerical Results

We now report on the results of using the generic HRTS approach described above for tackling the MaxSAT problem. We first assess our contribution by comparing HRTS to a static RTS strategy. We then examine how the parameters change over time in HRTS when solving several MaxSAT instances to better understand the inner workings of HRTS. Finally, we show how HRTS parameterizations can be used to augment maxroster [24], the winning solver at MSE'17 [1].

5.1   Benchmark and Evaluation Metric

We use data from the annual MaxSAT Evaluations [1,6]. In particular, we consider the scenario of a user who wants to build a superior MaxSAT solver after MSE'17. We train HRTS on the benchmarks from MSE'16. We tune multiple HRTS parameterizations using the GGA++ algorithm configurator [2,3], splitting the training data into 16 *families* for tuning based on the filenames of the instances. All tuning was performed with a 60 s target algorithm timeout due to limited computational resources. Testing is all performed with a 300 s timeout.

We build a portfolio of HRTS parameterizations in combination with maxroster. To test the performance of the resulting portfolios we evaluate on the MSE'17 data. Note that all solvers in the portfolios are built and tuned on pre-2017 benchmarks.

We use the same performance metric as at MSE'17, which considers the average gap in quality. That is, for each instance we compute the ratio of best known solution to an instance divided by the final assignment cost of the respective solver for that instance. As the best known solution, we use the better of the two solutions found when comparing two solvers head-to-head, or the best quality published by MSE results [1,6].

5.2   Tuning Setup

When optimizing the hyper-parameters, we need to define a function that establishes which parameterization is better than another when both have been evaluated on some MaxSAT instances. In the evaluation, we only run the algorithm to the best known solution quality and consider an instance "solved" if the solution was reached before the timeout. Parameterizations are first compared on average gap in quality. If that is equal, we next compare them based on the average time it took them to solve instances. Any remaining ties are broken randomly during tuning. As in [5], the surrogate model of GGA++ is trained by using relative ranks based on these pairwise comparisons.

We use a distributed version of GGA++ with 7 machines with 8 cores each and a memory limit of 32 GB each. The tuning uses a population size of 100 individuals and runs for 100 generations. In order to reduce experimental variance, each MaxSAT instance is evaluated with a common random seed.

To train the algorithm selector which picks the parameterizations at runtime, we use the cost-sensitive hierarchical clustering methodology (CSHC) [20]. We use the same 37 features from [4] and build 1,000 trees, whereby each is based on a sub sampling of the training instances with replacement with probability 0.7 and a subset of 6 features.

Experiments were run on a cluster containing Intel Xeon CPU E5-2670 processors at 2.6GHz running Scientific Linux 7.2.

5.3   Comparison with a Statically Tuned RTS

We create a static reactive tabu search variant, called SRTS, in which we only tune the full list of hyper-parameters that guide the reactive tabu lengths of the two tabu lists used in our approach. For all other tabu search parameters, we only tune the corresponding hyper-parameter w_0^p, which means that the corresponding search parameter is tuned yet stays fixed throughout the tabu search with no dynamically reactive behavior. This deprives the tuner from access to more dynamic search strategies, but on the other hand it has the advantage that the number of hyper-parameters to be tuned is lowered significantly.

Table 1 provides a comparison of HRTS and SRTS on several families of MSE'16 instances where local search traditionally works well. We consider an instance to be solved if the best known solution is reached. All runs are performed using a 300 s timeout.

Since the MSE'16 instances correspond to the training set, we use the results to assess the potential of the methods. We observe that HRTS provides a better or approximately the same score (the average ratio of best known quality over quality found by the solver) on five out of the six groups we consider. Overall, it therefore appears worthwhile to open more than just the tabu lengths for dynamic updates, even though the resulting tuning problem is much harder. As we see on family min-enc_warehouses, at times the tuning problem for the full HRTS may be so vast that an inferior parameterization is found. Overall, the existing state-of-the-art tuners (in this case GGA++ [2]) appear able to effectively tune even a fully hyper-parameterized version of tabu search, although we note that having enough instances available to tune is critical. With too few instances, HRTS is very prone to overfitting, although it is unfortunately not possible to exactly define "too few".

The configurations found for HRTS on these families are very competitive. On most families, HRTS outperforms the hyper-parameterized dialectic search [5] based portfolio in MSE'16 and works comparably well as human-developed MaxSAT solvers, like, e.g., ramp, the local search solver that maxroster employs. Note that the search guidance in HRTS knows nothing about MaxSAT itself, but is nonetheless competitive with state-of-the-art approaches.

5.4   Runtime Log Analysis

We now investigate how the parameters change over the course of a single run of HRTS, and split our analysis into looking at differences in the same parameters

Table 1. Average score and number of instances solved for HRTS and SRTS on MSE'16 instance families after 300 s. We also give number of solved instances by local search solver ramp and the dsat/wpm3 portfolio from MSE'16.

Family	Instances	HRTS		SRTS		ramp	dsat/wpm3
		Score	Solved	Score	Solved	Solved	Solved
auctions_auc-paths	20	100%	**20**	99.91%	12	18	17
auctions_auc-scheduling	20	100%	**20**	100%	**20**	**20**	19
frb	34	99.97%	29	99.98%	29	**32**	15
maxcut	48	100%	**48**	100%	**48**	**48**	47
min-enc_warehouses	18	77.64%	0	83.31%	1	1	**3**
spot5	42	99.01%	15	98.24%	13	28	**30**
Total	182	96.10%	132	96.91%	123	**147**	131

over different instances, and different parameters on the same instance. For the following experiments we use a 250 s timeout.

Dynamic Search Behavior Using the Same Parameterization: The scpnre instances are based on the well-known set covering problem. Figure 1 tracks eight different values over the course of solving three scpnre instances (SCPNRE1 (blue), SCPNRE2 (green), and SCPNRE3 (red)), when using the *same* hyperparameterization of HRTS.

We observe that, on all three instances, this parameterization makes quick progress in improving the solution quality, using a large neighborhood size for the first-improvement tabu greedy steps that are only diversified by making 5-7% of recently flipped variables and practically no individual solutions tabu in the beginning. This small diversification pressure is even further reduced by a very generous aspiration policy. After the initial phase of fast improvements, the aspiration criterion is made gradually more restrictive until only best solution improvements are allowed to override the tabu lists. In turn, however, the tabu variables list length decreases until this list is practically not used anymore at around half of the available computation time. Instead, the tabu solutions list length is increased, so that it has a total length of about 0.8 times the number of variables.

In the second half of the optimization, this list length is adjusted very dynamically, the aspiration is slightly relaxed again, and the probability for escape moves is increased, whereby the escape path length is controlled carefully. 0.4% of variables corresponds to roughly 20 variables on these instances. Furthermore, the average escape size (in two of the three instances) slowly ratchets up before smoothly declining. As a result of these changes in search behavior, we begin to see new and some abysmally bad solutions as evidenced by spikes in the average solution quality. This shows that the method is diversifying effectively, and the success of this strategy shows in the best solution quality: two of the three runs find improving solutions during this second half of the optimization.

Overall we see that the search is reactive, yet the overall search strategy (if one can call it that given its origins in tuning) is comparable on all three instances.

Comparison Of Multiple Parameterizations On the Same Instances:
Having just compared the dynamic search behavior of the same parameterization on different instances, in Fig. 2 we now compare two *different parameterizations* (solid and dashed lines) which we both run on the *same* instance, SCPNRG2. The dashed parameterization finds good solutions faster, as seen in the plot of the best value, although the solid parameterization does end up finding similar solutions later in its search.

The example shows beautifully how flexible the hyper-parameterized tabu search framework is. Let us look at the solid lines first. This is almost a pure tabu search strategy: The number of recently flipped variables that are tabu is kept firm at 5%, the tabu solutions list is not used. Furthermore, there are no escape moves at all, and almost all variables are open for flipping at any tabu best neighbor step. The only difference to a very traditional tabu search approach is the use of the aspiration criterion which allows practically any best neighbor to override the tabu variables list as long as the resulting quality is still below average cost. Only after about two thirds of the available runtime are exhausted, the aspiration criterion is made more restrictive.

Contrast this strategy with the one that leads to the dashed lines in our plots. Here, we observe that both tabu lists are used excessively: About 90% of recently flipped variables are tabu, and the history of "recently" visited solutions that is kept on file grows from 150 times the number of variables to 350 times the number of variables. Note further that this massive diversification pressure is not alleviated by a lax aspiration policy either. At the same time, only 50–60% of variables are considered for flipping during first improvement tabu steps. Obviously, this very aggressive, "go after the best you can find that is new" policy may lead to dead ends quickly. Consequently, the search restarts quickly, whereby, curiously, the method does actually *not* diversify at this point by using the hash table to find a good new starting point. The escape size is held firmly at zero. Merely, the restarts are used to clear the tabu lists only.

This is the quintessential take-away of hyper-parameterization: It allows the machine to drive the search in many different ways, explore vastly differing strategies, more than humans have considered and bothered to implement and test. Within the context of tuning and algorithm portfolios, it is *diversity* that makes all the difference. Any individual strategy is no longer required to work robustly across many different applications. As long as a certain policy can excel sometimes, it can become a valuable member of a portfolio of parameterizations.

5.5   Comparison With the State Of The Art

Finally, we show what effect the hyper-parameterized tabu search framework can have in a domain where regular solver competitions have driven the development

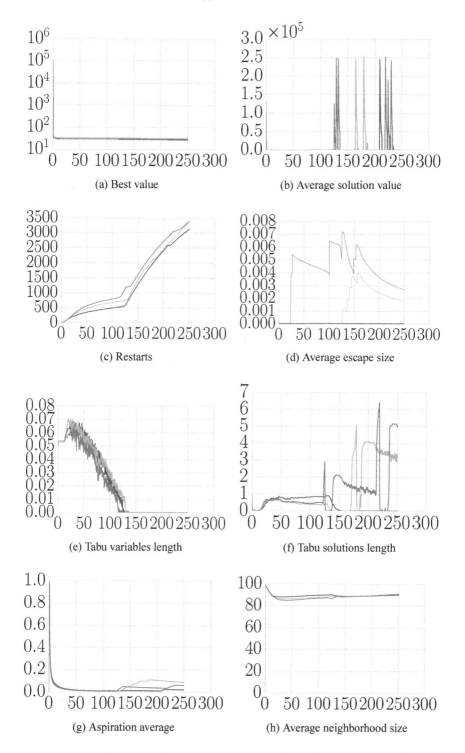

Fig. 1. Normalized characteristics over the course of running three SCPNRE instances, with seconds on the x axis.

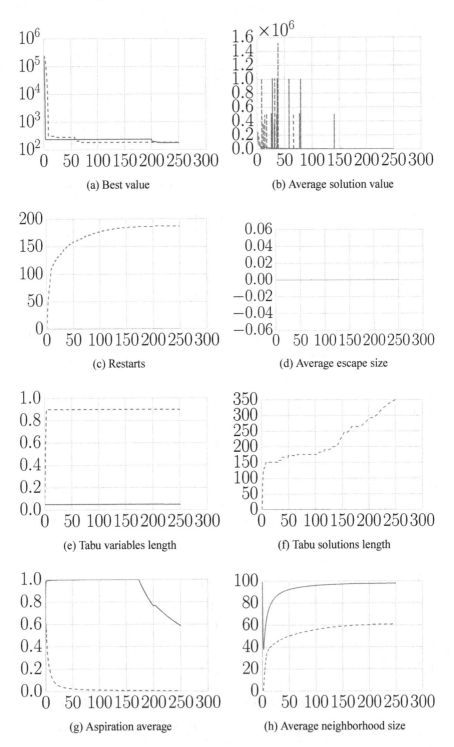

Fig. 2. Normalized characteristics over the course of running the SCPNRG2 instance with two parameterizations, with seconds on the x axis.

of incomplete methods for many years. We apply our framework to MaxSAT, trained on the MSE'16 benchmarks, and with automatically tuned hyper-reactive tabu parameterizations added, like the ones studied earlier, to a portfolio of solvers. In Table 2 we show the results when adding our parameterizations to maxroster, the winning solver at MSE'17, and compare to the virtual best solver (VBS).

In Table 2, we can see that entirely self-tuned tabu search hyper-parameterizations, which realize vastly differing search strategies, can effectively complement the current best MaxSAT solver, maxroster. The score (average ratio of the best known solution over solver quality) improves from 0.9721 to 0.9771 on the training set (we use the best quality in a portfolio of maxroster, wpm3 and HRTS parameterizations to define the instance best) and from 0.8311 to 0.8321 on the test set. This is achieved by improving the quality on 7 out of the 156 test instances, or about 4.5% of all instances, without ever decreasing it.

Table 2. Head-to-head comparison of maxroster and a maxroster+HRTS portfolio.

		maxr	maxr+HRTS
Training (961 instances)	Score	97.21%	**97.71%**
Test (156 instances)	Percent VBS	99.89%	**100%**
	Head-to-Head Wins	0	**7**
	Score	83.11%	**83.21%**
	Best Found	69	69

This leads us several important conclusions: First, hyper-parameterized tabu search can be tuned effectively. This is true even when the training set is not 100% indicative of the test set (the benchmark used at MSE'17 differed significantly from prior MaxSAT evaluations). Second, hyper-parameterized tabu search can realize search strategies that differ significantly from existing, human-incepted search methods, and thus offer the opportunity for improvements within a portfolio approach. This conclusion is analogue to the results presented in [17]. And third, as a consequence of the previous two points, even in entrenched, highly researched domains, hyper-parameterized tabu search can be applied effectively to achieve performance improvements automatically, without any need of the user to provide any domain knowledge. In our case, HRTS has access to an incremental cost evaluator for MaxSAT. That is all that the framework can exploit. Everything that regards search guidance, how to dynamically adapt, and what overall strategy to run, is entirely driven by the search experiences that are gathered during training, by monitoring the search features that we described in this paper and correlating these with the overall search performance.

6   Conclusion

We introduced a framework for the hyper-parameterization of tabu search. Our framework opens tabu search parameters such as escape rules, neighborhood sizes, and aspiration criterion for the dynamic adaptation during search, on top of the length of tabu lists which had been pioneered in [7]. The dynamic self-adaptation is based on various search features that we introduced. Experimental results show that portfolios of hyper-reactive tabu search parameterizations work competitively with human devised local search solvers for MaxSAT. Moreover, these parameterizations can even improve the most cutting edge MaxSAT solver maxroster in a joint portfolio approach.

Obvious extensions of this work are the consideration of other dynamic search features and other functional dependencies between search features and parameters. The most interesting new avenue, that we will consider in the future, is the integration of completely different local search frameworks by means of dynamic self-adaptation.

Acknowledgement. The authors would like to thank the Paderborn Center for Parallel Computation (PC^2) for the use of the OCuLUS cluster. This work was financially supported in part by TIN2016-76573-C2-2-P.

References

1. Ansótegui, C., Bacchus, F., Järvisalo, M., Martins, R.: MaxSAT Evaluation (2017). http://mse17.cs.helsinki.fi
2. Ansotegui, C., Malitsky, Y., Samulowitz, H., Sellmann, M., Tierney, K.: Model-based geneticalgorithms for algorithm configuration. In: IJCAI, pp. 733–739 (2015)
3. Ansotegui, C., Sellmann, M., Tierney, K.: A gender-based genetic algorithm for the automatic configuration of algorithms. In: CP, pp. 142–157 (2009)
4. Ansótegui, C., Gabas, J., Malitsky, Y., Sellmann, M.: Maxsat by improved instance-specific algorithm configuration. Artif. Intell. **235**, 26–39 (2016)
5. Ansótegui, C., Pon, J., Sellmann, M., Tierney, K.: Reactive dialectic search portfolios for maxsat. In: AAAI Conference on Artificial Intelligence (2017)
6. Argelich, J., Li, C., Manyà, F., Planes, J.: MaxSAT Evaluation (2016). www.maxsat.udl.cat
7. Battiti, R., Tecchiolli, G.: The reactive tabu search. ORSA J. Comput. **6**(2), 126–140 (1994)
8. Battiti, R., Brunato, M., Mascia, F.: Reactive Search and Intelligent Optimization, vol. 45. Springer Science & Business Media, Berlin (2008)
9. Burke, E., Kendall, G., Newall, J., Hart, E., Ross, P., Schulenburg, S.: Hyper-heuristics: an emerging direction in modern search technology. Handbook of meta-heuristics, pp. 457–474 (2003)
10. Burke, E.K., Gendreau, M., Hyde, M., Kendall, G., Ochoa, G., Özcan, E., Qu, R.: Hyper-heuristics: a survey of the state of the art. J. Oper. Res. Soc. **64**(12), 1695–1724 (2013)
11. Doerr, B., Doerr, C.: Optimal parameter choices through self-adjustment: Applying the 1/5-th rule in discrete settings. In: GECCO, pp. 1335–1342 (2015)

12. Doerr, B., Doerr, C.: Optimal static and self-adjusting parameter choices for the $(1 + (\lambda, \lambda))(1 + (\lambda, \lambda))$genetic algorithm. Algorithmica (2017)

13. Glover, F., Laguna, M.: Tabu search. In: Handbook of Combinatorial Optimization, pp. 3261–3362. Springer, Berlin (2013)

14. Glover, F., Laguna, M., Martí, R.: Principles of tabu search. In: Gonzalez, T. (ed.) Handbook of Approximation Algorithms and Metaheuristics (2007)

15. Kadioglu, S., Malitsky, Y., Sellmann, M., Tierney, K.: ISAC-instance-specific algorithm configuration. In: Coelho, H., Studer, R., Wooldridge, M. (eds.) ECAI. FAIA, vol. 215, pp. 751–756 (2010)

16. Kadioglu, S., Sellmann, M.: Dialectic search. CP, pp. 486–500 (2009)

17. KhudaBukhsh, A., Xu, L., Hoos, H., Leyton-Brown, K.: SATenstein: automatically building local search sat solvers from components. In: IJCAI, pp. 517–524 (2009)

18. Leventhal, D., Sellmann, M.: The accuracy of search heuristics: an empirical study on knapsack problems. In: Integration of AI and OR Techniques in Constraint Programming for Combinatorial Optimization Problems, pp. 142–157 (2008)

19. Leyton-Brown, K., Nudelman, E., Andrew, G., McFadden, J., Shoham, Y.: A portfolio approach to algorithm selection. In: IJCAI, pp. 1542–1543 (2003)

20. Malitsky, Y., Sabharwal, A., Samulowitz, H., Sellmann, M.: Algorithm portfolios based on cost-sensitive hierarchical clustering. In: IJCAI, pp. 608–614 (2013)

21. Mısır, M., Verbeeck, K., De Causmaecker, P., Berghe, G.V.: An intelligent hyper-heuristic framework for chesc 2011. In: Learning and Intelligent Optimization, pp. 461–466. Springer, Berlin (2012)

22. Özcan, E., Mısır, M., Ochoa, G., Burke, E.K.: A reinforcement learning: great-deluge hyper-heuristic. In: Modeling, Analysis, and Applications in Metaheuristic Computing: Advancements and Trends: Advancements and Trends, vol. 34 (2012)

23. Safarpour, S., Mangassarian, H., Veneris, A., Liffiton, M., Sakallah, K.: Improved design debugging using maximum satisfiability. In: Formal Methods in Computer Aided Design, pp. 13–19. IEEE (2007)

24. Sugawara, T.: Maxroster: solver description. In: MaxSAT Evaluation 2017, p. 12 (2017)

25. Vasquez, M., Hao, J.: A "logic-constrained" knapsack formulation and a tabu algorithm for the daily photograph scheduling of an earth observation satellite. Comput. Optim. Appl. **20**(2), 137–157 (2001)

26. Xu, H., Rutenbar, R., Sakallah, K.: sub-SAT: a formulation for relaxed boolean satisfiability with applications in routing. IEEE Trans. Comput.-Aided Des. Integr. Circuits Syst. **22**(6), 814–820 (2003)

27. Xu, L., Hoos, H., Leyton-Brown, K.: Hydra: automatically configuring algorithms for portfolio-based selection. In: AAAI, pp. 210–216 (2010)

Exact Algorithms for Two Quadratic Euclidean Problems of Searching for the Largest Subset and Longest Subsequence

Alexander Kel'manov[1,2], Sergey Khamidullin[1], Vladimir Khandeev[1,2(✉)], and Artem Pyatkin[1,2]

[1] Sobolev Institute of Mathematics, 4 Koptyug Ave., 630090 Novosibirsk, Russia
[2] Novosibirsk State University, 2 Pirogova St., 630090 Novosibirsk, Russia
{kelm,kham,khandeev,artem}@math.nsc.ru

Abstract. The following two strongly NP-hard problems are considered. In the first problem, we need to find in the given finite set of points in Euclidean space the subset of largest size such that the sum of squared distances between the elements of this subset and its unknown centroid (geometrical center) does not exceed a given percentage of the sum of squared distances between the elements of the input set and its centroid. In the second problem, the input is a sequence (not a set) and we have some additional constraints on the indices of the elements of the chosen subsequence under the same restriction on the sum of squared distances as in the first problem. Both problems can be treated as data editing problems aimed to find similar elements and removal of extraneous (dissimilar) elements. We propose exact algorithms for the cases of both problems in which the input points have integer-valued coordinates. If the space dimension is bounded by some constant, our algorithms run in a pseudopolynomial time. Some results of numerical experiments illustrating the performance of the algorithms are presented.

Keywords: Euclidean space · Largest set · Longest subsequence
Quadratic variation · NP-hard problem · Integer coordinates
Exact algorithm · Fixed space dimension · Pseudopolynomial time

1 Introduction

In this paper, we consider two discrete optimization problems that model the search for similar objects in measurement results. Our aim is to propose algorithms for these problems.

In the first problem, we need to find a subset of objects close to each other. In the second problem, we have time-ordered measurement results, i.e. we have a sequence. In addition, we have some restrictions on the choice of objects in this sequence. In both problems, one needs to find as much as possible similar objects.

© Springer Nature Switzerland AG 2019
R. Battiti et al. (Eds.): LION 12 2018, LNCS 11353, pp. 326–336, 2019.
https://doi.org/10.1007/978-3-030-05348-2_28

The study is motivated, on the one hand, by the insufficient studies on the problems, and on the other hand, by their importance for the applications, in particular, data editing, data cleaning, data mining, machine learning, etc (see the next section).

The paper is organized as follows. In Sect. 2, the mathematical statements of the considered problems are presented and the motivation is given. The statements of known problems close to them are also given. In the next section, some auxiliary algorithms are formulated. The exact algorithms and their analysis can be found in Sect. 4. Finally, Sect. 5 contains some results of numerical experiments illustrating the performance of the algorithms.

2   Problems Formulations, Their Complexity and Related Problems

Everywhere below $\mathbb{R}$ denotes the set of real numbers and $\|\cdot\|$ denotes the Euclidean norm.

The first problem under consideration is following

Problem 1. Given a set $\mathcal{Y} = \{y_1, \ldots, y_N\}$ of points in $\mathbb{R}^q$ and a number $\alpha \in (0, 1)$. *Find* a subset $\mathcal{C} \subset \mathcal{Y}$ of largest cardinality such that

$$S(\mathcal{C}) = \sum_{y \in \mathcal{C}} \|y - \overline{y}(\mathcal{C})\|^2 \leq \alpha \sum_{y \in \mathcal{Y}} \|y - \overline{y}(\mathcal{Y})\|^2, \tag{1}$$

where $\overline{y}(\mathcal{C}) = \frac{1}{|\mathcal{C}|} \sum_{y \in \mathcal{C}} y$ and $\overline{y}(\mathcal{Y}) = \frac{1}{|\mathcal{Y}|} \sum_{y \in \mathcal{Y}} y$ are the centroids of the subset $\mathcal{C}$ and the given set $\mathcal{Y}$, respectively.

In Problem 1, the goal is to find a subset $\mathcal{C}$ of the largest cardinality in the form of a spherical cluster, whose total quadratic spread with respect to the unknown centroid $\overline{y}(\mathcal{C})$ does not exceed α times the total quadratic spread of the input set $\mathcal{Y}$ with respect to its centroid $\overline{y}(\mathcal{Y})$. An example of the input set $\mathcal{Y}$ of points in $\mathbb{R}^2$ is shown in Fig. 1. One can easily find there a region where the points are placed more "compactly" than in the entire input set.

Problem 1 has the following interpretation. There is a set containing N measurements $y_1, \ldots, y_N$ of q numerical characteristics of some objects. Each measurement result has an error, and no correspondence is known between the elements of the set and the objects. Some of these objects have identical characteristics (or one can say that in the time series there are several measurements of one significant object). Other objects are distinguished and have different characteristics (or one can say that in the time series there are some outliers). The number M of measurements of identical objects is unknown. The characteristics of identical objects in contrast to the characteristics of other objects have an important information value. It is required to find the largest subset of measurements which correspond to the identical objects using the criterion of minimum sum of squared distances and to estimate the characteristics of these objects (taking into account the measuring errors in the data).

The second problem under consideration is following

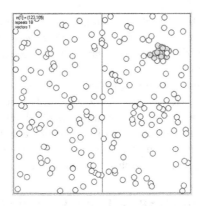

Fig. 1. An example of the input set with some concentrated points

Problem 2. Given a sequence $\mathcal{Y} = (y_1, \ldots, y_N)$ of points in $\mathbb{R}^q$, some positive integer numbers $T_{\min}$, $T_{\max}$, and a number $\alpha \in (0,1)$. *Find* a subset $\mathcal{M} = \{n_1, \ldots, n_M\} \subseteq \mathcal{N} = \{1, \ldots, N\}$ of largest size such that

$$T_{\min} \leq n_m - n_{m-1} \leq T_{\max} \leq N, \ m = 2, \ldots, M, \tag{2}$$

and

$$F(\mathcal{M}) = \sum_{j \in \mathcal{M}} \|y_j - \overline{y}(\mathcal{M})\|^2 \leq \alpha \sum_{j \in \mathcal{N}} \|y_j - \overline{y}(\mathcal{N})\|^2, \tag{3}$$

where $\overline{y}(\mathcal{M}) = \frac{1}{|\mathcal{M}|} \sum_{i \in \mathcal{M}} y_i$ and $\overline{y}(\mathcal{N}) = \frac{1}{N} \sum_{i \in \mathcal{N}} y_i$ are the centroids of the multisets $\{y_i \in \mathcal{Y} \mid i \in \mathcal{M}\}$ and $\{y_i \in \mathcal{Y} \mid i \in \mathcal{N}\}$ respectively.

In Problem 2, the goal is to find in a sequence $\mathcal{Y}$ a multisubset $\{y_i \in \mathcal{Y} \mid i \in \mathcal{M}\}$ of the maximum cardinality whose elements are well concentrated, i. e. the total quadratic spread of the points with respect to the unknown centroid $\overline{y}(\mathcal{M})$ is at most α times the total quadratic spread of the input sequence's points with respect to its centroid $\overline{y}(\mathcal{N})$. The inequalities (2) are the restrictions on the indices of elements chosen from the sequence $\mathcal{Y}$.

Fig. 2. An example of the input sequence containing a subsequence of concentrated points

The interpretation of Problem 2 is similar to the interpretation of Problem 1. But in Problem 2, in addition, it is known that the time interval between every two consequent results of measuring characteristics of the identical objects is bounded from above and below by some constants $T_{\max}$ and $T_{\min}$.

An example of the input sequence's points in $\mathbb{R}^2$ is shown in Fig. 2 in a tape form. Each point of the sequence $\mathcal{Y}$ corresponds to a vertical strip. The sets of points in Fig. 1 and 2 are the same.

The close in statement known problems are stated as follows.

Problem 3. Given a set $\mathcal{Y} = \{y_1, \ldots, y_N\}$ of points in $\mathbb{R}^q$ and a positive integer number M. *Find* a subset $\mathcal{C} \subset \mathcal{Y}$ of cardinality M such that

$$S(\mathcal{C}) \rightarrow \min.$$

Problem 4. Given a sequence $\mathcal{Y} = (y_1, \ldots, y_N)$ of points in $\mathbb{R}^q$ and positive integer numbers $T_{\min}$, $T_{\max}$ and $M > 1$. *Find* a subset $\mathcal{M} = \{n_1, \ldots, n_M\} \subseteq \mathcal{N} = \{1, \ldots, N\}$ of indices of the sequence elements such that

$$F(\mathcal{M}) \rightarrow \min,$$

subject to constraints (2).

The problems of search for a subset or a subsequence, similar to stated above Problems 1, 2, 3, 4, are typical for Data editing and Data cleaning problems (see, e.g., [13–15]). In problems of Machine learning and Data mining, cleaning the data from outliers is usually considered as a necessary element [16–26].

In Problems 3 and 4, in contrast to considered Problems 1 and 2, the sizes of the required subset and subsequence are known. The strong NP-hardness of Problems 3 and 4 was shown in [1] and [2] respectively. Some algorithmic results were obtained in [3–8] for Problem 3 and in [9–11] for Problem 4.

The strong NP-hardness of Problem 1 was recently shown in [12]. The strong NP-hardness of Problem 2 follows from this result, as Problem 1 is the special case of Problem 2 when $T_{\min} = 1$ and $T_{\max} = N$.

In [12], for the considered Problem 1 a polynomial-time 1/2-approximation algorithm was presented. This algorithm has $\mathcal{O}(N^2(q + \log N))$ running time. No other algorithmic results with guaranteed performance bounds are available now for the considered Problems 1 and 2.

In the current paper, we propose exact algorithms for the cases of both problems in which the input points have integer-valued coordinates. Both our algorithms implement the grid approach. The running time of the algorithm for Problem 1 is $\mathcal{O}(qN^2(2NB+1)^q)$, where B is the maximum absolute coordinate value of the input points. The algorithm for Problem 2 finds an optimal solution in $\mathcal{O}(N^2(N(T_{\max} - T_{\min} + 1) + q)(2NB + 1)^q)$ time. If the space dimension is bounded by some constant, our algorithms run in a pseudopolynomial time.

3   Algorithms Foundations

In order to substantiate our algorithms, we'll need algorithms for the special cases of Problems 3 and 4 and an auxiliary problem with an algorithm solving it.

First, we need the following algorithm for the special case of Problem 3 in which components of all points in the input set $\mathcal{Y}$ are integer. This algorithm is presented below.

Algorithm $\mathcal{A}_1$.

Input: a set $\mathcal{Y}$ and a positive integer M.

Step 1. Find the value of B and the size $|\mathcal{B}|$ of the set

$$\mathcal{B} = \{b \mid (b)^j = \frac{1}{M}(v)^j, \ (v)^j \in \mathbb{Z}, \ |(v)^j| \le MB, \ j = 1, \ldots, q\} \tag{4}$$

by formulas

$$B = \max_{y \in \mathcal{Y}} \max_{j \in \{1, \ldots, q\}} |(y)^j|, \tag{5}$$

$$|\mathcal{B}| = (2MB + 1)^q, \tag{6}$$

where $(x)^j$ denotes the j-th coordinate of a point x in $\mathbb{R}^q$. Construct the set $\mathcal{B}$ by formula (4).

Set $\mathcal{C}_A = \emptyset$, $G_{\min} = +\infty$, $i = 0$.

Step 2. $i := i + 1$; put $x = b_i$, where b_i is the i-th element of $\mathcal{B}$.

Step 3. Find the set $\mathcal{Y}(x)$ of M vectors nearest to x in the set $\mathcal{Y}$.

Step 4. Compute

$$G(x) = \sum_{y \in \mathcal{Y}(x)} \|y - x\|^2.$$

Step 5. If $G(x) \le G_{\min}$ then let $G_{\min} = G(x)$, $\mathcal{C}_A = \mathcal{Y}(x)$; else go to the next step.

Step 6. If $i < |\mathcal{B}|$ then go to Step 2 else go to the next step.

Output: the set $\mathcal{C}_A$.

Remark 1. It was proved in [5] that if components of all points in the set $\mathcal{Y}$ are integers in the interval $[-B, B]$ then the algorithm $\mathcal{A}_1$ finds an optimal solution of Problem 3 in $\mathcal{O}(qN(2MB + 1)^q)$ time.

Next, we need an exact polynomial-time algorithm for the following auxiliary

Problem 5. Given a sequence $\mathcal{Y} = (y_1, \ldots, y_N)$ of points in $\mathbb{R}^q$, a point $x \in \mathbb{R}^q$, and positive integer numbers $T_{\min}$, $T_{\max}$ and $M > 1$. Find a subset $\mathcal{M} = \{n_1, \ldots, n_M\} \subseteq \mathcal{N}$ of indices of the sequence elements such that

$$f^x(\mathcal{M}) = \sum_{i \in \mathcal{M}} \|y_i - x\|^2 \to \min,$$

while the elements of the tuple $(n_1, \ldots, n_M)$ satisfy the constraints (2).

The following dynamic programming scheme finds a solution $\mathcal{M}^x$ of Problem 5.

Algorithm $\mathcal{A}_2$.

Input: a sequence $\mathcal{Y}$, a point x, and numbers $T_{\min}$, $T_{\max}$ and M.

Step 1. Compute $g(n) = \|y_n - x\|^2$ for each $n \in \mathcal{N}$.

Step 2. Calculate the values of $f_m^x(n)$ for each $n \in \omega_m$ using the following recurrence formulas

$$f_m^x(n) = \begin{cases} g(n), & \text{if } n \in \omega_1, \ m = 1; \\ g(n) + \min\limits_{j \in \gamma_{m-1}^-(n)} f_{m-1}^x(j), & \text{if } n \in \omega_m, \ m = 2, \ldots, M, \end{cases}$$

where

$$\omega_m = \{n \mid 1 + (m-1)T_{\min} \le n \le N - (M-m)T_{\min}\},$$

$$\gamma_{m-1}^-(n) = \{j \mid \max\{1 + (m-2)T_{\min}, n - T_{\max}\} \le j \le n - T_{\min}\},$$

while $m = 1, \ldots, M$.

Step 3. Compute $f_{\min}^x = \min\limits_{n \in \omega_M} f_M^x(n)$ and find the tuple $\mathcal{M}^x = (n_1^x, \ldots, n_M^x)$ by formulas

$$n_M^x = \arg \min_{n \in \omega_M} f_M^x(n),$$

$$n_{m-1}^x = \arg \min_{n \in \gamma_m^-(n_m^x)} f_m^x(n), \quad m = M, M-1, \ldots, 2.$$

Output: the tuple $\mathcal{M}^x = (n_1^x, \ldots, n_M^x)$ and the value $f_{\min}^x$.

Remark 2. It was proved in [9, 27] that the algorithm $\mathcal{A}_2$ finds an optimal solution of Problem 5 in $\mathcal{O}(N(M(T_{\max} - T_{\min} + 1) + q))$ time.

Remark 3. In accordance with the statement of Problem 5, we have that $M \in \{2, \ldots, M_{\max}\}$, where

$$M_{\max} = \left\lfloor \frac{N-1}{T_{\min}} \right\rfloor + 1.$$

Finally, we need the following algorithm which is based on the algorithm $\mathcal{A}_2$ and allows to find a solution of Problem 4 in the case of integer coordinates of the input points of $\mathcal{Y}$.

Algorithm $\mathcal{A}_3$.

Input: a sequence $\mathcal{Y}$, numbers $T_{\min}$, $T_{\max}$ and M.

Step 1. Find the value of B and the size $|\mathcal{B}|$ of the set $\mathcal{B}$ by formulas (5) and (6). Construct the set $\mathcal{B}$ by formula (4). Put $\mathcal{M}_A = \emptyset$, $F_A = +\infty$, $i = 0$.

Step 2. $i := i + 1$; put $x = b_i$, where b_i is the i-th element of $\mathcal{B}$.

Step 3. Using the algorithm $\mathcal{A}_2$, find the value $f_{\min}^x$ and an optimal solution $\mathcal{M}^x = \{n_1^x, \ldots, n_M^x\}$ of Problem 5.

Step 4. If $F_A \ge f_{\min}^x$ then let $F_A = f_{\min}^x$, $\mathcal{M}_A = \mathcal{M}^x$; else go to the next step.

Step 5. If $i < |\mathcal{B}|$ then go to Step 2 else go to the next step.

Output: the tuple $\mathcal{M}_A$.

Remark 4. It was proved in [10] that if components of all points in the sequence $\mathcal{Y}$ are integers in the interval $[-B, B]$ then the algorithm $\mathcal{A}_3$ finds an optimal solution of Problem 4 in $\mathcal{O}(N(M(T_{\max} - T_{\min} + 1) + q)(2MB + 1)^q)$ time.

4   Exact Algorithms

We suggest the following approach for finding a solution of Problem 1. For each $M = 2, \ldots, N$, an auxiliary Problem 3 is solved by the algorithm $\mathcal{A}_1$. In a family of the found solutions, we choose the admissible (i.e. satisfying the inequality (1)) solution of the maximum size. The following algorithm realizes this approach.

Algorithm $\mathcal{A}_4$.

Input: a set $\mathcal{Y}$ and a number α.

Step 1. Compute $A = \alpha \sum_{y \in \mathcal{Y}} \|y - \overline{y}(\mathcal{Y})\|^2$.

Step 2. For every $M = 2, \ldots, N$ using the algorithm $\mathcal{A}_1$ find an exact solution $\mathcal{C}_A = \mathcal{C}_A(M)$ of Problem 3 and calculate for this solution the value of the goal function $S(\mathcal{C}_A(M))$.

Step 3. In the family $\{\mathcal{C}_A(M), M = 2, \ldots, N\}$ of the sets obtained in Step 2 find a set $\mathcal{C}_A$ of maximum cardinality for which $S(\mathcal{C}_A) \leq A$.

Output: the set $\mathcal{C}_A$.

The properties of this algorithm are established by the following

Theorem 1. *Suppose that the points of $\mathcal{Y}$ have integer components lying in $[-B, B]$. Then the algorithm $\mathcal{A}_4$ finds an optimal solution of Problem 1 in $\mathcal{O}(qN^2(2NB + 1)^q)$ time.*

Remark 5. If the space dimension q is bounded by some constant, then the algorithm's time complexity is $\mathcal{O}(N^2(NB)^q)$.

The algorithm for Problem 2 uses the similar approach as for Problem 1.

Namely, for each $M = 2, \ldots, N$ an auxiliary Problem 4 is solved by the algorithm $\mathcal{A}_3$. In the family of the found solutions, we choose the admissible (i.e. satisfying the inequality (3)) solution of the maximum length.

Algorithm $\mathcal{A}_5$.

Input: a sequence $\mathcal{Y}$ and numbers $T_{\min}, T_{\max}$, and α.

Step 1. Compute $A = \alpha \sum_{j \in \mathcal{N}} \|y_j - \overline{y}(\mathcal{N})\|^2$.

Step 2. For every $M = 2, \ldots, M_{\max}$ using the algorithm $\mathcal{A}_3$ find an exact solution $\mathcal{M}_A = \mathcal{M}_A(M) = (n_1, \ldots, n_M)$ of Problem 4 and calculate for this solution the value of the goal function $F(\mathcal{M}_A(M))$.

Step 3. In the family $\{\mathcal{M}_A(M), M = 2, \ldots, M_{\max}\}$ of the sets obtained in Step 2 find a set $\mathcal{M}_A$ of maximum cardinality for which $F(\mathcal{M}_A) \leq A$.

Output: the tuple $\mathcal{M}_A$.

The following theorem is true.

Theorem 2. *Suppose that the points of $\mathcal{Y}$ have integer components lying in $[-B, B]$. Then the algorithm $\mathcal{A}_5$ finds an optimal solution of Problem 2 in $\mathcal{O}(N^2(N(T_{\max} - T_{\min} + 1) + q)(2NB + 1)^q)$ time.*

Remark 6. If the space dimension q is bounded by some constant, then the algorithm's time complexity is $\mathcal{O}(N^4(NB)^q)$ since the value of $(T_{\max} - T_{\min} + 1)$ is at most N.

Since B is a numerical value given as an input, in the case of fixed space dimension q the algorithms are pseudopolynomial.

Some results of the numerical experiments for both algorithms are presented in the next section.

5 Examples of Numerical Simulation

The figures presented below show the robustness of the algorithm for the problem of search for a subset of similar elements in a data collection. An input set of integer-valued points (out of 500 points) is shown in Fig. 3 at the left-hand part. The subset of points found by the algorithm for $\alpha = 0.005$ and $B = 180$ is presented in the same Fig. 3 at the right-hand part (dark points). The found subset size is 198.

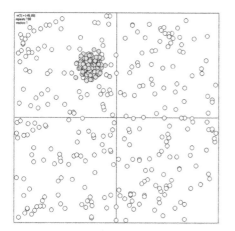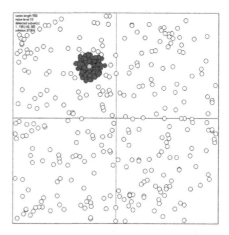

Fig. 3. The input of the algorithm $\mathcal{A}_4$ (left part) and the found subset (right part) of points (shown dark)

The next figures show the robustness of the algorithm for the problem of search for a subsequence. An input sequence of integer-valued points (out of 200 points) is shown in Fig. 4a (upper tape). Each point of the sequence corresponds to a vertical strip. The subsequence of points found by the algorithm for $\alpha = 0.0008$, $T_{\min} = 2$, $T_{\max} = 20$, $B = 180$ is presented in the same Fig. 4a on the lower tape. The found subsequence size is 19.

The points of the same input sequence are presented on a plane in Fig. 4b at the left-hand part. At the right-hand side, one can see a subset (dark points) corresponding to the subsequence presented in the Fig. 4a on the lower tape.

(a)

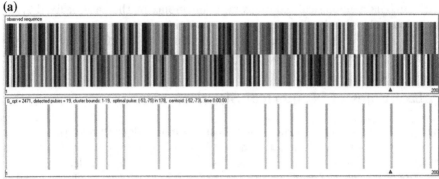

(b)

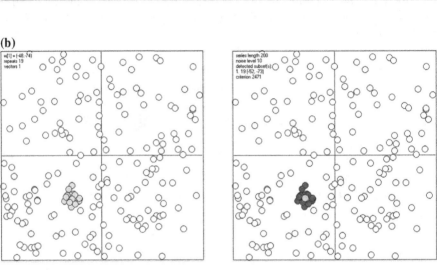

Fig. 4. (a) The input of the algorithm $\mathcal{A}_5$ (upper tape) and the found subsequence (lower tape). (b) The input of the algorithm $\mathcal{A}_5$ (left part) and the elements (shown dark) of the found subsequence (right part) shown on a plane

6   Conclusion

In this paper, we have considered two strongly NP-hard problems. The first one is the problem of searching for the subset of the largest size in a finite set of Euclidean points. The second one is the problem of searching for the longest subsequence (the subsequence of the largest size) in a finite sequence of Euclidean points. Exact algorithms for the cases of both problems in which the input points have integer-valued coordinates have been constructed. We have shown that these algorithms are pseudopolynomial if the space dimension is bounded by some constant. Thus, we have established the conditions under which the problems are solvable in a pseudopolynomial time.

In our opinion, the presented algorithms would be useful as tools for solving problems in applications related to Data editing, Data cleaning, Data mining, and Machine learning.

It is clear that the algorithms can be used to solve practical problems having integer instances of small dimensions. The development of approximation algorithms with guaranteed accuracy for these problems is of considerable interest.

Acknowledgments. The study presented in Sects. 2, 4 was supported by the Russian Science Foundation, project 16-11-10041. The study presented in Sects. 3, 5 was supported by the Russian Foundation for Basic Research, projects 16-07-00168 and 18-31-00398, by the Russian Academy of Science (the Program of basic research), project 0314-2016-0015, and by the Russian Ministry of Science and Education under the 5-100 Excellence Programme.

References

1. Kel'manov, A.V., Pyatkin, A.V.: NP-completeness of some problems of choosing a vector subset. J. Appl. Ind. Math. **5**(3), 352–357 (2011)
2. Kel'manov, A.V., Pyatkin, A.V.: On the complexity of some problems of choosing a vector subsequence. Zhurnal Vychislitel'noi Matematiki i Matematicheskoi Fiziki (in Russian) **52**(12), 2284–2291 (2012)
3. Aggarwal, A., Imai, H., Katoh, N., Suri, S.: Finding k points with minimum diameter and related problems. J. Algorithms **12**(1), 38–56 (1991)
4. Kel'manov, A.V., Romanchenko, S.M.: An approximation algorithm for solving a problem of search for a vector subset. J. Appl. Ind. Math. **6**(1), 90–96 (2012)
5. Kel'manov, A.V., Romanchenko, S.M.: Pseudopolynomial algorithms for certain computationally hard vector subset and cluster analysis problems. Autom. Remote Control **73**(2), 349–354 (2012)
6. Shenmaier, V.V.: An approximation scheme for a problem of search for a vector subset. J. Appl. Ind. Math. **6**(3), 381–386 (2012)
7. Kel'manov, A.V., Romanchenko, S.M.: An FPTAS for a vector subset search problem. J. Appl. Ind. Math. **8**(3), 329–336 (2014)
8. Shenmaier, V.V.: Solving some vector subset problems by voronoi diagrams. J. Appl. Ind. Math. **10**(2), 550–566 (2016)
9. Kel'manov, A.V., Romanchenko, S.M., Khamidullin, S.A.: Approximation algorithms for some intractable problems of choosing a vector subsequence. J. Appl. Ind. Math. **6**(4), 443–450 (2012)
10. Kel'manov, A.V., Romanchenko, S.M., Khamidullin, S.A.: Exact pseudopolynomial algorithms for some np-hard problems of searching a vectors subsequence. Zhurnal Vychislitel'noi Matematiki i Matematicheskoi Fiziki (in Russian) **53**(1), 143–153 (2013)
11. Kel'manov, A.V., Romanchenko, S.M., Khamidullin, S.A.: An approximation scheme for the problem of finding a subsequence. Numerical Anal. Appl. **10**(4), 313–323 (2017)
12. Ageev, A.A., Kel'manov, A.V., Pyatkin, A.V., Khamidullin, S.A., Shenmaier, V.V.: Approximation polynomial algorithm for the data editing and data cleaning problem. Pattern Recognit. Image Anal. **17**(3), 365–370 (2017)
13. de Waal, T., Pannekoek, J., Scholtus, S.: Handbook of Statistical Data Editing and Imputation. Wiley, Hoboken (2011)

14. Osborne, J.W.: Best Practices in Data Cleaning: A Complete Guide to Everything You Need to Do Before and After Collecting Your Data, 1st edn. SAGE Publication, Inc., Los Angeles (2013)
15. Farcomeni, A., Greco, L.: Robust Methods for Data Reduction. Chapman and Hall/CRC, Boca Raton (2015)
16. Hansen, P., Jaumard, B.: Cluster analysis and mathematical programming. Math. Program. **79**, 191–215 (1997)
17. Jain, A.K.: Data clustering: 50 years beyond k-means. Pattern Recognit. Lett. **31**(8), 651–666 (2010)
18. Shirkhorshidi, A.S., Aghabozorgi, S., Wah, T.Y., Herawan, T.: Big Data Clustering: A Review. LNCS. **8583**, 707–720 (2014)
19. Bishop, C.M.: Pattern Recognition and Machine Learning. Springer Science+Business Media, LLC, New York (2006)
20. James, G., Witten, D., Hastie, T., Tibshirani, R.: An Introduction to Statistical Learning. Springer Science+Business Media, LLC, New York (2013)
21. Hastie, T., Tibshirani, R., Friedman, J.: The Elements of Statistical Learning, 2nd edn. Springer, Berlin (2009)
22. Aggarwal, C.C.: Data Mining: The Textbook. Springer International Publishing, Berlin (2015)
23. Goodfellow, I., Bengio, Y., Courville, A.: Deep Learning (Adaptive Computation and Machine Learning series). MIT Press, Cambridge (2017)
24. Fu, T.: A review on time series data mining. Eng. Appl. Artif. Intell. **24**(1), 164–181 (2011)
25. Kuenzer, C., Dech, S., Wagner, W. (eds.): Remote Sensing Time Series. RSDIP, vol. 22. Springer, Cham (2015). https://doi.org/10.1007/978-3-319-15967-6
26. Liao, T.W.: Clustering of time series data – a survey. Pattern Recognit. **38**(11), 1857–1874 (2005)
27. Kel'manov, A.V., Khamidullin, S.A.: Posterior detection of a given number of identical subsequences in a quasi-periodic sequence. Comput. Math. Math. Phys. **41**(5), 762–774 (2001)

A Restarting Rule Based on the Schnabel Census for Genetic Algorithms

Anton V. Eremeev[1,2(✉)]

[1] Sobolev Institute of Mathematics SB RAS, Omsk, Russia
[2] Institute of Scientific Information on Social Sciences RAS, Moscow, Russia
eremeev@ofim.oscsbras.ru

Abstract. A new restart rule is proposed for Genetic Algorithms (GAs) with multiple restarts. The rule is based on the Schnabel Census method, transfered from the biometrics, where it was originally developed for the statistical estimation of a size of animal population. In this paper, the Schnabel Census method is applied to estimate the number of different solutions that may be visited with positive probability, given the current distribution of offspring. The rule consists in restarting the GA as soon as the maximum likelihood estimate reaches the number of different solutions observed at the recent iterations.We demonstrate how the new restart rule can be incorporated into a GA on the example of the Set Cover Problem. Computational experiments on benchmarks from OR-Library show a significant advantage of the GA with the new restarting rule over the original GA. On the unicost instances, the new rule also tends to be superior to the well-known rule, which restarts an algorithm when the current iteration number is twice the iteration number when the best incumbent was found.

Keywords: Maximum likelihood · Abundance of population
Set cover · Transfer of methods

1 Introduction

Genetic Algorithms (GAs) are the randomized search heuristics based on the biological analogy of selective breeding in nature, originating from the work of J. Holland [18]. A GA manipulates with a *population* of *individuals*, using the random operators that model mutation and crossover in nature. Suppose that a GA is applied to an optimization problem with the space of solutions D and the objective function $f : D \to \mathbb{R}$ to be maximized. An individual is a pair of *genotype* g and *phenotype* $x(g)$, where g is a fixed length string of symbols (called *genes*) from some finite alphabet, and $x(g)$ corresponds to a search point in the space of solutions D. The function $x(g)$ maps g to its phenotype $x(g) \in D$, thus defining a *representation* of solutions in a GA. The search in GAs is guided by the values of the *fitness* function $\Phi(g) = \phi(f(x(g)))$ on the genotypes of the current population Π^t on iteration t. Here $\phi : \mathbb{R} \to \mathbb{R}$ may be an identical mapping or a

© Springer Nature Switzerland AG 2019
R. Battiti et al. (Eds.): LION 12 2018, LNCS 11353, pp. 337–351, 2019.
https://doi.org/10.1007/978-3-030-05348-2_29

monotone function, chosen appropriately to intensify the search. The genotypes of the initial population Π^0 are generated according to some a priori defined probability distribution.

Due to randomness of initialization, selection, mutation and crossover operators, their behavior varies from run to run. In order to increase the probability of finding an optimal solution, it is a common practice to use multiple restarts of a GA. The choice of the iteration when the GA is stopped and restarted again (a restart rule) was considered in a number of papers [2,19,20,22]. A stopping criterion is also proposed for the multi-objective evolutionary algorithms in [23].

In this paper, a new restart rule is proposed for the GAs. The rule is based on the Schnabel Census method, transfered from the biometrics, where it was originally developed for the statistical estimation of a size of animal population [34], assuming that one takes repeated samples of size 1 (at suitable intervals of time) and counts the number of distinct animals seen. This method was used in computer science already to estimate the number of local optima on the basis of repeated local search [31]. Experiments [31,32] showed that the estimates based on this approach are adequate for the landscapes with uniform basin of attraction sizes, but have a negative bias when the basin sizes are significantly unequal.

Here we make a simplifying assumption that during the latest iterations, the GA population was generated according to the same distribution and the Schnabel Census method is applicable to estimate the number of different solutions that may be visited with positive probability in this distribution. The rule consists in restarting the GA as soon as a maximum likelihood estimate reaches the number of different solutions observed at the latest iterations. The rationale of this rule is that it stops the GA when, most likely, there are no more non-visited solutions in the area where the GA population spent the latest iterations. In such a case it would be more appropriate to restart the GA instead of waiting till it leaves the explored area by mutation and crossover.

There are two major GA outlines in use: the *elitist* GAs, which copy a certain number of the "most promising" individuals from the previous population to the next one, and the *non-elitist* GAs, which generate all individuals of a new population with the same probability distribution. The well-known Simple GA [18,35] is an example of non-elitist GAs. We expect that the new restart rule is sufficiently general to be applicable in both types of GAs, although in the present paper the rule is tested only with an elitist steady-state GA.

We demonstrate how the new restart rule can be incorporated into a GA [9] with Non-Binary Representation (NBGA) for the Set Cover Problem (SCP). Computational experiments on benchmarks from OR-Library show a significant advantage of the GA with the new restarting rule, compared to the original version of the GA [9]. In particular, given equal computational budget, in 35 out of 74 SCP instances the new version of the GA had a higher frequency of finding the optima and only in 5 out of 74 instances the new version showed inferior results. On the unicost SCP instances, the new rule also tends to be superior to the well-known rule [16], which restarts an algorithm when the current iteration number is twice the iteration number when the best incumbent was found.

The idea of using the Schnabel Census method for estimation of the number of unvisited solutions [31], as well as the basic ideas of GA [18], are the examples of transfer of ideas from biology into computer science. In the present paper, both methods are combined together. Schnabel Census, originally developed for estimation of animal population size, is not used for counting individuals here (the population size is a known parameter of a GA, kept constant throughout the run), but for estimation of the number of solutions which may be visited if the distribution of offspring in the GA remains unchanged.

In the next section, we briefly discuss the use of Schnabel Census in biology and computer science. Section 3 gives a motivation and a detailed description of the restart rule based on Schnabel Census. Section 4 briefly describes the GA considered in this paper. Section 5 presents the experimental evaluation of the proposed restart rule. Concluding remarks a given in Sect. 6.

2   Estimation of Animal Population Size and the Number of Local Optima

Schnabel Census method is developed in biometrics for statistical estimation of the size of animal populations [33,34]. According to this method, one takes repeated samples of size n_0 from a population and counts the number of *distinct* animals seen. Often it is assumed that the probability of catching any particular animal is the same. The sampled animals are marked, unless they were marked previously, and returned back into the population. Then a statistical estimate for the total number ν of individuals in population is computed on the basis of the total number of animals marked in all the samples.

This method, with the sample size $n_0 = 1$, was adapted in [31] to estimate the number of local optima in combinatorial optimization problems on the basis of repeated local search outcomes with random starting points. A number of other approaches have been proposed for estimation of the number of local optima. One may fit certain type of parametric distribution of the basin of attraction sizes (exponential, gamma, lognormal etc.) [14,32] to estimate this parameter. Nonparametric estimates, such as the bootstrap or the jackknife, can also be employed [11,27,32,34]. Assuming a particular type of distribution of basin sizes one can obtain the maximal likelihood estimate for the local optima number, or a confidence interval for it.

In what follows, we will apply the Schnabel Census method to estimate the number of values that a discrete random variable may take with non-zero probability. The other methods mentioned above could be applied to this problem as well, but unfortunately this problem does not have a satisfactory solution in the general case (see e.g. [22]), where different values of the random variable may have different probabilities.

3   Restart Rule Based on Schnabel Census

One of the theoretical approaches to understanding the Simple GA (and some of its its generalisations) as a dynamical system was suggested in [35,37]. Suppose that X is a finite genotypes space. In the dynamical system models of GAs, a *population vector* $\mathbf{p}$ is introduced which has a length $|X|$. A k-th component p_k of this vector is the proportion of the population Π that has a genotype $y_k \in X, k = 0, ..., |X| - 1$ (assuming that the genotypes in X are numbered in some standard order). Given a population vector $\mathbf{p}^t$, of the current population Π^t, a function $G(\mathbf{p}^t)$ produces a vector in $[0, 1]^{|X|}$, where the k-th component, $k = 1, ..., |X|$ equals the probability that an offspring computed for population Π^{t+1} will have the genotype y_k. The result of M.Vose [35] shows that as the population size tends to ∞, the sequence of population vectors of the Simple GA $\mathbf{p}^0, \mathbf{p}^1, ..., \mathbf{p}^t$ converges in probability to the sequence $\mathbf{p}^0, G(\mathbf{p}^0), ..., G^t(\mathbf{p}^0)$ for any finite t. This suggests that it may be helpful to consider the infinite-population GA as an approximation of a finite-population GA because on each iteration of an infinite-population GA the vector $\mathbf{p}^{t+1}$ is a deterministic function $G(\mathbf{p}^t)$ of the current population vector. It was shown in [36,37] that the fixed points of G (i.e. such population vectors $\mathbf{p}$ where $\mathbf{p} = G(\mathbf{p})$ holds) are crucial in the analysis of the infinite-population GA. The properties of a fixed point $\mathbf{p}$ of an infinite-population GA are determined by the eigenvalues associated with the differential of G at that point (see e.g. [36,37]). If all these eigenvalues belong to the interior of the unit disk, then the point $\mathbf{p}$ is called *stable*. It was shown in [35] that these stable points are *attractors* in the sense that from almost all initial population vectors $\mathbf{p}^0$ the infinite-population Simple GA converges to one of the stable points. The finite-population GAs demonstrate a similar *metastable* behavior, although in a randomized way: the probability distribution of their population remains "almost" stationary (close to a stable population vector $\mathbf{p}$) for a great number of iterations, until a seldom random event might shift it out into the basin of attraction of some other stable fixed point [24]. The elitist GAs were considered in [37] and shown to have a metastable behavior as well.

In [28], C. Reeves has shown that it is rather unlikely that a Hamming local optimum (w.r.t. Hamming neighborhood of radius 1 in the genotypes space) would not be a GA attractor as well. Experimental [28] and theoretical [36] studies have also revealed that there is a strong connection between attractors and local optima. With this in mind, it should be advisable to restart a GA, once its population has been trapped for a long time in one of the stable points. This idea is implemented in the new restart rule described below.

Let a parameter r define the length of the historical period considered for statistical analysis in the restart rule. Given a value of r, we assume that during the r latest iterations, all new offspring in the GA obeyed the same distribution and their genotypes may be treated as the sampled animals in the Schnabel Census method. Then we apply the Schnabel Census method in order to estimate the number ν of different solutions that may be visited with a positive probability, assuming that the current distribution of offspring remains unchanged.

In what follows, we assume that in the latest r iterations of a GA, the observed sample consists of r independent offspring solutions. Let us define the random variable K as the number of *distinct* solutions in this sample. We will make a simplifying assumption that all solutions, that may be generated in the current distribution, have equal probabilities. Then, as it was noticed in [7], for any fixed ν the value K has the following distribution:

$$\Pr\{K = k\} = \frac{\nu!}{(\nu - k)!} \frac{S(r, k)}{\nu^r},$$

where $S(r, k) = \frac{1}{k!} \sum_{s=0}^{k} (-1)^k \binom{k}{s}(k - s)^r$ is the Stirling number of the second kind. This distribution is also known as the Arfwedson distribution [21]. The maximum likelihood estimate $\hat{\nu}^{ML}$ for the unknown ν is

$$\hat{\nu}(r, k) = \text{argmax}\left\{\frac{\nu!}{(\nu - k)!\nu^r}\right\}, \tag{1}$$

where k is the number of different solutions actually generated on the latest r iterations of the GA. The value $\hat{\nu}^{ML} = \hat{\nu}(r, k)$ may be found from (1) by the standard one-dimensional optimization methods (see e.g. [31]).

The proposed rule restarts the GA as soon as the estimate $\hat{\nu}^{ML}$ becomes equal to k. The value of r is tuned adaptively during the GA execution. The rationale behind this rule is that once the equality $\hat{\nu}^{ML} = k$ is satisfied, most likely there are no more non-visited solutions in the area where the GA population spent the latest r iterations. In such a situation, it is more appropriate to restart the GA rather than to wait till the population distribution will significantly change by the evolutionary mechanisms.

4   The Genetic Algorithm for Set Cover Problem

The Set Cover Problem may be formulated as follows. *Consider a set $\mathcal{M} = \{1, \ldots, m\}$ and the subsets $\mathcal{M}_j \subseteq \mathcal{M}$, where $j \in \mathcal{N} = \{1, ..., n\}$. A subset $\mathcal{J} \subseteq \mathcal{N}$ is a cover of $\mathcal{M}$ if $\bigcup_{j \in \mathcal{J}} \mathcal{M}_j = \mathcal{M}$. For each $\mathcal{M}_j$, a positive cost c_j is assigned. The SCP is to find a cover of minimum summary cost.*

The SCP is a well-known NP-hard problem [13]. A number of heuristic algorithms are developed for approximate solving the SCP within relatively short running time: Lagrangian relaxation heuristics [6], neural networks [15], local search [38], GAs [4,9], ant colony algorithms [1] etc.

Here we will use the GA which was proposed in our earlier work without any restart rule [9]. This GA is based on the elitist steady-state population management. It is denoted as NBGA because of the non-binary representation of solutions [4,9], involving an alphabet with up to n symbols. The NBGA uses a problem-specific crossover operator based on the linear programming, the proportional selection operator and a mutation operator that makes random changes in every gene with a given probability p_m. The offsprings are improved by the means of different greedy procedures before they are added into the population.

5   Computational Experiments

The NBGA was implemented in Borland Delphi 5 and tested on Pentium-IV with 3 GHz CPU and 2 GB RAM, using the OR-Library [3] benchmark problem sets *4-6*, *A-H*, and two sets of combinatorial problems *CLR* and *Stein*. The sets *4-6* and *A-H* consist of randomly generated problems with costs c_j from 1,...,100, while *CLR* and *Stein* consist of combinatorial unicost problems, where $c_j = 1$ for all j. Dimensions of the problems and the number of instances in each randomly generated series are given in Table 1.

Table 1. Parameters of randomly generated instances

Series name	4	5	6	A	B	C	D	E	F	G	H
Rows (m)	200	200	200	300	300	400	400	500	500	1000	1000
Columns (n)	1000	2000	1000	3000	3000	4000	4000	5000	5000	10000	10000
Num. of probl.	10	10	5	5	5	5	5	5	5	5	5

We compared three modes of GA execution with equal computational budget:

- Mode A. Single run with no restarts.
- Mode B. Restart the GA as soon as the current iteration number becomes twice the iteration number when the best incumbent was found. This rule was used successfully by different authors to restart random hill-climbing method [16] and GAs [2,10]. To avoid early restarts, this rule is applied only after a certain number of iterations, denoted by t_{min}. We use t_{min} equal to the population size.
- Mode C. Restart the GA using the new rule proposed in Sect. 3. The value of parameter r is chosen adaptively as follows: *Whenever the best found solution is improved, r is set to be the population size. If the best incumbent was not improved during the latest $2r$ iterations, then the value of r is doubled.* We reset r to the population size when the best found solution is improved, assuming that whenever the best incumbent is improved, this means that the population has reached a new unexplored area and the length of the historic period for analysis should be reduced. To reduce the CPU cost, the termination condition is checked only when the value of r is updated.

A single experiment with a GA, given a certain computational budget we will call *a trial*. In the experiments, $N = 30$ independent trials of the GA in each of the three modes were carried out. The GA population size was set to 100. Let the statistic σ be the average relative error $\sum_{k=1}^{30} \frac{f_k - f^*}{30 f^*} \cdot 100\%$, where f_k is the cost of solution found in the k-th trial and f^* is the optimal cost. In what follows, F_{bst} will denote the frequency of obtaining a solution with the best known cost from the literature, estimated by 30 trials.

A statistical analysis of experimental data is carried out using the significance test from [5] (see Ch. 8, §2), which is used to compare two algorithms in terms of probability of finding an optimal or a best-known solution. Let P_1 and P_2 denote the probabilities of success for some Algorithms 1 and 2, respectively. The null hypothesis is that $P_1 = P_2$. Under the null hypothesis, the estimate of common success rate is $F = (F_1 + F_2)/2$, where F_1 denotes the frequency of success in N trials of Algorithm 1, and F_2 is the frequency of success in N trials of Algorithm 2. In our case $N = 30$. The difference $F_1 - F_2$ can be expressed in units of the standard deviation as the statistic $A = |F_1 - F_2|/\hat{SD}$, where $\hat{SD} = \sqrt{2F(1-F)/N}$ denotes the estimate of the standard deviation. It is appropriate to assume that A is normally distributed if the number of trials is sufficiently large. The null hypothesis may be tested at a confidence level p by comparison of the value of A to the quantile of the standard normal distribution $z_{p/2}$ (e.g. with $p = 0.05$, we have $z_{0.025} \approx 1.96$). If $A > z_{p/2}$, then the null hypothesis is rejected. In Tables 2, 3 , 4 and 6, the statistically significant difference (at level $p \le 0.05$) between the frequencies of finding optimal solutions is indicated by "$*$" when mode C is compared to mode A and by "$+$" when modes C and A are compared.

5.1 Experiments with Randomly Generated Problems

In the experiments described in this subsection, the total budget, counting all restarts during the GA trial, was set to 10000 tentative solutions. For all randomly generated problems, the mutation probability p_m is 0.1. Tables 2, 3 and 4 show the results of experiments with series *4-6*, *a,c,d* and *e,f,g,h*. The optimal solution values are known for all instances of these series (see e.g. [38]). The highest frequency of obtaining optima is indicated in bold. In series *b*, all runs yielded optimal solutions regardless of a restart rule, therefore we skip series *b* in Table 3. Tables 2, 3 and 4 also show the average number of iterations t_{avg} (over 30 runs) that were made until the restart rule terminated a GA. The symbol "–" indicates the cases where no restarts were made, until an optimum was found, or the total budget of 10000 iterations was reached.

Comparing the GA results in modes A and C reported in Tables 2, 3 and 4, one can see that among 37 instances, where these two modes yield different frequencies F_{bst}, mode C has a higher value F_{bst} in 31 cases and in 16 out of these 31 cases the difference is statistically significant. Mode A has a statistically significant advantage to mode C only on a single instance 404.

In these tables, modes B and C show different frequencies F_{bst} on 28 instances. On 16 of these 28 instances, mode C has a higher value F_{bst} than mode B and in 5 out of these 16 cases the difference is statistically significant. Mode B has a statistically significant advantage to mode C only on a single instance 502. In terms of percentage of deviation σ, averaged over all instances of series *4-6*, *a,c,d* and *e,f,g,h*, mode C gives the least error (see the row "average").

The restart mode C terminated the GA later than mode B in most of the cases. This seems to be natural because the restart rule of mode B is based solely on the values of objective function of the obtained solutions, while the new restart rule uses the information about the generated covers and their frequencies. In particular, in those cases where mode C yielded statistically greater frequency of finding the optima, compared to mode B, the average number of iterations t_{avg} in mode C was 4–8 times larger than in mode B.

Table 2. Relative errors and frequencies of finding optima in series *4,5* and *6*.

Instance	Single run mode A		Restart mode B			New restart mode C		
	σ	F_{bst}	σ	F_{bst}	t_{avg}	σ	F_{bst}	t_{avg}
401	0.117	0.967	0	1	537.6	0	1	760.8
402	0.703	0.833	0	1	537.6	0	1*	817.6
403	0.039	0.967	0	1	768.0	0	1	845.4
404	1.154	**0.767***	0.445	0.3	717.1	0.364	0.4	748.3
405	0.156	0.867	0	1	455.4	0.039	0.967	828.0
406	0.554	0.5	0	1	642.4	0	1*	770.0
407	0.186	0.867	0	1	488.9	0	1*	784.4
408	0.589	0.933	0	1	497.6	0	1	798.8
409	1.451	0.533	0.125	0.867	738.7	0.062	**0.933***	796.6
410	0.117	0.933	0	1	657.0	0	1	803.8
501	1.976	0.433	0	1	752.6	0.079	0.967*	803.8
502	1.987	0.667	0.066	**0.933+**	691.2	0.397	0.633	793.8
503	0	1	0	1	413.8	0	1	777.7
504	1.033	0.333	0.041	0.967	586.0	0	1*	839.0
505	4.171	0.333	2.559	0.267	373.4	2.417	**0.367**	684.9
506	0	1	0	1	249.1	0	1	759.7
507	0.785	0.233	0.307	0.7	889.4	0.273	**0.733***	792.7
508	0.486	0.767	0	1	518.0	0	1*	766.3
509	0.251	0.767	0	1	317.3	0.036	0.967*	770.4
510	1.245	0.567	1.132	**0.6**	457.6	1.396	0.5	765.3
601	0.87	0.867	0.145	**0.967**	174.9	0.29	0.933	711.0
602	0.89	0.667	0	1	165.9	0	1*	667.5
603	0	1	1.034	0.833	145.0	0	1+	661.9
604	0	1	0	1	127.8	0	1	1014.9
605	1.677	0.7	0	1	189.1	0	1*	664.5
Average	0.817	0.74	0.231	0.901	485.2	0.217	0.892	773.9

To evaluate the overall CPU cost of NBGA execution with the new restart rule (including the CPU time for Schnambel Census estimate $\hat{\nu}^{\text{ML}}$), we considered those instances, where the frequency of finding optima was equal to 100% (see Table 5). On average, in terms of the CPU time and in terms of the total number of tentative solutions made until first visiting an optimum, mode C tends to be more efficient than the other two. Here the CPU time is not proportional to the number of tentative solutions because the individuals of initial populations are computed faster than the offspring in the main loop of NBGA.

Table 3. Relative errors and frequencies of finding optima in series a, c and d.

Instance	Single run mode A		Restart mode B			New restart mode C		
	σ	F_{bst}	σ	F_{bst}	t_{avg}	σ	F_{bst}	t_{avg}
a1	1.344	**0.133**	1.067	0.1	501.9	1.146	0.033	773.1
a2	0.476	0.867	0.159	**0.933**	845.2	0.278	0.833	810.8
a3	1.207	0.2	0.862	0.333	541.8	0.819	**0.367**	756.3
a4	2.051	0.567	0.256	**0.8**	672.3	0.427	0.667	797.4
a5	0.89	0.3	0.466	0.633	465.9	0.381	**0.7***	752.9
c1	0.661	0.8	0.088	**0.933**	566.9	0.264	0.8	803.8
c2	0.868	0.467	0	1	523.9	0	1*	838.0
c3	1.811	0.533	0.288	**0.767**	763.7	0.37	0.7	889.0
c4	0.091	0.933	0	1	392.4	0	1	1017.0
c5	0.419	0.8	0	1	440.7	0	1*	839.4
d1	0	1	0	1	–	0	1	–
d2	0	1	0	1	120.9	0	1	–
d3	0.417	0.9	3.333	0.333	127.8	0	1^{+}	709.1
d4	0	1	0.806	0.833	108.3	0	1^{+}	932.4
d5	0	1	0	1	120.1	0	1	–
Average	0.682	0.7	0.488	0.778	–	0.246	0.807	–

5.2 Experiments on Combinatorial Unicost Problems

The three modes of running the GA were also tested on two series of combinatorial unicost SCP instances. In the experiments described in this subsection, the total budget, for each GA trial, was set to 10000 tentative solutions as in the previous subsection. The set of problems *clr* is derived from one of the well-known questions of P. Erdös [8], stated as a unicost SCP (see e.g [15]) which turns out to be very hard for exact solvers. The series of problems *Stein* arise from Steiner triple systems, and it was proposed in [12] as a set of examples of hard problems that can be used for evaluation of different algorithms. Table 6 shows the dimensions of the unicost SCP instances considered in this paper. Columns F_{bst} and t_{avg} have the same meaning as in tables above. To the best of our knowledge, optimality of the best-known solutions indicated in this table is proven only for *Stein* series [25]. Again we use the population size of 100 individuals and the total budget of GA iterations, equal to 10000 in each trial. The mutation probability p_{m} is set to 0.01 for all instances.

On the combinatorial unicost instances, restart mode C shows better or equal results compared to the other two modes, except for a single instance clr13. Column t_{avg} indicates that the restart rule of mode B triggers too early, precluding the GA from finding good solutions. This is more evident than in the case of randomly generated instances because the combinatorial unicost instances tend to have large plateaus of solutions with equal objective function values. On clr13,

Table 4. Relative errors and frequencies of finding optima in series e, f and h.

Instance	Single run mode A		Restart mode B			New restart mode C		
	σ	F_{bst}	σ	F_{bst}	t_{avg}	σ	F_{bst}	t_{avg}
e1	0	1	0	1	–	0	1	–
e2	0	1	0.667	0.933	114.5	0	1	662.7
e3	0	1	0	1	104.4	0	1	–
e4	0	1	0	1	102.3	0	1	–
e5	0	1	0	1	203.3	0	1	–
f1	0	1	0	1	101.0	0	1	–
f2	0	1	0	1	–	0	1	–
f3	0	1	0	1	101.7	0	1	473.1
f4	0	1	0	1	102.8	0	1	–
f5	22.308	**0.033**	23.077	0	110.9	22.308	**0.033**	579.9
g1	0.852	0.733	0	1	198.0	0	1*	816.0
g2	2.078	0.4	1.234	0.367	296.0	1.104	**0.433**	874.6
g3	2.831	0.033	2.53	0	268.6	1.747	**0.067**	892.3
g4	2.262	0.4	0.595	**0.833**	228.1	1.071	0.633	819.3
g5	0.952	0.667	0	1	271.6	0	1*	905.8
h1	4.762	0	4.762	0	136.3	4.762	0	777.0
h2	4.762	0	4.762	0	124.1	4.762	0	406.8
h3	0.169	**0.967**	3.898	0.233	142.0	0.508	0.9+	825.6
h4	0	1	2.759	0.467	139.9	0	1+	829.2
h5	0	1	0	1	111.0	0	1	–
Average	2.049	0.712	2.214	0.692	–	1.813	0.753	–

the best known solution was found only in mode A and it took more than 4000 iterations. Mode C was irrelevant on this instance, probably due to a negative bias of the maximum likelihood estimate $\hat{\nu}^{ML}$ from formula (1) (see [11,31,32]).

5.3 Reduced Iterations Budget

We repeated the experiments of Sects. 5.1 and 5.2 with the total iterations budget reduced to 5000 and obtained the following results.

Comparing the GA results on randomly generated problems in modes A and C, among 41 instances, where these two modes yielded different frequencies F_{bst}, mode C had a higher value F_{bst} in 29 cases and in 15 out of these 29 cases the difference was statistically significant. Mode A had a statistically significant advantage to mode C only on instances 404 and a4.

Modes B and C showed different frequencies F_{bst} on 35 randomly generated instances. On 21 of these instances, mode C had a higher value F_{bst} than mode B

Table 5. Average time (sec.) and number of covers generated till finding optima

Instance	Single run mode A		Restart mode B		New restart mode C	
	opt. time	tent. sol.	opt. time	tent. sol.	opt. time	tent. sol.
503	0.53	1364.0	1.09	1996.9	**0.39**	1256.0
603	0.09	2264.5	0.11	2962.6	**0.08**	1869.5
d1	0.04	129.5	0.04	136.1	**0.03**	127.9
d2	0.34	325.6	0.66	731.8	**0.23**	320.5
d5	0.05	174.7	0.05	166.5	**0.04**	144.4
e1	**0.01**	101.8	0.02	101.2	0.02	101.6
e3	**0.19**	249.0	0.35	558.1	0.24	293.5
e4	**0.18**	245.5	0.38	690.4	0.31	257.6
e5	**0.05**	137.7	**0.05**	137.3	**0.05**	132.6
f1	0.11	177.1	0.19	231.6	**0.06**	152.5
f2	**0.03**	107.3	**0.03**	107.4	**0.03**	107.6
f3	1.04	509.6	1.1	1596.8	**0.64**	594.3
f4	0.34	243.5	0.5	592.1	**0.20**	281.9
h5	0.31	235.0	0.45	423.4	**0.25**	211.3
Average	0.236	447.486	0.359	745.157	0.184	417.943

Table 6. Frequencies of finding optimal solutions in series *clr* and *Stein*.

Instance	Rows m	Columns n	Best known solution	Single run mode A	Restart mode B		New restart mode C	
				F_{bst}	F_{bst}	t_{avg}	F_{bst}	t_{avg}
clr10	511	210	25	0.2	0.033	119.8	**0.433**[*+]	746.2
clr11	1023	330	23	**0.433**	0	104.7	**0.433**[+]	724.0
clr12	2047	495	23	0.467	0.267	124.1	**0.567**	775.4
clr13	4095	715	23	**0.033**	0	113.9	0	899.7
Stein27	117	27	18	1	1	–	1	–
Stein45	330	45	30	0.867	0.967	101.0	**1**[*]	636.5
Stein81	1080	81	61	1	1	101.6	1	1328.5
Stein135	3015	135	103	0	0	111.6	0	684.8
Stein243	9801	243	198	0.833	0.667	111.7	**1**[*+]	680.0

and in 5 out of these 21 cases the difference was statistically significant. Mode B had a statistically significant advantage to mode C only on two instances 502 and a4. In terms of percentage of deviation σ, averaged over all instances of series *4-6*, *a,c,d* and *e,f,g,h*, mode C gave the least error.

On the combinatorial unicost instances, none of the modes significantly outperformed the other ones, presumably because 5000 iterations is a relatively short budget for these series and the restarts do not improve the results.

5.4   Increased Iterations Budget

We also repeated the experiments of Sects. 5.1 and 5.2 with the total iterations budget increased to 15000 and obtained the following results.

Comparing the GA results on randomly generated problems in modes A and C, among 38 instances, where these two modes yield different frequencies F_{bst}, mode C has a higher value F_{bst} in 35 cases and in 20 out of these 35 cases the difference is statistically significant. Mode A has a statistically significant advantage to mode C only on three instances 404, 502 and a1.

Modes B and C showed different frequencies F_{bst} on 22 randomly generated instances. On 14 of these instances mode C has a higher value F_{bst} than mode B and in 3 out of these 14 cases the difference is statistically significant. Mode B has a statistically significant advantage to mode C on two instances 502 and 507. In terms of percentage of deviation σ, averaged over all instances of series *4-6*, *a,c,d* and *e,f,g,h*, mode C gives the least error.

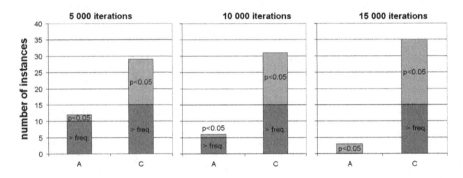

Fig. 1. Number of instances where modes A or C have a greater frequency of finding optima for randomly generated SCPs: The top segments show the instances with statistically significant difference in frequency

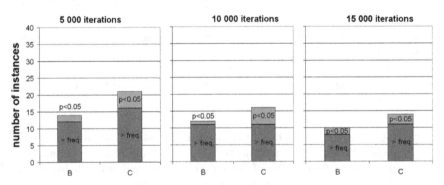

Fig. 2. Number of instances where modes B or C have a greater frequency of finding optima for randomly generated SCPs: The top segments show the instances with statistically significant difference in frequency

On the combinatorial instances, mode C shows better or equal results compared to the other two modes, except for a single instance Stein45, where it was outperformed by mode B but the difference in this case was not statistically significant. At the same time, mode C had a statistically significant advantage to mode B in 4 cases. Mode B performed poorly on these instances, showing a relative error on average by a factor of 1.5 greater than modes A and C.

Experiments with randomly generated SCPs are summarized in Figs. 1, 2.

6 Conclusions

A new restart rule is proposed for Genetic Algorithms, using the Schnabel Census method, originally developed for statistical estimation of size of animal populations. The performance of the new restart rule is demonstrated on a steady-state GA with non-binary representation for the Set Cover Problem. Computational experiments show a significant advantage of the GA with the new restarting rule over the GA without restarts and the GA restarted as soon as the current iteration number becomes twice the iteration number of the currently best incumbent.

Applying Schnabel Census in this research is an attempt to further benefit from the convergence between computer science and biology. That interface had been crucial for evolutionary computation to emerge, but was scarcely maintained afterwards [26]. As the present research shows, developing this transdisciplinary integration can be productive.

Further improvements of the restart strategy are expected via usage of less biased methods, developed for estimation of the number of local optima and for estimation of the abundance of populations, see e.g. [11,17]. The further research might address the usefulness of the proposed restart rule for other types of evolutionary algorithms and other optimization problems. The Schnabel Census method may be also applicable for a dynamical control of the mutation, if instead of restarting, the GA would increase the mutation probability by a certain factor.

Acknowledgments. This research is supported by the Russian Science Foundation grant 17-18-01536.

References

1. Alexandrov, D., Kochetov, Y.: Behavior of the Ant Colony Algorithm for the Set Covering Problem. In: Inderfurth, K., Schwödiauer, G., Domschke, W., Juhnke, F., Kleinschmidt, P., Wäscher, G. (eds.) Operations Research Proceedings 1999. ORP, vol. 1999, pp. 255–260. Springer, Heidelberg (2000). https://doi.org/10.1007/978-3-642-58300-1_38

2. Balas, E., Niehaus, W.: Optimized crossover-based genetic algorithms for the maximum cardinality and maximum weight clique problems. J. Heuristics 4(2), 107–122 (1998)

3. Beasley, J.E.: OR-Library: distributing test problems by electronic mail. J. Oper. Res. Soc. **41**(11), 1069–1072 (1990)
4. Beasley, J.E., Chu, P.C.: A genetic algorithm for the set covering problem. Eur. J. Oper. Res. **94**(2), 394–404 (1996)
5. Brown, B.W., Hollander, M.: Statistics: A Biomedical Introduction. Wiley, New York (1977)
6. Caprara, A., Fischetti, M., Toth, P.: Heuristic method for the set covering problem. Oper. Res. **47**(5), 730–743 (1999)
7. Craig, C.C.: Use of marked specimens in estimating populations. Biometrika **40**(1–2), 170–176 (1953)
8. Erdös, P.: On a combinatorial problem I. Nordisk Mat. Tidskrift **11**, 5–10 (1963)
9. Eremeev, A.V.: A genetic algorithm with a non-binary representation for the set covering problem. In: Proceedings of OR'98, pp. 175–181. Springer-Verlag (1999)
10. Eremeev, A.V., Kovalenko, Y.V.: Genetic algorithm with optimal recombination for the Asymmetric Travelling Salesman Problem. In: Lirkov, I., Margenov, S. (eds.) LSSC 2017. LNCS, vol. 10665, pp. 332–339. Springer, Cham (2018). https://doi.org/10.1007/978-3-319-73441-5_36
11. Eremeev, A.V., Reeves, C.R.: Non-parametric estimation of properties of combinatorial landscapes. In: Cagnoni, S., Gottlieb, J., Hart, E., Middendorf, M., Raidl, G.R. (eds.) EvoWorkshops 2002. LNCS, vol. 2279, pp. 31–40. Springer, Heidelberg (2002). https://doi.org/10.1007/3-540-46004-7_4
12. Fulkerson, D.R., Nemhauser, G.L., Trotter, L.E.: Two computationally difficult set covering problems that arise in computing the 1-width of incidence matrices of Steiner triple systems. Math. Program. Stud. **2**, 72–81 (1974)
13. Garey, M.R., Johnson, D.S.: Computers and Intractability. A Guide to the Theory of NP-Completeness. W.H. Freeman and Company, San Francisco (1979)
14. Garnier, J., Kallel, L.: How to detect all maxima of a function? In: Proceedings of the Second EVONET Summer School on Theoretical Aspects of Evolutionary Computing (Anvers, 1999), pp. 343–370 Springer, Berlin (2001)
15. Grossman, T., Wool, A.: Computational experience with approximation algorithms for the set covering problem. Eur. J. Oper. Res. **101**(1), 81–92 (1997)
16. Hampson, S., Kibler, D.: Plateaus and Plateau Search in Boolean Satisfiability Problems: When to Give Up Searching and Start Again, pp. 437–456. American Mathematical Society (1996)
17. Hernando, L., Mendiburu, A., Lozano, J.A.: An evaluation of methods for estimating the number of local optima in combinatorial optimization problems. Evol. Comput. **21**(4), 625–658 (2013)
18. Holland, J.: Adaptation in Natural and Artificial Systems. University of Michigan Press, Ann Arbor (1975)
19. Hulin, M.: An optimal stop criterion for genetic algorithms: a Bayesian approach. In: Proceedings of the Seventh International Conference on Genetic Algorithms (ICGA '97), pp. 135–143. Morgan Kaufmann (1997)
20. Jansen, T.: On the analysis of dynamic restart strategies for evolutionary algorithms. In: Guervós, J.J.M., Adamidis, P., Beyer, H.-G., Schwefel, H.-P., Fernández-Villacañas, J.-L. (eds.) PPSN 2002. LNCS, vol. 2439, pp. 33–43. Springer, Heidelberg (2002). https://doi.org/10.1007/3-540-45712-7_4
21. Johnson, N.L., Kotz, S.: Discrete Distributions. Wiley, New York (1969)
22. Luke, S.: When short runs beat long runs. In: Proceedings of he Genetic and Evolutionary Computation Conference (GECCO 2001), pp. 74–80. Morgan Kaufmann (2001)

23. Marti, L., Garcia, J., Berlanga, A. and Molina, J. M.: An approach to stopping criteria for multi-objective optimization evolutionary algorithms: the MGBM criterion. In: Proceedings of 2009 IEEE Congress on Evolutionary Computation, Trondheim, pp. 1263–1270. IEEE (2009)
24. Van Nimwegen, E., Crutchfield, J.P., Mitchell, M.: Finite populations induce metastability in evolutionary search. Phys. Lett. A **229**(2), 144–150 (1997)
25. Ostrowski, J., Linderoth, J.T., Rossi, F., Smriglio, S.: Solving Steiner triple covering problems. Oper. Res. Lett. **39**, 127–131 (2011)
26. Paixao, T., Badkobeh, G., Barton, N., et al.: Toward a unifying framework for evolutionary processes. J. Theor. Biol. **383**, 28–43 (2015)
27. Pledger, S.: The performance of mixture models in heterogeneous closed population capture-recapture. Biometrics **61**(3), 868–876 (2005)
28. Reeves, C.R.: The "crossover landscape" and the Hamming landscape for binary search spaces. In: Foundations of Genetic Algorithms, vol. 7, pp. 81–98. Morgan Kaufmann, San Francisco (2003)
29. Reeves, C.R.: Fitness landscapes and evolutionary algorithms. In: Fonlupt, C., Hao, J.-K., Lutton, E., Schoenauer, M., Ronald, E. (eds.) AE 1999. LNCS, vol. 1829, pp. 3–20. Springer, Heidelberg (2000). https://doi.org/10.1007/10721187_1
30. Reeves, C.R.: Landscapes, operators and heuristic search. Ann. Oper. Res. **86**, 473–490 (1999)
31. Reeves, C.R.: Direct statistical estimation of GA landscape properties. In: Foundations of Genetic Algorithms, vol. 6, pp. 91–108. Morgan Kaufmann, San Francisco (2001)
32. Reeves, C.R.: Estimating the number of optima in a landscape, part II: experimental investigations. Coventry University Technical Report SOR#01-04 (2001)
33. Schnabel, Z.E.: The estimation of the total fish population of a lake. Am. Math. Mon. **45**, 348–352 (1938)
34. Seber, G.A.F.: The Estimation of Animal Abundance. Charles Griffin, London (1982)
35. Vose, M.D.: The Simple Genetic Algorithm: Foundations and Theory. MIT Press, Cambridge (1999)
36. Vose, M.D., Wright, A.H.: Stability of vertex fixed points and applications. In: Foundations of Genetic Algorithms, vol. 3, pp. 103–114. Morgan Kaufmann, San Mateo, CA (1995)
37. Wright, A.H., Rowe, J.E.: Continuous dynamical systems models of steady-state genetic algorithms. In: Foundations of Genetic Algorithms, vol. 6, pp. 209–226. Morgan Kaufmann, San Francisco, CA (2001)
38. Yagiura, M., Kishida, M., Ibaraki, T.: A 3-flip neighborhood local search for the set covering problem. Eur. J. Oper. Res. **172**, 472–499 (2006)

Intelligent Pump Scheduling Optimization in Water Distribution Networks

Antonio Candelieri[1]([✉]), Riccardo Perego[1], and Francesco Archetti[1,2]

[1] Department of Computer Science, Systems and Communication, University of
Milano-Bicocca, 20126 Milan, Italy
antonio.candelieri@unimib.it
[2] Consorzio Milano Ricerche, 20125 Milan, Italy

Abstract. In this paper, the authors are concerned with Pump Scheduling
Optimization in Water Distribution Networks, targeted on the minimization of
the energy costs subject to operational constraints, such as satisfying demand,
keeping pressures within certain bounds to reduce leakage and the risk of pipe
burst, and keeping reservoir levels within bounds to avoid overflow. Urban
water networks are generating huge amounts of data from flow/pressure sensors
and smart metering of household consumption. Traditional optimization
strategies fail to capture the value hidden in real time data assets. In this paper
the authors are proposing a sequential optimization method based on Approximate Dynamic Programming in order to find a control policy defined as a
mapping from states of the system to actions, i.e. pump settings. Q-Learning,
one of the Approximate Dynamic Programming algorithms, well known in the
Reinforcement Learning community, is used. The key difference is that usual
Mathematical Programming approaches, including stochastic optimization,
requires knowing the water demand in advance or, at least, to have a reliable and
accurate forecasting. On the contrary, Approximate Dynamic Programming
provides a policy, that is a strategy to decide how to act time step to time step
according to the observation of the physical system. Results on the Anytown
benchmark network proved that the optimization policy/strategy identified
through Approximate Dynamic Programming is robust with respect to modifications of the water demand and, therefore, able to deal with real time data
without any distributional assumption.

Keywords: Reinforcement learning · Dynamic programming
Pump scheduling optimization

1 Introduction

Scarcity of water and the increasing awareness of the need to save energy in providing
good quality water to increasing numbers are driving the search for new ways to save
water as well as energy and improve the financials of water utilities. At the same time
the increasing "digitalization" of urban Water Distribution Networks (WDNs) is generating huge amounts of data from flow/pressure sensors and smart metering of
household consumption and enabling new ways to achieve more efficient operations.
Sequential decision models are offering an optimization framework more suitable to

© Springer Nature Switzerland AG 2019
R. Battiti et al. (Eds.): LION 12 2018, LNCS 11353, pp. 352–369, 2019.
https://doi.org/10.1007/978-3-030-05348-2_30

capture the value hidden in real time data assets. In this paper we propose a sequential optimization method based on Approximate Dynamic Programming (ADP). Initial computational results demonstrate that our methodology can reduce the electricity expenses while keeping the water pressure in a controlled range.

1.1 Pump Scheduling Optimization

Optimization of WDNs has been a very important field in the Operation Research community at least in the last 40 years and many tools from Mathematical Programming as well as metaheuristics have been proposed. An updated review on optimization of water distribution systems is given in [11], a very recent, wide and systematic survey where several classes of existing solutions including: linear programming, nonlinear programming, dynamic programming, metamodeling, heuristics, and metaheuristics are deeply analysed and referenced.

One of the major issues for water utilities is the excessive energy consumption due to non-optimal pumping scheduling. In this paper, the authors are concerned with energy-driven Pump Scheduling Optimization (PSO). A pump schedule defines which pumps are to be operated and with which settings at different periods of the day. PSO has to be performed by considering operational constraints, such as: satisfying demand, keeping pressures within certain bounds to reduce leakage and the risk of pipe burst, and keeping reservoir levels within bounds to avoid overflow.

One of the earliest approaches for PSO was based on Dynamic Programming (DP) [16]. However, since the number of states increases exponentially with the size of the WDN, this type of solutions has been usually considered impractical, due to the curse of dimensionality in DP: recently, [13] approximate methods has been proposed to overcome the curse of dimensionality.

A formulation closely related to DP is that based on a Markov Decision Process (MDP). In [9, 10], an MDP was used to model the PSO problem in the case of a simplified WDN with three water reservoirs. The disadvantage of using classical MDP is the need to control the size of the state space through coarser discretization, while a major advantage is that the solution provides full spectra of possible policies from whichever initial stage (i.e. defined through various discrete levels of reservoirs in the WDN). This is essentially a planning approach [17] that requires extensive simulations for building the state transition probability matrix. For instance, for each state-action pair in [9] more than a hundred random runs needed to be performed. A similar strategy was recently proposed in [6, 7], where the MDP is generated through an exhaustive interaction with the hydraulic simulation software and DP is used to solve the PSO problem.

This paper proposes an Approximate Dynamic Programming (ADP) framework to PSO based on MDP models but without the requirement of complete knowledge of transitional dynamics. Our basic idea is that, instead of generating the complete MDP of the WDN behaviour – and then solve the PSO using DP – we alternate exploration (i.e. try pumps configurations not yet evaluated with respect to the current state of the WDN) and exploitation (i.e. exploit available knowledge acquired so far, relatively to the current state of the WDN). Among ADP strategies we specifically selected Q-Learning, a well-known algorithm in the Reinforcement Learning (RL) community.

To test the proposed approach, we have used a well-known benchmark WDN, namely Anytown, first proposed for PSO in [12], and largely used in the literature.

The paper is organized as follows: Sect. 2 contains relevant background information; Sect. 3 illustrates how ADP, specifically Q-Learning, and hydraulic simulation can be used to optimize energy costs in a WDN; Sects. 4 and 5 describe our experiments and reports the results; and Sect. 6 summarizes our conclusions and points to directions for further research.

2 Markov Decision Processes

MDPs are a powerful framework that can be applied to model a variety of sequential optimization problems in different fields [14]. It works in a sequential process of decision epochs by performing actions that change the state at the next decision epoch, accordingly to a transition probability function representing the dynamics of the system, and that provides for the performed action [8]. Since the system is ongoing, the state of the system prior to next decision depends on the present decision/action. Therefore, the goal is to identify, for each state, the action that produces the highest expected reward in a long-time horizon and that will result in the system performing optimally with respect to some predetermined performance criterion.

In a formal way an MDP can be defined as a tuple $\langle S, A, T, R \rangle$ where S is a set of discrete states, A is a set of discrete actions, T is a state transition function, and R is a reward function [17]. The set of discrete states is defined as the finite set $S = \{s_1, s_2, \ldots, s_{N_S}\}$ where the size of the state space is N_S. Each state consists in N_V state variables, such that $s = \{\xi^1, \xi^1, \ldots, \xi^{N_V}\}$. Thus, the set of discrete states has cardinality N_S and each state is composed by the tuple of N_V state variables, as follows:

$$S = \{s_1, \ldots, s_{N_S}\} = \left\{ \left(\xi_1^1, \ldots, \xi_1^{N_V}\right), \ldots, \left(\xi_{N_S}^1, \ldots, \xi_{N_S}^{N_V}\right) \right\} \tag{1}$$

On the other hand, the actions allow for moving from one state to another. The actions set, A, can be defined as the union of all the subsets of allowed actions for every state s of the state set S:

$$A = \bigcup_{s \in S} A_s \tag{2}$$

By applying an action $a_t = a$ in the state $s_t = s$ at time step t the system makes a transition to the new state $s_{t+1} = s'$ based on a probability distribution T over the set of possible transitions:

$$T(s, a, s') = P(s_{t+1} = s' | s_t = s, a_t = a) \tag{3}$$

where:

$$\mathcal{T} : S \times \mathcal{A} \times S \rightarrow [0, 1], \quad \sum_{s' \in S} \mathcal{T}(s, a, s') = 1 \qquad (4)$$

and

$$0 \leq \mathcal{T}(s, a, s') \leq 1 \qquad (5)$$

Finally, the reward function specifies a value received for performing the action $a_t = a$ in the state $s_t = s$ at time step t and is defined as:

$$\mathcal{R}(s, a, s') = \{r_t | s_t = s, a_t = a, s_{t+1} = s'\} \qquad (6)$$

where $\mathcal{R} : S \times \mathcal{A} \times S \rightarrow \mathbb{R}$: while a positive $\mathcal{R}(s, a, s')$ may be regarded as a reward, a negative one can be considered a cost/punishment.

Solving an MDP consists in finding a control policy π, which is defined as a mapping from states to actions, $\pi : S \rightarrow \mathcal{A}$. Optimizing such policy corresponds to maximize the accumulated reward values received over a long-time horizon. To achieve this goal, the definition of value function or utility must be provided. The value function of a state $s_t = s$ at time step t under control policy π, denoted by $V^\pi(\sigma)$, is the expected return of rewards when starting in state s and by following, sequentially, the actions suggested by the policy π. The value function is therefore defined as:

$$V^\pi(s) = \mathbb{E}_\pi \left[\sum_{k=0}^{\infty} \gamma^k \cdot r_{t+k} \middle| s_t = s \right] \qquad (7)$$

where $\mathbb{E}_\pi$ is the expected value given by following policy π, and $\gamma \in [0, 1]$ is a discount factor that is used to balance current and future rewards. When γ is small the approach is said to be "myopic", which means that it is only concerned about immediate rewards, while, when γ is large future rewards become also important.

2.1 Dynamic Programming

One fundamental property of a value function is that it satisfies the Bellman Equation [1], that allows to break a dynamic optimization problem into simpler sub-problems, and can be defined recursively as follows:

$$\begin{aligned} V^\pi &= \mathbb{E}_\pi \{r_t + \gamma \cdot r_{t+1} + \gamma^2 \cdot r_{t+2} + \ldots | s_t = s\} \\ &= \mathbb{E}_\pi \{r_t + \gamma \cdot V^\pi(s') | s_t = s, s_{t+1} = s'\} \\ &= \mathcal{R}(s, \pi(s), s') + \gamma \sum_{s' \in S} \mathcal{T}(s, \pi(s), s') \cdot V^\pi(s') \end{aligned} \qquad (8)$$

An optimal policy, denoted by π^*, is such that $V^{\pi^*}(s) \geq V^\pi(s)$ for all $s \in S$ and all policies π. Thus, the optimal value function can be evaluated as follows:

$$V^*(s) = \max_{a \in A}\left[\mathcal{R}(s,a,s') + \gamma \sum_{s' \in S} \mathcal{T}(s,a,s') \cdot V^*(s')\right] \tag{9}$$

Finally, the optimal policy π^* consists in the optimal actions selected according to the optimal value function V^* and it is summarized by:

$$\pi^*(s) = \operatorname{argmax}_{a \in A}\left[\mathcal{R}(s,a,s') + \gamma \sum_{s' \in S} \mathcal{T}(s,a,s') \cdot V^*(s')\right] \tag{10}$$

There are two algorithms to compute the optimal policy π^* : *policy iteration* [8] and *value iteration* [17]. The former computes the utility of all states and improves the policy in each iteration until actions convergence to an optimal policy; the latter computes the expected utility of each state using the utilities of the neighbour states until the utilities for two consecutive steps are close enough.

Finally, independently on the specific algorithm, the best action will be the one with the highest expected value based on possible next states resulting from taking that action [18].

DP is also an effective approach for stochastic problems. The main difference is the need to model that information becomes available after the action a is performed, so uncertainty is both in next reached state s' and in the reward.

The deterministic form of the Bellman's equation can be adapted to the stochastic case by simply replacing the (deterministic) transition matrix with a probability transition matrix:

$$V^*(s) = \max_{a \in A}\left[\mathcal{R}(s,a,s') + \gamma \sum_{s' \in S} \mathbb{P}(s'|s,a) \cdot V^*(s')\right] \tag{11}$$

this is also known as the standard form of the Bellman's equation, used in almost every textbook on stochastic programming and dynamic programming.

Finally, if expected value is used instead of the sum of probabilities, an equivalent form – namely the expectation form of the Bellman's equation – is obtained, which is more appropriate for ADP:

$$V^*(s) = \max_{a \in A}[\mathcal{R}(s,a,s') + \gamma \mathbb{E}\{V(s')|s\}] \tag{12}$$

2.2   Approximate Dynamic Programming

ADP offers a powerful set of strategies for solving problems that are large as well as small but lacking a formal model, specifically the transition function. Most of relevant real-life problems belong to this class, also known as "information acquisition", where performing an action is the only way to obtain a better estimate of its value and increasing the knowledge about the system. While exact DP steps backward in time, computing the value function then used to produce optimal decisions, ADP steps forward in time, so an approximation of the value function updated and used to make decisions. Basically, going forward in time requires to alternate between:

- randomly generating a sample of what might happen
- making a decision about the action to perform.

According to [13], sampling can be performed in three different ways: from real data (e.g. real physical processes), via computer-simulation or sampling from a known distribution. The main benefit of ADP algorithms is the ability to solve problems without knowing the underlying probability distribution, which is usually the case of working on real data. On the other hand, when a probability model is available Monte Carlo simulation can be used to generate samples; another possibility is to use software simulation of the physical system to perform sampling. However, physical systems are usually complex and difficult to be mathematically modelled and probability distributions and/or simulation are not available: the system's behavior, in terms of transitions and reward, can only be observed by direct interaction, while optimization is performed accordingly to the expectation form of the Bellman's equation (12).

3   PSO as a Learning Problem

In this paper we focus on the value function approximations known as "lookup tables", which means we store a value $V(s)$ for each discrete state $s \in S$, in the case of state value function. Another possibility is to use the state-action value function, which represents how good is to perform action a in state s, for every pair state-action. More specifically, we adopt an ADP strategy of the latter type and known as Q-learning, a popular approach used for problems with small state and action spaces, where we do not have a mathematical model for how the system evolves over time, exactly the case study of this paper.

3.1   Q-Learning

Q-Learning takes its name from the variable $Q(s, a)$ which is the value of being in the state $s \in S$ and taking the action $a \in \mathcal{A}$ (i.e. state-action value function).

Another typical classification adopted in the RL community is between "learning" and "planning" algorithms. While planning algorithms have access to a model/simulator of the world, learning algorithms do not know anything about the system dynamics and must learn how to behave by direct experience with it (aka "environment"). Furthermore, RL algorithms can be used to address optimization problems by exploiting a "learning-by-doing" paradigm in two different ways:

- apply a planning algorithm after that a model of the world has been learned from experience (model-based approaches)
- learn and apply a policy or value function directly from experience (model-free).

Q-learning belongs to the latter category and, as an ADP algorithm, it works going forward in time, where the next action to perform is selected according to:

$$a^n = \operatorname{argmax}_{a \in A} \overline{Q}^{n-1}(s^n, a) \tag{13}$$

where $\overline{Q}^n(s^n, a)$ is an estimation of the true value of $Q(s, a)$ after n iterations, a^n and s^n are, respectively, the action to choose and the state the system is at the n-th iteration. Thus, at iteration n the estimation of $Q(s, a)$ at iteration $n - 1$, namely $\overline{Q}^{n-1}(s', a')$, is used.

After the action a^n, an immediate reward r is observed along with the new state s', so the value of the state-action pair, $\overline{Q}^n(s^n, a^n)$, is updated consequently:

$$\overline{Q}^n(s^n, a^n) = \overline{Q}^{n-1}(s^n, a^n) + \alpha \left[r + \gamma \max_{a' \in A} \overline{Q}^{n-1}(s', a') - \overline{Q}^{n-1}(s^n, a^n) \right] \tag{14}$$

where α is the learning rate – which set how much the old estimate of the Q-value has to change depending on the observed state and reward – and γ is the discount factor as defined in the Bellman's equation. Learning rate should be slightly decreased over time to guarantee convergence to the true Q-value.

In this updating rule, the immediate reward plus the discounted max Q-value in the observed next state is what satisfies the Bellman equation (expectation form Eq. 12).

Selecting the action a^n does not guarantee to reach an optimal solution: this is just one component of the Q-Learning algorithm which is basically associated to exploitation (i.e. making decisions depending on the knowledge acquired so far). As already mentioned, the goal of Q-Learning is to alternate optimization – thus exploitation – with learning something new about the environment – thus exploration. This is exactly the behavior required to an ADP approach to step forward in time balancing between: *(i)* randomly generating a sample of what might happen (exploration) and *(ii)* making a decision about the action to perform (exploitation).

In Q-Learning a typical strategy to manage this trade-off is known as ε-greedy policy, where the action a^n is randomly selected among all the possible actions, with probability ε, and "greedily" (according to Eq. 13), with probability $1 - \varepsilon$.

Solving the problem of when to explore and when to exploit is known as the exploration versus exploitation dilemma. This is a difficult problem and an active area of research. Not surprisingly, Q-learning is difficult to apply to problems with even modest state and action spaces, but its value lies in its ability to solve problems without a model and to work online with the system (learning-and-optimization).

3.2 An MDP for PSO: EPANET Simulator as an Environment

To solve PSO through RL, the first step consists in defining the underlying MDP. A similar approach has been proposed in [5] where the goal is to minimize energy cost over a certain time-period (horizon) under constraints on operation parameters.

We have decided to model the set of possible state, $s \in S$, through 2 variables: level of tank (h) and average pressure (p):

$$s = (h_t, p_t) \tag{15}$$

Differently from other approaches, such as in [6], we do not include time T as state variable. Another crucial difference is that

- in [6] the MDP is generated exhaustively by interacting with EPANET, so including time in the definition of state helps in modelling the deterministic transitions of the WDN system. Then DP – specifically, value iteration – is used to solve the MDP, assuming the learned policy is optimal for the WDN.
- in this paper, ADP is used and the MDP is not generated exhaustively (specifically the transition dynamics) but exploration and exploitation are performed to learn something more about the MDP by interacting with the system (exploration) while trying to generate the optimal strategy (exploitation).

Thus, while in [6] EPANET is used a simulator for generating the MDP, in this paper EPANET is assumed to be a real WDN, whose real data are acquired through sensors. Since Q-Learning is based on the expectation form of the Bellman's equation, the resulting policy should be more robust to uncertainty – basically related to water demand – with respect to the policy obtained by solving the deterministic MDP through DP.

Both the two state variables are continuous, and discretization is required to work with tabular MDP. In this paper, a discretization on 5 levels has been adopted for both h and p, resulting in $5 \times 5 = 25$ possible states. Discretized ranges are reported in Table 1.

Table 1. Discretization levels of continuous variables contained in the environment state.

Number level	Level of tank (h_t)	Average pressure (p_t)
1	[222.0; 227.6]	$(-\infty; 48.0]$
2	(227.6; 233.2]	(48.0; 72.0]
3	(233.2; 238.8]	(72.0; 96.0]
4	(238.8; 244.4]	(96.0; 120.0]
5	(244.4; 250.0]	$(120.0; +\infty)$

The set of actions $\mathcal{A}$ is represented by the variables that can affect the system's state. With respect to WDN, actions are represented by the status of the pumps, so that a generic action $a \in \mathcal{A}$ is defined as follows:

$$a = \{u_1, u_2, \ldots, u_{N_U}\} \tag{16}$$

where u_j is the status of the j-th pump in the system and $1 \le j \le N_U$ and N_U is the overall number of pumps (i.e. $N_U = 4$ in the case study considered). Since in this paper only on/off pumps are considered, this means that the possible values for every component of the action is $u_j = \{0, 1\}$, where 0 represents pump *off*, 1 represents pump *on*.

The transition function T in a WDN cannot be formally defined due to recurrence relation between the variables that compose the system. Thus, the transition function can be described as:

$$(h_{t+1}, p_{t+1}) = f(h_t, p_t, a_t) \tag{17}$$

In actual system h and p are provided by sensors, while in our experiments they are computed by EPANET. An important exogenous variable is the water demand, where d_t is the demand from t to $t+1$: it is unknown to the optimization algorithm. Water demand is a fundamental input for EPANET as it drives – exactly as a in real world WDN – the hydraulic behavior of the network. Since d_t can be observed only at $t+1$, it cannot be included in the definition of s_t; thus, reward associated to a_t is computed at $t+1$ and depends on s_t as well as the exogenous variable d_t and the action a_t.

Reward function is defined according to the PSO goal, which is to identify the pump schedule that minimizes the energy cost while satisfying the water demand. The objective function of the PSO problem, in the ON/OFF pumps setting, can be formulated as follows. Let denote with C_t the energy costs from $t-1$ to t, resulting from the configuration of pumps at $t-1$. Thus, C_t depends on the decision variables $a_{t-1} \in \{0, 1\}^{N_U}$, with N_U the number of pumps (maintaining the same notation used for actions). The final goal is to identify the actions a_t, with $t = 1, ., T$, that minimizes the total energy cost:

$$\min \sum_{t=1}^{T} C_t \tag{18}$$

To define C_t, is important to highlight that two consecutive time steps in the ADP algorithm can be related to many EPANET simulation time steps. For instance, in our case study simulation steps are on hourly basis while time steps for ADP are specific hours of the day when the action – i.e. possible modification in the activation of the pumps – is performed. So, we need to introduce an operator, $\varphi(t)$, which maps a time step of the ADP to a time step of the simulation. Just a simple example: suppose to have just two time steps in the ADP algorithm, which refer to the 6th and 18th time steps of the simulation, respectively, then $\varphi(t=1) = 6$ while $\varphi(t=2) = 18$. This notation allows for a definition of C_t which considers the simulation time resolution and, therefore, hourly variations in the price of energy.

$$C_t = \sum_{i=\varphi(t-1)}^{\varphi(t)} c_i \left(\sum_{j=1}^{N_U} Q_{i,j} \frac{H_{i,j}}{\eta_j} a_{t,j} \right) \tag{19}$$

Where:

- c_i is the energy price at time i (i.e., i-th hour of the day in a typical PSO setting)
- $Q_{i,j}$ is the quantity of water provided by the j-th pump at time i
- $H_{i,j}$ is the head loss of the j-th pump at time i
- η_j is the efficiency of the j-th pump (it does not depend on time)

- $a_{t,j}$ is the status of the j-th pump from $t-1$ to t (which is constant for every $i \in [\varphi(t-1); \varphi(t)]$)

In any case, the computation of C_t is performed by EPANET and the result is provided at the end of the simulation. Finally, the reward function $\mathcal{R}$ can be described as:

$$
\mathcal{R} = \begin{cases} r_t = 0 & \text{if } t = 0 \\ r_t = \left|\overline{C}_t - \overline{C}_{t-1}\right| & \text{if } (t > 0 \wedge C_{t-1} < 10^9 \wedge C_t < 10^9) \\ r_t = 10^9 & \text{otherwise} \end{cases} \tag{20}
$$

Where t is the time step over the optimization process, starting from $t = 1$ to $t = T$, C_t represents the energy cost at time t while $\overline{C}_t$ is the cost associated to the pump schedule up to t, as computed by EPANET.

4   Experimental Setting

4.1   Case Study

In this study the EPANET model of a benchmark WDN is considered, namely Anytown [4]. Anytown consists of 37 pipes, 19 nodes, 1 tank (having elevation and diameter equal to 65.53 m and 12.20 m, respectively) and 1 reservoir, with 4 pumps installed at the source of supply water. The storage tank keeps the pressure equalized in the system and helps to supply, jointly with the pumps, the water demand. The Anytown EPANET model is depicted in Fig. 1.

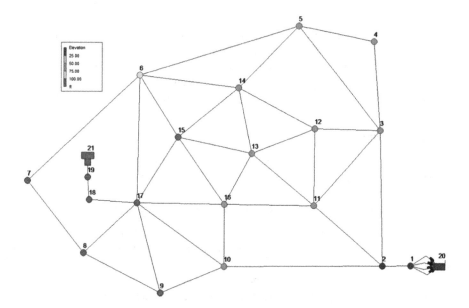

Fig. 1. Topology of the water distribution system Anytown.

One of the most recent works addressing pumps control on Anytown uses Harmony Search [4]: it is – according to the authors knowledge – the most recent, detailed and repeatable study on this benchmark. A penalty on the objective function is applied in the case errors or warnings occur during a EPANET simulation run but, differently from [4], the penalty value in this paper has been set to 10^9 instead of 9999. All the other settings are the same of those reported in [4]: minimum and maximum tank heads are set to 67.67 m and 76.20 m, respectively; hourly demand factors ranged from 0.4 to 1.2 and base demand is doubled from the original Anytown test case [12] - which was proposed for an optimal WDN design problem - to allow longer pump operations and tank storage, defining a total inflow between 161.51 lps and 484.53 lps. The daily water consumption pattern is reported in Fig. 2: it is possible to see two peaks during a 24-h period, one around 10:00 (morning peak) and another around 20:00 (night peak).

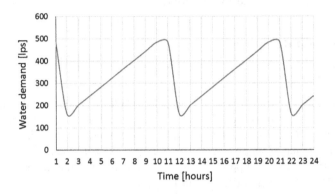

Fig. 2. The hourly water demand of Anytown network over 24 h.

Finally, the electricity tariff – reported in the following figure – consisting in a price of 0.0244 \$/kWh between 0:00–7:00 and 0.1194 \$/kWh between 8:00–24:00 (Fig. 3).

Usually PSO on Anytown has been targeted considering ON/OFF pumps with a time resolution of 1 h, leading to $4 \times 24 = 96$ decision variables. Thus, the total number of possible schedules is $2^{4 \times 24} = 2^{96} = 7.92 * 10^{28}$.

4.2   Experiments Setup

The experiments are organized in three consecutive steps to validate the proposed ADP approach:

4.2.1   First Experiment: Setting Up the ADP Framework for PSO

In this experiment the action space considered is very small with the aim to solve the PSO problem exhaustively and then estimate which is the more suitable Q-Learning to find the known optimum.

The overall time horizon (i.e. 24 h) has been divided in two time slots, reflecting the energy tariff (0:00–7:00 with 0.0244 \$/kWh and 8:00–23:00 with 0.1194 \$/kWh). Thus, there are only two time steps to make decisions about the activation of each one

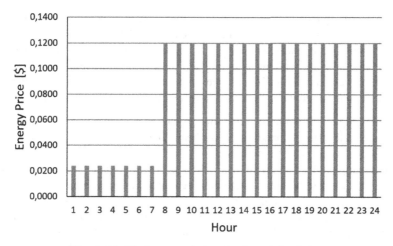

Fig. 3. Tariff of energy during the day of simulation

of the 4 ON/OFF pumps. This leads to $2^{4 \times 2} = 256$ possible pump schedules which can be easily evaluated through an exhaustive search to make the optimum known.

The following ε values, in the ε-greedy policy of Q-Learning, have been considered: 0.3, 0.4, 0.5. Learning rate and discount factor have been fixed to 0.1 and 0.9, respectively. To mitigate the impact of randomness, 30 different runs of Q-Learning have been performed for each ε value. Finally, every Q-Learning run consisted in 1000 episodes. Outcome of this experiment is to understand if an ADP approach can identify the known optimum with – possibly – a lower number of episodes than the function evaluations required by the exhaustive search.

4.2.2 Second Experiment: Testing the Set ADP Framework on a Larger Action Space

In this experiment the best Q-Learning configuration resulting from the previous step is used to solve PSO on a larger size problem, in terms of epochs (i.e. times to make decisions). In this case the overall time horizon (i.e., 24 h) has been divided in 4 time slots (0:00–7:00, 7:00–13:00, 13:00–19:00, 19:00–24:00), thus there are 4 different time steps to decide about the activation of the 4 ON/OFF pumps, leading to $2^{4 \times 4} = 65536$ possible pump schedules. Even in this case exhaustive search – even if largely more expensive with respect to the first experiment – has been performed to make the optimum known. The difference between the known optimum and the optimal solution identified through Q-Learning, configured according to the first experiment, is computed, along with the episodes needed to identify the optimum.

4.2.3 Third Experiment: Adaptivity of the ADP Learned Policy to Changing Water Demand

This experiment is aimed at validating the most promising property of the ADP approach, that is the ability to identify a policy robust to uncertainty, specifically the one affecting water demand. Indeed, exhaustive search, meta-heuristics algorithms and other simulation-optimization approaches can find an optimal solution to PSO after that demand is fixed in the EPANET simulator. DP and ADP, instead, learn a policy to make decisions online. ADP, specifically, as already highlighted in this paper, works going forward in time without assuming any probability distribution, mathematical model or software simulator. To evaluate how much the policy learned in the experiment 2 can adapt to new situations we have modified the water demand as follows:

- a value sampled from a normal distribution with mean 0 and variance equal to 10% of the overall water demand range has been added on each hourly water demand, separately (i.e. demand multipliers);
- a value sampled from a normal distribution with mean 0 and variance equal to 30% of the overall water demand range has been added on each hourly water demand, separately (i.e. demand multipliers);
- a value sampled from a normal distribution with mean 0 and variance changing from hour to hour according to the sequence 10%, 20%, 30%, has been added on each hourly water demand, separately (i.e. demand multipliers). This means that the (unknown) probability distribution of water demand changes form hour to hour, which is the most difficult case for traditional DP approaches but the most suitable for ADP strategies.

In the following figures the three modifications to the water demand multipliers are reported (Figs. 4, 5 and 6).

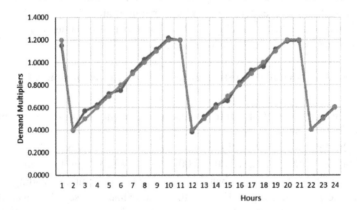

Fig. 4. Hourly water demand over 24 h: original, in orange, and with noise, in blue (noise is sampled from a normal distribution with mean 0 and variance 10% of the multipliers variation range).

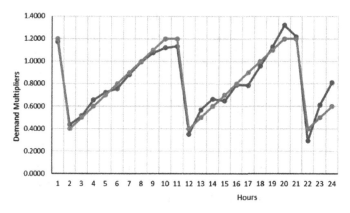

Fig. 5. Hourly water demand over 24 h: original, in orange, and with noise, in blue (noise is sampled from a normal distribution with mean 0 and variance 30% of the multipliers variation range).

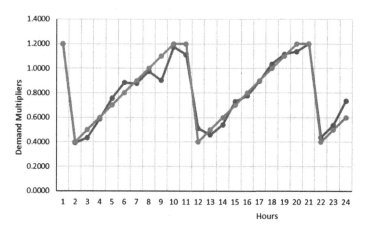

Fig. 6. Hourly water demand over 24 h: original, in orange, and with noise, in blue (noise is sampled from normal distributions with mean 0 and variance changing from hour to hour following the sequence 10%, 20% and 30% of the multipliers variation range).

5   Results

5.1   First Experiment

At the end of the exhaustive search, the optimum pump schedule has an energy cost of 1454.32 \$/day. It is important to highlight that only 80 out of the 256 (31.25%) possible pump schedules are hydraulically feasible, so many evaluations are useless. One of the expected benefits of ADP is that it should throw unfeasible schedules away, quite quickly, by interacting with the WDN. Results are summarized in Table 2.

Finally, the best configuration of Q-Learning is obtained for $\varepsilon = 0.4$, as it found the optimum before 256 episodes in 63.3% of the runs. When only these runs are

Table 2. Results for different values of ε over 30 runs of Q-Learning

ε	Optimum found at episode (avg [min; max])	#Runs with optimum found before 256 episodes	Optimum found at episode - before 256 (avg [min; max])
0.3	277 [16; 844]	15/30 (50.0%)	154 [16; 247]
0.4	246 [1; 821]	19/30 (63.3%)	101 [1; 226]
0.5	294 [4; 883]	16/30 (53.3%)	140 [4; 251]

considered, the algorithm was able to find the optimum by using, on average, only the 40% of all the possible schedules, meaning that the proposed ADP approach has learned a policy able to move quickly towards effective and efficient pump schedules.

5.2 Second Experiment

For the second experiment, related to a larger size problem, in terms of time steps for making decisions (i.e. epochs), only the Q-Learning with $\varepsilon = 0.4$ has been used, as it has been identified as the most promising configuration according to the previous experiment. Learning rate and discount factor are again fixed to 0.1 and 0.9, respectively. Even in this case the exhaustive search has been performed to make the optimum known: the best pump schedule has a cost of 1267.78 \$/day. It is important to highlight that in this case the percentage of hydraulically feasible pump schedules is even lower than the previous experiment (17.19%), since the larger number of possible pump schedules (65536).

According to the results from the previous experiment, on 60% of runs Q-Learning should be able to identify the optimum in a number of episodes lower than the number of possible pump schedules, approximately by exploring 40% of them.

Thus, we have decided to perform only 3 runs of the best Q-Learning configuration with 26000 episodes each (approximately 40% of 65536 possible schedules). Our idea is that Q-Learning can identify the (known) optimum – or at least a close value – within the given number of episodes.

ADP learned policy was not able to identify the optimum but a very close solution – the pump schedule associated to the second lowest cost – in 2 out of 3 runs, by using just the 40% of the overall search space (Table 3).

Table 3. Results of three runs of Q-Learning with $\varepsilon = 0.4$ and 26000 episodes

Run	Minimum energy cost [rank]	Gap to the optimum	Optimum found at episode
#1	1310.62 [5th]	42.84 \$/day	7597/26000
#2	1282.98 [2nd]	15.20 \$/day	25931/26000
#3	1282.98 [2nd]	15.20 \$/day	20545/26000

5.3   Third Experiment

The first important result, when some uncertainty is added to the demand, is that the optimum significantly changes. This means that solutions identified through exhaustive search as well as meta-heuristics and simulation-optimization approaches are not valid any longer and should be recomputed.

The following table summarizes which is the result provided by the best schedule of the second experiment identified through exhaustive search when applied to solve PSO on Anytown with modified demand. Only on 1 out of 3 scenarios the best pump schedule remains valid; in the other two cases – even if related to slight modifications of the water demand pattern – this schedule becomes hydraulically unfeasible (Table 4).

Table 4. Application of the best pump schedule obtained through exhaustive search to solve PSO on Anytown with three different water demand modifications

Demand variation	Energy Cost [$/day]
Scenario #1	1264.85
Scenario #2	Unfeasible
Scenario #3	Unfeasible

On the contrary, the policy learned through ADP is more robust to the modifications applied to the water demand and provides feasible schedules by adapting the actions to perform. More specifically, the best result is provided by the policy learned from Q-Learning in the run #1 of the second experiment (which has identified, as optimal solution, the pump schedule associated to the 5th lowest energy cost).

For reducing computational burden, we have not performed exhaustive search for this experiment: we have just considered the reduction with respect to the highest energy cost possible, that is the one associated to the schedule having all the pumps always active over the time horizon (24 h). The following table summarizes the costs reduction obtained by using the ADP learned policy (Table 5).

The most important result is that all the schedules generated by the ADP approach are feasible. Moreover, the fact that the best result in this experiment is provided by the policy learned by the Q-Learning which has provided just the 5th lowest energy cost on the second experiment could suggest that an "under-fitted" policy should be less sensitive to uncertainty with respect to one "over-specialized" on it.

Table 5. Results of the application of the ADP learned policy (run#1 of the second experiment) on the three demand modification scenarios

Demand variation	Energy cost [$/day]	Max cost [$/day]	Cost reduction [$/day]
Scenario #1	1416.13	3925.52	2509.39 (63.92%)
Scenario #2	1414.75	3889.70	2474.95 (63.63%)
Scenario #3	1421.87	3959.58	2537.71 (64.09%)

6 Conclusions

The increasing "digitalization" of urban water distribution networks is generating huge amounts of real time data from flow/pressure sensors and smart metering of household consumption and enabling new ways to achieve more efficient operations. Approximate Dynamic Programming has the potential to leverage the value hidden in real time data assets into energy driven PSO which satisfies the operational constraints.

The main result is that ADP strategies are robust with respect to different demand level. Thus, while traditional approaches require to know the water demand in advance or, at least, to have a reliable and accurate forecasting, ADP provides a policy, that is a strategy to decide how to act, as new sensor data become available.

Future work will have to explore other Reinforcement Learning approaches also for continuous state and action spaces, hyperparameter optimization in particular learning rate and discount factor, and a principled way to identify effective reward functions.

References

1. Bellman, R.E.: Dynamic Programming. Princeton University Press, Princeton (1957)
2. Candelieri, A., Soldi, D., Archetti, F.: Short-term forecasting of hourly water consumption by using automatic metering readers data. Procedia Eng. **119**, 844–853 (2015)
3. Candelieri, A.: Clustering and Support Vector Regression for Water Demand Forecasting and Anomaly Detection. Water **9**(3), 224 (2017)
4. De Paola, F., Fontana, N., Giugni, M., Marini, G., Pugliese, F.: An application of the Harmony-Search Multi-Objective (HSMO) optimization algorithm for the solution of pump scheduling problem. Procedia Eng. **162**, 494–502 (2016)
5. Ertin, E., Dean, A.N., Moore, M.L., Priddy, K.L.: Dynamic optimization for optimal control of water distribution systems. Applications and science of computational intelligence IV. Proc. SPIE **4390**, 142–149 (2001)
6. Fracasso, P.T., Barnes, F.S., Costa, A.H.R.: Energy cost optimization in water distribution systems using Markov decision processes. In: International Green Computing Conference Proceedings, Arlington, pp. 1–6 (2013)
7. Fracasso, P.T., Barnes, F.S., Costa, A.H.R.: Optimized Control for Water Utilities. Procedia Eng. **70**, 678–687 (2014)
8. Howard, R.A.: Dynamic Programming and Markov Processes. MIT Press, Cambridge (1960)
9. Ikonen, E., Bene, J.: Scheduling and disturbance control of a water distribution network. In: Proceedings of 18th World Congress of the International Federation of Automatic Control (IFAC 2011), Milano, Italy (2011)
10. Ikonen, E., Selek, I., Tervaskanto, M.: Short-term pump schedule optimization using MDP and neutral GA. IFAC Proc. Vol. **43**(1), 315–320 (2010)
11. Mala-Jetmarova, H., Sultanova, N., Savic, D.: Lost in optimization of water distribution systems? A literature review of system operations. Environ. Model. Softw. **93**, 209–254 (2017)
12. Pasha, M.F.K., Lansey, K.: Optimal pump scheduling by linear programming. In: Proceedings of World Environmental and Water Resources Congress 2009 - World Environmental and Water Resources Congress 2009: Great Rivers, vol. 342, pp. 395–404 (2009)

13. Powell, W.B.: Approximate Dynamic Programming: Solving the Curses of Dimensionality. Wiley, New York (2007)
14. Puterman, M.: Markov Decision Processes. Discrete Stochastic Dynamic Programming. Wiley, New York (1994)
15. Shabani, S., Candelieri, A., Archetti, F., Naser, G.: Gene expression programming coupled with unsupervised learning: a two-stage learning process in multi-scale, short-term water demand forecasts. Water 10(2), 142 (2018)
16. Sterling, M.J.H., Coulbeck, B.: A dynamic programming solution to the optimization of pumping costs, in Hybrid genetic algorithm in the optimization of energy costs in water supply networks. ICE Proc. 59(2), 813–818 (1975)
17. Sutton, R.S., Barto, A.G.: Reinforcement Learning: An introduction – Adaptive Computation and Machine Learning. MIT Press, Cambridge (1998)
18. Wiering, M., Van Otterlo, M.: Reinforcement Learning - State-of-the- Art, 1st edn. Springer, Berlin (2012)

Detecting Patterns in Benchmark Instances of the Swap-Body Vehicle Routing Problem

Dimitris Souravlias$^{(\boxtimes)}$ and Sandra Huber

Logistics Management Department, Helmut-Schmidt University,
Holstenhofweg 85, 22043 Hamburg, Germany
{dsouravl,sandra-huber}@hsu-hh.de

Abstract. The present study aims at identifying possible relations between solution features and different characteristics of the Swap-body Vehicle Routing Problem. For this purpose, an investigation has been conducted on two established benchmark sets. Our analysis reveals the existence of hidden patterns with respect to various aspects of the corresponding problem instances. The detected patterns are then used to formulate problem-specific properties, which hold for the majority of the instances under consideration. Our findings may be employed as guidelines in the design of algorithmic components, such as new selection techniques that choose only a subset of specific nodes of interest from the vehicle routing network. Also, our work sheds further light on the effect of various problem characteristics on the structure of their best known solutions.

Keywords: Swap-body vehicle routing problem · Benchmarks
Patterns

1 Introduction

Vehicle routing has been long established as a crucial part of smart cities project having significant economic and environmental impact. An abundance of vehicle routing problem types has been studied in recent years. The Swap-body Vehicle Routing Problem (SB-VRP) is a very challenging problem of this type. Its significance stems from the economic, environmental, and social impact that accompanies the efficient distribution of goods.

The SB-VRP is a generalization of the well-known Capacitated Vehicle Routing Problem. It is also related to the Truck and Trailer Routing Problem, which has been the subject of relevant research [4,7]. SB-VRP was introduced in the VeRoLog Implementation Challenge 2014. Since then, it has received considerable research attention [1,6,8–10]. All relevant studies have been focused on the development of effective solution methods aiming at high-quality solutions. In

© Springer Nature Switzerland AG 2019
R. Battiti et al. (Eds.): LION 12 2018, LNCS 11353, pp. 370–385, 2019.
https://doi.org/10.1007/978-3-030-05348-2_31

this context, ad hoc (meta-)heuristics have been frequently employed, specifically tailored to the requirements of the problems at hand. To the best of our knowledge, post-optimization analysis has not been conducted in existing works in order to gain profound knowledge from the acquired results with respect to problem-specific properties of the SB-VRP.

The contribution of this paper lies in the detection of possible hidden relational patterns between characteristics of various instances of the SB-VRP and features of their corresponding solutions. Regarding the solution features, our study is focused on three main aspects, namely (i) the use of swap locations, (ii) combination of swap actions, and (iii) impact of the tactical fleet management. The term *swap locations* refers to network nodes where the vehicles can modify their configurations by performing specific actions, such as parking or picking up a swap-body, also called *swap actions*. An interesting hypothesis is that specific characteristics of the problem can be related to high-quality solutions. This is a first attempt to put this hypothesis under scrutiny. For this purpose, appropriate frequency measures are introduced for the aforementioned aspects of interest. These measures can be highly useful in assessing solution quality, and may complement the role of the objective function towards this direction. The outcome of the present study is expected to be beneficial for the design of powerful algorithmic approaches. For example, the provided insights may contribute in developing suitable selection techniques that manage the use of swap locations under the presence of particular problem characteristics.

Our study is based on two sets of benchmark instances. The first one was released in the VeRoLog challenge [5] and consists of 18 instances, while the second one [10] has been recently made available and includes 100 problem instances. The best known solutions of the problems are publicly available. The rest of the paper is structured as follows: Sect. 2 provides a brief description of the SB-VRP. Details about the benchmark sets are provided in Sect. 3, followed by our analysis in Sect. 4. Section 5 concludes the paper.

2 Problem Description

The SB-VRP is a similar to the widely studied Truck and Trailer Routing Problem (TTRP) [3], which is an NP-hard problem [2]. In TTRP a fleet of truck vehicles is considered, each one combined with a trailer to increase capacity. The use of this combination is not allowed for some customers that can be visited only by a truck without a trailer. This can be attributed to possible access limitations of specific customer locations, and difficult maneuvering due to inadequate space. In SB-VRP the fleet consists of trucks and trailers pulling swap-bodies (see Fig. 1a). A swap-body is a type of freight container that can be (dis)connected to the vehicle according to the transportation needs.

The SB-VRP is defined on a directed, asymmetric graph $G(V, A)$, where V is the set of nodes and A is the set of arcs. The set V comprises the subset of n customers, the subset of m swap locations, and the depot, denoted as V_c, V_s, and v_0, respectively. Demand values $q_i \geqslant 0$, $i \in V_c$, are known for each customer

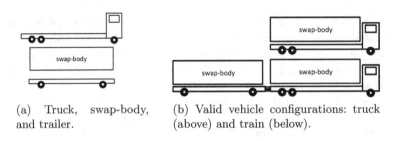

(a) Truck, swap-body, and trailer.

(b) Valid vehicle configurations: truck (above) and train (below).

Fig. 1. Types of fleet vehicles in SB-VRP.

and shall be satisfied in a single vehicle visit. The fleet consists of an infinite number of vehicles, each one traversing a distance denoted as d_{ij}, $i, j \in V$, and consuming driving times t_{ij}. Three counterparts comprise each vehicle, namely truck, trailer, and swap-bodies. Two vehicle modifications are allowed, namely (i) a truck that carries one swap-body, and (ii) a truck pulling a trailer, each one loaded with a swap-body. Henceforth, the first combination will be called a *truck*, while the second one will be called a *train*. Figure 1b illustrates the two valid vehicle combinations.

Swap locations are nodes of the graph where the current configuration of a train can be altered. At swap locations four operations are allowed, namely *park*, *pickup*, *swap*, and *exchange*, which are illustrated in Fig. 2. Let SB_1 and SB_2 be the swap-bodies carried by the truck and the trailer of a train, respectively. The park action occurs when the train parks its trailer loaded with swap-body SB_2 and continues its trip carrying only swap-body SB_1. Pickup takes place when a truck visits the swap location and gets attached to a trailer that is located there. The swap action involves switching the current swap-body of the truck with another one, previously parked at the swap location. In the exchange operation, the train exchanges SB_1 that was loaded to its truck with SB_2 that was loaded to its trailer. Then, it parks the trailer and continues its trip as a truck carrying SB_2. Note that a train is restricted to carry the same swap-bodies when it leaves and reaches the depot, although not necessarily with the same ordering.

Actions at swap locations require relevant handling times, e.g., parking typically takes less time than swapping. There are no limitations on the maximum number of actions taken at a swap location or the number of vehicles that can access it. Also, no restrictions arise regarding the maximum number of swap locations during the trip of a train. Although, driving regulations impose a maximum driving time DT for all drivers. Moreover, the capacity of each swap-body is equal to QT, while exchanges of (parts of) loads between swap-bodies is not allowed. The costumers are distinguished in three different types, namely (i) train customers that can be visited by either a truck or a train, (ii) mandatory train customers that can be reached only by a train since their demand is higher than QT, and (iii) truck customers that can be served only by a truck. The goal in solving the SB-VRP is, in tactical level, to determine the appropriate vehicle

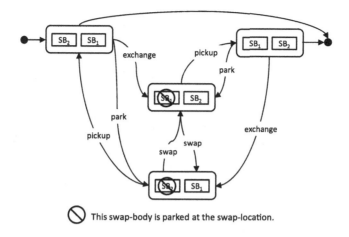

Fig. 2. Valid transitive operations at swap locations.

configurations of the fleet and, in operational level, to determine a minimum-cost set of tours. Routing costs include truck-related costs, such as fixed costs incurred by the use of trucks and variable costs related to the traveled distances. Additional fixed and variable costs arise if trailers are employed in the solution. A mixed integer programming model of the SB-VRP was introduced in [9] and experimentally verified in [6].

3 Benchmark Sets

3.1 VeRoLog Challenge Problem Instances

In the VeRoLog Implementation Challenge, a number of real-world instances was released by VeRoLog and the PTV Group. These are available at the VRP repository platform (http://www.vrep-rep.org/datasets.html). Characteristics of these problems instances are reported in the left part of Table 1. They are organized with respect to type of the customers, demand, as well as some additional features. The costumers are distinguished in train customers (V_{tt}^k), mandatory train customers (V_{tr}^k), and truck customers (V_{ck}^k), for each instance $k \in \{1, 2, \ldots, 18\}$. Also, the average (Avg) and the maximum (Max) demand is reported per problem instance. Additional information is provided on the number of available swap locations (V_s^k) and the maximum capacity (QT) of the used swap-bodies. In the right part of Table 1, various features of the best known solution of each instance are presented. Details about these features are exposed in Sect. 4.1.

Each instance belongs to a group $g \in \{1, 2, \ldots, 6\}$. Every group consists of three types of problem instances, namely *normal*, *all with trailer*, and *all without trailer*. The first problem type comprises all customer types, the second problem type consists of train and mandatory train customers, and the latter type includes only truck customers. Additionally, three different problem sizes with

Table 1. Analysis of best known solutions of VeRoLog problem instances [6,10].

Instance			Customers type			Demand		Features		Swap locations		Routes			Swap actions		
g	k	Name	V_{tt}^k	V_{tr}^k	V_{ck}^k	Avg	Max	QT	V_s^k	SL_{dist}^k	SL_{tot}^k	R_{tot}^k	R_{trc}^k	R_{trn}^k	NS^k	PP^k	PSP^k
Small																	
1	1	All with trailer	56	1	0	149	995	500	20	0	0	11	4	7	7	0	0
	2	All without trailer	0	0	57	140	500	500	20	5	5	12	7	5	0	0	5
	3	Normal	41	1	15	149	995	500	20	2	2	12	6	6	4	2	0
Medium																	
2	4	All with trailer	206	0	0	121	720	1,000	41	1	1	17	8	9	8	1	0
	5	All without trailer	0	0	206	117	500	1,000	41	6	7	18	11	7	0	0	7
	6	Normal	186	0	20	121	720	1,000	41	1	1	18	10	8	7	0	1
Large1																	
3	7	All with trailer	548	0	0	119	995	1,000	99	1	1	43	17	26	25	0	1
	8	All without trailer	0	0	548	116	500	1,000	99	13	20	46	26	20	0	0	20
	9	Normal	498	0	50	119	995	1,000	99	2	2	44	20	24	22	1	1
Large2																	
4	10	All with trailer	550	0	0	107	960	1,000	101	0	0	51	35	16	16	0	0
	11	All without trailer	0	0	550	107	960	1,000	101	7	8	57	49	8	0	0	8
	12	Normal	500	0	50	107	960	1,000	101	3	3	53	40	13	10	3	0
Final																	
5	13	All with trailer	549	0	0	108	960	1,000	102	0	0	38	15	23	23	0	0
	14	All without trailer	0	0	549	108	960	1,000	102	15	17	45	28	17	0	0	17
	15	Normal	499	0	50	108	960	1,000	102	3	3	40	19	21	18	2	1
Final random																	
6	16	All with trailer	549	0	0	587	999	1,000	102	0	0	168	1	167	167	0	0
	17	All without trailer	0	0	549	594	1,000	1,000	102	66	178	179	1	178	0	0	178
	18	Normal	499	0	50	604	999	1,000	102	27	28	171	1	170	142	11	17

respect to the number of customers are considered. *Small* instances assume a maximum number of 57 customers, *medium* instances include up to 206 customers, and larger instances consist of approximately 550 customers. Differences among the instances occur with respect to the levels of demand as well as the maximum capacity of the swap-bodies. As we see in Table 1, the maximum demand is higher than 900 for most problem instances, with the exceptions of medium instances as well as the *all without trailer* versions of small and larger instances. For small instances, a total of 20 swap locations can be used, whereas for medium instances 41 swap locations are available. Larger instances allow the use of approximately 100 swap locations.

3.2   Additional Problem Instances

In [10] a set of 100 additional problem instances is provided, and it is available at http://benchmark.gent.cs.kuleuven.be/sbvrp/en/problem/. This benchmark set has been generated by using instances of the Capacitated Vehicle Routing Problem (CVRP) [11], which assume that customers are located within a 1000×1000 grid. In order to produce SB-VRP problems, a number of 4, 20, and 100 swap locations was added to the CVRP instances. Swap locations were generated as follows: a customer was selected and a random point was taken within a radius of 100 from the location of the selected customer. If this point was located within the grid, it was labeled as the selected swap location. The procedure was repeated for the desirable number of swap locations. Moreover, the cost of a truck is equal to 1.0 per distance unit, whereas for the cost of a train three different values are possible that is $c^k \in \{1.2, 1.4, 1.6\}$.

An overview of the characteristics of the new problem instances is presented in Table 2, where the columns are distinguished with respect to customers type, demand, and some additional features. The columns of the dataset represent the type of customers, where V_c^k is the number of customers and V_{tt}^k is the number of customers, who can be served by truck or train (no mandatory train customers exist). Moreover, *Place* refers to the placement of the customers, which may be random (R), in n clusters $(C(n))$, or randomly clustered $(RC(n))$. Also, the dataset reports the demand D^k, which may be unitary (U) where all demands are equal to 1.0, uniformly distributed within a range $[n_1, n_2]$, quadrant-based (Q) or "many small and few large values" (SL). Other features include the maximum capacity of each swap-body (QT), the number of available swap locations (V_s^k) and the cost of a train (c^k). Moreover, the positioning of the depot is referred as *Depot*, which can be centered (C) at the point $(500, 500)$ of the grid, eccentric (E) at the point $(0, 0)$ or random (R) at a random point of the grid.

4   Pattern Detection in Benchmark Instances

The SB-VRP has been the subject of intense research efforts [1,6,8,10]. Relevant works are mostly focused on developing effective solution methods that can provide high-quality solutions. Although this is an essential research goal, post-optimization analysis of acquired results may also be highly beneficial for identifying possible hidden patterns that correlate problem characteristics with solution features. The present work is a first step towards this direction, aiming at investigating the impact of using swap locations, the relation between different combinations of swap actions, and fleet management on solution quality. Exploring these aspects can offer ample ground in gaining profound knowledge that enables the development of better algorithms.

4.1   Case of VeRoLog Challenge Problems

In the first part of our analysis, the best known solutions of the 18 VeRoLog challenge instances are taken into account. A number of 15 solutions was found using the approach in [10], and they are available in the SB-VRP repository platform. The remaining 3 solutions, namely for the *final random normal*, *medium normal*, and *small all without trailer* instances were detected by the solution method proposed in [6]. Each solution comprises several routes starting from the depot and returning back to it. A route consists of a number of intermediate visits to network locations (swap locations or customers). For each route, certain information are stored including the vehicle configuration at the time of the visit, e.g., the vehicle type and, the delivery quantities.

The analysis of the best solutions is performed mainly with respect to three types of solution characteristics presented in the right part of Table 1. The first type refers to swap locations involved in the solution of instance $k \in \{1, 2, \ldots, 18\}$. These are distinguished in the number of distinct swap locations, SL_{dist}^k, and the number of total swap locations, SL_{tot}^k. SL_{dist}^k is the number of each distinct swap location used by the vehicles to perform a swap action. Since each swap location can be employed more than once in a solution, SL_{tot}^k maintains the number of multiple instances of the same swap location.

The second type of solution characteristics refers to the routes involved in the solution of instance k. These include the total number of routes, R_{tot}^k, the number of truck routes, R_{trc}^k, and the number of train routes, R_{trn}^k. A route is a *truck route*, if it is traversed by a truck carrying a single swap-body, or a *train route*, if it is used by a train loaded with two swap-bodies. Across train routes, the vehicle may use swap locations to conduct actions in order to modify its configuration.

In the third type of solution characteristics, the number of different combinations of swap actions is considered. Specifically, two combinations of actions are present in the solution of instance k, i.e., *Park-Pickup* (denoted as PP^k) and *Park-Swap-Pickup* (denoted as PSP^k). The existing combinations consist of three types of swap actions, namely *Park*, *Swap*, and *Pickup*. The action *Exchange* is not present in the acquired solutions, which might be due to the occurred time restrictions [6]. Moreover, it is possible that no swap actions are performed across a train route. The number of train routes that include no swap locations is denoted as NS^k.

A close inspection of the right part of Table 1 reveals several meaningful properties. Firstly, we observe that only a small fraction of the available swap locations is used in the best solution of each instance. This observation can be exploited to design appropriate methods for the selection of the set of most promising swap locations. Also, the detected solutions for nearly all *with trailer* instances make no use of swap locations. An exception to this pattern is the *large1 all with trailer* instance, where only a single swap location is used. This strong evidence suggests the neglection of swap locations in the presence of no truck customers. Additionally, the data suggest the existence of a common pattern that is present in all groups of instances. Specifically, the number of distinct swap locations is higher for the *all without trailer* instances than the *normal* ones, which in turn use a higher or equal number of swap locations than the *all with trailer* instances.

More formally, let $SL^g_{dist,with}$, $SL^g_{dist,without}$, $SL^g_{dist,normal}$, be the number of distinct swap locations used in the solutions of *with trailer*, *without trailer*, and *normal* instance, which belong to the group $g \in \{1, 2, \ldots, 6\}$. Then, it holds that

$$SL^g_{dist,with} \leqslant SL^g_{dist,normal} < SL^g_{dist,without}, \quad \forall g \in \{1, 2, \ldots, 6\} \qquad (1)$$

Obviously, it can be observed that the property of Eq. (1) is valid for the number of total swap locations used in the solutions, namely,

$$SL^g_{tot,with} \leqslant SL^g_{tot,normal} < SL^g_{tot,without}, \quad \forall g \in \{1, 2, \ldots, 6\}, \qquad (2)$$

where $SL^g_{tot,with}$, $SL^g_{tot,without}$, $SL^g_{tot,normal}$, stand for the total number of swap locations used in the solutions of *with trailer*, *without trailer*, and *normal* instance, which belong to the group g.

Also, let $R^g_{tot,with}$, $R^g_{tot,without}$, $R^g_{tot,normal}$, be the total number of routes that constitute a solution of *with trailer*, *without trailer*, the *normal* instance of the group g. Then, the data reveal that

$$R^g_{tot,with} < R^g_{tot,normal} \leqslant R^g_{tot,without}, \quad \forall g \in \{1, 2, \ldots, 6\}, \qquad (3)$$

A similar pattern appears for the routes that are traversed by a truck. Let $R^g_{trc,with}$, $R^g_{trc,without}$, $R^g_{trc,normal}$, be the total number of truck routes that constitute a solution of *with trailer*, *without trailer* and *normal* instance, respectively, of group g. Then, it holds that

$$R^g_{trc,with} \leqslant R^g_{trc,normal} \leqslant R^g_{trc,without}, \quad \forall g \in \{1, 2, \ldots, 6\}. \qquad (4)$$

Regarding the combination of swap actions used in the solutions, an additional pattern can be detected. In all solutions of *all without trailer* instances, only the combination of *Park-Swap-Pickup* actions is present. This is an essential pattern that reveals the potential of using the specific combination of actions instead of *Park-Pickup* for this type of instances.

4.2 Case of Additional Problem Instances

The best solutions of the new SB-VRP instances have been detected by using the method presented in [10], and they can be downloaded from http://benchmark. gent.cs.kuleuven.be/sbvrp/en/results/. Each row of Table 2 corresponds to several characteristics of an instance followed by statistics about its best known solution similarly to the VeRoLog instances. The employed notation is the same with that of Sect. 4.1. Our efforts are again focused on finding correlations between instance and solution features. However, for this dataset, it proved to be a harder task since each instance does not belong to a clearly defined group of instances as in the VeRoLog benchmarks. Also, the higher number of characteristics in these instances and their larger domains makes our attempt even more difficult.

A careful inspection of Table 2 brings to light various patterns. In particular, the data indicate that the number of train customers has an impact on the number of swap locations used in the solutions. When no truck customers exist, a low number of swap locations appears in the corresponding solutions. The existence of this pattern in this dataset is also verified by the data of the VeRoLog instances. Formally, let $r_k = V_{tt}^k / V_c^k$ be the fraction of the number of train customers to the total number of customers for instance $k \in \{1, 2, \ldots, 100\}$. Also, let SL_{dist}^k be the number of distinct swap locations used by the best known solution of instance k. Then, it holds that

$$r_k = 1.0 \overset{\sim}{\Longrightarrow} SL_{dist}^k \leq 10, \quad \forall k \in \{1, 2, \ldots, 100\}. \tag{5}$$

The symbol $\overset{\sim}{\Longrightarrow}$ does not imply a strong mathematical implication. Instead, it stands for a relaxed *if...then* (causal) relation, where the *if* part refers to an observation in the characteristics of the instances, while the *then* part refers to a finding in the solution features.

As we observe in the data exposed in Table 2, Eq. (5) holds for 20 instances. Among them, 12 instances have solutions without swap locations. The solutions of the remaining 8 instances use a small number of swap locations (up to 10). This is a helpful observation, which offers a useful insight that endorses the use of a small number of swap locations when the condition $r_k = 1.0$ is met.

The fact that only 12 out of 100 instances neglect swap locations suggests that their use is associated with high solution quality. For this reason, our efforts are concentrated on detecting patterns in the data that encourage the existence of such network locations. The first pattern is formulated as

$$r_k < 1.0 \overset{\sim}{\Longrightarrow} SL_{dist}^k > 0, \quad \forall k \in \{1, 2, \ldots, 100\}. \tag{6}$$

This indicates that if at least one truck customer exists ($r_k < 1.0$), then at least one swap location should be used. This property is met in 80 out of 100 instances. However, recall that a total of 88 instances needs the presence of at least one swap location in their solutions. Therefore, the remaining 8 instances have solutions that use swap locations, but for them it holds that $r_k = 1.0$.

The second pattern considers the demand distribution. Specifically, when the distribution of the demand, D^k, is unitary (U), at least one swap location exists in the solution of instance k. It holds that

$$D^k = U \overset{\sim}{\Longrightarrow} SL^k_{dist} > 0, \quad \forall k \in \{1, 2, \ldots, 100\}. \tag{7}$$

Notice that the pattern exposed in Eq. (7) holds for a total of 11 instances without any exceptions.

The third pattern is based on costs when a train vehicle is used. We observe that in case of cost $c^k = 1.6$, at least one swap location is used by the solution of instance k. More formally:

$$c^k = 1.6 \overset{\sim}{\Longrightarrow} SL^k_{dist} > 0, \quad \forall k \in \{1, 2, \ldots, 100\}. \tag{8}$$

This pattern appears in 33 instances with solutions employing at least one swap location. However, there exists one exception where the train cost is equal to 1.6, but no swap locations are used in its solution.

Next, our analysis is on correlations between characteristics of solutions. Henceforth, the symbol $\overset{\sim}{\Longrightarrow}$ will refer to *if... then* relations with respect to the features of the best known solutions. One important finding is that the use of swap locations promotes *Park-Swap-Pickup* combinations of actions in almost all best known solutions. It holds that:

$$SL^k_{dist} > 0 \overset{\sim}{\Longrightarrow} PSP^k > 0, \quad \forall k \in \{1, 2, \ldots, 100\}. \tag{9}$$

This is valid for 86 out of 88 instances with solutions including swap locations. Consequently, this is a strong indication that correlates *Park-Swap-Pickup* combinations with high-quality solutions for the instances of the benchmark set under investigation.

Another pattern in the data verifies the correlation between the combination of actions *Park-Pickup* and *Park-Swap-Pickup*. Let PP^k be the number of times that *Park-Pickup* is present in the solution of instance k. The existence of both combinations in a solution is rather frequent. Moreover, the existence of *Park-Pickup* entails the presence of *Park-Swap-Pickup* in the same solution of instance k, i.e.,

$$PP^k > 0 \overset{\sim}{\Longrightarrow} PSP^k > 0, \quad \forall k \in \{1, 2, \ldots, 100\}. \tag{10}$$

This specific pattern holds for 37 cases overall, while only solutions of two instances use the *Park-Pickup* combination, but not the *Park-Swap-Pickup* one.

5   Conclusions

In this paper, we analyzed the relations between solution features and characteristics of the SB-VRP. Two sets of benchmarks with different characteristics were considered along with their best known solutions. Our post-optimization analysis showed that for a plethora of instances, the number and the selection

Table 2. Detailed analysis of best known solutions of additional instances [10].

Instance		Customers			Demand	Other features				Swap locations		Routes			Swap actions		
k	Name	V_c^k	V_{tt}^k	Place	D^k	QT	Depot	V_s^k	c^k	SL_{dist}^k	SL_{tot}^k	R_{tot}^k	R_{trc}^k	R_{trn}^k	NS^k	PP^k	PSP^k
1	sbvrp-n101-s100	100	4	RC(7)	1–100	206	R	100	1.2	13	13	14	1	13	0	0	13
2	sbvrp-n106-s20	105	0	C(3)	50–100	600	E	20	1.6	6	7	7	0	7	0	0	7
3	sbvrp-n110-s4	109	52	R	5–10	66	C	4	1.4	3	6	7	1	6	0	1	5
4	sbvrp-n115-s100	114	102	R	SL	169	C	100	1.4	5	5	5	0	5	0	1	4
5	sbvrp-n120-s4	119	119	RC(8)	U	21	E	4	1.6	1	3	3	0	3	0	1	2
6	sbvrp-n125-s20	124	107	C(5)	Q	188	R	20	1.2	5	5	15	0	15	10	2	3
7	sbvrp-n129-s4	128	0	RC(8)	1–10	39	E	4	1.6	3	9	9	0	9	0	0	9
8	sbvrp-n134-s100	133	11	C(4)	Q	643	R	100	1.4	6	6	7	1	6	0	0	6
9	sbvrp-n139-s20	138	138	R	5–10	106	C	20	1.2	1	1	5	0	5	4	0	1
10	sbvrp-n143-s100	142	61	R	1–100	1190	E	100	1.4	3	3	4	1	3	0	0	3
11	sbvrp-n148-s4	147	18	RC(7)	1–10	18	R	4	1.6	4	15	31	16	15	0	1	14
12	sbvrp-n153-s20	152	152	C(3)	SL	144	C	20	1.2	0	0	11	0	11	11	0	0
13	sbvrp-n157-s20	156	142	C(3)	U	12	R	20	1.6	6	6	7	1	6	0	0	6
14	sbvrp-n162-s100	161	82	RC(8)	51–100	1174	C	100	1.4	5	5	6	1	5	0	0	5
15	sbvrp-n167-s4	166	0	R	5–10	133	E	4	1.2	2	5	5	0	5	0	0	5
16	sbvrp-n172-s100	171	171	RC(5)	Q	161	C	100	1.2	0	0	26	0	26	26	0	0
17	sbvrp-n176-s4	175	22	R	SL	142	E	4	1.6	3	10	16	6	10	0	0	10
18	sbvrp-n181-s20	180	81	C(6)	U	8	R	20	1.4	9	11	12	1	11	0	1	10
19	sbvrp-n186-s4	185	0	R	50–100	974	R	4	1.2	2	7	8	1	7	0	0	7
20	sbvrp-n190-s100	189	173	C(3)	1–10	138	E	100	1.6	4	4	4	0	4	0	0	4
21	sbvrp-n195-s20	194	167	RC(5)	1–100	181	C	20	1.4	9	9	27	2	25	16	1	8
22	sbvrp-n200-s100	199	0	C(8)	Q	402	R	100	1.6	16	18	18	0	18	0	0	18
23	sbvrp-n204-s4	203	101	RC(6)	50–100	836	C	4	1.2	4	7	10	1	9	2	0	7
24	sbvrp-n209-s20	208	208	R	5–10	101	E	20	1.4	0	0	8	0	8	8	0	0
25	sbvrp-n214-s100	213	23	C(4)	1–100	944	C	100	1.6	5	5	6	1	5	0	0	5

continued

Table 2. continued

Instance		Customers			Demand	Other features				Swap locations		Routes			Swap actions		
k	Name	V_c^k	V_{tt}^k	Place	D^k	QT	Depot	V_s^k	c^k	SL_{dist}^k	SL_{tot}^k	R_{tot}^k	R_{trc}^k	R_{trn}^k	NS^k	PP^k	PSP^k
26	sbvrp-n219-s4	218	17	R	U	3	E	4	1.2	4	35	38	3	35	0	0	35
27	sbvrp-n223-s20	222	105	RC(5)	1–10	37	R	20	1.4	9	15	17	0	17	2	3	12
28	sbvrp-n228-s20	227	0	C(8)	SL	154	R	20	1.2	7	11	12	1	11	0	0	11
29	sbvrp-n233-s4	232	212	RC(7)	Q	631	C	4	1.4	3	4	9	1	8	4	2	2
30	sbvrp-n237-s100	236	236	R	U	18	E	100	1.6	7	7	7	0	7	0	3	4
31	sbvrp-n242-s20	241	0	R	1–10	28	E	20	1.2	13	24	24	0	24	0	0	24
32	sbvrp-n247-s100	246	215	C(4)	SL	134	C	100	1.4	9	9	24	1	23	14	2	7
33	sbvrp-n251-s4	250	29	RC(3)	5–10	69	R	4	1.6	3	7	21	14	7	0	0	7
34	sbvrp-n256-s4	255	121	C(8)	50–100	1225	C	4	1.6	4	7	9	2	7	0	0	7
35	sbvrp-n261-s100	260	260	R	1–100	1081	E	100	1.2	0	0	7	1	6	6	0	0
36	sbvrp-n266-s20	265	0	RC(6)	5–10	35	R	20	1.4	14	29	30	1	29	0	0	29
37	sbvrp-n270-s20	269	133	RC(5)	50–100	585	C	20	1.2	11	14	18	0	18	4	2	12
38	sbvrp-n275-s4	274	274	C(3)	U	10	R	4	1.4	3	5	14	0	14	9	1	4
39	sbvrp-n280-s100	279	27	R	SL	192	E	100	1.6	8	8	9	1	8	0	0	8
40	sbvrp-n284-s100	283	261	C(8)	1–10	109	R	100	1.6	7	7	8	1	7	0	2	5
41	sbvrp-n289-s20	288	288	RC(7)	Q	267	E	20	1.4	0	0	31	1	30	30	0	0
42	sbvrp-n294-s4	293	33	R	1–100	285	C	4	1.2	4	19	32	13	19	0	0	19
43	sbvrp-n298-s100	297	0	R	1–10	55	R	100	1.4	14	15	16	1	15	0	0	15
44	sbvrp-n303-s4	302	278	C(8)	1–100	794	C	4	1.6	2	5	11	1	10	5	1	4
45	sbvrp-n308-s20	307	140	RC(6)	SL	246	E	20	1.2	5	6	7	1	6	0	1	5
46	sbvrp-n313-s4	312	154	RC(3)	Q	248	R	4	1.2	4	15	45	17	28	13	0	15
47	sbvrp-n317-s100	316	316	C(4)	U	6	E	100	1.4	4	4	27	1	26	22	2	2
48	sbvrp-n322-s20	321	286	R	50–100	868	C	20	1.6	6	13	14	0	14	1	1	12
49	sbvrp-n327-s4	326	0	RC(7)	5–10	128	R	4	1.2	3	10	10	0	10	0	0	10
50	sbvrp-n331-s100	330	32	R	U	23	E	100	1.4	7	7	8	1	7	0	0	7

continued

Table 2. continued

Instance		Customers			Demand	Other features				Swap locations		Routes			Swap actions		
k	Name	V_c^k	V_{tt}^k	Place	D^k	QT	Depot	V_s^k	c^k	SL_{dist}^k	SL_{tot}^k	R_{tot}^k	R_{trc}^k	R_{trn}^k	NS^k	PP^k	PSP^k
52	sbvrp-n344-s4	343	27	RC(7)	5-10	61	C	4	1.2	4	16	27	11	16	0	0	16
53	sbvrp-n351-s20	350	0	C(3)	1-100	436	C	20	1.6	7	20	21	1	20	0	0	20
54	sbvrp-n359-s100	358	358	RC(7)	1-10	68	E	100	1.4	0	0	15	1	14	14	0	0
55	sbvrp-n367-s20	366	329	C(4)	SL	218	R	20	1.6	5	7	9	1	8	1	1	6
56	sbvrp-n376-s100	375	344	R	U	4	E	100	1.4	13	14	47	0	47	33	4	10
57	sbvrp-n384-s4	383	383	R	50-100	564	R	4	1.2	0	0	26	0	26	26	0	0
58	sbvrp-n393-s100	392	0	RC(5)	5-10	78	C	100	1.6	15	19	19	0	19	0	0	19
59	sbvrp-n401-s20	400	37	C(6)	Q	745	E	20	1.4	11	14	15	1	14	0	0	14
60	sbvrp-n411-s4	410	209	C(5)	SL	216	R	4	1.2	4	9	10	1	9	0	2	7
61	sbvrp-n420-s4	419	198	RC(3)	1-10	18	C	4	1.2	4	31	75	19	56	25	4	27
62	sbvrp-n429-s100	428	37	R	50-100	536	R	100	1.4	29	31	31	0	31	0	0	31
63	sbvrp-n439-s20	438	438	C(4)	U	12	C	20	1.6	10	13	19	1	18	5	4	9
64	sbvrp-n449-s100	448	407	R	1-100	777	E	100	1.2	5	5	15	1	14	9	5	0
65	sbvrp-n459-s20	458	0	C(4)	Q	1106	C	20	1.4	6	13	13	0	13	0	0	13
66	sbvrp-n469-s4	468	421	R	50-100	256	E	4	1.6	4	13	70	1	69	56	1	12
67	sbvrp-n480-s4	479	229	C(8)	5-10	52	R	4	1.6	3	15	45	20	25	10	3	12
68	sbvrp-n491-s20	490	490	RC(6)	1-100	428	R	20	1.2	0	0	30	1	29	29	0	0
69	sbvrp-n502-s100	501	0	C(3)	U	13	E	100	1.4	19	19	20	1	19	0	0	19
70	sbvrp-n513-s100	512	47	RC(4)	1-10	142	C	100	1.6	6	10	11	1	10	0	0	10
71	sbvrp-n524-s4	523	268	R	SL	125	R	4	1.4	4	39	77	11	66	27	0	39
72	sbvrp-n536-s20	535	58	C(7)	Q	371	C	20	1.2	19	48	50	2	48	0	0	48
73	sbvrp-n548-s100	547	494	R	U	11	E	100	1.2	7	7	25	0	25	18	3	4
74	sbvrp-n561-s4	560	0	RC(7)	1-10	74	C	4	1.6	2	21	21	0	21	0	0	21
75	sbvrp-n573-s20	572	572	C(3)	SL	210	E	20	1.4	2	2	15	0	15	13	0	2

continued

Table 2. continued

k	Name	V_c^k	V_{tt}^k	Place	D^k	QT	Depot	V_s^k	c^k	SL_{dist}^k	SL_{tot}	R_{tot}^k	R_{trc}^k	R_{trn}^k	NS^k	PP^k	PSP^k
76	sbvrp-n586-s20	585	65	RC(4)	5–10	28	R	20	1.2	18	80	80	0	80	0	3	77
77	sbvrp-n599-s4	598	598	R	50–100	487	R	4	1.6	0	0	47	1	46	46	0	0
78	sbvrp-n613-s100	612	0	R	1–100	523	C	100	1.4	29	31	32	1	31	0	0	31
79	sbvrp-n627-s4	626	549	C(5)	5–10	110	E	4	1.6	3	7	24	5	19	12	7	0
80	sbvrp-n641-s20	640	335	RC(8)	50–100	1381	E	20	1.4	11	17	18	1	17	0	0	17
81	sbvrp-n655-s100	654	347	C(4)	U	5	C	100	1.2	38	43	66	1	65	22	4	39
82	sbvrp-n670-s4	669	592	R	SL	129	R	4	1.2	4	9	64	1	63	54	2	7
83	sbvrp-n685-s20	684	684	RC(6)	Q	408	C	20	1.6	6	12	38	0	38	26	3	9
84	sbvrp-n701-s100	700	79	RC(7)	1–10	87	E	100	1.4	21	21	23	2	21	0	0	21
85	sbvrp-n716-s20	715	0	C(3)	1–100	1007	R	20	1.4	10	17	18	1	17	0	0	17
86	sbvrp-n733-s4	732	71	R	1–10	25	C	4	1.6	4	37	122	84	38	1	0	37
87	sbvrp-n749-s100	748	664	C(8)	1–100	396	R	100	1.2	13	13	49	0	49	36	3	10
88	sbvrp-n766-s100	765	0	RC(7)	SL	166	E	100	1.6	25	35	36	1	35	0	0	35
89	sbvrp-n783-s4	782	782	R	Q	832	R	4	1.4	0	0	24	0	24	24	0	0
90	sbvrp-n801-s20	800	401	R	U	20	E	20	1.2	12	20	20	0	20	0	2	18
91	sbvrp-n819-s100	818	427	C(6)	50–100	358	C	100	1.4	51	69	88	3	85	16	6	63
92	sbvrp-n837-s4	836	759	RC(7)	5–10	44	R	4	1.2	3	9	72	2	70	61	2	7
93	sbvrp-n856-s20	855	0	RC(3)	U	9	C	20	1.6	13	47	48	1	47	0	0	47
94	sbvrp-n876-s20	875	100	C(5)	1–100	764	E	20	1.6	11	29	30	1	29	0	0	29
95	sbvrp-n895-s4	894	894	R	50–100	1816	R	4	1.2	0	0	19	0	19	19	0	0
96	sbvrp-n916-s100	915	915	RC(6)	5–10	33	E	100	1.4	0	0	104	1	103	103	0	0
97	sbvrp-n936-s4	935	0	R	SL	138	C	4	1.2	4	76	85	9	76	0	0	76
98	sbvrp-n957-s100	956	92	RC(4)	U	11	R	100	1.4	36	43	44	1	43	0	0	43
99	sbvrp-n979-s20	978	483	C(6)	Q	998	E	20	1.6	10	28	30	2	28	0	5	23
100	sbvrp-n1001-s4	1000	894	R	1–10	131	R	4	1.6	4	9	22	1	21	12	2	7

of swap locations is important. Also, it resulted in the detection of various patterns used to formulate useful problem-specific properties, which are valid for the majority of the instances under consideration. Specifically, it was observed that the absence of any truck customers in an instance may imply the use of no swap locations in its solution. Regarding the tactical fleet management, our study revealed that trains are more preferable than trucks for a multitude of best known solutions of the additional instances. For the same benchmark set, the combination of swap actions *Park-Swap-Pickup* is promoted for almost all best known solutions, where at least one swap location is exploited. On the contrary, in the solutions of both datasets, the combination *Park-Pickup* is rarely present. Additionally, there is the case that no swap locations are visited by a train, which is employed due to the high demand of the customers. Future work involves the use of the detected motifs in the design of promising solution methods for the considered problem.

References

1. Absi, N., Cattaruzza, D., Feillet, D., Housseman, S.: A relax-and-repair heuristic for the Swap-Body Vehicle Routing Problem. Ann. Oper. Res. **253**(2), 957–978 (2017)
2. Chao, I.M.: A tabu search method for the truck and trailer routing problem. Comput. Oper. Res. **29**(1), 33–51 (2002). https://doi.org/10.1016/S0305-0548(00)00056-3
3. Derigs, U., Pullmann, M., Vogel, U.: Truck and trailer routing - problems, heuristics and computational experience. Comput. Oper. Res. **40**(2), 536–546 (2013). https://doi.org/10.1016/j.cor.2012.08.007, http://www.sciencedirect.com/science/article/pii/S0305054812001724
4. Drexl, M.: Branch-and-price and heuristic column generation for the generalized truck-and-trailer routing problem. J. Quant. Methods Econ. Bus. Adm. **12**, 5 (2011)
5. Heid, W., Hasle, G., Vigo, D.: VeRoLog solver challenge 2014 VSC2014 problem description. Technical report (2014). http://verolog.deis.unibo.it/news-events/general-news/verolog-solver-challenge-2014
6. Huber, S., Geiger, M.J.: Order matters - a variable neighborhood search for the swap-body vehicle routing problem. Eur. J. Oper. Res. **263**(2), 419–445 (2017). https://doi.org/10.1016/j.ejor.2017.04.046, http://www.sciencedirect.com/science/article/pii/S0377221717303934
7. Lin, S.W., Yu, V.F., Chou, S.Y.: Solving the truck and trailer routing problem based on a simulated annealing heuristic. Comput. Oper. Res. **36**(5), 1683–1692 (2009)
8. Miranda-Bront, J.J., Curcio, B., Méndez-Díaz, I., Montero, A., Pousa, F., Zabala, P.: A cluster-first route-second approach for the swap body vehicle routing problem. Ann. Oper. Res. **253**(2), 935–956 (2017)
9. Todosijević, R., Hanafi, S., Urošević, D., Jarboui, B., Gendron, B.: A general variable neighborhood search for the swap-body vehicle routing problem. Comput. Oper. Res. **78**, 468–479 (2017). https://doi.org/10.1016/j.cor.2016.01.016, http://www.sciencedirect.com/science/article/pii/S0305054816300120

10. Toffolo, T.A., Christiaens, J., Malderen, S.V., Wauters, T., Vanden Berghe, G.: Stochastic local search with learning automaton for the swap-body vehicle routing problem. Comput. Oper. Res. **89**, 68–81 (2018)

11. Uchoa, E., Pecin, D., Pessoa, A., Poggi, M., Vidal, T., Subramanian, A.: New benchmark instances for the capacitated vehicle routing problem. Eur. J. Oper. Res. **257**(3), 845–858 (2017)

Evolutionary Deep Learning for Car Park Occupancy Prediction in Smart Cities

Andrés Camero$^{(\boxtimes)}$, Jamal Toutouh, Daniel H. Stolfi, and Enrique Alba

Departamento de Lenguajes y Ciencias de la Computación,
Universidad de Málaga, 29071 Málaga, Spain
andrescamero@uma.es, {jamal,dhstolfi,eat}@lcc.uma.es

Abstract. This study presents a new technique based on Deep Learning with Recurrent Neural Networks to address the prediction of car park occupancy rate. This is an interesting problem in smart mobility and we here approach it in an innovative way, consisting in automatically design a deep network that encapsulates the behavior of the car occupancy and then is able to make an informed guess on the number of free parking spaces near to the medium time horizon. We analyze a real world case study consisting of the occupancy values of 29 car parks in Birmingham, UK, during eleven weeks and compare our results to other predictors in the state-of-the-art. The results show that our approach is accurate to the point of being useful for being used by citizens in their daily lives, as well as it outperforms the existing competitors.

Keywords: Deep neuroevolution · Deep learning
Evolutionary algorithms · Smart cities · Car park occupancy

1 Introduction

Nowadays, most of the world population lives in urban areas, and it is expected that the number of inhabitants in cities will be 75% of the world's population by 2050 [3]. Thus, a wide range of challenges have to be faced by the different city stakeholders in order to mitigate the negative effects of a very fast growth of such urban areas. With the application of new types of computing and the technological innovation of critical infrastructure and services, the concept of *Smart City* emerges as a means to efficiently address big city challenges.

One of the main concerns in modern cities is mobility. The vast increment in the volume of urban road traffic experienced during the last decades causes serious issues that have be confronted with new tools. Traffic jams bother the daily life of the population, mainly because traffic congestion causes longer trip times and a larger associated pollution, not to mention the economic losses due to delays and other transport problems. Thus, great efforts are being made to develop along one of the dimensions of the Smart City initiative: Smart Mobility, which focuses on providing sustainable transport systems and logistics to allow

© Springer Nature Switzerland AG 2019
R. Battiti et al. (Eds.): LION 12 2018, LNCS 11353, pp. 386–401, 2019.
https://doi.org/10.1007/978-3-030-05348-2_32

a smooth urban traffic and commuting by mainly applying information and communication technologies [4, 7, 19, 22, 33].

The search of free parking spaces is an important activity that negatively affects the road traffic flows: it is responsible for up to 40% of the total traffic within cities [10]. This is mainly due to the fact that drivers often do not take the most efficient parking decisions, because those are based basically on partial on-road perceptions and past personal experiences. During the last few years, a number of systems have emerged with the aim of simplifying the search of free parking spaces [19]. These systems have a positive impact on traffic operations in urban areas. The main advantages of using such systems are [34]:

- they reduce the driver's frustration since they increase the probability of finding free parking spaces,
- they improve the global road traffic efficiency (e.g., fuel/energy consumption, generated gas emissions, travel times, etc.) because they reduce the total distance traveled by vehicles,
- and they help road users optimize their trips by taking into account the expected free parking space information to decide where to park in advance.

This type of systems involve learning, predicting, and exploiting cloud based architectures. In this article, we aim to provide a new technique for learning and predicting car park systems, which can be also applied to other forecast problems. Indeed, a new Deep Learning (DL) technique based on Recurrent Neural Networks (RNN) and Evolutionary Algorithms is proposed, a deep neuroevolutionary hyper-parameter architecture optimization. We compare the performance of our proposal against other machine learning (ML) techniques applied to predict the occupancy rate of several car parks in Birmingham (U.K.) [32].

When dealing with DL, the efficiency is an important issue, mainly due to the required computational cost, which hinders the use of such a powerful tool. Using a neural network is often a solution to many practical problems similar to this one. But one of the main aspects that affects the efficiency is the appropriateness of the used neural network architecture [24], that is, how good is the defined set of layers and links, and which method is to be used for its best design.

In this study, we define an optimization problem consisting in determining the optimal design of an RNN, obtaining an architecture that minimizes the required computational costs while keeping a high accuracy. As a global statement, an optimization problem is defined by a search space and a quality or fitness function. The search space defines (and then restricts) the possible configurations of a solution vector, which is associated with a numerical cost by the fitness function. Thus, solving an optimization problem consists in finding the least-cost configuration of a solution vector (assuming minimization).

The vast number of possible RNN architectures that can be defined makes this task very hard. Then, the use of automatic intelligent tools seems a mandatory requirement when addressing them. In this sense, Evolutionary Algorithms (EA) [2, 11] emerge as an efficient stochastic techniques able to solve optimization problems. Indeed, these algorithms are currently employed in a multitude of hard-to-solve problems, e.g., in the domain of Smart City [20, 33], showing a successful performance. Nevertheless, the use of such a methodology in the domain of DL is still limited [21].

Therefore, the main contributions of this study are two: first, defining a general technique to automatically design efficient RNN by using EAs, and second, applying such tool for developing an efficient parking prediction system to be applied in *smart urban areas*. The remainder of this paper is organized as follows. Next section reviews similar works solving the prediction of car park occupancy rates. Section 3 presents our proposal. Section 4 presents the experiments carried out, the results, a benchmark, and the analysis. Finally, conclusions and future work are considered in Sect. 5.

2   Related Work

The prediction of car park availability is a demanded subject that is being studied in the context of smart cities [19], especially now when most car parking spaces have a sensing infrastructure connected to a cloud based system. Smart parking services based on parking prediction allow drivers to organize their transports before departures or during their trips. This type of services is a common way of forecasting the occupancy rate of parking spaces, and even they could provide the possibility of booking a parking spot in advance.

Some of the most popular prediction approaches assume that vehicles arrive to car park spaces following a Poison distribution. Then, they predict their capacity by using a Markov Chain [16,25]. However, the efficacy of these methods is limited because the demand of the parking spaces depends on different factors, including the time of the day, the day of the week, weather conditions, etc., which are not considered by these proposed models.

Smart City projects are promoting the collection of large-scale car park information in urban areas. Therefore, researchers and practitioners have access to realistic car park data sets. The analysis of *SmartSantander* on-street parking data shown that the occupancy and parking periods of different parking areas followed a Weibull distribution [34]. In [37], three different ML methods were applied to predict the car park occupancy rate, over two data sets from San Francisco (*SFpark*) and Melbourne: a regression tree, a neural network, and supported vector machines, in order to show their relative strengths and weaknesses. SFpark was also used as a use case to compare a number of spatio-temporal clustering strategies [29]. These methods reduced the storage required by other prediction methods, providing similar accurate fitting than seven-day models. A multi-variate regressive model to predict spatial and temporal correlations of car park availability has been also applied using real-time data from San Francisco and Los Angeles [26,27].

In this study, we focus on the use of DL based on a special type of neural networks, RNN. Neural networks have been applied by different authors in this domain. The main idea is to study the relation between aggregating parking lots and predicting car park availability by applying feed-forward networks [18]. These type of approaches are improved when they are used together with Internet of Things (IoT) systems, because they can continuously improve the accuracy of the occupancy predictions by using back-propagation [34,35].

Like the aforementioned studies, this work considers real world data retrieved by our smart-parking data-collecting system [33] from Birmingham (U.K.). Thus, we analyze the application of DL for predicting the occupancy rate and compare the performance against other previously used ML methods [33].

DL allows the application of learning processes that can be done with multiple layers of abstraction, by taking the advantage of the currently available, high computational resources [12]. DL has dramatically improved the results provided by ML in the past [17]. Nevertheless, the efficient design of DL methods is still an open problem and there is room for improvement [24]. In RNN, the issue of efficiency is even harder, because they are updated or rebuilt repeatedly to capture the temporal structure of the data. Thus, special care must be taken with the efficiency of the training process, the efficacy of the training method, and the appropriateness of the architecture.

In order to deal with the issue of finding a suitable architecture to the car park prediction problem (a very influent decision for the final quality of the prediction), we propose the application of an automatic and intelligent procedure based on EAs. A few number of authors have already studied *Neuroevolution*, i.e., the optimization of artificial neural networks by using EAs [1,36]. However, their solutions cannot directly be applied on DL due to the high complexity of the neural networks used in DL [24]. Therefore we propose to improve and extend these ideas to DL, giving rise to deep neuroevolution.

3   Evolutionary-Based Architecture Optimization

In this section we present the details of our proposal. First, we comment on the training of the RNN, then we introduce the *fitness*, and finally we outline two evolutionary approaches to optimize the architecture of the RNN.

3.1   Learning

Artificial neural networks (ANN) are computational models inspired by the human brain, and as our brain does, they are capable of *learning*. In our particular problem, we are interested in an iterative type of learning process referred as *supervised learning* [13]. Supervised learning consists in supplying training data of N input-output pairs (X,Y). Then, for each X the ANN produces an output Z, which is compared to Y using an error (cost or distance) function. Finally, this error is minimized in an iterative manner.

Minimizing the error is a tough task. There have been several approaches proposed for this, including gradient descent-based [13] and metaheuristic-based ones [1,24], but up-to-date the most used method is a first-order gradient descent algorithm named *backpropagation* [30] (BP). It is very important to notice that in order to use BP to train a RNN, the network has to be *unfold* [15] (so the error is propagated backwards). This means that the network is *copied* and *connected* in series, building an unrolled or unfold version of the RNN. The number of times that the RNN is unfold is usually referred as to *look back*, because we are allowing the RNN to look back a finite number of times.

Once an ANN is trained, we are interested in its generalization capability, i.e. its proficiency to offer general solutions rather than overfitting to a training data set. This issue is usually tackled by adding a stochastic component in the training process, a technique called dropout [31]. This technique has proved to be very useful to reduce overfitting, however Reed et al. [28] showed that "large networks generally learn rapidly", but "they tend to generalize poorly", suggesting that we might consider the architecture to improve this capability.

Therefore, we propose to look for an RNN that better adapts to our problem by optimizing its generalization ability, including its architecture (the number of hidden layers and neurons), and its unfolded representation.

Since training an RNN is costly (in terms of computational resources) and the number of RNN architectures is infinite (or extremely large if we impose restrictions to the number of hidden layers or neurons), we are enforced to define a smart search strategy to find an optimal RNN.

Among the many potential optimization techniques, metaheuristics are well know because of their ability to combine the exploration and exploitation strategies. Thus, they are suitable to address complex, nonlinear, and non differentiable problems [2,24]. Having said that, we decided to adapt two metaheuristic algorithms to solve our problem of finding an RNN architecture that allow us to minimize the generalization error.

3.2   Fitness

To evaluate the fitness of an RNN architecture (for solving our problem), we propose to measure its generalization capability. Specifically, we decided to use the *mean absolute error* (MAE) of the predicted car park occupancy versus the observed value. Considering that in this particular problem we would like to forecast a whole day (or week) in advance (not only the next value of the time series), we defined the *fitness* of the RNN as the MAE of the prediction of multiple (future) observations, based on the already predicted values. In other words, the predicted values are used as the input for future predictions.

$$\text{minimize} \quad \text{Fitness} = \frac{1}{N} \sum_{i}^{N} MAE(z_i, y_i) \tag{1}$$

$$\text{subject to} \quad H \leq max\_hidden\_layers \tag{2}$$

$$L \leq max\_neurons\_per\_layer \tag{3}$$

$$\hat{x}_i = \begin{cases} x_0 & \text{if } i = 0 \\ z_{i-1} & \text{if } i > 0 \end{cases} \tag{4}$$

Equation 1 presents the problem of minimizing the fitness, where N corresponds to the number of samples in the testing data set (X,Y), z_i stands for the predicted occupancy of the i-th sample, and y_i corresponds to the ground truly occupancy of the i-th sample. Notice that the RNN is fed with already predicted data $\hat{x}$, and that the architecture is constraint by H and L.

3.3   GA-Based Approach

In order to find a RNN architecture that is *fitted* to the series, we designed a genetic algorithm (GA) [14] that *evolves* an initial population of RNN candidates (Algorithm 1). Particularly, we encoded (i.e. the **representation**) the architecture of the RNN and two training features (*solution*) as an integer vector of variable length. The first position of the solution corresponds to the *dropout* (a learning parameter that avoids over-fitting), the second, to the *look back* (how many times the net is unfold during the training), and the third and successive positions correspond to the number of neurons of the i-th hidden layer (the architecture, properly). Thus, the number of hidden layers is defined by the length of the solution. Note that the input and output layers are defined by the time series, therefore we did not include them as part of the solution.

First, an initial population of *pop size* individuals (**Initialize**) is created randomly, and evaluated (**Evaluate**), i.e. each solution is decoded into an RNN architecture, then it is trained (using a training data set), and finally the *fitness* is computed over a test data set.

Then, while the number of evaluations is smaller than *max_evaluations*, the evolution of the population is performed by selecting a subset of *parents* using the *binary tournament* (**Selection**). After this, the parents are recombined into an offspring (of size equal to *offspring_size*) using the *single point crossover* (**Recombination**) with *cx_prob*. It is important to remark that with 1-*cx_prob* probability one of the parents is returned unmodified.

Once the recombination is done, each offspring is mutated by a two phase process (**Mutation**). In the first step, a uniformly distributed value in the range $[1, mut\_max\_step]$ is added to or subtracted from the i-th component of the solution with probability *mut_prob*. In the second step, a hidden layer is added to (copied) or removed from the solution with probability *mut_x_prob*.

Finally, the *offspring* is evaluated, and the *offspring_size* worst solutions (in terms of fitness) of the population are replaced by the offspring (**Replace**).

Algorithm 1 Pseudo-code of the GA-based RNN architecture optimizer.

1: *population* ← Initialize()
2: Evaluate(*population*)
3: **while** evaluations ≤ max evaluations **do**
4:     *parents* ← Selection(*population*)
5:     *offspring* ← Recombination(*parents*)
6:     *mutated* ← Mutation(*offspring*)
7:     Evaluate(*mutated*)
8:     Replace *mutated* **in** *population*
9: **end while**

3.4   ES-Based Approach

In line with the GA-based RNN optimizer, we designed an RNN optimizer based on an $(1 + 1)$ Evolutionary Strategy (ES) [2] approach. Note that we selected small μ and λ because of the high computational cost of training an RNN. Our proposal (Algorithm 2) evolves a single *solution* using the encoding and mutation already defined in Sect. 3.3, and a *plus* replacement criteria, i.e. if the fitness of the new candidate solution (mutated) is at least as good as the *old* solution, the new candidate replaces the old one.

To improve the performance of the ES-based algorithm we included a procedure to self-adjust the parameters [8]. Particularly, if the new candidate solution (*mutated*) improves the fitness in regard to the old solution, then the *mut_prob* and *mut_x_prob* values are multiplied by 1.5, otherwise these values are divided by 4. Therefore, while the solution is improving, we widen the local search.

Algorithm 2 Self Adjusted (1+1)ES-based RNN architecture optimizer.

1: *solution* ← Initialize()
2: Evaluate(*solution*)
3: **while** evaluations ≤ max evaluations **do**
4:     *mutated* ← Mutation(*solution*)
5:     Evaluate(*mutated*)
6:     **if** Fitness(*mutated*) ≤ Fitness(*solution*) **then**
7:         *solution* ← *mutated*
8:     **end if**
9:     SelfAdjust()
10: **end while**

---

4   Experimental Study

We implemented the algorithms[1] in Python 3, using the deep learning hyperparameter optimization library **dlopt** [6], the evolutionary computation framework **deap** [9], and the DL frameworks **keras** and **tensorflow**. Then, we (*i*) selected a data set to test our proposal, (*ii*) performed a preliminary study on the use of the algorithms, (*iii*) analyzed the performance of the optimized RNNs against expert defined architectures, and (*iv*) benchmarked our predictions against the state-of-the-art of car park occupancy predictors.

4.1   Data Set: Birmingham Car Park Occupancy

The data set analyzed in this article is the one used in [32], comprising valid occupancy rates of 29 car parks operated by NCP (National Car Parks) in the city of Birmingham in the U.K.

---

[1] https://github.com/acamero/dlopt.

Birmingham, is a major city in the West Midlands of England, standing on the small River Rea. It is the largest and most populous British city outside London, with an estimated population of 1,124,569 as of 2016 [23].

Several cities in the U.K. have been publishing their open data to be used, not only by researchers and companies, but also for citizens for better know the place where they live. The Birmingham data set is licensed under the Open Government License v3.0 and it is updated every 30 minutes from 8:00 to 16:30 (18 occupancy values per car park and day). In our study, we worked with data collected from Oct 4, 2016 to Dec 19, 2016 (11 weeks).

Figure 1a shows the occupancy data available for all the car parks and dates. Figure 1b presents a box plot showing the data distribution of the car park occupancy by weekdays. We can see in the former that almost all car parks begin the day with at least 50% of free spaces and that they are progressively occupied during the day with a clear peak between 13:00 and 14:00 h. Finally, the latter figure shows that car parks have more available spaces on Sundays and Saturdays which was to be expected as they are not working days.

The numeric characteristics of the data set are: 77 days of occupancy data of 29 car parks which account for 33,292 values for training plus 3,425 for testing.

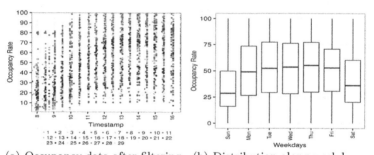

(a) Occupancy data after filtering. (b) Distribution along weekdays.

Fig. 1. Occupancy data of the 29 car parks and their distribution on weekdays.

4.2 Preliminary Study

To begin with our experimentation we defined an initial setup for the training process (Table 1). These values were taken from the related literature, our experience, and from a brief analysis of a small set of RNN training logs.

The GA-based approach for RNN architecture optimization (refer to Sect. 3.3) requires the setup of several parameters. Therefore, we performed the hyper-parametrization of the algorithm using **hyperopt**, a Python library introduced by Bergstra et al. [5]. Particularly, we used the *Tree of Parzen Estimators* algorithm to optimize the set of parameters (in regard to the fitness defined in Sect. 3.2, computed over the test data set defined in Sect. 4.1), and the parameters values from the ranges defined in Table 2. Note that we included in the hyper-parameterization process three *training* parameters (*batch size, min*

Table 1. Default training parameters.

Parameter	Value	Parameter	Value	Parameter	Value
Epoch	10000	Max evaluations	100	Patience	500
Learning rate	5e-5	Max hidden layers	8	Dropout	[0,100]
Min delta	1e-6	Max neurons per layer	64	Look back	[1,30]
Validation split	0.3				

delta, and *patience*), because we wanted to explore the impact of discarding an RNN earlier (i.e. increasing the *min delta* and decreasing the *patience*) and augmenting the amount of data available on each BP iteration.

Table 2. Parameters search space.

Parameter	Min	Max	Type	Parameter	Min	Max	Type
Mut max step	1	10	Discrete	Offspring size	1	10	Discrete
Cx prob	0	1	Continuous	Batch size	10	100	Discrete
Mut prob	0	1	Continuous	Min delta	10e-6	10e-4	Continuous
Mut x prob	0	1	Continuous	Patience	10	500	Discrete
Pop size	1	10	Discrete				

The best configuration found (using the referred method) for the GA-based is shown in Table 3. We inferred from the configuration that discarding solutions a 'little bit earlier' than the original configuration improves the final result.

Table 3. Best GA configuration found.

Parameter	Value	Parameter	Value	Parameter	Value
Batch size	80	Mut max step	2	Offspring size	1
Cx prob	0.73	Mut prob	0.48	Pop size	4
Min delta	5.9e-5	Mut x prob	0.51	Patience	283

On the other hand, the ES-based approach does not require an hyper-parameterization, because it is a self-adjusting algorithm. Hence, we initialize the algorithm using an off-the-shelve configuration (presented in Table 4).

4.3 RNN Optimization Performance

In order to benchmark the performance of the optimized RNN against an *expert defined* architecture, we compare the predictions for both approaches.

Table 4. Off-the-shelve ES parameters configuration.

Parameter	Value
Mut max step	3
Mut prob	0.2
Mut x prob	0.2

First, we executed 30 independent runs of both RNN optimization algorithms (GA and ES-based), using the parameters configuration defined in Tables 3 and 4, the data set defined in Sect. 4.1, and the fitness defined in Sect. 3.2.

Then, we trained three *expert defined* RNN architectures (refer to Table 5 for details), the training parameters defined in Table 1, and the same data set and fitness measured mentioned above. Note that these configurations are based on Google Tensorflow sample models[2].

Table 5. Expert defined RNN architectures.

Parameter	Small	Medium	Large
Hidden layers	2	2	2
Hidden size	200	650	1500

Table 6 summarizes the results of the optimization benchmark. The table includes the number of *hidden layers* (H. Layers), the number of *trainable parameters* (Tr. Params., i.e. the number of weights in the net), and the *fitness* computed over the test data set.

From the latter results we concluded that optimizing the architecture of an RNN is a very influential task, not only because it helps to improve the accuracy of the predictions, but also because it could help to reduce the training time dramatically. Moreover, we observed that increasing the number of neurons (*trainable parameters*) is not as important as arranging them adequately. Due to this finding, we decided to briefly analyze all the architectures evaluated by both optimizers, with the aim of unveiling useful information for future design of optimization algorithms.

First, we sorted the architectures evaluated by both algorithms according to their fitness in ascending order and split them into deciles. Then, we computed the smooth density estimates for each decile, grouping them by the number of layers and by the number of neurons (refer to Fig. 2). The density estimates show that the best solutions (first decile) are mainly consisting in architectures of eight layers, arranging 250 to 350 neurons.

---

[2] https://github.com/tensorflow/models.

Table 6. Optimization experimental results.

Architecture	H. Layers	Tr. Params.	Fitness
Small	2	511,228	0.34
Medium	2	5,171,428	0.32
Large	2	27,234,028	0.84
Mean GA-based	6	53,696	0.08
Max GA-based	7	100,964	0.09
Min GA-based	5	7,684	0.08
SD GA-based	0.8	42,546	0.01
Mean ES-based	6	7,136	0.08
Max ES-based	7	8,452	0.10
Min ES-based	5	5,820	0.08
SD ES-based	1.4	1,861	0.01

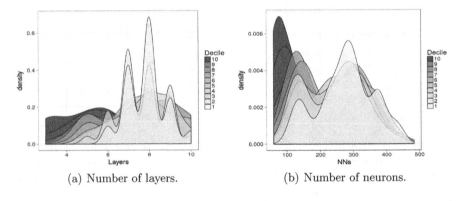

(a) Number of layers.   (b) Number of neurons.

Fig. 2. Number of neurons/layers by fitness decile.

To continue with our analysis, we study the architectures of the first decile. Figure 3 shows the fitness of the *best* architectures evaluated. A smaller fitness (blacker dots) implies a more accurate prediction.

The results suggest that, for this particular problem, there is an archetype (i.e. specific shape) of RNN architecture, because the majority of the architectures that belong to the first decile present a similar number of neurons and hidden layers. Therefore, we envision that this kind of information has to be considered for the design of future architecture optimization algorithms.

4.4   Prediction Benchmark

Since the final goal of our work is to obtain an accurate prediction of the car park occupancy, we selected the *best* of the trained RNN (including both algorithms) and used it to predict the car park occupancy on the *testing* data set. Then, we

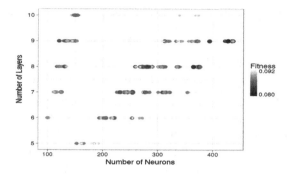

Fig. 3. Fitness of the first decile. A lower fitness is desirable.

compared these predictions against the ones presented by Stolfi et al. [32]. In that article, the authors trained six predictors to forecast the future occupancy rates of the same data set analyzed here. Concretely, they used polynomials (P), Fourier series (F), k-means clustering (KM), polynomials fitted to the k-means' centroids (KP), shift and phase modifications to KP polynomials (SP), and time series (TS).

Table 7 presents the MAE measured for each car park (over the entire predicted period), as well as the summarized statistics. The results of the RNN do not exceed all its competitors (in terms of the mean or median), however there is no significant difference between its predictions and the ones made using *polynomials* (Wilcoxon test p-value=0.096) or *time series* (Wilcoxon test p-value=0.099).

Therefore, we consider that the results are useful, not only because there is no significant difference between the RNN forecast values and polynomials or time series, but also because the predictions were made by only one predictor, while all its competitors consist on multiple predictors (one per each car park and day, i.e. 203 predictors!). Moreover, the predictions were made based on already predicted data, hence the results could be improved in a real situation by updating the forecast periodically with updated data.

5   Conclusions and Future Work

RNN architecture optimization is a promising field of study, because most efforts have been made in regard to feedforward architectures. Our results show that the overall performance of an RNN could be improved by selecting an appropriate configuration. Moreover, adding hidden layers or neurons arbitrarily to an RNN do not guarantee an improvement in its generalization capability, thus defining a search strategy is mandatory.

We introduced two metaheuristics approaches to RNN architecture optimization: a GA-based and an ES-based. Both strategies proved to be suitable for solving the problem, and not only that: they also provided useful information to improve our understanding of the RNN architecture optimization problem.

Table 7. Mean absolute error of the predicted occupancy.

Parking lot	P	F	KM	KP	SP	TS	RNN
BHMBCCMKT01	0.041	0.053	0.087	0.086	0.059	0.067	0.063
BHMBCCPST01	0.076	0.072	0.148	0.149	0.083	0.111	0.137
BHMBCCSNH01	0.132	0.141	0.15	0.148	0.139	0.069	0.117
BHMBCCTHL01	0.122	0.142	0.134	0.131	0.123	0.08	0.103
BHMBRCBRG01	0.101	0.148	0.148	0.149	0.133	0.095	0.123
BHMBRCBRG02	0.087	0.116	0.122	0.122	0.097	0.088	0.112
BHMBRCBRG03	0.068	0.085	0.113	0.112	0.074	0.059	0.076
BHMEURBRD01	0.044	0.057	0.087	0.085	0.036	0.042	0.077
BHMEURBRD02	0.072	0.078	0.064	0.063	0.067	0.068	0.062
BHMMBMMBX01	0.063	0.067	0.074	0.072	0.084	0.129	0.084
BHMNCPHST01	0.060	0.079	0.13	0.127	0.073	0.034	0.050
BHMNCPLDH01	0.030	0.034	0.087	0.084	0.036	0.072	0.072
BHMNCPNHS01	0.072	0.084	0.082	0.078	0.06	0.082	0.102
BHMNCPNST01	0.085	0.083	0.15	0.154	0.117	0.074	0.124
BHMNCPPLS01	0.078	0.088	0.067	0.066	0.08	0.058	0.076
BHMNCPRAN01	0.083	0.094	0.143	0.14	0.084	0.055	0.089
Broad Street	0.047	0.057	0.073	0.071	0.041	0.034	0.064
Bull Ring	0.088	0.119	0.113	0.112	0.101	0.074	0.100
NIA Car Parks	0.033	0.033	0.048	0.049	0.028	0.054	0.033
NIA South	0.040	0.036	0.064	0.064	0.031	0.078	0.053
Others-CCCPS105a	0.032	0.050	0.119	0.121	0.05	0.072	0.065
Others-CCCPS119a	0.090	0.092	0.081	0.081	0.089	0.095	0.091
Others-CCCPS133	0.083	0.108	0.093	0.092	0.087	0.061	0.091
Others-CCCPS135a	0.057	0.075	0.078	0.076	0.058	0.029	0.049
Others-CCCPS202	0.016	0.024	0.074	0.075	0.025	0.023	0.033
Others-CCCPS8	0.038	0.055	0.081	0.079	0.04	0.047	0.061
Others-CCCPS98	0.092	0.089	0.177	0.179	0.101	0.097	0.092
Shopping	0.035	0.054	0.065	0.066	0.032	0.032	0.037
Median	0.070	0.078	0.087	0.086	0.074	0.069	0.077
Mean	0.067	0.079	0.102	0.101	0.073	0.067	0.079
Max	0.132	0.148	0.177	0.179	0.139	0.129	0.137
Min	0.016	0.024	0.048	0.049	0.025	0.023	0.033
Sd	0.029	0.033	0.035	0.035	0.033	0.026	0.028

We believe that these insights will help us to define more adequate optimization strategies, as well as to improve the overall performance of RNN.

As future work we propose to use the proposed strategies to solve problems of other domains, such as load forecast (energy) and artificial intelligence for gaming, among others.

Acknowledgements. This research was partially funded by Ministerio de Economía, Industria y Competitividad, Gobierno de España, and European Regional Development Fund grant numbers TIN2014-57341-R (http://moveon.lcc.uma.es), TIN2016-81766-REDT (http://cirti.es), and TIN2017-88213-R (http://6city.lcc.uma.es). Daniel H. Stolfi is supported by a FPU grant (FPU13/00954) from the Spanish Ministry of Education, Culture and Sports. Universidad de Málaga. Campus Internacional de Excelencia, Andalucía TECH.

References

1. Alba, E., Martí, R.: Metaheuristic Procedures for Training Neural Networks, vol, p. 35. Springer Science & Business Media, Berlin (2006)
2. Back, T.: Evolutionary Algorithms in Theory and Practice: Evolution Strategies, Evolutionary Programming, Genetic Algorithms. Oxford University Press, Oxford (1996)
3. Bakici, T., Almirall, E., Wareham, J.: A smart city initiative: the case of Barcelona. J. Knowl. Econ. **4**(2), 135–148 (2013). https://doi.org/10.1007/s13132-012-0084-9
4. Benevolo, C., Dameri, R.P., D'Auria, B.: Smart mobility in smart city. In: Torre, T., Braccini, A.M., Spinelli, R. (eds.) Empowering Organizations, pp. 13–28. Springer International Publishing (2016)
5. Bergstra, J., Yamins, D., Cox, D.: Making a science of model search: Hyperparameter optimization in hundreds of dimensions for vision architectures. In: International Conference on Machine Learning, pp. 115–123 (2013)
6. Camero, A., Toutouh, J., Alba, E.: DLOPT: deep learning optimization library (2018). arXiv:1807.03523
7. Cintrano, C., Stolfi, D.H., Toutouh, J., Chicano, F., Alba, E.: Ctpath: A real world system to enable green transportation by optimizing environmentaly friendly routing paths. In: Alba, E., Chicano, F., Luque, G. (eds.) Smart Cities, pp. 63–75. Springer International Publishing (2016)
8. Doerr, C.: Non-static parameter choices in evolutionary computation. In: Genetic and Evolutionary Computation Conference, GECCO 2017, Berlin, Germany, July 15–19, 2017, Companion Material Proceedings. ACM (2017). https://doi.org/10.1145/3067695.3067707
9. Fortin, F.A., De Rainville, F.M., Gardner, M.A., Parizeau, M., Gagné, C.: DEAP: Evolutionary algorithms made easy. J. Mach. Learn. Res. **13**, 2171–2175 (2012)
10. Giuffré, T., Siniscalchi, S.M., Tesoriere, G.: A novel architecture of parking management for smart cities. Procedia - Soc. Behav. Sci. **53**, 16–28 (2012) (sIIV-5th Intl. Congress - Sustainability of Road Infrastructures 2012)
11. Goldberg, D.E., Holland, J.H.: Genetic algorithms and machine learning. Mach. Learn. **3**(2), 95–99 (1988)
12. Goodfellow, I., Bengio, Y., Courville, A.: Deep Learning. The MIT Press (2016)
13. Haykin, S.: Neural Networks and Learning Machines, vol. 3. Pearson (2009)
14. Holland John, H.: Adaptation in Natural and Artificial Systems: An Introductory Analysis with Applications to Biology, Control, and Artificial Intelligence. University of Michigan, USA (1975)

15. Jaeger, H.: Tutorial on Training Recurrent Neural Networks, Covering BPPT, RTRL, EKF and the Echo State Network Approach, vol. 5. GMD (2002)
16. Klappenecker, A., Lee, H., Welch, J.L.: Finding available parking spaces made easy. Ad Hoc Netw. **12**, 243–249 (2014)
17. LeCun, Y., Bengio, Y., Hinton, G.: Deep learning. Nature **521**(7553), 436 (2015)
18. Lin, T.: Smart parking: network, infrastructure and urban service. Ph.D. thesis, Lyon, INSA (2015)
19. Lin, T., Rivano, H., Mouël, F.L.: A survey of smart parking solutions. IEEE Trans. Intell. Transp. Syst. **18**(12), 3229–3253 (2017). https://doi.org/10.1109/TITS.2017.2685143. Dec
20. Massobrio, R., Toutouh, J., Nesmachnow, S., Alba, E.: Infrastructure deployment in vehicular communication networks using a parallel multiobjective evolutionary algorithm. Int. J. Intell. Syst. **32**(8), 801–829 (2017). https://doi.org/10.1002/int.21890
21. Morse, G., Stanley, K.O.: Simple evolutionary optimization can rival stochastic gradient descent in neural networks. In: Proceedings of the Genetic and Evolutionary Computation Conference 2016, pp. 477–484. GECCO '16, ACM (2016)
22. Nesmachnow, S., Rossit, D., Toutouh, J.: Comparison of multiobjective evolutionary algorithms for prioritized urban waste collection in montevideo, uruguay. Electron. Notes Discret. Math. (2018) (in press)
23. Office for National Statistics: Population Estimates for UK. http://www.nomisweb.co.uk/articles/747.aspx (2016). Accessed 16 Dec 2017
24. Ojha, V.K., Abraham, A., Snášel, V.: Metaheuristic design of feedforward neural networks: a review of two decades of research. Eng. Appl. Artif. Intell. **60**, 97–116 (2017)
25. Pullola, S., Atrey, P.K., Saddik, A.E.: Towards an intelligent GPS-based vehicle navigation system for finding street parking lots. In: 2007 IEEE International Conference on Signal Processing and Communications, pp. 1251–1254 (2007). https://doi.org/10.1109/ICSPC.2007.4728553
26. Rajabioun, T., Foster, B., Ioannou, P.A.: Intelligent parking assist. In: Control & Automation (MED), 2013 21st Mediterranean Conference, pp. 1156–1161. IEEE (2013)
27. Rajabioun, T., Ioannou, P.A.: On-street and off-street parking availability prediction using multivariate spatiotemporal models. IEEE Trans. Intell. Transp. Syst. **16**(5), 2913–2924 (2015). Oct
28. Reed, R., Marks, R., Oh, S.: Similarities of error regularization, sigmoid gain scaling, target smoothing, and training with jitter. IEEE Trans. Neural Netw. **6**(3), 529–538 (1995)
29. Richter, F., Martino, S.D., Mattfeld, D.C.: Temporal and spatial clustering for a parking prediction service. In: 2014 IEEE 26th International Conference on Tools with Artificial Intelligence, pp. 278–282 (2014)
30. Rumelhart, D., Hinton, G.E., Williams, R.j.: Learning internal representations by error propagation. Technical Report No. ICS-8506, California University San Diego La Jolla Inst for Cognitive Science (1985)
31. Srivastava, N., Hinton, G., Krizhevsky, A., Sutskever, I., Salakhutdinov, R.: Dropout: a simple way to prevent neural networks from overfitting. J. Mach. Learn. Res. **15**(1), 1929–1958 (2014)
32. Stolfi, D.H., Alba, E., Yao, : X.: Predicting car park occupancy rates in smart cities, pp. 107–117. Springer (2017)

33. Stolfi, D.H., Armas, R., Alba, E., Aguirre, H., Tanaka, K.: Fine tuning of traffic in our cities with smart panels: the Quito city case study. In: Proceedings of the Genetic and Evolutionary Computation Conference 2016, pp. 1013–1019. GECCO '16, ACM (2016)
34. Vlahogianni, E.I., Kepaptsoglou, K., Tsetsos, V., Karlaftis, M.G.: A real-time parking prediction system for smart cities. J. Intell. Transp. Syst. **20**(2), 192–204 (2016). https://doi.org/10.1080/15472450.2015.1037955
35. Vlahogianni, E., Kepaptsoglou, K., Tsetsos, V., Karlaftis, M.G.: Exploiting new sensor technologies for real-time parking prediction in urban areas. In: Transportation Research Board 93rd Annual Meeting Compendium of Papers, pp. 14–1673 (2014)
36. Yao, X.: Evolving artificial neural networks. Proc. IEEE **87**(9), 1423–1447 (1999). https://doi.org/10.1109/5.784219
37. Zheng, Y., Rajasegarar, S., Leckie, C.: Parking availability prediction for sensor-enabled car parks in smart cities. In: 2015 IEEE 10th International Conference on Intelligent Sensors, Sensor Networks and Information Processing (ISSNIP), pp. 1–6 (2015)

Asymptotically Optimal Algorithm for the Maximum m-Peripatetic Salesman Problem in a Normed Space

E. Kh. Gimadi[1,2] and O. Yu. Tsidulko[1,2](✉)

[1] Sobolev Institute of Mathematics SB RAS, Novosibirsk, Russia
gimadi@math.nsc.ru
[2] Department of Mechanics and Mathematics, Novosibirsk State University,
Novosibirsk, Russia
tsidulko@math.nsc.ru

Abstract. The maximum m-Peripatetic Salesman Problem (m-PSP) consists of determining m edge-disjoint Hamiltonian cycles of maximum total weight in a given complete weighted n-vertex graph. We consider a geometric variant of the problem and describe a polynomial time approximation algorithm for the m-PSP in a normed space of fixed dimension. We prove that the algorithm is asymptotically optimal for $m = o(n)$.

Keywords: Maximum traveling salesman problem · Maximum m-peripatetic salesman problem · Normed space · Asymptotically optimal algorithm

1 Introduction

Given a complete edge-weighted n-vertex graph, the m-Peripatetic Salesman Problem (m-PSP) is to find m edge-disjoint Hamiltonian cycles of minimum or maximum total weight. The problem was first introduced by Krarup in [8] and is known to be NP-hard [4]. In this paper we consider the maximum geometric version of this problem (maximum m-PSP).

Suppose that the vertices of the graph G belong to a normed space $\mathbb{R}^k$ and the weight of an edge (x, y) is equal to $\|x - y\|$, where $\|\cdot\|$ is a given norm on $\mathbb{R}^k$.

Following Shenmaier [11] we define a concept of an angle in an arbitrary normed space, setting the remote angle α between two vectors x and y to be equal to the distance between the vector $x/\|x\|$ and the nearest vector $\pm y/\|y\|$, that is $\alpha(x, y) = \min\{\|x/\|x\| - y/\|y\|\|, \|x/\|x\| + y/\|y\|\|\}$. For $x = \lambda y$ or if the norm of one of the vectors equals zero, the remote angle between x and y is assumed to be zero.

Sections 1 and 2 are supported by the RFBR (project 16-07-00168), by the Russian Ministry of Science and Education under the 5-100 Excellence Programme and by the program of fundamental scientific researches of the SB RAS I.5.1. Sections 3 and 4 are supported by Russian Science Foundation (project 16-11-10041)

R. Battiti et al. (Eds.): LION 12 2018, LNCS 11353, pp. 402–410, 2019.
https://doi.org/10.1007/978-3-030-05348-2_33

In the case of $m = 1$ the maximum m-PSP is the well-known maximum Traveling Salesman Problem (maximum TSP) [3,9], which is NP-hard even in Euclidean k-dimensional space, $k \geq 3$ [2]. In [11] for the maximum TSP in a normed space a polynomial asymptotically exact algorithm is described. This algorithm uses ideas of known algorithms [6,10] for the TSP in a Euclidean space. The algorithm is based on the following geometric facts:

Lemma 1 ([11]). *Among any t vectors ($t < n/2$) in a normed space $\mathbb{R}^k$ there exist two vectors such that the remote angle between them doesn't exceed the value*

$$\alpha(k,t) = \frac{2k}{\lfloor (2t-1)^{1/k} \rfloor} \,. \tag{1}$$

Lemma 2 ([11]). *Let AB and CD be two intervals in $\mathbb{R}^k$ and α be the remote angle between them. Then*

$$\frac{\max(\|AC\| + \|BD\|, \|AD\| + \|BC\|)}{\|AB\| + \|CD\|} \geq 1 - \alpha/2 \,.$$

Baburin and Gimadi [1] constructed an asymptotically optimal algorithm for the maximum m-PSP in Euclidean space of fixed dimension, for the case of $m = o(n)$. Shenmaier [11] showed how to obtain an asymptotically optimal algorithm for the maximum m-PSP in a polyhedral space, if $f \cdot m = o(n)$, where f is the number of facets in a unit ball, and m is the number of desired Hamiltonian cycles.

In this paper, we combine the asymptotically optimal approach to the maximum TSP in a normed space with technique from [1], that leads to an asymptotically optimal polynomial-time algorithm for the maximum m-PSP in *an arbitrary normed space.*

Definition 1. *An approximation algorithm A for some maximization problem has a* guaranteed relative error ε, *if the algorithm is a $(1 - \varepsilon)$-approximation. Namely,*

$$\frac{OPT(X) - F_A(X)}{OPT(X)} \leq \varepsilon \,,$$

for any input X, where $OPT(X)$ is the optimal value of the objective function for the input X and $F_A(X)$ is the value of the objective function obtained by the algorithm A.

Consider an optimization problem on a graph $G = (V, E)$ with $|V| = n$. We will specify the dependence of the relative error of an algorithm A on the number of vertices n of an instance and use the notation $\varepsilon_A(n)$.

Definition 2. *Algorithm A solving an optimization problem on a graph is* asymptotically optimal *if $\varepsilon_A(n) \to 0$ as the number of vertices n of the input graph indefinitely increases.*

2   Description of the Algorithm $\mathcal{A}$ for the Maximum m-PSP in a Normed Space

Let $w(u, v)$ be the weight of edge $e = (u, v) \in E$ and $W(G') = \sum e \in E' w(e)$ be the total weight of a subgraph $G' = (V; E')$ of the initial graph $G = (V; E)$ with the set of edges $E' \subseteq E$. The goal of the algorithm $\mathcal{A}$ for the maximum m-PSP is to find a subset of edges $\widetilde{E} \subset E$, consisting of m edge-disjoint Hamiltonian cycles H_i, where $i = \overline{1, m}$. At the beginning of the algorithm, $\widetilde{E}$ is empty.

Let $\mathcal{M}^* = \{I_1, \ldots, I_\mu\}$ be the set of edges (intervals in $\mathbb{R}^k$) of a maximum weight matching in G; $\mu = \lfloor n/2 \rfloor$.

Definition 3. *Two edges $e_1, e_2 \in E$ are linked (with respect to set $\widetilde{E} \subseteq E$), if there exists an edge $e \in E$ ($e \in \widetilde{E}$), that connects the end vertices of e_1 and e_2.*

Definition 4. *An I-chain is a sequence of edges, where each two neighboring edges are linked.*

Definition 5. *Two I-chains are linked (with respect to set $\widetilde{E}$) if their end edges are linked.*

We declare one of the end edges of an I-chain to be the master edge and the other one to be the inferior edge.

Definition 6. *An α-chain is an I-chain, where the remote angle between any two neighboring edges of the chain is at most α.*

Now let's describe the approximation algorithm $\mathcal{A}$.

2.1   Algorithm $\mathcal{A}$

Preliminary Steps.

In a given complete graph G find a matching $\mathcal{M}^* = \{I_1, \ldots, I_\mu\}$ of maximum weight, where $\mu = \lfloor n/2 \rfloor$ is the number of its edges (intervals).

Set $\widetilde{E} = \emptyset$ and fix a parameter $t \leq \mu/2$. Sort the edges of $\mathcal{M}^*$ in the non-increasing order with respect to their weights. We will now refer to the first $(\mu - t)$ edges of $\mathcal{M}^*$ as heavy edges, and the last t edges as light.

Phase $i = 1, \ldots, m$

Phase i constructs a Hamiltonian cycle H_i.

Stage 1.

Define angle $\alpha = \alpha(k, t)$ according to the relation (1):

$$\alpha(k, t) = \begin{cases} \frac{2k}{\lfloor (2t-1)^{1/k} \rfloor}, & \text{if } i = 1; \\ \frac{2k}{\lfloor (2t/(2i-2)-1)^{1/k} \rfloor}, & \text{if } 1 < i \leq m, \end{cases} \tag{2}$$

where t is the number of available edges.

We are going to build a set $\mathcal{I}$ of α-chains, $|\mathcal{I}| = t$. Recall that we defined the $(\mu - t)$ heaviest edges of $\mathcal{M}^*$ as the heavy edges. Each α-chain will consist only of the heavy edges. Note that an edge is a one-element α-chain.

We start with $\mathcal{I}_0$ consisting of the first t heaviest edges of $\mathcal{M}^*$: $\mathcal{I}_{\prime} = \{I_1, I_2, \ldots, I_t\}$.

Set $j = t$.

In the current t-set $\mathcal{I}_i$, find a pair of non-linked with respect to the set $\widetilde{E}$ α-chains such that the angle between their master edges is at most α.

Join thees chains into one α-chain by setting their master edges to be neighbors and assign one of the end edges of the joined chain (one of the former inferior edges) to be the new master edge. Set $j := j + 1$. If $j < \mu - t$, then append one more heavy edge I_j to the current set $\mathcal{I}$ and repeat Stage 1.

Otherwise, we have obtained a sequence $\mathcal{C} = \{C_1, \ldots, C_t\}$ of t α-chains such that each chain consists of a sequence of heavy edges, where the angle between any consecutive (neighboring) pair of edges is at most $\alpha = \alpha(k, t)$.

Stage 2. Let's regard the sequence $\mathcal{C}$ as a cycle, i.e. the α-chain C_t is followed by the α-chain C_1. Let the edges of the α-chains $C_1, \ldots, C_t$ be enumerated so that $C_r = \{I_{\nu_{r-1}+1}, \ldots, I_{\nu_r-1}\}$, $1 \le r \le t$, where $\nu_1 < \nu_2 < \ldots < \nu_t$ are the numbers reserved for the remaining light edges of the maximum weight matching $\mathcal{M}^*$ ($\nu_0 = 0$, $\nu_t = \mu$).

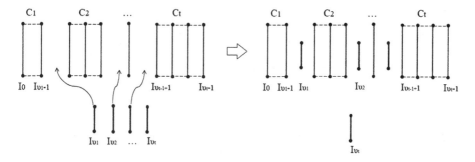

Fig. 1. The bold lines correspond to the edges of $\mathcal{M}^*$, while the dashed lines indicates the α-chains. The t light edges of $\mathcal{M}^*$ were placed to the positions between the α-chains. The last light edge I_{ν_t} is placed between the first and the t-th α-chains.

Place t light edges of the maximum weight matching $\mathcal{M}^*$ to the positions $\nu_1, \nu_2, \ldots, \nu_t = \mu$, such that no light edge is linked with respect to $\widetilde{E}$ to the neighboring end edges of the α-chains. This can be done by solving a corresponding assignment problem.

Finally, we have a sequence $\mathcal{S}_i = \{C_1, I_{\nu_1}, C_2, I_{\nu_2}, \ldots, C_t, I_{\nu_t}\}$ of t α-chains consisting of heavy edges, which alternate with the t light edges (Fig. 1.).

Stage 3. Construct a Hamiltonian cycle H_i that is edge-disjoint with $H_1, H_2, \ldots, H_{i-1}$ in the following way.

We assume that the sequence of edges of the maximum weight matching $\mathcal{M}^* = \{I_1, I_2, \ldots, I_\mu\}$ is given according to their order in the sequence $\mathcal{S}$, $I_j =$

(x_j, y_j), $j = 1, \ldots, \mu$. Now we are going to construct a partial tour T consisting of the end vertices of the light edge $I_\mu = (x_\mu, y_\mu)$ and of two $(\mu - 1)$-vertex paths.

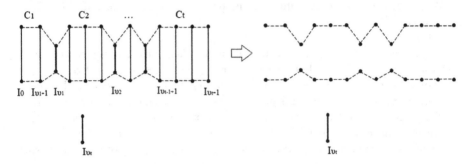

Fig. 2. The bold lines correspond to the edges of $\mathcal{M}^*$, the dashed lines correspond to the partial tour. We connect each two neighboring edges of $\mathcal{M}^*$ with the edges of the partial tour in a best possible way. Note that the edges of the maximum weight matching are not included to the partial tour.

Step 1. Set $T = x_\mu \cup y_\mu$; $u_1 := x_1$; $v_1 := y_1$ and $j = 2$.
Step 2 While $1 < j < \mu$ do:
If

$$w(u_{j-1}, x_j) + w(v_{j-1}, y_j) \geq w(u_{j-1}, y_j) + w(v_{j-1}, x_j),$$

then set $u_j = x_j$; $v_j = y_j$; otherwise, set $u_j = y_j$ and $v_j = x_j$. Supplement the partial tour T with a pair of new edges (Fig. 2.):

$$T = T \cup (u_{j-1}, u_j) \cup (v_{j-1}, v_j).$$

Increase j by 1.
Step 3. At the previous step we have obtained a partial tour consisting of end vertices of the light edge $I_\mu = (x_\mu, y_\mu)$ and two non-intersecting paths $(u_1, u_2, \ldots, u_{\mu-1})$ and $(v_1, v_2, \ldots, v_{\mu-1})$:

$$T = (u_1, u_2, \ldots, u_{\mu-1}) \cup (v_1, v_2, \ldots, v_{\mu-1}) \cup \{x_\mu\} \cup \{y_\mu\}.$$

Close T (Fig. 3.) into a 2μ-vertex cycle by adding a pair of two-edge chains $(u_{\mu-1}, y_\mu, v_1) \cup (v_{\mu-1}, x_\mu, u_1)$ or $(u_{\mu-1}, x_\mu, v_1) \cup (v_{\mu-1}, y_\mu, u_1)$ with the greatest total weight

$$\delta W = \max \begin{cases} W((u_{\mu-1}, y_\mu, v_1) \cup (v_{\mu-1}, x_\mu, u_1)); \\ W((u_{\mu-1}, x_\mu, v_1) \cup (v_{\mu-1}, y_\mu, u_1)), \end{cases}$$

In the case of even n, the obtained cycle is Hamiltonian. If n is odd, there exists a vertex x_0 that is not in $\mathcal{M}^*$. In this case replace an edge (x, y) of the constructed $(n-1)$-vertex cycle by the pair of edges $(x_0; x)$ and (x_0, y) so that none of these edges intersects the set $\widetilde{E}$. The triangle inequality guarantees that the weight of the cycle will not decrease.

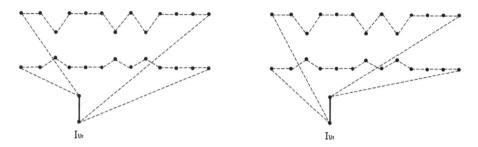

Fig. 3. We close the partial tour into a Hamiltonian cycle, choosing the best of two possible variants

Step 4. Append the edges of the obtained Hamiltonian cycle H_i to the set $\widetilde{E}$. The description of the algorithm $\mathcal{A}$ is complete.

The algorithm $\mathcal{A}$ produces m Hamiltonian cycles $H_1, H_2, \ldots, H_m$, and since each time we arrange the edges in $\mathcal{S}$ so that they are not linked with respect to the edges of $\widetilde{E}$, the obtained Hamiltonian cycles are edge-disjoint. The running-time of the algorithm is determined by the time one needs to construct a maximum weight matching, which is $O(n^3)$. The correctness of the algorithm follows from [1].

3   Analysis of Algorithm $\mathcal{A}$

Note that the total weight $W(\widetilde{\mathcal{M}^*})$ of the first $(\mu - t)$ heaviest edges of $\mathcal{M}^*$ satisfies the inequality [1]:

$$W(\widetilde{\mathcal{M}^*}) \geq W(\mathcal{M}^*)\left(1 - \frac{t}{\mu}\right). \tag{3}$$

Therefore, the weight of a Hamiltonian tour H_i satisfies

$$W(H_i) \geq 2W(\mathcal{M}^*)\left(1 - \frac{t}{\mu}\right)(1 - \alpha_i/2),$$

where the angle α_i is defined by (2). Thus,

$$\frac{W(H_i)}{W(H^*)} \geq 1 - \frac{2t + 1}{n} - k\left(2\tau_i - 1\right)^{-1/k},$$

where H^* is the solution of the maximum TSP in the given graph and

$$\tau_i = \begin{cases} t, & \text{if } i = 1; \\ t/(2i - 2) & \text{for } 1 < i \leq m. \end{cases} \tag{4}$$

Using (3) and (4), for the approximation ratio $\varepsilon_{\mathcal{A}}(n)$ of algorithm $\mathcal{A}$ we have

$$\varepsilon_{\mathcal{A}}(n) = 1 - \frac{W_{\mathcal{A}}}{OPT} \leq 1 - \frac{W_{\mathcal{A}}}{mW(H^*)} = \tag{5}$$

$$= 1 - \frac{W(H_1) + \ldots + W(H_m)}{mW(H^*)} \leq \frac{2t+1}{n} - \frac{k}{m} \sum_{i=1}^{m} (2\tau_i - 1)^{-1/k}.$$

Lemma 3. *For $m \leq t$, $k \geq 2$ the following inequality holds:*

$$\sum_{i=1}^{m} (2\tau_i - 1)^{-1/k} \leq 2m \left(\frac{m}{t}\right)^{\frac{1}{k}}.$$

Proof. Using the definition (4) of τ_i, we have:

$$\sum_{i=1}^{m} (2\tau_i - 1)^{-1/k} \leq m \left(\frac{t}{m} - 1\right)^{-1/k} \leq m \frac{(m/t)^{1/k}}{(1 - m/t)^{1/k}} \leq m \frac{(m/t)^{1/k}}{1 - m/(kt)} \leq 2m \left(\frac{m}{t}\right)^{1/k}.$$

Due to Eq. (5) together with Lemma 3, we have proved the following theorem.

Theorem 1. *Algorithm $\mathcal{A}$ gives approximate solutions to the maximum m-PSP in a normed space of fixed dimension with relative error*

$$\varepsilon_{\mathcal{A}}(n) \leq \frac{2t+1}{n} + 2k \left(\frac{m}{t}\right)^{1/k}. \tag{6}$$

Theorem 2. *Algorithm $\mathcal{A}$ with parameter $t^* = m\left(\frac{n}{m}\right)^{1/k}$ gives asymptotically optimal solutions for the maximum m-PSP in a normed space $\mathbb{R}^k$, if the dimension k is fixed and $m = o(n)$.*

Proof. Setting $t = t^*$ in (6), we obtain

$$\varepsilon_{\mathcal{A}}(n) \leq \frac{1}{n} + 2(k+1) \left(\frac{m}{n}\right)^{1/(k+1)} \to 0$$

as $n \to \infty$.

4   The m-Capacitated Peripatetic Salesman Problem

In this short section we show how the results obtained in the previous section can be simply extended to the case of maximum m-Capacitated Peripatetic Salesman problem (m-CPSP). Given a complete graph $G(V, E)$, a capacity function of edges $c : E \to \{1, 2, \ldots, m\}$ and a weight function of edges $w : E \to \mathcal{R}^+$, the m-CPSP is to find m Hamiltonian cycles of maximum total weight, such that each edge e belongs to at most $c(e)$ of these cycles. If all the edge capacities are equal to 1, the m-CPSP is the m-PSP. The problem was first introduced in [5]. Only a few papers on the problem are known in the literature. The branch-and-cut and branch-and-price algorithms for the m-CPSP are constructed in [5]. Paper [7] presents polynomial ρ-approximation algorithms for some special cases of 2-CPSP.

Corollary 1. *Algorithm $\mathcal{A}$ with parameter $t^* = m\left(\frac{n}{m}\right)^{1/k}$ gives asymptotically optimal solutions for the maximum m-Capacitated Peripatetic Salesman problem in a normed space $\mathbb{R}^k$ of a fixed dimension k, if $m = o(n)$.*

Proof. Let $W_{\mathcal{A}}$ be the weight of the solution obtained by algorithm $\mathcal{A}$, $W(m\text{-PSP})$ be the weight of the maximum m-PSP solution, $W(m\text{-CPSP})$ be the weight of the maximum m-CPSP solution, and $W(H^*)$ be the weight of the maximum TSP solution. Obviously, a feasible solution of the m-PSP is a feasible solution for the m-CPSP. It is also easy to see that for any given capacity function

$$mW(H^*) \geq W(m\text{-CPSP}) \geq W(m\text{-PSP}) \geq W_{\mathcal{A}}.$$

Thus, for the relative error $\varepsilon'_{\mathcal{A}}$ of algorithm $\mathcal{A}$ solving the m-CPSP, we have:

$$\varepsilon'_{\mathcal{A}}(n) = 1 - \frac{W_{\mathcal{A}}}{W(m\text{-CPSP})} \leq 1 - \frac{W_{\mathcal{A}}}{mW(H^*)},$$

which is exactly the same expression (5) we were estimating in the previous section.

5   Conclusion

In this paper using the definition and useful properties of the remote angle between two vectors in the normed space introduced in [11], we were able to extend an approach to the maximum m-PSP, so it can give asymptotically optimal solution for the problem in an arbitrary normed space of fixed dimension. We also showed that the same algorithm gives asymptotically optimal solutions for the capacitated version of the problem.

For the further research it is interesting to apply this approach and ideas to different modifications of the considered problem. For example, it is of natural interest to construct an asymptotically optimal algorithm for the m-CPSP that would use the features of the problem statement and would have a better relative error or less tight condition for the number m of Hamiltonian cycles.

References

1. Baburin, A.E., Gimadi, EKh: On the asymptotic optimality of an algorithm for solving the maximum m-PSP in a multidimensional Euclidean space. Proc. Steklov Inst. Math. **272**(1), 1–13 (2011)
2. Barvinok, A.I., Fekete, S.P., Johnson, D.S., Tamir, A., Woeginger, G., Woodroofe, R.: The geometric maximum traveling salesman problem. J. ACM **50**(5), 641–664 (2003)
3. Barvinok, A.I., Gimadi, E.Kh., Serdyukov, A.I.: In: Punnen, A., Gutin, G. (eds.) The Maximum TSP. In: The Traveling Salesman Problem and Its Variations, pp. 585–608. Kluwer Academic Publishers, Dordrecht (2002)
4. De Kort, J.B.J.M.: Upper bounds and lower bounds for the symmetric K peripatetic salesman problem. Optimization **23**(4), 357–367 (1992)
5. Duhenne, E., Laporte, G., Semet, F.: The undirected m-capacitated peripatetic salesman problem. Eur. J. Oper. Res. **223**(3), 637–643 (2012)
6. Gimadi, E.Kh.: A new version of the asymptotically optimal algorithm for solving the Euclidean maximum traveling salesman problem (in Russian). Proc. 12th Baykal International Conference 2001, Irkutsk, vol. 1, pp. 117–123 (2001)

7. Gimadi, EKh, Istomin, A.M., Rykov, I.A.: On m-capa itated peripatetic salesman problem with capaity restritions. J. Appl. Ind. Math. **8**(1), 1–15 (2014)

8. Krarup, J.: The peripatetic salesman and some related unsolved problems. Comb. Program.: Methods Appl. NATO Adv. Study Inst. Ser. **19**, 173–178 (1975)

9. Lawler, E.L., Lenstra, J.K.: Rinnooy Kan, A.H.G., Shmoys, D.: The Traveling Salesman Problem: A Guided Tour of Combinatorial Optimization. Wiley, Chichester (1985)

10. Serdyukov, A. I: An asymptotically optimal algorithm for the maximum traveling salesman problem in Euclidean space (in Russian). Upravlyaemye Sistemy 27, Novosibirsk, pp. 79–87 (1987)

11. Shenmaier, V.V.: Asymptotically optimal algorithms for geometric Max TSP and Max m-PSP. Discret. Appl. Math. **163**(2), 214–219 (2014)

Computational Intelligence for Locating Garbage Accumulation Points in Urban Scenarios

Jamal Toutouh[1](✉), Diego Rossit[2], and Sergio Nesmachnow[3]

[1] Universidad de Málaga, Málaga, Spain
jamal@lcc.uma.es
[2] Universidad Nacional del Sur-CONICET, Bahía Blanca, Argentina
diego.rossit@uns.edu.ar
[3] Universidad de la República, Montevideo, Uruguay
sergion@fing.edu.uy

Abstract. This article presents computational intelligence methods for solving the problem of locating garbage accumulation points in urban scenarios, which is a relevant problem in nowadays smart cities to optimize budget and reduce negative environmental and social impacts. The problem model considers reducing the investment costs, enhance the proportion of citizens served by bins, and the accessibility to the system. A family of heuristics based on the generic PageRank schema and a mutiobjective evolutionary algorithm are proposed. Experimental evaluation performed on real scenarios on the city of Montevideo, Uruguay, demonstrates the effectiveness of the proposed approaches. The methods allow computing plannings with different trade-off between the problem objectives and improving over the current planning.

Keywords: Computational intelligence · Waste management
Smart cities

1 Introduction

Solid waste management is an important issue in modern cities. It refers to the process of collecting, treating, and disposing of solid material that is discarded by citizens. The current increase of garbage generation rate in urban areas [14] has led to initiatives for encouraging efficient and sustainable practices. In fact, solid waste management is one of the main challenges for local governments in order to mitigate environmental and social impacts, especially in highly populated cities.

A specific problem related to solid waste management in urban scenarios refers to find a proper location for *garbage accumulation points* (GAPs) or *collection sites*, where a set of waste bins are to be installed. This is a relevant problem to guarantee a good service to citizens, because a paltry spatial distribution of waste bins in the city may lead to not fulfilling the needs of residents and/or affecting the quality of the service (e.g., people must walk long

© Springer Nature Switzerland AG 2019
R. Battiti et al. (Eds.): LION 12 2018, LNCS 11353, pp. 411–426, 2019.
https://doi.org/10.1007/978-3-030-05348-2_34

distances for garbage disposal, or certain waste bins fill quickly while others remain empty). Finding appropriate locations for waste bins is a variation of the Facility Location Problem, which is proved to be NP-hard [18]. In this context, where exact optimization methods require long execution times for solving realistic instances, heuristics and metaheuristics [21] are viable options to find good-quality approximate solutions, allowing to analyze different configurations for waste bins and also different scenarios. These capabilities are important when planning solid waste management in nowadays smart cities, especially considering that the final location plan for GAPs usually considers several criteria (thus, the underlying problem is multiobjective), it must fulfill a set of hard constraints, and also that it is desirable that waste bin locations change periodically in order to not disturb the same citizens. This last feature is closely related with the 'not in my backyard' phenomenon of semi-obnoxious facilities, such as waste bins [16]. Despite knowing that GAPs must be located somewhere, citizens are reluctant to have a waste bin very close to their homes, since it is generally linked to undesirable aspects, e.g., bad smell, visual pollution, disturbing noises, and heavy traffic associated with collecting vehicles.

The problem model proposed in this article considers performing the GAP distribution along the city while taking into account different criteria: reduce the expenses of installing bins, facilitate the accessibility to the system by reducing the walking distance of the citizens to the bins and serve as many people as possible, maximizing the total amount of waste collected. Two computational intelligence approaches are applied to solve the GAP location problem: three Pagerank heuristics and a Multiobjective Evolutionary Algorithm (MOEA). PageRank provides a simple approach for planning, while the MOEA allows exploring solutions that account garbage accumulation points configurations with different trade-off between the problem objectives. The experimental evaluation performed on real scenarios on Montevideo, Uruguay, demonstrated that significant improvements are obtained over the current solution implemented in the city (up to 90% in distance and 31% in cost).

The article is structured as follows. Section 2 introduces the problem. The proposed methods for solving the problem are described in Sect. 3. The experimental evaluation is reported in Sect. 4. Finally, Sect. 5 presents the main conclusions and formulates the main lines for future work.

2   Problem Description

The problem addressed in this paper aims to select the best locations of GAPs from a predefined set of potential places and determine the number and type of waste bins (indicated, hereafter, as "bins") that are to be install in each chosen GAP, according to three different criteria: (i) maximize the total amount of waste collected; (ii) minimize the installation cost of bins; and (iii) minimize the average distance between the garbage generators and the assigned bins as a metric of the Quality of Service (QoS) offered to the users.

2.1   Mathematical Formulation

The mathematical formulation of the GAP location problem applying a Mixed Integer Programming (MIP) model considers the following elements:

- A set $I = \{i_1, \ldots, i_M\}$ of potential GAPs for bins. Each GAP i has an available space S_i for installing bins.
- A set $P = \{p_1, \ldots, p_N\}$ of generators. Following a usual approach in the related literature, nearby generators are grouped in clusters, assuming a similar behavior between elements in each cluster. The amount of waste produced by generator p (in volumetric units) is b_p. The distance from generator p to GAP $i \in I$ is $d_{p,i}$, and the maximum distance between any generator in P and its assigned GAP (in meters) is D.
- A set $J = \{j_1, \ldots, j_H\}$ of bin types. Each type has a given purchase price c_j, capacity C_j, and required space for its installation e_j. The maximum number of bins of type j available is MB_j.

The model is described in Eqs. 1–12, using the following variables: $t_{j,i}$ is the number of bins of type j installed in GAP i, $x_{p,i}$ is 1 if generator p is assigned to GAP i and 0 otherwise, and $f_{p,i}$ is the fraction of the waste produced by generator p that is deposited in GAP i.

$$\max \sum_{p\in P,\ i\in I} f_{p,i} \times b_p \tag{1}$$

$$\min \sum_{p\in P,\ i\in I} \frac{d_{p,i} \times f_{p,i}}{|P|} \tag{2}$$

$$\min \sum_{j\in J,\ h\in H,\ i\in I} t_{j,h,i} \times c_j \tag{3}$$

subject to

$$\sum_{i\in I} (f_{p,i}) \leq 1 \qquad\qquad \forall\, p \in P \tag{4}$$

$$\sum_{j\in J,\ h\in H} (t_{j,i} e_j) \leq S_i \qquad\qquad \forall\, i \in I \tag{5}$$

$$\sum_{p\in P} (b_p f_{p,i}) \leq \sum_{j\in J} (C_j t_{j,i}) \qquad\qquad \forall\, i \in I \tag{6}$$

$$\sum_{h\in H,\ i\in I} t_{j,h,i} \leq MB_j \qquad\qquad \forall\, j \in J \tag{7}$$

$$d_{p,i} x_{p,i} \leq D \qquad\qquad \forall\, p \in P,\ i \in I \tag{8}$$

$$f_{p,i} \leq x_{p,i} \leq k f_{p,i} \qquad\qquad \forall\, p \in P,\ i \in I \tag{9}$$

$$0 \leq f_{p,i} \leq 1 \qquad\qquad \forall\, p \in P,\ i \in I \tag{10}$$

$$x_{p,i} \in [0,1] \qquad\qquad \forall\, p \in P,\ i \in I \tag{11}$$

$$t_{j,i} \in \mathbb{Z}_0^+ \qquad\qquad \forall\, j \in J,\ i \in I \tag{12}$$

Three objective functions are proposed: the total waste that generators are able to dispose (Eq. 1); the average distance between generators and the assigned GAPs, weighted according to the waste fraction that is deposited in each GAP (Eq. 2); and the total investment cost (Eq. 3). Regarding the problem constraints, the sum of relative waste fractions of a generator must be less than one (Eq. 4); the space occupied with bins in a GAP must not be larger than the available space of the GAP (Eq. 5); the waste volume assigned to one GAP must not be larger than the storage capacity installed in that GAP, i.e., S_i (Eq. 6); the number of bins of each type must be smaller than the maximum for that type (Eq. 7); the maximum distance between a generator and any assigned GAP must be smaller than the threshold D (Eq. 8); variable $x_{p,i}$ is one if and only if some of the waste produced by generator p is deposited in GAP i, where k is a positive integer constant such that $k \times f_{p,i} \geq 1 \; \forall \; f_{p,i} > 0$ (Eq. 9); the continuous variable $f_{p,i}$ is defined between zero and one (Eq. 10), variable $x_{p,i}$ is binary (Eq. 11), and variable $t_{j,i}$ is a non-negative integer (Eq. 12).

The model considers that each block in the city contains a number of potential GAPs for bins. Each GAP has a limited available space (S_i) that restricts the number of bins that can be installed. Conversely to other approaches in the literature, the model considers the possibility that a generator can deposit its waste in several different GAPs. This is a rather realistic feature that is highly probable to occur in everyday life when a generator finds a GAP that has all its bins full. Moreover, taking into account that generators are grouped in clusters, instead of considering standalone individuals, this feature allows expressing that not all generators of a given cluster shall deposit their garbage in the same GAP.

2.2 Related Work

Bautista and Pereira [3] modeled the problem as a minimal set covering/maximum satisfiability problem, and proposed a genetic algorithm and a GRASP metaheuristic for solving real instances in Barcelona, Spain. Other authors have applied integral approaches to solve the bins location problem and the routing/collection problem simultaneously. For example, Chang and Wei [6] used a fuzzy MOEA to solve the problem for a scenario in Kaohsiung, Taiwan. The model considered the percentage of population served, the average walking distance between users and their assigned GAP, and the approximate length of the routes of collecting vehicles as objectives. Hemmelmay et al. [13] introduced the Waste Bin Allocation and Routing Problem, which was solved applying different methodologies that combine sequential and simultaneous strategies; the allocation was solved either with an exact or an heuristic method, while the routing was solved using Variable Neighborhood Search.

A similar problem was addressed by Ghiani et al. [10], who proposed a constructive heuristic for solving large instances in Nardò, Italy, that cannot be properly handled by exact methods implemented in CPLEX. Later [11], the heuristic was modified to bound posterior routing costs, e.g., not allowing the installation in a same GAP of bins that require different type of collecting vehicles. Di Felice [8] proposed a two-phase heuristic for the problem, for a real case in L'Aquila, Italy. The first phase solved the location of the GAPs thorough a constructive heuristic while the second determines the quantity and size of bins needed for each GAP according to the number of generators served by that GAP. A similar heuristic was applied by Boskovic and Jovici [4] for Kragujevac, Serbia, using the ArcGIS Network Analyst. Since bins location is a problem that uses spatial information, other authors relied on Geographic Information Systems to gather and analyze data. For example, Valeo et al. [24] used a constructive heuristic to establish the GAPs sequentially according to some priorities in order to cover an studied area in Dundas, Canada.

Regarding bibliographic reviews about the studied topic, Purkayastha et al. [23] made an effort to review the different applications related to bins location problems, concluding that the bibliography is rather scarce compare to the potential benefits that smart solutions in this problem can produce. Goulart et al. [12] reviewed multicriteria approaches in municipal solid waste (MSW), emphasizing that there are again few applications that take advantage of the potential of multiobjective optimization in this field.

Our research contributes with a novel model to solve the GAPs location problem, and determine which bins are to be established in each GAP, that allows that a generator can visit several bins. This is a realistic aspect that is not common in the literature. A mathematical formulation of the problem and two heuristic approaches –PageRank and MOEA–, are proposed. Experiments performed on real cases in Montevideo show the competitiveness of the model and the proposed computational intelligence algorithms.

3 Computational Intelligence Methods for the GAP Location Problem

This section describes the computational intelligence approaches developed for solving the GAP location problem: three Pagerank heuristics and a MOEA.

3.1 PageRank Algorithms

PageRank is a well-known voting algorithm to compute the relevance of web pages in Internet taking into account inbound and outbound links [15]. It has been successfully used to solve location problems that can be mathematically defined over graphs in the field of *smart cities*, e.g. by modeling road networks as weighted graphs considering road-traffic information, and applying the voting algorithm to rank the potential locations to install infrastructure [5,17].

Weighted PageRank is applied to a given directed graph $G = (V, E)$ defined by V (a set of vertices) and E (a set of edges). The algorithm starts by initializing the PageRank value of each vertex v_i to a fixed value d, i.e., $PR^W(v_i) = d$, $\forall v_i \in V$. d is known as the *dumping parameter* and its default value is 0.85. Then, an iterative process is performed until a stop condition is reached (the convergence value is below a given threshold or a maximum number of iterations performed). $PR^W(v_i)$ is computed according to Eq. 13, where $In(v_i)$ is the set of vertices that point to it (*predecessors*), and $Out(v_i)$ is the set of vertices that v_i points to (*successors*), and w_{ij} is the weight that for the edge that connects v_i and v_j.

$$PR^W(v_i) = (1 - d) + d \times \left(\sum_{v_j \in In(v_i)} w_{ij} \times \frac{PR^W(v_j)}{\sum_{v_k \in Out(v_j)} w_{jk}} \right) \quad (13)$$

In the GAP location problem model, information about waste generators and collection points is modeled as a fully connected weighted graph $G = (V, E)$. G is defined by the set of waste generators P and the set of edges E. The weight of each edge $w_{jk} = \frac{b_j + b_k}{d_{j,k}}$ relates the impact of the waste generated in both generators and their distance. Thus, the tentative locations of GAPs are ranked in a sorted vector I^{PR} in which $i_j^{PR}, i_k^{PR} \in I, j < k \Leftrightarrow PR^W(i_j^{PR}) > PR^W(s_k^{PR})$.

Once the GAPs are sorted in I^{PR}, a constructive heuristic is applied to select a collection point configuration and locate it. The heuristic iterates over the sorted vector I^{PR} starting by the best ranked element (i_1^{PR}). For each collection point $i_j^{PR} \in I^{PR}$, each of the three metrics evaluated (volume of the waste collected, distance from the generators to the assigned collection points, and installation costs) are evaluated for each possible collection point configuration.

Three constructive heuristics were defined by prioritizing one of the three objectives of the GAP location problem:

- *Pagerank-Vol*, selects the configuration that collects the maximum volume of waste from the nearby generators. If more than one GAP configuration collect the same maximum of waste, the one with the cheapest installation cost is selected;
- *Pagerank-Dist*, considers the solutions that collects at least all the generated waste by the nearest generator, then it selects the one that provokes the users walk the shortest distance. If more than one configuration have the same minimum distance, the one that collects the maximum volume of waste is selected.
- *Pagerank-Cost*, evaluates the solutions that collects at least all the generated waste by the nearest generator, then it considers the one with the cheapest installation costs. If more than one configuration have the same minimum cost, the one that collects the maximum volume of waste is selected.

3.2   Multiobjective Evolutionary Algorithm

The MOEA applied in the study is *Non-dominated Sorting Genetic Algorithm, version II* (NSGA-II) [7]. NSGA-II is a state-of-the-art MOEA characterized by an evolutionary search using a non-dominated elitist ordering that diminishes the complexity of the dominance check, a crowding technique for diversity preservation, and a fitness assignment method considering dominance ranks and crowding distance values. It has successfully been applied in other waste management problems [22]. The main details of the proposed NSGA-II for the GAP location problem are presented next.

Solution encoding. Solutions are encoded as a vector of integers in a given range $[0, Z-1]$. Each index in the vector represent a possible GAP location, and the corresponding integer value represents one of the Z possible configurations (number of bins for each bin type) for locating bins in each GAP, according to the problem model and constraints for a given scenario.

Initialization. The population is initialized by applying a random procedure that selects a configuration for each GAP, according to a uniform distribution over the Z configurations defined for the problem scenario.

Selection, replacement, and fitness assignment. NSGA-II applies the $(\mu+\lambda)$ evolution model. Tournament selection is applied, with tournament size of two solution representations (*individuals*). The tournament criteria is based on dominance, and if the two compared individuals are non-dominated, the selection is made based on crowding distance. Fitness assignment is performed considering Pareto dominance rank and crowding distance values.

Evolutionary operators. The recombination operator is the standard two points crossover applied over two selected individuals with probability p_C. The mutation operator is based on modifying specific attributes (configurations) in an individual. Elements in a solution encoding are replaced by an integer value uniformly selected in the range $[0, Z-1]$, with probability p_M.

Solution feasibility. Feasibility of the solutions generated by applying the evolutionary operators is guaranteed by applying a greedy heuristic for deciding the GAP where citizens dispose their waste. Based on the intuitive behavior of citizens, also confirmed by surveys performed in the city [2], people always try to dispose waste in the nearest GAP, while there is space available in the located bins. When there is no free space on the nearest GAP, the remaining waste is disposed on the second near GAP, and so on.

4   Experimental Evaluation

This section reports the experimental analysis of the proposed computational intelligence methods to solve the real instances of the GAP location problem.

4.1   Problem Instances

The experimental evaluation was performed over two different real urban areas in Montevideo, Uruguay (Trouville and Villa Española neighborhoods). For each area, three different scenarios are considered according to variations of the waste generation rate along the year: a *normal demand* scenario, with the average generation rate estimated by Uruguayan authorities [19], a *high demand* scenario, and a *low demand* scenario, with generation rates 20% larger and smaller than the one from the *normal demand* scenario, respectively. This percentage variation is in line with surveys carried out to practitioners about what may occur along different periods of the year [2]. Trouville is an inner-city area that is densely populated (on average, 200 inhabitants per ha). It includes 82 generators and the same number of potential GAPs. Villa Española is a more sparsely populated suburban area (on average, 100 inhabitants per ha) that includes 70 generators and potential GAPs. The maximum distance between any generator in P and its assigned GAP is $D = 300$m, in line with suggestions for the maximum distance for accessing to public services.

Considering real information about bins used in the city, from the government of Montevideo [1], three bin types (j_1, j_2, and j_3) according to the normal usage in Montevideo. The values of parameters c_j, C_j, and e_j are: 1000 monetary units (m.u.), 1 m^3 and 1 m^2 for bin type j_1, 2000 m.u., 2 m^3 and 2 m^2 for bin type j_2, and 3000 m.u., 3 m^3 and 3 m^2 for bin type j_3. A set of 12 GAP configurations (number of bins of each type) were defined for the problem instances in Montevideo considering an available space for each GAP 5 m^2 (S_i). Their details, including the required space for installation, total cost, and capacity, are presented in Table 1.

Table 1. Feasible configurations according to the problem definition and GAP constraints ($S_i = 5$ for all collection points).

Config. id	Number of bins			Required space (m^2)	Installation cost (m.u.)	Maximum capacity (m^3)
	j_1	j_2	j_3			
0	0	0	0	0	0	0
1	1	0	0	1	1000	1
2	2	0	0	2	2000	2
3	3	0	0	3	3000	3
4	4	0	0	4	4000	4
5	5	0	0	5	5000	5
6	1	1	0	3	3000	3
7	1	2	0	5	5000	5
8	1	0	1	4	4000	4
9	0	1	0	2	2000	2
10	0	1	1	5	5000	5
11	0	0	1	3	3000	3

4.2 Methodology

Development and execution platform. The proposed methods were implemented in Python, by using Distributed Evolutionary Algorithms in Python (deap) framework [9]. The experimental evaluation was performed on a Dell PowerEdge M620 (Intel Xeon E5-2680 processor at 2.50GHz, 24 cores and 32 GB RAM) from Cluster FING, the High Performance Computing platform from Universidad de la República, Uruguay [20].

Metrics. Results computed by the proposed MOEAs are evaluated considering the *relative hypervolume* (RHV) metric, standard metrics for multiobjective optimization. RHV is defined as the ratio between the volumes (in the objective functions space) covered by the computed Pareto front and the true Pareto front. The ideal value for RHV is 1. In the analysis, the true Pareto front—unknown for the problem instances studied—is approximated by all non-dominated solutions found for each instance in each execution.

4.3 Numerical Results

Parameters Calibration. A set of parametric setting experiments were performed to determine the best parameter values for the proposed MOEA. The parameter setting analysis were made over three instances (different from validation scenarios) to avoid bias in the results.

The population size ($\#p$) and the maximum number of generations ($\#g$) were calibrated in preliminary experiments. The analysis confirmed that using $\#p = 100$ and $\#g = 1000$ provided a good exploration pattern, which allowed computing the best results. For the crossover probability (p_C) and the mutation probability (p_M), three different candidate values were defined and all combinations of p_C and p_M were studied on 30 independent executions performed for the proposed MOEA. Candidate values were $p_C \in \{0.5, 0.7, 0.9\}$ and $p_M \in \{0.01, 0.05, 0.1\}$. The result distributions obtained using each configuration were analyzed by applying the non-parametric Friedman rank statistical test to determine the configuration that allowed computing the best results.

Table 2 reports hypervolume values and the Friedman rank for each parameter configuration. The three best configurations were (0.9, 0.01), (0.7, 0.01), and (0.5, 0.01), with $\chi^2 = 231.9$ and p-value $<10^{-10}$. A post-hoc analysis applying the Wilcoxon rank test for pairwise comparisons reported that $p_C = 0.9$, $p_M = 0.01$ is the best configuration with statistical confidence of 0.97.

Table 2. Hypervolume results and Friedman rank for different parameter configurations of the proposed MOEA

p_C	p_M	Friedman rank	Minimum	Median	Maximum
0.5	0.01	7.30	0.954960	0.974518	0.954960
0.5	0.05	4.27	0.911992	0.945057	0.911992
0.5	0.1	1.50	0.857651	0.893930	0.857651
0.7	0.1	2.10	0.865003	0.899056	0.865003
0.7	0.01	7.70	0.969649	0.978279	0.969649
0.7	0.05	5.03	0.916363	0.953600	0.916363
0.9	**0.01**	**8.47**	0.969440	**0.982455**	0.969440
0.9	0.05	6.13	0.947243	0.962788	0.947243
0.9	0.1	2.50	0.879467	0.905054	0.879467

Multiobjective Optimization Analysis. Figure 1 presents sample Pareto fronts (3D view) computed by the proposed MOEA for Trouville and Villa Española scenarios with normal waste generation, which are representative of the results computed for the other waste generation rates. The results computed by the Pagerank heuristics are also reported for comparative purposes.

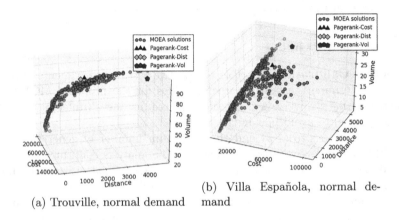

(a) Trouville, normal demand

(b) Villa Española, normal demand

Fig. 1. Sample MOEA Pareto fronts and PageRank solutions

Table 3. Hypervolume results for different proposed problem instances

Scenario	Waste generation	Minimum	Median	Maximum
Trouville	Low demand	0.848981	0.915520	0.959794
	Normal demand	0.835673	0.918477	0.971029
	High demand	0.856235	0.924728	0.973995
Villa Española	Low demand	0.917414	0.938922	0.960938
	Normal demand	0.911611	0.943481	0.969071
	High demand	0.875456	0.941840	0.965635

Results in Fig. 1 indicate that the proposed MOEA is able to accurately sample the set of trade-off solutions for the problem instances studied. Pagerank-Dist and Pagerank-Cost computed accurate solutions regarding both distance and cost objectives, while PageRank-Vol computed significantly worst solutions (i.e., with the worst cost and QoS). The proposed MOEA takes advantage of the evolutionary search to compute better solutions than PageRank regarding distance and cost, while also sampling properly the Pareto front of the problem.

Table 3 reports the hypervolume results obtained by the proposed MOEA for each of the problem instances studied. The median value is used as estimator, because results do not follow a normal distribution, as confirmed by the Kolmogorov-Smirnov statistical test applied to analyze the results.

Results in Table 3 indicate that the proposed MOEA is able to consistently compute accurate solutions for all problem instances studied. Hypervolume results were over 0.91 for all cases. The best hypervolume results computed were 0.96 or superior, demonstrating the robustness of the proposed evolutionary search. Overall, results show that Trouville scenarios are more difficult to solve the Villa Española scenarios, mainly because they involve a larger number of generators and potential locations for GAPs. On the other hand, waste generation rates did not affect significantly the hypervolume results.

Comparative Analysis. Table 4 reports the improvements of the MOEA results over the PageRank heuristics. The reported values accounts for the average and best improvements in each one of the three problem objectives (distance, cost, and volume) over each Pagerank solution, computed over those solutions in the Pareto front that dominate the corresponding PageRank solution in distance and cost objectives, and has up to 10% difference on the volume of the collected waste. Given that most of the waste is collected in the solutions computed by all PageRank heuristics and the MOEA, the analysis is focused on the benefits for both citizens (i.e., QoS, given by the average distance they must walk to dispose the waste) and the city administration (evaluating the cost of implementing a certain GAP planning).

Table 4. Improvements of the MOEA solutions over the PageRank solutions

Scenario	Waste generation	Baseline	Average improvement			Best improvement	
			Distance	Cost	Volume	Distance	Cost
Trouville	Low demand	PageRank-Cost	6.0%	8.0%	8.7%	15.6%	13.6%
		PageRank-Dist	9.9%	7.3%	8.9%	33.5%	14.1%
		PageRank-Vol	44.0%	17.7%	5.1%	79.5%	37.2%
	Normal demand	PageRank-Cost	16.8%	9.4%	6.7%	38.0%	20.0%
		PageRank-Dist	18.0%	9.9%	6.6%	36.5%	26.6%
		PageRank-Vol	44.8%	23.8%	4.7%	76.3%	44.4%
	High demand	PageRank-Cost	18.1%	10.4%	8.5%	33.3%	20.2%
		PageRank-Dist	14.8%	10.5%	8.4%	22.1%	21.0%
		PageRank-Vol	47.5%	18.9%	4.0%	80.7%	35.5%
Villa Española	Low demand	PageRank-Cost	9.3%	0.0%	9.1%	9.3%	0.0%
		PageRank-Dist	16.8%	12.1%	4.4%	36.6%	37.1%
		PageRank-Vol	23.7%	14.0%	3.8%	59.1%	33.3%
	Normal demand	PageRank-Cost	2.3%	9.2%	7.1%	4.07%	18.37%
		Pagerank-Dist	19.8%	10.8%	6.0%	40.0%	22.2%
		PageRank-Vol	31.8%	13.0%	4.8%	66.0%	25.6%
	High demand	PageRank-Cost	10.0%	7.4%	6.2%	11.7%	7.4%
		PageRank-Dist	16.8%	12.4%	5.6%	31.7%	25.3%
		PageRank-Vol	36.3%	19.8%	5.1%	69.7%	37.0%

The analysis of results in Table 4 allows concluding that the proposed MOEA is able to compute solutions that account for significant improvements over the PageRank algorithms. Regarding the distance objective, the average

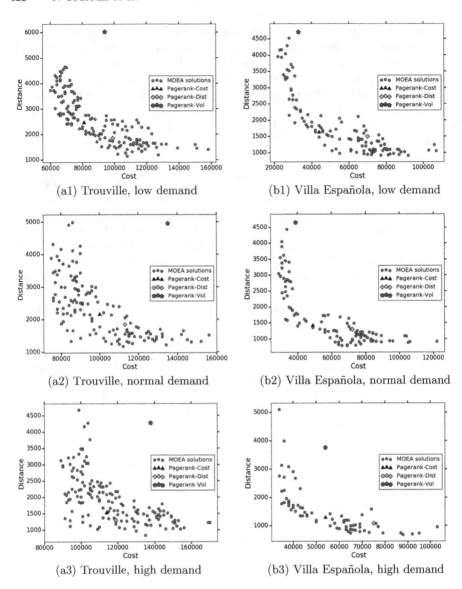

(a1) Trouville, low demand   (b1) Villa Española, low demand

(a2) Trouville, normal demand   (b2) Villa Española, normal demand

(a3) Trouville, high demand   (b3) Villa Española, high demand

Fig. 2. 2D cuts of the Pareto fronts (distance and cost objectives)

improvement over PageRank-Dist was up to 12.5% (Villa Española, high demand scenario) and the best improvement was up to 40.0% (Villa Española, normal demand scenario). Regarding the cost objective, the average improvement over PageRank-Cost was up to 10.4% (Trouville, high demand scenario) and the best improvement was up to 38.0% (Trouville, normal demand scenario). Overall, the solutions computed by the proposed MOEA outperformed the PageRank solutions in all but one case (PageRank-Cost, Villa Española, low demand scenario).

Figure 2 presents 2D cuts of the Pareto fronts obtained by the proposed MOEA and the results computed by the Pagerank heuristics. Results correspond to those solutions in the Pareto front that has a ±10% difference on the volume of the collected waste. Results in Fig. 2 demonstrate that the proposed MOEA is able to properly sample the Pareto front of the problem, computing solutions that improve over PageRank algorithms. Significant improvements are obtained regarding both distance and cost objectives, especially for scenarios with normal and high waste demand.

The superiority of Pagerank-Dist and Pagerank-Cost over PageRank-Vol solutions is also shown clearly. The proposed MOEA is able to compute solutions that dominate the best PageRank solution, improving both distance and cost, for all studied scenarios.

Comparison with Current GAP Locations in Montevideo. The results computed by the proposed computational intelligence methods were also compared with the current location planning for GAPs according to the government of Montevideo [1]. Table 5 reports the improvements of the best compromise solution (the closest solution to the ideal vector [7]) computed by the MOEA and the solutions of PageRank-Cost and PageRank-Dist. Results are reported for each scenario, considering different waste generation rates and the real GAP location and configuration in Montevideo, as of February, 2018.

Table 5. Comparison of the proposed algorithms with the real solutions

		Trouville			Villa Española		
		Low demand	Normal demand	High demand	Low demand	Normal demand	High demand
MOEA	Distance	84.6%	84.2%	84.2%	86.0%	86.3%	87.2%
	Cost	31.4%	23.6%	10.7%	2.3%	−4.4%	4.1%
	Volume	4.5%	4.0%	3.3%	4.1%	3.3%	4.8%
PageRank-Cost	Distance	84.8%	85.4%	87.6%	89.5%	90.1%	90.7%
	Cost	38.2%	23.7%	13.0%	24.1%	15.5%	6.9%
	Volume	0.0%	0.0%	0.0%	0.0%	0.0%	0.0%
PageRank-Dist	Distance	88.6%	87.5%	89.4%	90.3%	90.9%	92.2%
	Cost	24.4%	13.7%	9.2%	−20.7%	−24.1%	−29.3%
	Volume	0.0%	0.0%	0.0%	0.0%	0.0%	0.0%

Regarding the distance objective, the solutions computed by the proposed approaches account for significant improvements over the current GAP location in Montevideo. For Trouville scenarios, with the current GAP location a citizen must walk between 150.81–199.75 m to dispose the waste, while in the MOEA solution the distance is reduced to 28.75–23.84m (average reduction of 84%). For the Villa Española scenario, distances are reduced 86% on average, from 167–188 m to 21–26 m. The cost objective is also improved in most cases, although there are certain exceptions, such as the normal demand case in the Villa Española scenario for the MOEA and the same scenario with any demand patrons for the PageRank-Dist. The PageRank algorithms collect all the available waste while the MOEA leaves on average 4% uncollected.

5   Conclusions and Future Work

This article introduced an optimization model for locating GAPs in an urban area aimed at maximize the collected waste and the QoS, and minimize the investment cost. This is a relevant problem in modern smart cities as the initial stage in MSW system, having an important impact in both the QoS provided to the citizens and the budget expenses of the city authorities.

Computational intelligence techniques, namely three PageRank greedy algorithm and a MOEA, were proposed to solve the problem. Three PageRank methods were devised, focusing on the different objectives considered in the problem model. The MOEA applies a Pareto-based evolutionary search, allowing to compute solutions with different trade-off between the problem objectives.

The experimental evaluation was performed on real scenarios of Montevideo, Uruguay, built over real geographic locations (Trouville and Villa Española) and using real data about waste bin types used in the city and their features. Three waste generation patterns (normal, low, and high demand) were also considered to represent the expected fluctuations of waste generation rate along the year. Results showed that MOEA outperformed PageRank in all the studied scenarios regarding both distance and cost objectives, which account for the main QoS metric from the point of view of the citizens (the average distance they must walk to dispose the waste) and the main metric for the city administration (installation cost of GAPs). Regarding distance, MOEA improvements over PageRank were 12.5% in average, and up to 40.0%, in Villa Española scenarios. Regarding cost, MOEA improvements over PageRank were 10.4% in average, and up to 38.0%, in Trouville scenarios.

The Pareto front analysis confirmed that the proposed MOEA was able to compute accurate solutions, with different trade-off values between the problem objectives. MOEA dominated PAgeRank in all but one scenario. The proposed computational intelligence approaches also improve over the current solution applied by the authorities. Regarding the distance objective, the best MOEA results computed allow reducing from 150m to 25m the average distance that a citizen must walk to dispose the waste, improving in 6.9% the installation cost and maintaining the waste volume collected. This is an important result that suggests specific benefits for the city in order to apply a smart MSW system.

The main lines for future work are related to extend the experimental evaluation of the proposed algorithms on other realistic scenarios and include specific features such as the aleatory nature of the generation waste, following a more integral approach through stochastic programming. The inclusion of uncertainty in the decision making process will enhance the robustness of the solutions.

Acknowledgments. The work of J. Toutouh has been partially funded by MINECO and FEDER projects TIN2014-57341-R, TIN2016-81766-REDT, and TIN2017-88213-R, Spain. University of Malaga. International Campus of Excellence Andalucia TECH. The work of S. Nesmachnow is partly supported by ANII and PEDECIBA, Uruguay. The work of D. Rossit is partly funded by the Department of Engineering of Universidad Nacional del Sur, Argentina.

References

1. Arriba un montevideo más limpio (2018). http://www.montevideo.gub.uy/arriba-un-montevideo-mas-limpio
2. Compromise for recycling (2017). http://www.cempre.org.uy
3. Bautista, J., Pereira, J.: Modeling the problem of locating collection areas for urban waste management. An application to the metropolitan area of Barcelona. Omega **34**(6), 617–629 (2006)
4. Boskovic, G., Jovicic, N.: Fast methodology to design the optimal collection point locations and number of waste bins: A case study. Waste Manag. Res. **33**(12), 1094–1102 (2015)
5. Brahim, M.B., Drira, W., Filali, F.: Roadside units placement within city-scaled area in vehicular ad-hoc networks. In: 3rd International Conference on Connected Vehicles and Expo, pp. 1–7 (2014)
6. Chang, N.B., Wei, Y.: Siting recycling drop-off stations in urban area by genetic algorithm-based fuzzy multiobjective nonlinear integer programming modeling. Fuzzy Sets Syst. **114**(1), 133–149 (2000)
7. Deb, K.: Multi-objective Optimization Using Evolutionary Algorithms. Wiley, New York (2001)
8. Di Felice, P.: Integration of spatial and descriptive information to solve the urban waste accumulation problem: a pilot study. Procedia - Soc. Behav. Sci. **147**, 592–597 (2014). https://doi.org/10.1016/j.sbspro.2014.07.636
9. Fortin, F.A., De Rainville, F.M., Gardner, M.A., Parizeau, M., Gagné, C.: DEAP: evolutionary algorithms made easy. J. Mach. Learn. Res. **13**, 2171–2175 (2012)
10. Ghiani, G., Laganà, D., Manni, E., Triki, C.: Capacitated location of collection sites in an urban waste management system. Waste Manag. **32**(7), 1291–1296 (2012)
11. Ghiani, G., Manni, A., Manni, E., Toraldo, M.: The impact of an efficient collection sites location on the zoning phase in municipal solid waste management. Waste Manag. **34**(11), 1949–1956 (2014). https://doi.org/10.1016/j.wasman.2014.05.026
12. Goulart Coelho, L.M., Lange, L.C., Coelho, H.M.: Multi-criteria decision making to support waste management: a critical review of current practices and methods. Waste Manag. Res. **35**(1), 3–28 (2017). https://doi.org/10.1177/0734242X16664024
13. Hemmelmayr, V.C., Doerner, K.F., Hartl, R.F., Vigo, D.: Models and algorithms for the integrated planning of bin allocation and vehicle routing in solid waste management. Transp. Sci. **48**(1), 103–120 (2013)
14. Hoornweg, D., Bhada-Tata, P., Kennedy, C.: Peak waste: when is it likely to occur? J. Ind. Ecol. **19**(1), 117–128 (2015)
15. Langville, A.N., Meyer, C.D.: Google's PageRank and Beyond: The Science of Search Engine Rankings. Princeton University Press, Princeton (2011)
16. Lindell, M.K., Earle, T.C.: How close is close enough: public perceptions of the risks of industrial facilities. Risk Anal. **3**(4), 245–253 (1983)

17. Massobrio, R., Toutouh, J., Nesmachnow, S., Alba, E.: Infrastructure deployment in vehicular communication networks using a parallel multiobjective evolutionary algorithm. Int. J. Intell. Syst. **32**(8), 801–829 (2017)
18. Megiddo, N., Tamir, A.: On the complexity of locating linear facilities in the plane. Oper. Res. Lett. **1**(5), 194–197 (1982)
19. Ministerio de Vivienda, Ordenamiento Territorial y Medio Ambiente: Plan de gestión de Montevideo para la recuperación de residuos de envases no retornables. Government of Uruguay (2012)
20. Nesmachnow, S.: Computación científica de alto desempeño en la Facultad de Ingeniería, Universidad de la República. Revista de la Asociación de Ingenieros del Uruguay **61**(1), 12–15 (2010)
21. Nesmachnow, S.: An overview of metaheuristics: accurate and efficient methods for optimisation. Int. J. Metaheuristics **3**(4), 320–347 (2014)
22. Nesmachnow, S., Rossit, D., Toutouth, J.: Comparison of multiobjective evolutionary algorithms for prioritized urban waste collection in Montevideo, Uruguay. Electron. Notes Discret. Math. **69**, 93–100 (2018) (in press)
23. Purkayastha, D., Majumder, M., Chakrabarti, S.: Collection and recycle bin location-allocation problem in solid waste management: a review. Pollution **1**(2), 175–191 (2015)
24. Valeo, C., Baetz, B.W., Tsanis, I.K.: Location of recycling depots with GIS. J. Urban Plan. Dev. **124**(2), 93–99 (1998). https://doi.org/10.1061/(ASCE)0733-9488(1998)124:2(93)

Fully Convolutional Neural Networks for Mapping Oil Palm Plantations in Kalimantan

Artem Baklanov[1,2]([⊠])(iD), Michael Khachay[3,4](iD), and Maxim Pasynkov[3]

[1] National Research University Higher School of Economics, St. Petersburg, Russia
[2] International Institute for Applied Systems Analysis (IIASA), Laxenburg, Austria
baklanov@iiasa.ac.at
[3] Krasovsky Institute of Mathematics and Mechanics, Ekaterinburg, Russia
{mkhachay,pmk}@imm.uran.ru
[4] Ural Federal University, Ekaterinburg, Russia

Abstract. This research is motivated by the global warming problem, which is likely influenced by human activity. Fast-growing oil palm plantations in the tropical belt of Africa, Southeast Asia and parts of Brazil lead to significant loss of rainforest and contribute to the global warming by the corresponding decrease of carbon dioxide absorption. We propose a novel approach to monitoring of the development of such plantations based on an application of state-of-the-art Fully Convolutional Neural Networks (FCNs) to solve Semantic Segmentation Problem for Landsat imagery.

Keywords: Remote sensing · Mapping · Landsat · Fully convolutional neural network

1 Introduction

Oil palm is a dominant source of vegetable oil, which causes degradation to the environment [2] and exceedingly threatens forest conservation in the tropics. To support sustainability of Oil Palm Plantations (OPPs), one should be able to monitor their development. Remote sensing is the most promising way to allow for independent monitoring in all oil producing regions. As suggested in [2], there are following applications of remote sensing in OPP monitoring: land-cover classification; change detection; tree counting; estimation of yield, age, biomass, carbon production; pest and disease detection. To study complex issues, e.g., how patterns of oil palm driven deforestation change due to zero-deforestation commitments, a fusion of the above applications is required [1].

There are basically two approaches to map OPPs: visual interpretation by experts and semi-automated methods. Visual interpretation of Landsat imagery was successfully used in [1,5,12] and relies on a rather easily detectable pattern:

This research was supported by Russian Science Foundation, grant no. 14-11-00109.

R. Battiti et al. (Eds.): LION 12 2018, LNCS 11353, pp. 427–432, 2019.
https://doi.org/10.1007/978-3-030-05348-2_35

OPPs are mostly *"organized in rectilinear patterns and co-occur with context indicators such as road networks"* [1].

Two approaches are suggested to map OPPs: visual interpretation ('manually' performed by experts) and semi-automated methods based on supervised learning procedures. Visual interpretation of Landsat imagery was successfully used in [1,5,12] and relies on a rather easily detectable pattern: OPPs are mostly '*organized in rectilinear patterns and co-occur with context indicators such as road networks*' [1].

The conventional semi-automated methods rely on classification algorithms to assign every pixel in an image to a specific class; these are so-called per-pixel classifiers. For these methods, human expertise is needed to generate training examples (reference data) for automated models to learn; see [2,12] for a review of the most recent related results for OPPs mapping. The limitation of existing semi-automated methods for OPPs mapping becomes obvious if we consider to which extent they are used. Namely, for high-resolution imagery, these methods are applied only to small regions, e.g., 637.2 km^2 [11], 1020 km^2 [9]. We are not aware of any OPPs map with resolution 30 m or better that covers entire countries and was obtained by semi-automated methods. Basically, the only sources for high-resolution OPPs maps for entire countries are infrequent publications with maps obtained by visual interpretation. In contrast, the situation with deforestation mapping is absolutely different: Hansen's deforestation map [7] is available for the entire world and updated yearly thanks to a semi-automated implementation. A conceptually similar map product for OPPs monitoring can be highly beneficial for sustainability science. The paper relies on the recent advances in deep convolutional networks to make the first step towards this product.

Convolutional Neural Networks (CNNs) are a widely known tool for solving problems of image analysis. Although CNNs are state of the art for *scene classification*, their applicability for *pixel-wise classification* (segmentation problems) appears to be limited; see discussion in [3, p. 5]. On the other hand, Fully Convolutional Neural Networks (FCNs) show promising results [3,13] by assigning to an image the collection of *heat map* with values equal to probabilities of classes. Thus, FCNs classify every pixel separately.

We propose the novel semi-automated method for OPPs monitoring based on the reduction of the initial problem to the well-known Semantic Segmentation Problem and subsequent solution of the latter with FCNs. In particular, we use the state-of-the-art FCN-8s network, showed one the best performance on the famous PASCAL-VOC competition [13]. This approach shows results that are promising for development of the advanced map product for OPPs.

Overview of the Method. The proposed method consists of the following stages:

(i) to train a FCN to make a segmentation on historical Landsat images, whose experts-made labeling is taken as the ground truth; in the simplest case, several types of plantations do not differ from each other, and we solve two-class semantic segmentation problem;

(ii) to make forecasts for the future (or past), we compare segmentation results of the trained network on query images with the expert mask.

The method is based on the recent advances in semantic segmentation by FCNs [4] and the assumption that visual cues (e.g. textures) of OPPs on Landsat imagery of a fixed area change slowly. Generally speaking, we may assign a separate visual class to any type of industrial plantations. In this paper, we focus only on a single foreground class C_1, *OPP* featuring an authentic visual pattern, and train an FCN to discriminate it against the background class C_0, *non-OPP*; see Fig. 1 for an example of semantic segmentation for class *OPP*.

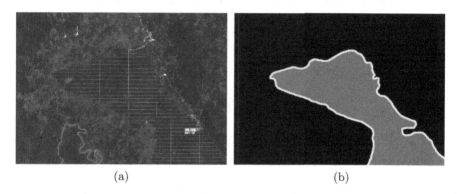

(a)   (b)

Fig. 1. The initial scene (a) and the corresponding true segmentation of the *OPP* (b)

Data. We illustrate our approach on an area in Kalimantan (Indonesia). To train and test the FCN, we used a cloud-free Landsat false color composite (5-4-3) acquired in 2014 [7] and publicly available[1] at Google Earth Engine (GEE). Each band has 30m resolution. The area of study is the rectangle[2] with the total area equal to 61170 km^2. As a picture, the rectangle's dimensions are 12246 × 5572 pixels. The true segmentation that we used for training and performance evaluation was the result of visual interpretation[3] [1] and is publicly available[4]. In the specified rectangle, this map reports that OPPs occupy 16630 km^2 meaning that accuracy of the naive classifier is equal to 0.728.

---

[1] Image ID: UMD/hansen/global_forest_change_2015_v1_3; Bands: 'last_50', 'last_40', 'last_30'.

[2] In GEE, rectangles are represented by the minimum and maximum corners as the list of four numbers in the order $xMin, yMin, xMax, yMax$. In our case $(xMin, yMin, xMax, yMax) = (110.2, -3, 113.5, -1.5)$.

[3] As reported in [1, Appendix A. SI Table 3], the overall accuracy, sensitivity (recall), and precision of the produced map are equal to $0.911, 0.969$, and 0.84, respectively.

[4] http://pure.iiasa.ac.at/14829/.

Performance Metrics. Performance of the algorithms was evaluated using the map [1] as the ground truth. For a two-class classification problem, Table 1 shows the confusion matrix helping to define the following performance metrics related to a particular class i: precision (P_i), recall (R_i), producer's accuracy (PA_i), and user's accuracy (UA_i); overall accuracy (OA) is an aggregate metric. Note that the notions of producer's and user's accuracy correspond to common definitions in machine learning:

$$P_i = UA_i = \frac{p_{i,i}}{p_{i,\cdot}}, \quad R_i = PA_i = \frac{p_{i,i}}{p_{\cdot,i}}, \quad OA = \frac{p_{1,1} + p_{0,0}}{p_{\cdot,\cdot}}, \quad i = 0, 1.$$

Table 1. Confusion matrix for a two-class classification problem

		Ground truth		
		Class C_1	Class C_0	Total
Prediction	Class C_1	$p_{1,1}$	$p_{1,0}$	$p_{1,\cdot}$
	Class C_0	$p_{0,1}$	$p_{0,0}$	$p_{0,\cdot}$
	Total	$p_{\cdot,1}$	$p_{\cdot,0}$	$p_{\cdot,\cdot}$

2   Methods and Results

For the semantic segmentation, we employ FCN-8s, which takes RGB 500×500 images as inputs. Therefore, the initial high-resolution image was sliced into more than 400 square tiles (of this size) with 250 pixels overlap. Since the conventional transfer learning approach (see, e.g., [8]) led to poor accuracy, we train the network from the scratch starting with a small number of instances that admitted an obvious segmentation and increasing the number of training samples gradually. Eventually, we trained 6 networks, for different intervals of foreground / background disbalance: *first100* containing starting tiles of fg/bg balance about 50%, *sel45-55*, *sel40-60*, *sel35-65*, *sel30-70*, and *sel25-75*, whose names encode the corresponding interval.

For all networks, UA_1 and PA_1 evaluated on the training samples exceed 97%. The values obtained for the entire rectangle are presented in Table 2.

Table 2. Accuracy assessment on testing data (%)

	first100	sel45-55	sel40-60	sel35-65	sel30-70	sel25-75
UA_1 (precision)	83.7	84.2	88.7	87.8	93.2	95.0
PA_1 (recall)	89.0	91.2	91.5	95.2	93.8	96.4

3   Conclusion and Future Work

We proposed the novel semi-automated method for OPPs monitoring based on the state-of-the-art network FCN-8s. The achieved performance is comparable with UAs and PAs reported in [1,12] for the visual interpretation method. The following steps may push the performance higher. First, we plan to utilize more bands available for Landsat imagery. Namely, we will test different combinations of 7 bands available in 30m resolution and one (panchromatic) band available in 15 m resolution. We will also explore if pan-sharpening may improve the results of mapping. Then we will produce OPPs map for entire Indonesia and compare it with existing products. Moreover, we aim to test thoroughly our approach as a tool for yearly (monthly) monitoring by evaluating how well our algorithm detects expansion of existing plantations and new ones. Figure 2 shows the application of the trained networks to the imagery obtained in 2016; the OPP expansion was successfully detected.

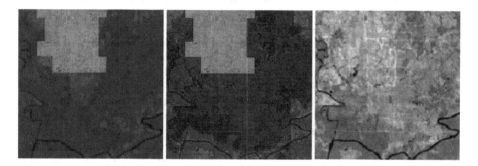

Fig. 2. The left, center, and right images are acquired in 2014, 2016, and 2016, respectively. The green region corresponds to the OPP detected in 2014. The blue region depicts the correctly detected expansion of the OPP as of 2016. The right image (false color composite: 5-4-3) demonstrates the rectilinear patterns of OPPs.

Further, we are interested to apply our methods to other regions. For example, in Peru, large-scale OPPs have the similar structure (see [6, Fig. 2]) to the one we observed in Kalimantan. Thus, one may expect that the trained net should perform well even without additional learning. In some other region, the complete retraining of the net or transfer learning may be needed. Based on the obtained results, we conjecture that the recent advances of FCNs make it feasible not only to develop the net that successfully detects more types of industrial plantations (e.g., coffee, rubber, and eucalyptus) but apply FCNs to a wider context of data mining in agriculture [10].

References

1. Austin, K., Mosnier, A., Pirker, J., McCallum, I., Fritz, S., Kasibhatla, P.: Shifting patterns of oil palm driven deforestation in Indonesia and implications for zero-deforestation commitments. Land Use Policy **69**, 41–48 (2017). https://doi.org/10. 1016/j.landusepol.2017.08.036, http://www.sciencedirect.com/science/article/pii/ S0264837717301552

2. Chong, K.L., Kanniah, K.D., Pohl, C., Tan, K.P.: A review of remote sensing applications for oil palm studies. Geo-Spat. Inf. Sci. **20**(2), 184–200 (2017). https:// doi.org/10.1080/10095020.2017.1337317

3. Fu, G., Liu, C., Zhou, R., Sun, T., Zhang, Q.: Classification for high resolution remote sensing imagery using a fully convolutional network. Remote Sens. **9**(5) (2017). https://doi.org/10.3390/rs9050498, http://www.mdpi.com/2072-4292/9/ 5/498

4. Garcia-Garcia, A., Orts-Escolano, S., Oprea, S., Villena-Martinez, V., Jose Garcia-Rodriguez, V.: A review on deep learning techniques applied to semantic segmentation. Manuscript 1 (2017)

5. Gaveau, D.L.A., et al.: Rapid conversions and avoided deforestation: examining four decades of industrial plantation expansion in Borneo. Sci. Rep. **6** (2016). https://doi.org/10.1038/srep32017

6. Gutiérrez-Vélez, V.H., DeFries, R.: Annual multi-resolution detection of land cover conversion to oil palm in the Peruvian Amazon. Remote Sens. Environ. **129**, 154–167 (2013). https://doi.org/10.1016/j.rse.2012.10.033, http://www.sciencedirect. com/science/article/pii/S003442571200421X

7. Hansen, M.C., et al.: High-resolution global maps of 21st-century forest cover change. Science **342**(6160), 850–853 (2013). https://doi.org/10.1126/science. 1244693, http://science.sciencemag.org/content/342/6160/850

8. Huang, Z., Pan, Z., Lei, B.: Transfer learning with deep convolutional neural network for SAR target classification with limited labeled data. Remote Sens. **9**(9) (2017). https://doi.org/10.3390/rs9090907, http://www.mdpi.com/2072-4291/9/ 9/907

9. Lee, J.S.H., Wich, S., Widayati, A., Koh, L.P.: Detecting industrial oil palm plantations on Landsat images with Google Earth Engine. Remote Sens. Appl.: Soc. Environ. **4**, 219–224 (2016). https://doi.org/10.1016/j.rsase.2016.11.003, https:// www.sciencedirect.com/science/article/pii/S235293851630129X

10. Mucherino, A., Papajorgji, P.J., Pardalos, P.M.: Data Mining in Agriculture, 1st edn. Springer Publishing Company, Incorporated (2009)

11. Nooni, I., Duker, A., Van Duren, I., Addae-Wireko, L., Osei Jnr, E.: Support vector machine to map oil palm in a heterogeneous environment. Int. J. Remote Sens. **35**(13), 4778–4794 (2014). https://doi.org/10.1080/01431161.2014.930201

12. Petersen, R., et al.: Mapping tree plantations with multispectral imagery: preliminary results for seven tropical countries. Technical report, World Resources Institute (2016). www.wri.org/publication/mapping-tree-plantations

13. Shelhamer, E., Long, J., Darrell, T.: Fully convolutional networks for semantic segmentation. IEEE Trans. Pattern Anal. Mach. Intell. **39**, 640–651 (2017). https:// doi.org/10.1109/TPAMI.2016.2572683

Calibration of a Water Distribution Network with Limited Field Measures: The Case Study of Castellammare di Stabia (Naples, Italy)

Armado Di Nardo[1], Michele Di Natale[1], Anna Di Mauro[1], Giovanni Francesco Santonastaso[1(✉)], Andrea Palomba[2], and Stefano Locoratolo[2]

[1] Dipartimento di Ingegneria, Università della Campania, via Roma 29, 81031 Aversa, Italy
giovannifrancesco.santonastaso@unicampania.it
[2] GORI Spa Gestione Ottimale Risorse Idriche, Via Trentola 211, 80056 Ercolano, Italy

Abstract. The great amount of big data provided to the water utilities from the application of smart technologies represents a key role for the future of water analysis and management. Water big data (WBD) can offer various ways of employment to support water network analysis and to achieve a smart management of the system. In this regard, one of the main application of WBD is the implementation and calibration of the hydraulic model of a water distribution network (WDN). Yet, to date, WDNs are still not fully-equipped and, consequently, WBD and IoT in water sectors are still limited. The paper presents a case study of calibration of Castellammare di Stabia, with limited field measures and a high level of water losses, highlighting the low calibration reliability without the availability of WBD about water demand and water losses.

Keywords: Hydraulic model · Calibration · Water demand · Water losses
Water big data · Smart water network

1 Introduction

The increasing of the technological development, the application of Information and Communication Technologies (ICT) and the diffusion of the Internet of Thing (IoT) made available a huge amount of data, Big Data (BD), thanks to which large amounts of information can be collected and updated in a short time [1].

In the water sector, more and more attention is being paid to the integration of innovative "smart technologies" (sensors, electronic meters, meters, etc.) to transform the traditional water networks in Smart Water Networks (SWAN) to promote smart water management solutions and improve decision-making methodologies.

In this regard, the availability of an hydraulic model represents an effective tool to support the management decisions about the hydraulic infrastructure. Anyway, the numerical model of a water distribution network (WDN) must be properly calibrated to reproduce operational conditions of the real systems [2].

© Springer Nature Switzerland AG 2019
R. Battiti et al. (Eds.): LION 12 2018, LNCS 11353, pp. 433–436, 2019.
https://doi.org/10.1007/978-3-030-05348-2_36

As known, the calibration process of a WDN model can be carried out with different algorithms [3, 4]. To this purpose, water big data (WBD) can offer a precious way to improve the calibration of network hydraulic models. Unfortunately, to date, the availability of WBD is poor and, in most cases the field data is limited to few sensors. This process is still ongoing due to the high cost of technologies, the problems related to sensors placement and communications and to the difficult in the connection between operational data and network models. Thus, in this work, a preliminary attempt to calibrate the hydraulic model of the WDN of Castellammare di Stabia, with a limited number of field measures and a high level of water losses, is investigate. The approach is developed also in order to suggest to Water Utility how to increase data collection – with the best location of flow and pressure sensors in the network – to improve the model calibration.

2 Case Study and Results

The water supply system of Castellammare di Stabia (NA), a city in the South of Italy with about 66,000 inhabitants, consists of 118 km in pipes that discharge an average flow equal to 450 l/s. The hydraulic model was build starting from topological and hydraulic data provided by GORI Spa (the Water Utility) that manages the water system. The available field measures are only the values of flow, measured downstream the feeding pipes of the network (the collected data concerns the time interval between 17/05/2016 and 17/05/2017).

Since the water balance of the water network of Castellammare, assessed according to Italian law, has shown a large percentage of leakage (about 68% of the total supplied volume), the water demand Q_i of i-th node was modeled as follows:

$$Q_i = d_i + l_i \tag{1}$$

where d_i and l_i are respectively the water supplied to the users and the physical leakage in the i-th node; specifically, as known, the physical losses can be described as function of the local pressure h_i of the i-th node [5]:

$$l_i = c \cdot h_i^{\gamma} \tag{2}$$

in which the values of the coefficients c and γ depend on site conditions.

The spatial distribution of water consumption was obtained from the analysis of billing data provided by the GORI Spa. Figure 1a shows the water network of Castellammare and the positioning of the few flow measuring stations. In this work, the problem of calibration was approached as an optimization problem, using different algorithms [6, 7], aimed to minimize the deviation between measures and values computed by the hydraulic simulations.

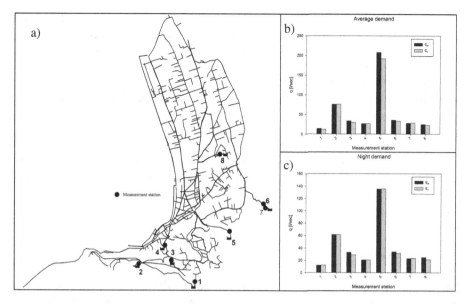

Fig. 1. a) Castellammare di Stabia water network; b) Comparison between measured and computed flows in the average demand; c) Comparison between measured and computed flows in night demand

To this purpose, the following objective function (*OF*) was used:

$$OF = \sqrt{\sum_{j=1}^{k} \left(q_{m,j} - q_{c,j}\right)^2} \tag{3}$$

In Eq. 3, k is the number of measurement stations, $q_{m,j}$ and $q_{c,j}$ are the measured and computed values of *j-th* flow, respectively. The minimization of *OF* was carried out by different heuristic optimization methods. In this short paper, the solutions carried out with a Genetic Algorithm (*GA*) [6] by adjusting, step by step, the pipe roughness and the spatial distribution of losses, were showed. Specifically, the calibration strategy proposed is based on:

(a) pipes were grouped according to the material and a roughness coefficient was assigned to each group;
(b) the evaluation of the spatial distribution of losses was obtained specifying the value of the coefficient c for each node: c = 0 for no leakage in the node, c $\neq$ 0 for the presence of leakage in the node. The coefficients of Eq. 2 were considered as constant values for the Castellammare di Stabia water network ($c = 0.05$ and $\gamma = 0.5$);
(c) to take into account the regulation of the outlet flow from Suppezza reservoir during the night, two scenarios were considered: Average demand and Night demand; d) in order to satisfy a good level of service for the users in terms of supplied demand, different values of minimum pressure were ensured for the different areas of Castellammare WDN.

The hydraulic simulation of the network was carried out with the solver EPANET 2 [8]. In Fig. 1b, c, the comparison between measured flow, q_m, and the computed value, q_c, provided by the hydraulic simulation, are reported in both scenarios. The maximum error between measured and computed values is 15.63 l/s for the average demand and 4.25 l/s for the night demand.

Although the results seem significant they cannot be considered reliable due to few measurement stations, no pressure measurements in the water network and scarce information about demand and leakage model.

The last aspect is crucial; in fact, the results of the case study show the possibility to minimize the error even using few measures but it does not allow to evaluate the reliability of the model calibration. In this regard, a data collection of fine-resolution demand [7] and water losses distribution at nodes or demand points [9] is essential to improve the model reliability.

3 Conclusions

The study analyzed the possibility to calibrate the hydraulic model of Castellammare di Stabia with few data of flows. The results, obtained with *GA*, in terms of errors between measured and computed values appear satisfactory with very few field measurements. Nevertheless, the reliability of the calibration can be very low because the knowledge about water demand and water losses models are too limited. In this regard, the availability of WBD of the water systems is crucial to improve the calibration of network hydraulic models.

Further studies will investigate the effectiveness of other heuristic algorithms but, above all, the improvement of the calibration reliability in case of WBD are available

References

1. Candelieri, A., Archetti, F.: Smart water in urban distribution networks: limited financial capacity and big data analytics. Wit Trans. Built. Env. **139**, 63–73 (2014)
2. Alvesa, Z., Muranhob, J., Albuquerquee, T., Ferreira, A.: Water distribution network's modeling and calibration. A case study based on scarce inventory data. In: 12th International Conference on Computing and Control for the Water Industry, CCWI2013, pp. 31–40 (2013)
3. Lingireddy, S., Ormsbee, L.E.: Hydraulic network calibration using genetic optimization. Civ. Eng. Environ. Syst. **19**(1), 13–39 (2002)
4. Reddy, P.V.N., Sridharan, K., Rao, P.V.: WLS method for parameter estimation in water distribution networks. J. Water Resour. Plan. Manag. **122**(3), 157–164 (1996)
5. Jowitt, P.W., Xu, C.: Optimal valve control in water distribution networks. J. Water Resour. Plan. Mange. **116**(4), 455–472 (1990)
6. Goldberg, D.E.: Genetic Algorithms in Search, Optimization and Machine Learning. Addison-Wesley Longman Publishing Co., Inc., Boston (1989)
7. Cominola, A., Spang, E.S., Giuliani, M., Castelletti, A., Lund, J.R., Logebdz, F.J.: Segmentation analysis of residential water-electricity demand for customized demand-side management programs. J. Clean. Prod. **172**, 1607–1619 (2018)
8. Rossman, L.A.: EPANET2 users manual. US EPA, Cincinnati (2000)
9. Di Nardo, A., Di Natale, M., Gisonni, C., Iervolino, M.: A genetic algorithm for demand pattern and leakage estimation in a water distribution network. J. Water Supply Res. T, 35–46 (2015)

Combinatorial Methods for Testing Communication Protocols in Smart Cities

Dimitris E. Simos[1]([✉]), Ludwig Kampel[1], and Murat Ozcan[2]

[1] SBA Research, 1040 Vienna, Austria
{dsimos,lkampel}@sba-research.org
[2] Siemens Building Technologies CPS Software House, Chicago, IL, USA
murat.ozcan@siemens.com

Abstract. In this paper, we conduct a feasibility study for combinatorial methods applied to widely used communication protocols in smart city environments. Even though, our initial results reveal no failures in the involved products, the approach looks promising in terms of software reuse and test efficiency.

Keywords: Combinatorial testing · BACnet protocol · Smart cities

1 Introduction

A report conducted by the Centre of Regional Science at the Vienna University of Technology identified six main axes (dimensions) along which a ranking of 70 European middle size cities can be made [3]. These axes are: a smart economy; smart mobility; a smart environment; smart people; smart living; and, finally, smart governance.

Moreover, a study documented in [6] shows that Americans spend 87% of their time indoors. This gives us sufficient justification to consider smart buildings a main factor of smart cities. In modern smart buildings various hardware pieces and processes, such as lights, air vants, air condition, fire alarm, water etc., have to be properly coordinated (and centrally controlled). To be able to facilitate the communication between these interacting parts, they also have to be compliant to a common communication protocol.

In this paper, we focus on the Building Automation and Control Networks Protocol (BACnet) which is an interoperable communication protocol for building automation and control networks. When a piece of hardware or software is BACnet compliant, this means it can communicate with the hardware and software of any other BACnet compliant vendor. BACnet compliance is an essential aspect of every research and development project at major corporations like Siemens Building Technologies, Honeywell, Johnson Controls, Schneider Electric and others.

Our motivation for this work is to propose a combinatorial method for testing communication protocols in buildings, as a means to provide an intelligent

© Springer Nature Switzerland AG 2019
R. Battiti et al. (Eds.): LION 12 2018, LNCS 11353, pp. 437–440, 2019.
https://doi.org/10.1007/978-3-030-05348-2_37

way which could increase the quality of smart buildings in smart cities. As a proof-of-concept of this early stage devised methodology we develop a combinatorial model for the widely used APOGEE Insight® workstation which acts as a BACnet client.

2   Combinatorial Models of the BACnet Protocol

For BACnet testing the standard processes so far has evolved around exhaustive testing, and user-defined specification testing (e.g. by contractual work with third parties). Hence our first steps for proposing a thorough testing methodology for the BACnet protocol is novel in that sense and also could aid the practitioners by utilizing specific advantages of combinatorial testing, which are explained below.

Introduction to Combinatorial Testing. Combinatorial testing (CT) is a highly sophisticated testing methodology, that is capable of producing comparable small test sets (e.g. when compared to exhaustive testing), while at the same time providing guarantees of (certain) input space coverage. To apply CT to a system under test (SUT), it is necessary to have an *input parameter model* (IPM) of the SUT [5]. To devise an IPM for an SUT it is necessary to identify *input parameters* and their respective *values*, such that an input to such a model can be represented by parameter-value assignments. Engineering an IPM can be a tedious and time intensive task itself [2]. The underlying mathematical primitives of CT are covering arrays (CAs), which are discrete structures appearing in *combinatorial design theory*, and can be represented as matrices with specific coverage properties [7]. To further apply CT, the parameters of the IPM are matched with the columns of an appropriate CA, such that a row of the CA can be interpreted as an assignment of values to the parameters of the IPM. Translating all rows of a CA in such a way, the mathematical properties of CAs guarantee that the generated test set is a *t-way test set*, i.e. a test set which ensures that all *t*-way combinations of parameter-value assignments are tested, once all tests have been executed [7]. Note that the general problem of constructing a t-way test set is believed to be NP-hard as it is tightly coupled with hard combinatorial optimization problems (see [4]). A study from the National Institute of Standards and Technology (NIST) [7] shows that in all tested software products all faults rely on the interaction of at most six input parameters. This means that all faults in the tested software products can be triggered using a 6-way interaction test set, which is generally much smaller than an exhaustive test set, but yet achieving the same testing quality.

Application to the BACnet Protocol via the APOGEE Insight® Product. We propose to apply CT to the BACnet client as this is utilized by the APOGEE Insight® workstation, a commercial tool manufactured by Siemens Inc., in order to reduce the time and costs for the testing cycle. Considerably, as the APOGEE Insight® has to be tested newly for every vendor the test setup needs a lot of time and thus, resources.

A *BACnet Event* can be characterized as any change in the value of any property of any object that meets a particular criteria. The purpose of an Event Enrollment Object (EEO) is to define an event and offer the engineer an association with the occurring event and the transmission of notification messages.

Vendors and building control products of those vendors may have varying implementations for EEO configurations in the user interface (UI). Until recently, the Siemens legacy building control product, Siemens APOGEE Insight® Workstation, could have above 5 million ways to configure an EEO. Given that one configuration takes one second to execute (a conservative estimate), the total effort spent towards exhaustive testing makes apparent the need for more intelligent testing methodologies. Due to its complicated nature the EEO configuration was previously left open for the user and most of these configurations could be invalid. If the configuration was invalid at the workstation, once the EEO was downloaded to the field panel, it would show a configuration error for the EEO. The EEO would then need to be modified at the field panel or it would have to be modified at the workstation and re-downloaded to the panel.

For practical testing, to be able to execute the generated test set, one needs to be able to input EEOs into the System. This is however enabled by the Harmonization tool of APOGEE Insight®. The steps below, describe a high-level view of a CT methodology for testing the EEOs.

1. Devise IPMs for the EEO configurations.
2. Generate t-way test sets using an intelligent CT generation algorithm (e.g. the IPO strategy as this is implemented by the widely used ACTS tool [8]).
3. Then the test sets are fed to the Harmonization Tool.
4. The EEO configurations are downloaded to the field panels, where the field panels should show configuration OK if the EEOs were configured correctly.
5. Finally the EEO configurations can be bulk-uploaded to the workstation from the panel. EEOs should show configuration OK if they were configured correctly.

Due to space constraints the devised IPMs for the EEOs and the derived 2-way test sets are available upon request. We briefly, sketch below how an IPM for a reduced EEO configuration looks like together with a (reduced) test set:

An EEO can be modelled consisting of the following parameters that can be configured assigning the respective subsequent values to them:

- *ObjectType* (OT): Analog-Input (AI), Analog-Output (AO)
- *EventType* (ET): Floating-Limit (FL), Out-of-Range (OoR)
- *SetPointType* (SPT): Schedule (S), Trend-Log (TL)
- *SetPointProperty* (SPP): Start-Time (ST), Present-Value (PV)

A 2-way test set for this IPM of the reduced EEO can be attained by computing a (binary) covering array (of *strength* 2) with four columns and replacing the entries in the columns with the values of the corresponding parameter. The resulting test set covers all 2-way combinations of parameter-value assignments, see Fig. 1.

c_1	c_2	c_2	c_2		OT	ET	SPT	SPP
0	0	0	1		AI	FL	S	PV
0	1	1	1		AI	OoR	TL	PV
1	0	1	1		AO	FL	TL	PV
1	1	0	0		AO	OoR	S	ST
0	0	1	0		AI	FL	TL	ST

Fig. 1. On the left hand side a covering array (of strength 2), and on the right hand side the derived 2-way test set for the IPM of the reduced EEO.

3   Further Remarks

The initial challenge in this study was how to best test EEO configurations in the most efficient manner. We have conducted sample experiments with the derived test sets from our combinatorial models that are still ongoing. So far approximately 1000 test cases have been executed versus the APOGEE Insight®, however an initial investigation of the test results revealed no faults. Nevertheless, a final conclusion can only be made once the whole testing cycle is completed.

In addition, as future work we want to devise real world tests, also for other tools besides APOGEE Insight® for the BACnet protocol, and shift focus to security in smart cities [1].

Acknowledgment. This research was carried out in the context of the Austrian COMET K1 program and publicly funded by the Austrian Research Promotion Agency (FFG) and the Vienna Business Agency (WAW).

References

1. Baig, Z.A., et al.: Future challenges for smart cities: cyber-security and digital forensics. Digit. Investig. **22**, 3–13 (2017)
2. Bartholomew, R., Collins, R.: Using combinatorial testing to reduce software rework. CrossTalk **23**, 23–26 (2014)
3. Centre of Regional Science: Smart cities, Ranking of European medium-sized cities (2018). http://www.smart-cities.eu/download/smart_cities_final_report.pdf. Accessed 10 Feb 2018
4. Cheng, C.T.: The test suite generation problem: optimal instances and their implications. Discret. Appl. Math. **155**(15), 1943–1957 (2007)
5. Grindal, M., Offutt, J.: Input parameter modeling for combination strategies. In: Proceedings of the 25th Conference on IASTED International Multi-Conference: Software Engineering, pp. 255–260. SE'07, ACTA Press, Anaheim, CA, USA (2007)
6. Klepeis, N.E., et al.: The national human activity pattern survey (NHAPS): a resource for assessing exposure to environmental pollutants. J. Expo. Sci. Environ. Epidemiol. **11**(3), 231 (2001)
7. Kuhn, D., Kacker, R., Lei, Y.: Practical combinatorial testing. NIST Special Publication, pp. 800–142 (2010)
8. Yu, L., Lei, Y., Kacker, R.N., Kuhn, D.R.: ACTS: A combinatorial test generation tool. In: 2013 IEEE Sixth International Conference on Software Testing, Verification and Validation, pp. 370–375. IEEE (2013)

Pseudo-pyramidal Tours and Efficient Solvability of the Euclidean Generalized Traveling Salesman Problem in Grid Clusters

Michael Khachay[1,2,3]([✉]) [iD] and Katherine Neznakhina[1,2] [iD]

[1] Krasovsky Institute of Mathematics and Mechanics, Ekaterinburg, Russia
{eneznakhina,mkhachay}@imm.uran.ru
[2] Ural Federal University, Ekaterinburg, Russia
[3] Omsk State Technical University, Omsk, Russia

Abstract. Generalized Traveling Salesman Problem (GTSP) is a well-known combinatorial optimization problem having numerous applications in operations research. For a given edge-weighted graph and a partition of its nodeset onto k (disjoint) clusters it is required to find a minimum cost cyclic tour visiting all the clusters once. The problem is strongly NP-hard even in the Euclidean plane provided the number of clusters is a part of the instance. Recently we proposed efficient optimal algorithms for GTSP based on quasi- and pseudo-pyramidal tours. As a consequence, we proved polynomial time solvability of the Euclidean GTSP in Grid Clusters defined by a grid of height at most 2. In this short paper, we show how to extend this result to the case defined by grids of an arbitrary fixed height.

Keywords: Generalized traveling salesman problem
Pseudo-pyramidal tour · Polynomial time solvability

1 Introduction

The Generalized Traveling Salesman Problem (GTSP) is a widely known extension of the famous Traveling Salesman Problem (TSP) (see, e.g. [4]). In addition to a complete edge-weighted graph $G = (V, E, c)$, any instance of the GTSP is defined by some partition $V_1 \cup \ldots \cup V_k = V$ of its nodeset V. The goal is to find a minimum cost cyclic tour visiting each cluster V_i at a single node. Enclosing the TSP, the GTSP is also strongly NP-hard even on the Euclidean plane provided the number of clusters k belongs to the instance. On the other hand, for any fixed k, the GTSP can be solved to optimality in time $O((k-1)!n^3)$ by exhaustive search over all permutations of the clusters [3].

This research was supported by Russian Science Foundation, grant no. 14-11-00109.

R. Battiti et al. (Eds.): LION 12 2018, LNCS 11353, pp. 441–446, 2019.
https://doi.org/10.1007/978-3-030-05348-2_38

We consider a geometric setting of the GTSP known as the Euclidean Generalized Traveling Salesman Problem in Grid Clusters (EGTSP-GC), which was introduced recently in [1] by Bhattacharya et al. In EGTSP-GC, nodes of the graph G are points on the plane, edges are weighted by Euclidean distances between the incident nodes, and clusters are defined implicitly by non-empty cells of the unit grid of some height h and width w (Fig. 1).

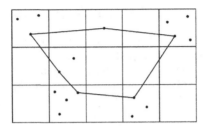

Fig. 1. EGTSP-GC: an instance for $n = 17$ and $k = 6$, and one of its optimal solutions

As the GTSP, the problem under consideration is strongly NP-hard in the general case, when k belongs to the input. In [1], for any $\varepsilon > 0$, a $(1.5 + 8\sqrt{2} + \varepsilon)$-approximation algorithm was proposed for this case of the EGTSP-GC. Augmented by some additional constraints, the problem may become approximable much better. For instance, the results of [2] imply that, for any instance defined by a grid with a fixed height h, such that the set of non-empty cells is connected, 2-approximate solution can be found in a polynomial time.

In [5,6], three polynomial time approximation schemes for slow and fast growing dependence of the number of clusters k on the size n of the node-set were proposed. Actually, first two of them have time complexity bounds of $O(k^2 O(1/\varepsilon)^{2k}) + O(n)$ and $2^{O(k)} k^4 (\log k)^{O(1/\varepsilon)} + O(n)$, respectively, and remain PTAS for $k = O(\log n)$. The last one, for any $\varepsilon > 0$, provides a $(1 + \varepsilon)$-approximate solution in time of $(n/k)^k (\log k)^{O(1/\varepsilon)}$ depending on n polynomially for $k = n - O(\log n)$.

In [9], we claimed the conjecture *'for any fixed h, the special case of the EGTSP-GC defined[1] by grids with height of at most h has an optimal algorithm with time complexity bound depending on n and k polynomially'*. Employing the recently proved accuracy guarantee for univariate k-medians clustering [7] and our FPT algorithm for the general GTSP finding a quasi-pyramidal tour of minimum cost, we obtained a partial proof of this conjecture for $h \leq 2$. Actually, we showed in [8,9] that any instance of EGTSP-GC(2) can be solved to optimality in time of $O(n^3)$. In this paper, we extend this result to the case of any fixed $h \geq 1$.

---

[1] We call it EGTSP-GC(h).

2   Short Overview of the Result

Our approach is based on the following two points

(i) Recent results show that, for any fixed natural l and any instance of the general GTSP, a minimum cost l-pseudo-pyramidal tour can be found in polynomial time on n and k;

(ii) Theorem 2 of this paper claiming that, for any fixed h, there exists the number $l = l(h)$ such that, for any instance of the EGTSP-GC(h), at least one minimum cost tour is $l(h)$-pseudo-pyramidal.

As a consequence, we obtain an exact algorithm for the EGTSP-GC(h) with complexity bound of $poly(n, k)$.

3   Pseudo-pyramidal Tours

We proceed with some necessary definitions and notation. Let an instance of the GTSP be defined by a complete edge-weighted graph $G = (V, E, c)$, $c \colon E \to \mathbb{R}_+$, and partition $V_1 \cup \ldots \cup V_k = V$ of the nodeset V onto k non-empty disjoint clusters V_i. The feasible solutions set is exhausted by cyclic k-tours $\tau = v_{i_1}, \ldots, v_{i_k}$ visiting each cluster V_j once, at node v_j. We call such tours *Clustered Hamiltonian tours* or briefly CH-tours. It is required to find a minimum cost CH-tour, where, for any CH-tour τ, its cost $C(\tau)$ is defined by the equation $C(\tau) = \sum_{j=1}^{k-1} c(\{v_{i_j}, v_{i_{j+1}}\}) + c(\{v_{i_k}, v_{i_1}\})$.

The natural ordering of clusters $V_1, \ldots, V_k$ induces the partial order on the nodeset V as follows: for any $u \in V_i$ and $v \in V_j$, $u \prec v$ iff $i < j$. We consider a special type of CH-tours that are consistent with this order and are referred to as *pseudo-pyramidal tours*.

Definition 1. *A CH-tour* $\tau = v_1, v_{i_1}, \ldots, v_{i_r}, v_k, v_{j_{k-r-2}}, \ldots, v_{j_1}$ *is called an l-pseudo-pyramidal tour, if* $i_p - i_{p+1} \le l$ *and* $j_q - j_{q+1} \le l$ *for any* $1 \le p \le r - 1$ *and* $1 \le q \le k - r - 3$.

Actually, any l-pseudo-pyramidal tour consists of two chains

$$\tau^+ = v_1, v_{i_1}, \ldots, v_k \text{ and } \tau^- = v_k, \ldots, v_{j_1}, v_1$$

that are almost monotonous with respect to the order defined above. Namely, suppose u and v are two arbitrary successive vertices of the chain τ^+ of some l-pseudo-pyramidal CH-tour τ (the case of τ^- can be considered similarly), such that $u \in V_i$ and $v \in V_j$. If the chain τ^+ would be truly monotonic, then $j - i > 0$. By definition, for any u and v, the monotonicity can be violated, but anyway $j - i \ge -l$.

Similarly to the classic pyramidal tours, pseudo-pyramidal tours of minimum (or maximum) cost can be found efficiently. The following theorem is valid.

Theorem 1. *For any instance of the GTSP with an arbitrary non-negative cost function, a minimum cost l-pseudo-pyramidal CH-tour can be found in* $O(k \cdot l \cdot n^{O(l)})$ *time.*

4   Optimal $l(h)$-Pseudo-pyramidal Tours of EGTSP-GC(h)

Let an instance of the EGTSP-GC(h) be given by a complete edge-weighted graph $G = (V, E, c)$, where $V \subset \mathbb{R}_2$, cost function c be defined by the Euclidean distance, and clustering $V_1 \cup \ldots \cup V_k$ be induced by a unit grid of height h and width w. On the set of clusters, we define an order as follows: we number all non-empty cells of the grid (and the appropriate clusters) lexicographically, from left to right and from bottom to top. The following theorem is valid. In this short paper, we present the sketch of its proof postponing the full version to the forthcoming paper.

Theorem 2. *For any natural number h, each optimal CH-tour is l-pseudo-pyramidal, where $l = l(h) = 15h^3 + 2h$.*

Proof. Consider an arbitrary optimal CH-tour τ of the given instance of EGTSP-GC(h). Evidently, τ is l-pseudo-pyramidal for some $l \leq k - 3$. We show that $l \leq 15h^3 + 2h$, since, otherwise, there exists the cheaper CH-tour τ', for which $C(\tau') < C(\tau)$ that contradicts the optimality of τ.

Indeed, assume that τ is l-pseudo-pyramidal for $l > 15h^3 + 2h$. W.l.o.g. we can assume that there exists an edge u, v, such that

$$u \in V_{i_1}, v \in V_{i_2} \text{ and } i_1 - i_2 \geq l > 15h^3 + 2h, \tag{1}$$

and $\{u, v\}$ belongs to the chain τ^+. Moreover, we can assume that the subchain of τ^+ containing the edge $\{u, v\}$ has the form as presented in Fig. 2.

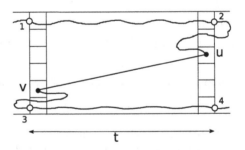

Fig. 2. Subgrid T containing the edge $\{u, v\}$

Introduce some necessary notation. By $\tau^+(1, 2)$ and $\tau^+(3, 4)$ we denote the segments of τ^+ connecting the point 1 with the point 2 and the point 3 with the point 4, respectively. By s we denote a length of the horizontal projection of the edge $\{u, v\}$. It can be easily shown that (1) implies $s > 15h^2$.

We split this projection into five equal parts producing the correspondent vertical stripes $S_1, \ldots, S_5$ of width greater than 3. Further, for any j and line segment $[p, q]$, we define the clipped segment $[\bar{p}, \bar{q}] = [p, q] \cap S_j$ and the value $\text{loss}_j(p, q)$ by the following equation

$$\text{loss}_j(p, q) = |[\bar{p}, \bar{q}]| - (x(p) - x(q)), \tag{2}$$

where $x(p)$ and $x(q)$ are horizontal coordinates of the points p and q.

It can be shown that the segments $\tau^+(1, 2)$ and $\tau^+(3, 4)$ have edges $\{p_1, q_1\}$ and $\{p_2, q_2\}$ intersecting the stripes S_2 and S_4, respectively, such that the following equation

$$\min\{\text{loss}_2(p_1, q_1), \text{loss}_4(p_2, q_2)\} \geq 1/4$$

is valid (Fig. 3a).

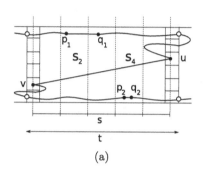 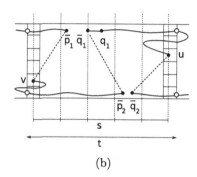

(a)   (b)

Fig. 3. (a) the stripes $S_1, \ldots, S_5$; (b) shortcutting the route τ

Then, we obtain the CH-tour τ' by exclusion from τ the clipped segments $[\bar{p}_1, \bar{q}_1]$ and $[\bar{p}_2, \bar{q}_2]$ together with the edge $\{u, v\}$ and linking $\bar{p}_1$ with v, $\bar{p}_2$ with $\bar{q}_1$, and u with $\bar{q}_2$ directly (Fig. 3b).

The last point is to show that $C(\tau') < C(\tau)$. Denoting by $\alpha_1 s, \alpha_2 s$ and $\alpha_3 s$ horizontal projections of the segments $[\bar{p}_1, v]$, $[\bar{p}_2, \bar{q}_2]$, and $[u, \bar{q}_2]$, respectively, we obtain $s = s\sum_{i=1}^{3}\alpha_i - (x(\bar{p}_1) - x(\bar{q}_1)) - (x(\bar{p}_2) - x(\bar{q}_2))$, where $\alpha_i \geq 1/5$, by construction. Therefore, since $s > 15h^2$,

$$C(\tau') - C(\tau) \leq \sum_{i=1}^{3}\sqrt{(\alpha_i s)^2 + h^2} - s - |[\bar{p}_1, \bar{q}_1]| - |[\bar{p}_2, \bar{q}_2]|$$

$$\leq \sum_{i=1}^{3}\left(\sqrt{(\alpha_i s)^2 + h^2} - \alpha_i s\right) - \text{loss}_2(p_1, q_1) - \text{loss}_4(p_2, q_2)$$

$$\leq \sum_{i=1}^{3}\frac{h^2}{\sqrt{(\alpha_i s)^2 + h^2} + \alpha_i s} - \frac{1}{2} \leq \frac{1}{2}\left(\sum_{i=1}^{3}\frac{h^2}{2\alpha_i s} - 1\right) \leq \frac{1}{2}\left(\frac{15h^2}{s} - 1\right) < 0$$

Theorem is proved.

Together with Theorems 1 and 2 implies the proof of the aforementioned conjecture.

5   Conclusion

In this short paper, we outlined an approach to proving the efficient solvability for the cases of Generalized Traveling Salesman Problem defined on the Euclidean plane. Obviously, this result can be easily downgraded to the case of the classic Euclidean TSP on the plane, which seems interesting in comparison with the famous Papadimitriou result [10]. As for the future work, the bound $l(h)$ obtained in Theorem 2 seems to be unattainable and possibly can be improved. Further, it would be interesting to extend the results presented to other distances, e.g. l_1 metric.

References

1. Bhattacharya, B., Ćustić, A., Rafiey, A., Rafiey, A., Sokol, V.: Approximation algorithms for generalized MST and TSP in grid clusters. In: Lu, Z., Kim, D., Wu, W., Li, W., Du, D.-Z. (eds.) COCOA 2015. LNCS, vol. 9486, pp. 110–125. Springer, Cham (2015). https://doi.org/10.1007/978-3-319-26626-8_9

2. Feremans, C., Grigoriev, A., Sitters, R.: The geometric generalized minimum spanning tree problem with grid clustering. 4OR **4**(4), 319–329 (2006). https://doi.org/10.1007/s10288-006-0012-6

3. Fischetti, M., González, J.J.S., Toth, P.: A branch-and-cut algorithm for the symmetric generalized traveling salesman problem. Oper. Res. **45**(3), 378–394 (1997). https://doi.org/10.1287/opre.45.3.378

4. Gutin, G., Punnen, A.P.: The Traveling Salesman Problem and Its Variations. Springer, US, Boston (2007)

5. Khachai, M.Y., Neznakhina, E.D.: Approximation schemes for the generalized traveling salesman problem. Proc. Steklov Inst. Math. **299**(1), 97–105 (2017). https://doi.org/10.1134/S0081543817090127

6. Khachay, M., Neznakhina, K.: Towards a PTAS for the generalized TSP in grid cluster. AIP Conf. Proc. **1776**(1), 050003 (2016). https://doi.org/10.1063/1.4965324

7. Khachay, M., Pankratov, V., Khachay, D.: Attainable best guarantee for the accuracy of k-medians clustering in $[0, 1]$. In: CEUR Workshop Proceedings 1987, pp. 322–327 (2017). urn:nbn:de:0074-1987-8

8. Khachay, M., Neznakhina, K.: Generalized pyramidal tours for the generalized traveling salesman problem. Lecture Notes in Computer Science, vol. 10627, pp. 265–277. Springer International Publishing, Cham (2017). https://doi.org/10.1007/978-3-319-71150-8_23

9. Khachay, M., Neznakhina, K.: Polynomial time solvable subclass of the generalized traveling salesman problem on grid clusters. In: van der Aalst, W.M.P., Ignatov, D.I., Khachay, M., Kuznetsov, S.O., Lempitsky, V., Lomazova, I.A., Loukachevitch, N., Napoli, A., Panchenko, A., Pardalos, P.M., Savchenko, A.V., Wasserman, S. (eds.) AIST 2017. LNCS, vol. 10716, pp. 346–355. Springer, Cham (2018). https://doi.org/10.1007/978-3-319-73013-4_32

10. Papadimitriou, C.: Euclidean TSP is NP-complete. Theoret. Comput. Sci. **4**, 237–244 (1977)

Constant Factor Approximation for Intersecting Line Segments with Disks

Konstantin Kobylkin[1,2]([✉])

[1] Institute of Mathematics and Mechanics, Ural Branch of RAS,
Sophya Kovalevskaya str. 16, 620990 Yekaterinburg, Russia
[2] Ural Federal University, Mira str. 19, 620002 Yekaterinburg, Russia
kobylkinks@gmail.com
http://wwwrus.imm.uran.ru

Abstract. Fast constant factor approximation algorithms are devised for a problem of intersecting a set of n straight line segments with the smallest cardinality set of disks of fixed radii $r > 0$ where the set of segments forms a straight line drawing $G = (V, E)$ of a planar graph without edge crossings. Exploiting its tough connection with the geometric Hitting Set problem we give $(100 + \varepsilon)$-approximate $O(n^4 \log n)$-time and $O(n^2 \log n)$-space algorithm based on the modified Agarwal-Pan reweighting algorithm, where $\varepsilon > 0$ is an arbitrary small constant. Moreover, $O(n^2 \log n)$-time and $O(n^2)$-space 18-approximation is designed for the case where G is any subgraph of a Gabriel graph.

Keywords: Computational geometry · Approximation algorithms Hitting set problem · Epsilon nets · Line segments

1 Introduction

Design of fast algorithms with low approximation factors is an important task for many NP-hard geometric combinatorial optimization problems among which are numerous special settings of the known geometric HITTING SET problem on the plane which itself has close links to optimal geometric coverage and piercing problems. The classical HITTING SET problem is to find the smallest cardinality subset H of some given set $Y \subseteq \mathbb{R}^2$ such that $H \cap R \neq \varnothing$ for any $R \in \mathcal{R}$,[1] where $\mathcal{R}$ is some given family of subsets of $\mathbb{R}^2$ also called *objects* in the sequel. Moreover, a pair $(Y, \mathcal{R})$ is associated with each instance of the HITTING SET problem called a *range space*. In this paper fast constant factor approximation algorithms are constructed for special settings of the geometric HITTING SET problem related to network analysis.

INTERSECTING PLANE GRAPH WITH DISKS (IPGD): given a straight line drawing (or a plane graph) $G = (V, E)$ of an arbitrary simple planar graph without

---

This work was supported by Russian Science Foundation, project 14-11-00109.

[1] A subset $H \subset Y$ is named a *hitting set* for $\mathcal{R}$ if $H \cap R \neq \varnothing$ for every $R \in \mathcal{R}$.

© Springer Nature Switzerland AG 2019
R. Battiti et al. (Eds.): LION 12 2018, LNCS 11353, pp. 447–454, 2019.
https://doi.org/10.1007/978-3-030-05348-2_39

edge crossings and a constant $r > 0$, find the smallest cardinality set $C \subset \mathbb{R}^2$ of points (disk centers) such that each edge $e \in E$ is within Euclidean distance r from some point $c = c(e) \in C$ or, equivalently, the disk of radius r centered at c intersects e.

The IPGD abbreviation is used below in our paper to denote the problem for simplicity of presentation.

The IPGD problem could be of interest in network security, sensor network deployment and facility location. For example, in [8] its sensor deployment applications are reported for road networks. Namely, for a road network possible intrusion activity must be monitored by deploying sensor devices near its roads which have uniform circular sensing area. As full network surveillance might be costly, it may be sufficient to guarantee that the intruder movement is detected at least once somewhere along any its road. Minimizing total cost of the deployed identical sensors can be modeled in the form of the IPGD problem assuming that network roads are modeled by interior disjoint straight line segments. Moreover, another network security application may be of interest of the IPGD problem for optical fiber networks following the approach of [1].

As far as it is known, special settings are first considered close to the IPGD problem in [8] for cases of (possibly overlapping) axis-parallel and bounded length straight line segments. Of course, the general IPGD problem can be considered as the special geometric intersection problem which is to find the smallest cardinality subset of radius r disks whose union has nonempty intersection with each segment from E. Besides, it coincides with the HITTING SET problem in which $Y = \mathbb{R}^2$, $\mathcal{R} = \mathcal{N}_r(E) = \{\{x \in \mathbb{R}^2 : d(x, e) \le r\} : e \in E\}$ and $d(x, e)$ is Euclidean distance between a point $x \in \mathbb{R}^2$ and a segment $e \in E$.

Moreover, the IPGD problem generalizes a classical NP-hard disk covering problem on the case of non-zero length segments. In the latter problem one needs to cover a given finite point set on the plane with the smallest cardinality set of equal disks. Generally, the IPGD problem is quite different from the HITTING SET problem for families of identical disks (which is equivalent to the disk covering problem) and other bounded aspect ratio objects as segment lengths are not assumed to be bounded from above by any linear function of r. Surprisingly, it admits fast constant factor approximation algorithms whose factors are modest.

In the recent years a significant reduction is achieved in complexity of those approximation algorithms for several geometric HITTING SET problems which use ideas of local search [3,7]. Local search is initially used to design PTAS [11] for geometric HITTING SET problems. More precisely, local search approach has become quite competitive for designing low constant factor approximations for geometric HITTING SET problems for families of disks [7] and pseudo-disks[2] [3]. As straight line segments from E intersect at most at their endpoints, objects

---

[2] A set of geometric objects on the plane is called a set of pseudo-disks if any two its objects have their boundaries intersecting at most twice and both their differences are connected.

from $\mathcal{N}_r(E)$ form a family of pseudo-disks without loss of generality[3]. Therefore an algorithm from [3] yields 4-approximate solution for the IPGD problem in $O(n^{18})$-time, being too complicated, where $n = |E|$.

In this paper another approach is taken which involves using epsilon nets [2]. It gives faster algorithms at the expense of larger constants in their approximation factors. More precisely, $\left(50 + 52\sqrt{\frac{12}{13}} + \varepsilon\right)$-approximation algorithm is proposed for the IPGD problem working in $O\left(n^4 \log n\right)$ time, where $\varepsilon > 0$ is an arbitrary small constant. Moreover, an $O(n^2 \log n)$-time 18-approximation is designed for the case where G is any subgraph of a Gabriel graph. Within epsilon net approach the only relevant constant factor approximation is known for the HITTING SET problem for sets of pseudo-disks [12] with similar time complexity bounds, giving huge constant approximation factors. Thus, our algorithms give a competitive tradeoff of approximation factor and time complexity being compared with known local search and epsilon net based approximation algorithms designed for similar geometric HITTING SET problems on sets of pseudo-disks.

There are a few restricted settings of the IPGD problem for which low constant factor approximation algorithms are designed of small complexity. If E contains only zero length segments, a 4-approximation is known [4] which works in $O(n \log n)$ time and $O(n)$ space. Moreover, when E consists of axis-parallel segments, 8-approximation algorithm is given [8] taking $O(n \log n)$ time and $O(n \log n)$ space.

2 Constant Factor Approximation for the IPGD Problem

2.1 Algorithmic Ideas: Epsilon Nets and Independent Sets

As reported in the Introduction, the IPGD problem is equivalent for a set E of straight line segments to the HITTING SET problem on a set $\mathcal{N}_r(E)$ of Euclidean r-neighbourhoods of segments of E (which are often called hippodromes). To get constant factor approximation algorithms for the IPGD problem improved versions of two approaches are used from [2,12], which are devised for the general geometric HITTING SET problem.

To outline basic concepts behind our algorithms for the IPGD problem, we start with the following idea which underlies low constant factor approximation algorithms for cases of the IPGD problem with either zero-length [4] or axis-parallel [8] segments. Say, if segments from E are all of zero-length, the idea is to apply a "divide-and-conquer" approach by finding a maximal (with respect to inclusion) independent set $\mathcal{I}$ of radius r disks within $\mathcal{N}_r(E)$, i.e. a maximal subset of pairwise non-intersecting disks. As 7 radius r disks are sufficient to cover $2r$ radius disk, 7-point set S_I can be constructed such that $S_I \cap M \neq \varnothing$

---

[3] Indeed, segments of E can be slightly shifted to become pairwise disjoint while keeping all nonempty intersections of subsets of objects from $\mathcal{N}_r(E)$ with some slightly larger r.

for any $M \in \mathcal{N}_I = \{N \in \mathcal{N}_r(E) : N \cap I \neq \varnothing\}$ and $I \in \mathcal{I}$. Therefore a set $\bigcup_{I \in \mathcal{I}} S_I$ gives a 7-approximate solution.

In our paper a similar but more relaxed and indirect approach is employed which is inspired by work [12]. This is because the aforementioned idea can not be applied to the general IPGD problem more or less directly. Indeed, its application amounts to finding constant cardinality hitting sets for sets $\mathcal{M}_I$ of objects, being subsets of $\mathcal{N}_I$, associated with some object $I \in \mathcal{N}_r(E)$. But there can be no such hitting sets for $\mathcal{M}_I$ that have their sizes uniformly (within $\mathcal{I}$) bounded from above by any absolute constant. This is because, first, $I = \{x \in \mathbb{R}^2 : d(x, e(I)) \leq r\}$ for some straight line segment $e(I) \in E$ whose Euclidean length is not assumed to be bounded from above by any linear function of r. Second, the number of distinct orientations of segments from E can be arbitrarily large. Therefore, it is not possible to partition segments of E into constant number of groups of segments with similar orientations and apply the aforementioned idea in each group as done in [8].

Instead, one can generalize object independency notion by allowing intersection of objects in maximal independent set, but control amount of their overlapping in some way. Following approach of [2], one can get an approximate solution to the IPGD problem by a compound algorithm which combines two different subalgorithms, first of which exercises a similar "divide-and-conquer" scheme, seeking those generalized independent sets and respective partitioning of the family $\mathcal{N}_r(E)$.

First Subalgorithm. Let $0 < \varepsilon < 1$. Assuming that a positive weight map $w : Y_0 \to \mathbb{R}_+$ is defined on some finite point set $Y_0 \subset \mathbb{R}^2$ with OPT $=$ OPT$(\mathbb{R}^2, \mathcal{N}_r(E)) = $ OPT$(Y_0, \mathcal{N}_r(E))$,[4] first subalgorithm returns a hitting set for a subset of those objects from $\mathcal{N}_r(E)$ whose weight is at least εth fraction of $w(Y_0)$, where $w(N) = \sum_{y \in N \cap Y_0} w(y)$ for $N \subseteq \mathbb{R}^2$ and OPT$(Y, \mathcal{R})$ is the optimum of the HITTING SET problem for a given range space $(Y, \mathcal{R})$. Hitting set has a special name of the kind, which is output by the first subalgorithm.

Definition 1. *Let $0 < \varepsilon < 1$ and $w : Y_0 \to \mathbb{R}_+$. A subset $Y' \subset \mathbb{R}^2$ is called a (weighted) weak ε-net [9] for a range space $(Y_0, \mathcal{N}_r(E), w)$ if $Y' \cap N \neq \varnothing$ for any $N \in \mathcal{N}_r(E)$ with $w(N) > \varepsilon w(Y_0)$.*

To build a weak ε-net for space $(Y_0, \mathcal{N}_r(E), w)$, first subalgorithm initially seeks a maximal subset $\mathcal{I}$ of *almost* non-intersecting objects within the subset $\mathcal{N}_\varepsilon = \{N \in \mathcal{N}_r(E) : w(N) > \varepsilon w(Y_0)\}$. To control amount of overlap of objects from $\mathcal{I}$ special parameter δ and a weight map w on Y_0 are used as follows [12].

Definition 2. *Given a subset $\mathcal{P} \subseteq \mathcal{N}_r(E)$, a parameter $\delta > 0$ and a weight map $w : Y_0 \to \mathbb{R}_+$, a subset $\mathcal{I} = \mathcal{I}(\delta) \subseteq \mathcal{P}$ is called a maximal (with respect to*

---

[4] Set Y_0 can be constructed by computing vertices of arrangement of boundaries of objects from $\mathcal{N}_r(E)$.

inclusion) δ-*independent for a range space* $(Y_0, \mathcal{P}, w)$ *if*

$$w(I \cap I') \leq \delta w(Y_0)$$

for any distinct $I, I' \in \mathcal{I}$ *and for any* $N \in \mathcal{P}$ *there is some* $I = I(N) \in \mathcal{I}$ *with* $w(N \cap I) > \delta w(Y_0)$.

Below a pseudo-code is presented of the first subalgorithm. It involves using an auxiliary procedure to seek hitting sets $H_I(\delta)$ of size at most $\frac{c_1}{\delta} + c_2$ for subsets $\mathcal{N}_I(\delta) \subseteq \{N \in \mathcal{N}_r(E) : w(N \cap I) > \delta w(I)\}$ for a given $I \in \mathcal{N}_r(E)$ and a parameter $0 < \delta < 1$, where c_1 and c_2 are some constants. Such a procedure is named a *hitting set finder* below whereas sets $\mathcal{N}_I(\delta)$ are referred to as *sets of dependent objects*. It is assumed to exist (see the Lemma 2 below).

FIRST SUBALGORITHM.

Input: a range space $(Y_0, \mathcal{N}_r(E), w)$ and parameters $0 < \varepsilon, \theta_0 < 1, \alpha, \beta, \tau, c_1$ and c_2;

Output: a weak ε-net for $(Y_0, \mathcal{N}_r(E), w)$.

1. set $\delta := \theta_0 \varepsilon$, find a maximal δ-independent set $\mathcal{I} \subset \mathcal{N}_\varepsilon$ for $(Y_0, \mathcal{N}_\varepsilon, w)$ and form disjoint sets $\mathcal{N}_{\varepsilon,I} \subseteq \{N \in \mathcal{N}_\varepsilon : w(N \cap I) > \delta w(Y_0)\}, I \in \mathcal{I}$, with $\bigcup_{I \in \mathcal{I}} \mathcal{N}_{\varepsilon,I} = \mathcal{N}_\varepsilon$;

2. for each $I \in \mathcal{I}$ compute a hitting set H_I for $\mathcal{N}_{\varepsilon,I}$ with $|H_I| \leq \frac{c_1 w(I)}{\delta w(Y_0)} + c_2$ by applying the aforementioned auxiliary procedure with its parameter equal to $\frac{\delta w(Y_0)}{w(I)}$;

3. return the set $H_{\theta_0} = \bigcup_{I \in \mathcal{I}} H_I$.

Bounds for Size of Epsilon Nets, Given by the First Subalgorithm. The following lemma summarizes on the length of the weak ε-net H_{θ_0} returned by the first subalgorithm and provides a way to choose its parameters α, β, τ and θ_0 to guarantee $O\left(\frac{1}{\varepsilon}\right)$ bounds for $|H_{\theta_0}|$. It inherits main ideas of work [12] but establishes better upper bounds for $|H_{\theta_0}|$ by at least factor of 2.

Lemma 1. *Suppose there are absolute constants* α, β, τ *and a graph* $G_{\mathcal{I}} = (\mathcal{I}, U)$ *for any* $\mathcal{I} \subseteq \mathcal{N}_r(E)$ *such that* $|U| \leq \beta |\mathcal{I}|$ *and* $m_{\mathcal{I}}(y) \geq \alpha n_{\mathcal{I}}(y) - \tau$ *for every* $y \in Y_0$, *where* $n_{\mathcal{I}}(y) = |\{I \in \mathcal{I} : y \in I\}|$ *and* $m_{\mathcal{I}}(y) = |\{\{I, I'\} \in U : y \in I \cap I'\}|$. *Then a weak* ε-*net* H_{θ_0} *can be constructed for a range space* $(Y_0, \mathcal{N}_r(E), w)$ *by the first subalgorithm for any* $0 < \varepsilon < 1$ *of size at most*

$$|H_{\theta_0}| \leq \left[\left(1 + \frac{1}{\sqrt{1 + \frac{c_2 \alpha}{c_1 \beta}}}\right)\left(\frac{2c_1 \tau \beta}{\alpha^2} + \frac{c_2 \tau}{\alpha}\right) + \frac{c_2 \tau}{\alpha \sqrt{1 + \frac{c_2 \alpha}{c_1 \beta}}}\right]\frac{1}{\varepsilon},$$

where $\theta_0 = \dfrac{\frac{\alpha}{\beta}}{1 + \sqrt{1 + \frac{c_2 \alpha}{c_1 \beta}}}$.

Definition 3. *A map is called a structural map for the range space* $(Y_0, \mathcal{N}_r(E))$, *which assigns a graph* $G_{\mathcal{I}}$ *for each* $\mathcal{I} \subseteq \mathcal{N}_r(E)$ *as defined in the Lemma 1, where the constants* α, β *and* τ *are named as structural parameters for that space.*

Second Subalgorithm. Second subalgorithm adjusts the parameter ε and computes a weight map w to get $|H_{\theta_0}| = O(\text{OPT})$. Namely, it computes point weights $w(y)$, $y \in Y_0$, to get the inequality

$$w(N) > \varepsilon w(Y_0) \tag{1}$$

hold for all $N \in \mathcal{N}_r(E)$, where ε is also found such that $\varepsilon = \frac{1}{\lambda_0 \text{OPT}}$ for some $\lambda_0 \approx 1$. To get a suitable weight map (i.e. satisfying the inequality (1) for all $N \in \mathcal{N}_r(E)$) and compute right ε a modification of the known fast Agarwal-Pan algorithm [2] is applied which is similar to the one in [6].

Main Algorithm. Finally, the compound main algorithm works as follows: second subalgorithm runs to compute a suitable weight map w_0 and gets a value $\varepsilon_0 = \frac{1}{\lambda_0 \text{OPT}}$ of the parameter ε whereas first (weak epsilon net seeking) subalgorithm constructs a hitting set for $\mathcal{N}_r(E)$ being applied for found ε_0 and w_0, thus, arriving at the constant factor approximation due to the Lemma 1. Namely, a constant C in the bound $|H_{\theta_0}| \leq \frac{C}{\varepsilon}$ (see the Lemma 1) provides an upper bound on the approximation factor of the compound algorithm.

2.2 Geometric Constructions

Hitting Set Finders for Sets of Dependent Objects. In order for the first subalgorithm to be working, a special procedure should be designed which returns hitting sets $H_I(\delta)$ of length at most $\frac{c_1}{\delta} + c_2$ for sets $\mathcal{N}_I(\delta) \subseteq \{N \in \mathcal{N}_r(E) : w(N \cap I) > \delta w(I)\} \cup \{I\}$, where $I \in \mathcal{N}_r(E)$, $0 < \delta < 1$ and $c_1, c_2 > 0$ are some absolute constants. Such a special procedure is described below.

Some notations are first given to simplify our work with machinery of straight line segments and their Euclidean r-neighbourhoods. Let $N_r(e) = \{x \in \mathbb{R}^2 : d(x, e) \leq r\}$ for a straight line segment e on the plane; particularly, for $x \in \mathbb{R}^2$ $N_r(x)$ denotes a radius r disk centered at x. Moreover, for $N \in \mathcal{N}_r(E)$ a segment $e(N)$ is such that $N = N_r(e(N))$ whereas $E(\mathcal{N}) \subseteq E$ is the segment set with $\mathcal{N} = \mathcal{N}_r(E(\mathcal{N}))$. Set $z_e(e') = \{x \in e' : d(x, e) \leq 2r\}$ and let $l(e)$ be a straight line through a non-zero length segment e, $h_1(e)$ and $h_2(e)$ be positive and negative halfplanes respectively, whose boundary coincides with $l(e)$; here orientation is chosen arbitrarily for $l(e)$. The set $\text{bd} \, N_r(e)$ can be represented in the form of a union of two halfcircles and two segments $f_1(e)$ and $f_2(e)$, where $f_i(e) \subset \text{int} \, h_i(e)$, $i = 1, 2$. Let $l_i(e)$ be the straight line through $f_i(e)$.

The procedure below seeks hitting sets for sets of dependent objects in the case of interior disjoint segments. It is based on finding hitting sets for sets of 1-dimensional r-neighbourhoods of (interval) projections of segments $\{z_e(e')\}_{e' \in E'}$, $E' \subseteq E$, onto straight lines $l_i(e)$. Let $N_{ir}(f) = \{x \in l_i(e) : d(x, f) \leq r\}$ for an arbitrary interval $f \subset l_i(e)$, $i = 1, 2$.

HITTING SET FINDER FOR DEPENDENT OBJECTS.

Input: a parameter $0 < \delta < 1$ and a set $\mathcal{N}_I(\delta)$, where $I \in \mathcal{N}_r(E)$ and $\mathcal{N}_I(\delta) \subseteq \{N \in \mathcal{N}_r(E) : w(N \cap I) > \delta w(I)\} \cup \{I\}$ for some weight map $w : Y_0 \to \mathbb{R}_+$;

Output: a hitting set $H_I(\delta) \subset \mathbb{R}^2$ for $\mathcal{N}_I(\delta)$.

---

1. set $\{v_1, v_2\} = l(e(I)) \cap \operatorname{bd} I$ and

$$\mathcal{P} := \mathcal{N}_I(\delta) \backslash \{N \in \mathcal{N}_r(E) : N \cap \{v_1, v_2\} \neq \varnothing\};$$

2. form sets $Z_i = \{z_{e(I)}(e) : e \in E(\mathcal{P}), z_{e(I)}(e) \subset h_i(e(I))\}$, $i = 1, 2$;
3. form a set P_i of orthogonal projections of segments from Z_i onto the straight line $l_i(e(I))$ and construct sets $P_i(r) = \{N_{ir}(p) : p \in P_i\}$, $i = 1, 2$;
4. find the minimum cardinality hitting set $H_i \subset l_i(e(I))$ for $P_i(r)$, $i = 1, 2$;
5. for each $x_0 \in H_i$ and $i = 1, 2$ construct a set $S(x_0)$ of 4 points such that $N_{\sqrt{2}r}(x_0) \subset \bigcup_{x \in S(x_0)} N_r(x)$ and return a set $H_I(\delta) = \{v_1, v_2\} \cup \bigcup_{x_0 \in H_i, i=1,2} S(x_0)$.

The following lemma summarizes on the procedure performance.

Lemma 2. *Let* $n = |\mathcal{N}_I(\delta)|$. *The* HITTING SET FINDER FOR DEPENDENT OBJECTS *procedure returns a hitting set* $H_I(\delta)$ *for* $\mathcal{N}_I(\delta)$ *of size at most* $\frac{8}{\delta} + 2$ *in* $O(n \log n)$ *time and* $O(n)$ *space.*

Structural Map for the IPGD Problem. According to the Lemma 1, to guarantee $O\left(\frac{1}{\varepsilon}\right)$ bounds for size of weak epsilon nets, obtained from the first subalgorithm, a structural map should be identified for $(Y_0, \mathcal{N}_r(E))$. In other words, one needs to build a structural map by assigning a graph for each subset $\mathcal{N} \subseteq \mathcal{N}_r(E)$ or, equivalently, for each subset $E' \subseteq E$. It can be established that Delaunay triangulation (planar) graph [5] of the segment set E' turns out to be the sought graph for which ratios $\frac{\beta}{\alpha}$ and $\frac{\tau}{\alpha}$ are small.

Lemma 3. *There is a structural map for* $(Y_0, \mathcal{N}_r(E))$ *with* $\beta = 3$ *and* $\alpha = \tau = 1$.

2.3   Performances of Our Approximation Algorithms

Our main result is on constant factor approximation for the IPGD problem.

Theorem 1. *Let* $n = |E|$. *There is a* $\left(50 + 52\sqrt{\frac{12}{13}} + \nu\right)$*-approximation for the* IPGD *problem, which works in*

$$O\left(\left(n^2 + \frac{n \log n}{\nu^2} + \frac{\log n}{\nu^3}\right) n^2 \log n\right)$$

time and $O\left(\frac{n^2 \log n}{\nu}\right)$ *space for any small* $\nu > 0$.

Let S be a plane point set in general position no 4 of which are cocircular.

Definition 4. *A plane graph* $G = (S, E)$ *is called a Gabriel graph when* $[u, v] \in E$ *iff the disk having* $[u, v]$ *as its diameter does not contain any other points of* S *distinct from* u *and* v.

In [10] it is shown that the IPGD problem is NP-hard for the case where E is edge set of a Gabriel graph. The theorem below reports an approximation algorithm to exist for sets of segments given by edge sets of Gabriel graphs.

Theorem 2. *If G is subgraph of a Gabriel graph, there is a 18-approximation algorithm, which works in $O(n^2 \log n)$ time and $O(n^2)$ space.*

3   Conclusion

Approximation algorithms with constant factors are proposed for a problem of intersecting a set of n straight line segments, overlapping at most at their endpoints, with the least cardinality set of equal disks. Namely, a $(100 + \varepsilon)$-approximation is given which works in $O(n^4 \log n)$ time and $O(n^2 \log n)$ space. Also 18-approximation is reported to exist with $O(n^2 \log n)$ time and $O(n^2)$ space complexities for the case where segments form edge set of a Gabriel graph.

References

1. Agarwal, P.K., Efrat, A., Ganjugunte, S.K., Hay, D., Sankararaman, S., Zussman, G.: The resilience of WDM networks to probabilistic geographical failures. IEEE/ACM Trans. Netw. **21**(5), 1525–1538 (2013)
2. Agarwal, P., Pan, J.: Near-linear algorithms for geometric hitting sets and set covers. In: Proceedings of 30th Annual Symposium on Computational geometry, pp. 271–279 (2014)
3. Antunes, D., Mathieu, C., Mustafa, N.: Combinatorics of local search: an optimal 4-local Hall's theorem for planar graphs. In: Proceedings of 25th Annual European Symposium on Algorithms (ESA), vol. 87, pp. 8:1–8:13 (2017)
4. Biniaz, A., Liu, P., Maheshwari, A., Smid, M.: Approximation algorithms for the unit disk cover problem in 2D and 3D. Comput. Geom. **60**, 8–18 (2017)
5. Brévilliers, M., Chevallier, N., Schmitt, D.: Triangulations of line segment sets in the plane. In: Arvind, V., Prasad, S. (eds.) FSTTCS 2007. LNCS, vol. 4855, pp. 388–399. Springer, Heidelberg (2007). https://doi.org/10.1007/978-3-540-77050-3_32
6. Bus, N., Mustafa, N., Ray, S.: Practical and efficient algorithms for the geometric hitting set problem. Discret. Appl. Math. **240**, 25–32 (2018)
7. Bus, N., Garg, S., Mustafa, N., Ray, S.: Limits of local search: quality and efficiency. Discret. Comput. Geom. **57**(3), 607–624 (2017)
8. Dash, D., Bishnu, A., Gupta, A., Nandy, S.C.: Approximation algorithms for deployment of sensors for line segment coverage in wireless sensor networks. Wirel. Netw. **19**(5), 857–870 (2013)
9. Haussler, D., Welzl, E.: ε-nets and simplex range queries. Discret. Comput. Geom. **2**(2), 127–151 (1987)
10. Kobylkin, K.: Stabbing line segments with disks: complexity and approximation algorithms. In: van der Aalst, W.M.P., et al. (eds.) AIST 2017. LNCS, vol. 10716, pp. 356–367. Springer, Cham (2018). https://doi.org/10.1007/978-3-319-73013-4_33
11. Mustafa, N., Ray, S.: Improved results on geometric hitting set problems. Discret. Comput. Geom. **44**(4), 883–895 (2010)
12. Pyrga, E., Ray, S.: New existence proofs for ε-nets. In: Proceedings of the 24th Annual Symposium on Computational Geometry, pp. 199–207 (2008)

Scheduling Deteriorating Jobs and Module Changes with Incompatible Job Families on Parallel Machines Using a Hybrid SADE-AFSA Algorithm

Yuwei Sun, Xiaofei Qian$^{(\boxtimes)}$, and Siwen Liu$^{(\boxtimes)}$

School of Management, Hefei University of Technology, Hefei 230009, China
{qianxiaofei888, liusiwen67}@126.com

Abstract. This research is motivated by a scheduling problem found in the special steel industry of continuous casting processing, where the special steel is produced on the parallel machines, i.e., the continuous casting machine, and each machine can produce more than one types of special steel. Usually, different types of special steel have diversity alloy content, which generates distinct cooling requirements. Consequently, the job families are incompatible, different types of special steel cannot be continuous process. This indicates that the machine will pause for a period of time to execute the module change activity between two adjacent job families. In this context, we attempt to investigate a parallel machine scheduling problem with the objective of minimizing the makespan, i.e., the completion time of the last job. The effect of deterioration, incompatible job families, and the module change activity are taken into consideration simultaneously, and the actual processing time of each job depends on its starting time and normal processing time. A hybrid SADE-AFSA algorithm combining Self-Adaptive Differential Evolution (SADE) and Artificial fish swarm algorithm (AFSA) is proposed to tackle this problem. Finally, the computational experiments are conducted to evaluate the performance of the proposed algorithm.

Keywords: Scheduling · Deteriorating jobs · Incompatible job family
Module change · SADE-AFSA

1 Introduction

An increasing number of companies start to adopt the strategy of high-quality products with various families and specifications to satisfy the diversified requirements of the customer order. Among them, the special steel industry is a representative industry, which has the features of multi-species and small-batch. The continuous casting machine is one of the crucial accomplishments of special steel processing, on which various types of special steel can be manufactured in a continuous manner, and the production efficiency is improved significantly. In this paper, we focus on the scheduling problem of the cooling operation, which plays an essential role in special steel continuous casting. According to the continuous casting processing depicted in [1], during the cooling operation as shown in Fig. 1, the liquid steel is first infused into the ladle, which is placed on the end of the turntable. On top of that, it flows through the tundish and then pours into a water-cooled copper sheath, which is the major module in cooling operation of the continuous casting machine. After the casting processing of one ladle is completed, the turntable rotates, and another ladle of the same family begins to process in succession. Owing to the diversity alloy content, different types of special steel should be cooled with the different-sized water-cooled copper sheath, the job family is incompatible. And there exists a module change activity between two adjacent job families, and due to the duration of the module change activity is generally longer compared with the job processing time, effective scheduling of it has an important impact on achieving good manufacturing performance. Furthermore, the build-up of oxide inclusion, the aging effect of machines and operator condition recovery will likely result in a delay during the casting processing, which cannot be ignored. Therefore, it is significant to study the scheduling problems by considering the module change activity, the effect of deterioration and the completion time-related objective simultaneously.

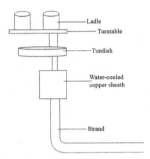

Fig. 1. Continuous casting machine

The deteriorating effect is firstly introduced to the scheduling model by Browne and Yechiali [2], in which the job processing time is assumed as a non-decreasing, start-time dependent linear function. Since then, the machine scheduling problems with deteriorating jobs draws more and more attention [3–6].

In traditional scheduling problems, the machine is usually assumed to be continuously available. Nevertheless, there are many reasons which will result in a temporarily inoperative period, and one of the important reasons is the module change activity. Over the last two decades, researchers have shown an increased interest in the scheduling problem with the module change activity. Akturk et al. [7], Akturk et al. [8], Chen [9], Xu et al. [10] have concerned the module change scheduling problem on a single machine with constant job processing time. However, there exist limited studies considering the deteriorating jobs on parallel machines

In terms of the parallel machine scheduling problem, many researchers have attempted to employ meta-heuristic algorithms to find the near-optimal solutions, which are proved to have excellent performance. Guo et al. [11] proposed a hybrid discrete cuckoo search algorithm to gain the near-optimal solution on the parallel machines. With the restriction that the maximum lateness had an upper bound, Wu et al. [12] applied a tabu search (TS) algorithm to find the near-optimal solutions of the two-agent single-machine scheduling problem. Wu et al. [13] refined the Ant colony algorithm (ACO) and the TS algorithm to minimize the makespan respectively. In addition, PSO has been regarded as an effective algorithm which was widely applied to solve several scheduling problems [14, 15].

We have investigated the scheduling problem by considering the linear deteriorating effect in our previous research [16–23], distinguish from which, this paper focuses on the practical problems with incompatible job families and the module change activity, which has not been studied before. There exist two main differences between this paper and our previous research. On one hand, this paper introduces the module change activity and the family attribute of jobs, while our previous studies assume that jobs are same job family which can be processed continuously. On the other hand, some structural properties, as well as a novel hybrid SADE-AFSA algorithm, are proposed to solve this problem. As shown in Table 1, we compare this study with our previous research.

Table 1. Key comparisons of our previous work, existing literature, and current study

Publication	Effect		Objective	Machine type	Jobs processing	JF and CH	Algorithm
	LE	DE					
Pei et al. [16]		$\checkmark$	C_{max}, T_{max}, E_{max} & L_{max}	Single	Serial-batching	Only JF	Heuristic
Pei et al. [17]		$\checkmark$	C_{max}	Single	Serial-batching		Heuristic
Pei et al. [18]		$\checkmark$	C_{max} & $\sum_{i=1}^{n} U_i$	Single	Serial-batching		Heuristic
Pei et al. [19]	$\checkmark$	$\checkmark$	C_{max}	Single	Serial-batching		Heuristic
Pei et al. [20]		$\checkmark$	C_{max}	Multiple	Serial-batching		Heuristic & BA-VNS
Liu et al. [21]		$\checkmark$	C_{max}	Multiple	Parallel-batching		Heuristic & VNS-HS
Pei et al. [22]	$\checkmark$	$\checkmark$	C_{max}, E_{max} & $\sum_{i=1}^{n} U_i$	Single	Serial-batching		Heuristic
Fan et al. [23]	$\checkmark$	$\checkmark$	C_{max}	Single	Serial-batching		Heuristic & VNS-ASHLO
Current study		$\checkmark$	C_{max}	Multiple	General	$\checkmark$	SADE-AFSA

In Table 1, LE and DE denote learning effect and deteriorating effect respectively, JF and CH denote job family and the module change activity respectively, and C_{max}, T_{max}, E_{max}, L_{max}, $\sum_{i=1}^{n} C_i$, $\sum_{i=1}^{n} U_i$ denote the makespan, the maximum tardiness, the maximum lateness, the maximum earliness and the total completion time of all jobs and the number of tardy jobs in a schedule, respectively.

The contributions of this paper can be summarized as follows:

(i)   A parallel-machine scheduling problem is investigated, and the features of deteriorating jobs, incompatible job families, and the module change activity are taken into consideration simultaneously.

(ii) Under the special situation that the jobs are assigned on a single machine, some structural properties are identified, which provides the basis of making the optimal policy to the objective of minimizing the makespan.

(iii) Based on the proposed structural properties, a hybrid SADE-AFSA algorithm which combines self-adaptive differential evolution (SADE) and artificial fish swarm algorithm (AFSA) are designed to solve the parallel-machine scheduling problem.

The rest of this paper is organized as follows. In Sect. 2, the notations used in this paper are introduced and the problem is described in detail. Some structural properties and the SADE-AFSA are proposed in Sect. 3 and 4, respectively. Computational experiments are presented in Sect. 5 for depicting the performance of the proposed algorithm. Finally, Sect. 6 summarizes this paper and points out the future research.

2   Notations and Problem Formulation

Notations used in this paper are illustrated in Table 2.

There are n families to be processed on m parallel machines, each job belongs to a certain family. All jobs and machines are non-resumable and available at time t_0 in manufacturing processing. J_{ki} has a job-dependent normal processing time p_{ki}, the actual job processing time is defined as in [24].

$$p_{ki}^{A} = p_{ki}(1 + at) \tag{1}$$

Where a is the common deteriorating rate shared by all jobs.

According to the requirement of matching job families and modules with different sizes, machines must undergo mandatory module change activities when different families are processed in succession, except for the last family. During the module change activity, no job can be performed. We assume that the duration of the module change activity is a machine-dependent constant $s_j, j = 1, 2, 3, \ldots, m$. In this paper, we need to make the decisions of family split, job sequencing within the same family and families sequencing to minimize the makespan on parallel machines. Following the three field notation in [25], the problem can be denoted as $Pm|nr, ch, p_{ki}^{A} =$

Table 2. The list of notations

Notation	Definition
n	The number of families
m	The number of machines
F_i	Family i, $i = 1, 2, 3, \ldots, n$
M_j	Machine j, $j = 1, 2, 3, \ldots, m$
F_{ij}	family i on M_j
n_j	The number of families on M_j, $j = 1, 2, 3, \ldots, m$
N	The total number of jobs
N_{ij}	The number of jobs in F_{ij}, $i = 1, 2, 3, \ldots, n$
s_j	The changeover time of machine M_j
J_k	Job k, $k = 1, 2, \cdots, n$
J_{ki}	Job k in F_i, $k = 1, 2, 3, \ldots, N_i$
p_k	The normal processing time of J_k, $i = 1, 2, \cdots, N$
p_{ki}	The normal processing time of J_{ki}
p_{ki}^A	The actual processing time of J_{ki}
J_{kij}	Job k in F_{ij}, $k = 1, 2, 3, \ldots, N_{ij}$
p_{kij}	The normal processing time of J_{kij}
p_{kij}^A	The actual processing time of J_{kij}
θ_{hj}	The total processing time for F_{hj}
$C(F_{hj})$	The completion time of F_{hj}
$C(M_j)$	The makespan on M_j
C_{max}	The makespan

$p_{ki}(1 + at)|C_{max}$, where nr and ch in the second field denote non-resumable and the module change activity respectively.

Some assumptions for the problem are listed as follows:

(1) All jobs and machines are available at time zero and machine idle-time is not allowed.
(2) There exists no idle-time between processing and the module change activity.
(3) Preemption is prohibited. Once a job starts on a machine, any other jobs cannot be processed on the machine until the job is finished.

3   Structural Properties

In this section, the problem of minimizing the makespan for single-machine case $1|nr, ch, p_{ki}^A = p_{ki}(1 + at)|C_{max}$ is studied, and some structural properties for the optimal schedule are given.

Lemma 1. *[24] For the problem* $1\left|p_{ki}^A = p_{ki}(1+at)\right|C_{max}$, *the completion time of the kth job in the sequence* $\{J_{1i}, J_{2i}, \ldots, J_{ki}, \ldots, J_{Ni}\}$ *is formulated in Eq. (2) when the first job is started at time* t_0.

$$C(J_{ki}) = \left(t_0 + \frac{1}{a}\right) \prod_{z=1}^{k} (1 + ap_{ki}) - \frac{1}{a} \tag{2}$$

We follow the demonstration of jobs and families sequences as shown in [20, 21], the lemmas 2, 3, 4 are proved.

Lemma 2. *For any given schedule* $\omega = \{J_{1i}, J_{2i}, \ldots, J_{ki}, J_{(k+1)i} \ldots, J_{Ni}\}$, *the completion time of this schedule remains unchanged when* J_{ki} *and* $J_{(k+1)i}$ *swapped,* $k = 1, 2, 3, \ldots, N_i - 1, i = 1, 2, 3, \ldots, n$.

Proof. We omit the proof, which is easy to obtain. $\square$

Lemma 3. *For the problem* $1\left|nr, ch, p_{ki}^A = p_{ki}(1+at)\right|C_{max}$, *if all jobs in the identical family are processed successively, the processing order of the families is given. With the first family* F_{1j} *starting at time* $t_0 > 0$, *the makespan of schedule is*

$$C(M_j) = \left(s_0 + \frac{1}{a}\right) \prod_{h=1}^{n_j} \prod_{k=1}^{N_{hj}} (1 + ap_{khj}) + s_j \sum_{h=1}^{n_j-1} \prod_{i=1}^{h} \prod_{k=1}^{N_{(n_j-i+1)j}} \left(1 + ap_{k(n_j-i+1)j}\right) - \frac{1}{a} \tag{3}$$

Proof. It is easy to understand, we do not to prove it here. $\square$

Lemma 4. *For the problem* $1\left|nr, ch, p_{ki}^A = p_{ki}(1+at)\right|C_{max}$, *considering two adjacent families* F_{ej} *and* $F_{(e+1)j}$, *if* $\theta_{ej} \le \theta_{(e+1)j}$, *where* $\theta_{ej} = \prod_{k=1}^{N_{ej}} (1 + ap_{kej})$, $e = 1, 2, \cdots, n_j - 1$, *then* $F_{(e+1)j}$ *is processed before* F_{ej} *in the optimal schedule.*

Proof. It is easy to obtain, we do not to prove it here. $\square$

Without the assumption that jobs in the same family are processed continuously, we allow families to be split and processed separately. The schedule can be expressed as a series of the fundamental runs, a fundamental run consists of some jobs which are processed between the adjacent module change activities. For a given schedule, let η_j denote the total number of fundamental runs on M_j, let r_{ij} denote the ith fundamental run on M_j, let β_{ij} denote the number of jobs in r_{ij}, let φ_{ij} denote the processing time of r_{ij}, then we have $\iota_{ij} = \prod_{z=1}^{\beta_{ij}} (1 + ap_{zij})$. Based on Lemma 4, the optimal schedule can be obtained by the non-increasing order of ι_{ij}.

Corollary 1. *There is an optimal schedule in which jobs from the same family are processed consecutively on a single machine and the families are processed in the non-increasing order of* θ_{ij}, *where* $\theta_{ij} = \prod_{k=1}^{N_{ij}} (1 + ap_{kij})$, $i = 1, 2, \cdots, n_j$

Proof. r_{pj} and r_{qj} are two fundamental runs from the same family with $\iota_{pj} > \iota_{qj}$. From Lemma 4, we can get the optimal processing sequence when the fundamental runs are sorted by the non-increasing order of ι_{ij}. Assume that there exist no other fundamental runs from the same family as r_{pj} and r_{qj} before the position of r_{qj}, we can get a new schedule by combining r_{pj} and r_{qj} as a new fundamental run $r_{\delta j}$, and then according to the non-increasing order of ι_{ij}, $r_{\delta j}$ will be inserted into the kth position (Fig. 2). The two schedules are compared as follows.

Since we have

$$\varphi_{\delta j} = \left(s_0 + \frac{1}{a}\right)^{\beta_{pj}} \prod_{z=1}(1 + ap_{zpj}) \prod_{z=1}^{\beta_{qj}}(1 + ap_{zqj}) - \frac{1}{a},$$

For ω, the completion time of r_{qj} is

$$C_\omega(r_q) = \left(s_0 + \frac{1}{a}\right) \prod_{h=1}^{q-1}\prod_{z=1}^{\beta_{hj}}(1 + ap_{zhj}) + s_j \sum_{h=1}^{q-1}\prod_{i=1}^{h} \iota_{(n-i+1)j} - \frac{1}{a},$$

For ω', the completion time of $r_{(q-1)j}$ is

$$C'_\omega(r_q) = \left(s_0 + \frac{1}{a}\right) \prod_{h=1}^{q-1}\prod_{z=1}^{\beta_{hj}}(1 + ap_{zhj}) + s_j \sum_{h=1}^{q-2}\prod_{i=1}^{h} \iota_{(n-i+1)j} - \frac{1}{a},$$

Then,

$$C_\omega(r_q) - C_{\omega'}(r_q) = s_j(\iota_q - 1)\sum_{h=1}^{q-p-1}\prod_{i=1}^{h}\iota_{q-i} + s_j\prod_{i=1}^{q-p-1}\left(\sum_{h=q-p-1}^{q-k-2}\left(\iota_q \prod_{i=q-p-1}^{h}\iota_{q-i-1} - \prod_{i=q-p-1}^{h}\iota_{q-i-2}\right)\right) \geq 0$$

We can obtain that the schedule ω' has a later starting time than the schedule ω, and the makespan of the fundamental runs under ω' is not less than that under ω, which proves the corollary. And based on the structural properties proposed before, an algorithm is proposed as follows

Algorithm 1	
Step 1.	Index jobs on M_j such that $p_1 \geq p_2 \geq \cdots \geq p_n$ and get a job list
Step 2.	Group jobs belonging to the same family together, and calculate $p_{ki}^A = p_{ki}(1 + at)$, $k=1,2,3,\ldots,N_i$, $i=1,2,3,\ldots,n$.
Step 3.	Calculate the actual processing time θ_{ij} of each family.
Step 4.	Schedule all families in non-increasing order of θ_{ij}, i.e., $\theta_{1j} \leq \theta_{2j} \leq \cdots \leq \theta_{n_jj}$.
Step 5.	Output the solutions of job scheduling on M_j

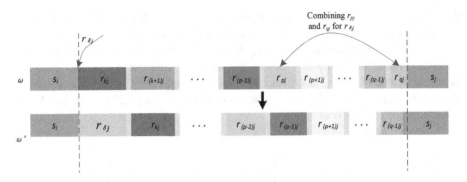

Fig. 2. The combinatorial process of two adjacent fundamental runs

4 The Hybrid Algorithm

The problem of minimizing the makespan under parallel machine $P_m||C_{max}$ is NP-hard, it has been proved by Garey and Johnson [26]. In this paper, considering that $a = 0$ and $s_i = 0$, then the problem is reduced to $P_m||C_{max}$. Hence, the problem $Pm|nr, ch, p_{ki}^A = p_{ki}(1 + at)|C_{max}$ is also NP-hard. In this section, the hybrid SADE-AFSA algorithm is put forward to solve the parallel machine scheduling problem with the objective of minimizing the makespan.

4.1 Key Procedures of SADE-AFSA

Artificial fish swarm algorithm (AFSA) was first proposed by Li, X. L. and Shao, Z. J [27] as a novel swarm intelligent algorithm to solve the optimization problem. This algorithm is used frequently in various fields, like a prediction of stock indices [28], wireless sensor network [29], scheduling [30].

Differential Evolution (DE) is one of the Evolutionary Algorithms for global optimization [31]. It is a powerful stochastic real-parameter algorithm. DE has been widely applied to solve various optimization problems, such as biology [32], power systems optimization [33] and scheduling [34]. Brest et al. [35] proposed a new adaptive DE, Self-Adaptive DE (SADE), which improved DE by adjusting the control parameters F_i and CR_i with individual characteristics.

AFSA only uses target function in the algorithm to ensure the adaptive ability and it has been proved insensitive to initial values. These characteristics guarantee the good performance of AFSA in global optimization. However, on account of that AF perceives the concentration of food, it is prone to converge prematurely, and the performance is quite dependent on the setting of parameters. AFSA easily trap into the local optimum during the later period. In order to overcome the shortcomings, the SADE algorithm is applied as the local search strategy to strengthen the ability of specific search.

Generate Initial Populations. The hybrid SADE-AFSA algorithm for the scheduling problem of parallel machines was initialized by generating a population randomly. On account of the coding rule, we can transform random permutation into an integer sequence, get the job sets on each machine, and obtain the makespan depending on the normal processing time of each job by Algorithm 1.

Behaviors of SADE-AFSA. Suppose in an objective searching space of n-dimension, there is a population composed of a set of alternative solutions, which are called artificial fishes, AFs. The current state and the food concentration of the uth AF, AF_u can be represented as vector $X_u = \{x_u(1), x_u(2), x_u(3), \dots, x_u(n)\}$ and Y_u respectively, the distance between AF_u and AF_r is indicated by $Dis_{ur} = ||X_u - X_r||$, $X_{u \setminus next}$ implies the next state of the uth AF. $Visual_g$ and $step_g$ represent the scope of sight field, and the maximum length of each movement at the gth generation respectively, which shall be neither too large nor too small. δ is the threshold of crowded degree, AT_{max} denotes the maximum number of attempts ().

(1) **AF_Prey:** Prey is a kind of basic behavior that AF will perceive a state with more food concentration [27]. This paper improves the prey-behavior by introducing mutation and crossover operations of SADE, which is used to update the current state of X_u. The algorithm framework of improved prey behavior is shown in Fig. 3.

The improved prey-behavior
1. Define $visual_g$, $step_g$, the amplification factor of the difference vector F_g, the crossover control parameter CR_g, the number of maximum attempt times AT_{max}, and set attempt times $k = 1$
2. Select a state X_r randomly in $visual_g$, $X_r(i) = X_u(i) + visual_g * Rand(-1,1)\, i = 1,2,\dots,n$
3. Calculate the food concentration of X_r
4. If $(Y_r \geq Y_u)$
5.   If $(k < AT_{max})$ then
6. $k++$, go to line 2
7.   Else
8.   Execute $X_{u/next}(i) = X_u(i) + step_g * Rand(-1,1)\, i = 1,2,\dots,n$ for X_u to step randomly in $visual_g$
9.   End if
10. Else if $(Y_r < Y_u)$
11.   If $(Rand(0,1) > CR_g)$ then
12.   Select three state (X_{u0}, X_{u1}, X_{u2}) in $visual_g$ randomly, calculate $V_u(i) = X_{u0}(i) + F_g * (X_{u1}(i) - X_{u2}(i)) + F_g * (X_{best}(i) - X_u(i)), i = 1,2,\dots,n$
13.   Set $X_{u'}(i) = V_u(i)$
14.   Else
15.   Set $X_{u'}(i) = X_r(i)$
16.   End if
17. End if
18. $X_{u/next}(i) = X_u(i) + g * Rand(-1,1) * (X_{u'}(i) - X_u(i))/Dis_{uu'}$
19. Return X_{d1}

Fig. 3. Description of improved prey behavior

X_{best} is the state of AF with the best food concentration in the current generation, X_{u0}, X_{u1}, and X_{u2} are three states in $visual_g$, which is calculated based on [27].

$$X_r(i) = X_u(i) + visual_g * Rand(-1, 1) \tag{4}$$

(2) **AF_Swarm:** Swarm behavior means that the fish will assemble in shoal and move in the direction of the center position near its fellow, meanwhile the shoal avoids being crowed [27]. Let $X_u = X_{u/next}$ denotes the current state of AF_u, $f_{u/center}$ denotes the number of fellows around AF_u in $visual_g$, meanwhile the position of AF_u is regarded as the center. If $S_u = \{X_r|||X_u - X_r|| \le visual_g\}$ is not empty, X_c, the center position of S_u will be calculated as[27]

$$X_c = \frac{\sum_{r=1}^{f_c} X_r}{f_{u/center}} \tag{5}$$

If $Y_c < Y_u$ and $\frac{Y_c}{f_{u/center}} < Y_u * \delta$, then the state of the current AF is updated as follows [27].

$$X_{u/next}(i) = X_u(i) + step_g * Rand(-1, 1) * (X_c(i) - X_u(i))/Dis_{cu} \tag{6}$$

Otherwise, execute prey behavior.

(3) **AF_Follow:** In the moving process, the fish will follow the most active fellows in visual which finds the best food concentration [27]. Let $X_u = X_{u/next}$ denotes the current state of AF_u, X_b denotes the fellow state which has the best food concentration in neighborhood ($Dis_{ub} \le visual_g$). $f_{b/center}$ denotes the number of fellows around X_b in $visual_g$, meanwhile the position of X_b is regarded as the center. If $Y_b < Y_u$ and $\frac{Y_b}{f_{b/center}} < Y_u * \delta$, the state of the current AF is updated as [27]

$$X_{u/next}(i) = X_u(i) + step_g * Rand(-1, 1) * (X_b(i) - X_u(i))/Dis_{bu} \tag{7}$$

Otherwise, execute prey behavior.

(4) **Bulletin Board:** At each generation, after all the behaviors have completed, the best food concentration will be compared with the record in the bulletin. If the current food concentration is better, then the record will be replaced.

(5) **Selection of Parameters:** According to Li et al. [27], the performance of AFSA is sensitive to the choice of $visual_g$ and $step_g$. In classical AFSA, $visual_g$ and $step_g$ are identified in advance and remain unchanged throughout the entire iterative process. Different from the constant parameters, this paper designs an adaptive approach which is adjusted by the convergence characteristic h. The parameters are calculated as

$$visual_g = \frac{visual_i}{g_{total}} * g * Gaussian\left(0, e^{h+1}\right) \tag{8}$$

$$Step_g = step_l + \frac{step_u - step_l}{g_{total}} * g * e^{\beta(h+1)} \tag{9}$$

h implies the number of continuous non-convergent generations at generation g, β is a positive constant between 0 and 1, g_{total} implies the total number of iteration generations. $visual_i$ denotes the initial value of $visual_g$, $step_{max}, step_{min}$ determimed the range of $step_g$ respectively. Besides, according to Brest et al. [35], the values of F_g and CR_g are calculated as

$$F_g = \begin{cases} F_l + Rand(0,1) * F_u, & if\ Rand(0,1) < \tau_1 \\ F_{g-1}, & otherwise \end{cases} \tag{10}$$

$$CR_g = \begin{cases} Rand(0,1), & if\ Rand(0,1) < \tau_2 \\ CR_{g-1}, & otherwise \end{cases} \tag{11}$$

τ_1, τ_2 represent the probabilities to adjust factor F_g and CR_g respectively, F_l and F_u determine the range of F_g. We have made additional experiments to take the best value of τ_1, τ_2, F_l and F_u, $visual_i$ and $step_u, step_l$ with multiple optimization runs, we set $F_l = 0.1, F_u = 0.9, visual_i = 0.83, step_u = 0.5, step_l = 0.2$. Besides, we notice that there are no obvious differences in results by changing τ_1, τ_2, and then we set $\tau_1 = \tau_2 = 0.1$.

4.2 Framework of the Hybrid SADE- AFSA Algorithm

In the hybrid SADE-AFSA algorithm, we not only imitate the behavior of fish but also apply the core operators of SADE. The critical parameters are dynamically changed according to the convergence condition. The algorithm framework and the flow chart of SADE-AFSA is described in Fig. 4 and 5.

5 Computational Experiments and Comparison

5.1 Experimental Design

In this section, a number of categories are proposed to test the performance of the proposed SADE-AFSA, to cover various types of problems, four factors are defined: the number of jobs, the number of machines, the number of families, and the normal processing time of jobs. A summary of all experimental factors and levels is shown in Table 3. The normal processing time of each job is generated from a discrete uniform distribution $[0.05, 0.35]$. Combining with different numbers of jobs, machines, and families, 64 categories are generated in our experiments and each category is represented with a code *JiMjFk*. For example, a problem category with 100 jobs, 3 machines, and 5 families is represented by *J2M1F3*.

Artificial fish swarm algorithms and Differential Evolution (SADE- AFSA)
1.   Define $visual_i$ and $step_u$, $step_l$, δ, n, $AFNum$(the number of AFs) and other parameters.
2.   Generate the initial AF swarm randomly in the search space; set iterations $g = 1, h = 0$.
3.   Calculate the current food concentration of each AF
4.   While $g < g_{total}$ do
5.   For $u = 1: AFNum$ do
6.   Assign the value of the best AF to bulletin board;
7.   If the value of the best AF is not changing, $h + +$,
8.   else $h = 0$
9.   End if
10.   Execute prey behavior
11.   If the set of AFs in the $visual_g$ of X_u is not null then
12.   Calculate the center position X_c in $visual_g$
13.   If($Y_c < Y_u$ && $\dfrac{Y_c}{f_{u/center}} < Y_u * \delta$)
14.   Set $X_{u/next} = X_u(i) + step_g * Rand(-1,1) * (X_c(i) - X_u(i))/Dis_{uc}$
15.   Else execute prey behavior
16.   End if
17.   Else execute prey behavior
18.   End if
19.   If the set of AFs in the neighborhood of X_u is not null then
20.   Choose the AF which has best food concentration in the neighborhood of X_u
21.   If ($Y_b < Y_u$ && $\dfrac{Y_b}{f_{b/center}} < Y_u * \delta$) then
22.   Set $X_{u/next} = X_u(i) + step_g * Rand(-1,1) * (X_b(i) - X_u(i))/Dis_{ub}$
23.   Else execute prey behavior
24.   End if
25.   Else execute prey behavior
26.   End if
27.   go to line 6
28.   End for
29.   Update $visual_g, Step_g, CR_g$, F_g
30. $g + +$, go to line 5
31.   End while
32.   End

Fig. 4. Pseudocode of the SADE-AFSA algorithm

According to the recommend in [27], we choose the initial value of the crowing index as $\delta = 0.618$. The number of AFs is set as 30 and the maximum iteration is 200. To avoid multi-factor interference, each experiment executes 20 times and its average value is taken. All algorithms are coded in java language and their codes are run on an Inter Core 5, 3.4 GHZ PC with 8 GB of RAM.

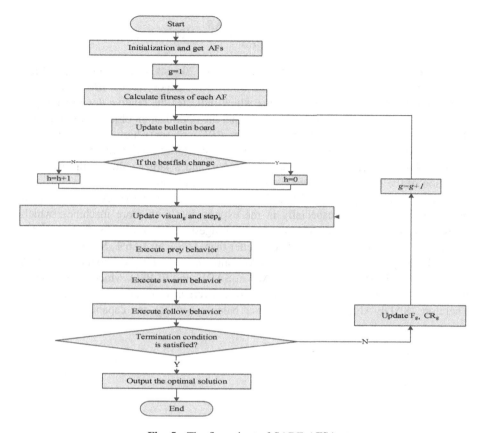

Fig. 5. The flow chart of SADE-AFSA

Table 3. Summary of experimental factors

Factors	Levels
Number of jobs	60,100, 150, 200
Number of machines	3, 5, 6, 7
Number of families	1,3,5,7
Normal processing time of jobs	$Uniform[0.05, 0.35]$

5.2 Experiment Results and the Analyses

SADE [35], AFSA [27], and Enhanced leader PSO (ELPSO) [36] are adopted to compare the solution obtained by SADE-AFSA. ELPSO is a hybrid algorithm which combines PSO and DE. In order to identify the different performance of these algorithms, the Relative Error (RE) is represented as *Eq.* (12).

$$RE = \frac{C_{max}(Alg) - BKS}{BKS} \times 100\% \tag{12}$$

Where $C_{max}(Alg)$ denotes the mean makespan of 20 tests by algorithm Alg and BK denotes the best-known solution obtained from each category. Thus, BRE (RE of the best solution) should be a positive value which indicates the relative deviation of the $C_{max}(Alg)$ over the best one. On account of that the optimization object is to minimize the makespan of parallel machines, a smaller BRE represents higher performance of algorithm Alg. The value of ARE (RE of the average solutions) is an arbitrary value that could be both positive value and negative value, and a larger value of the negative deviation of the solution signifies better convergence efficiency.

In Fig. 6, it can be observed that the BRE values of SADE-AFSA are lower than the compared algorithm, especially in the experiments with more machines, which shows that the improved algorithm has the strongest search ability for finding better solutions. For example, SADE-AFSA is better than AFSA by 12.6%, SADE by 13.6% and ELPSO by 21.04% in terms of the BRE value for $J3M2F2$. Additionally, the ARE gained by SADE-AFSA is lower than AFSA, SADE, and ELPSO, which demonstrates that the SADE-AFSA has the best quality of convergence for all of these categories. Therefore, SADE-AFSA achieves a better balance between the capability of exploration and the converging speed.

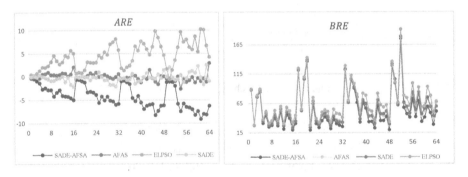

Fig. 6. The ARE and BRE of SADE -ADSA, AFSA, SADE, ELPSO for each category

Figure 7 shows the convergence behaviors of SADE-AFSA, AFSA, SADE, and ELPSO for some typical categories. For each category, we plot the average value of the best solution among the twenty running times at each iteration. It should be remarked that for some small categories, AFSA usually converges sharply at the early iterations, however, it cannot avoid trapping in local optimum, and usually convergent to a worse solution than SADE.

Figure 8 contains three stochastic box plots of the modified proportional, from which we can be see that the hybrid SADE-AFSA algorithm has better stability than other algorithms. Compared with AFSA, SADE, and ELPSO, the hybrid SADE-AFSA can not only find the global local optimum solution sufficiently fast but also improve the quality of the final solutions moderately. Based on the experimental results and our

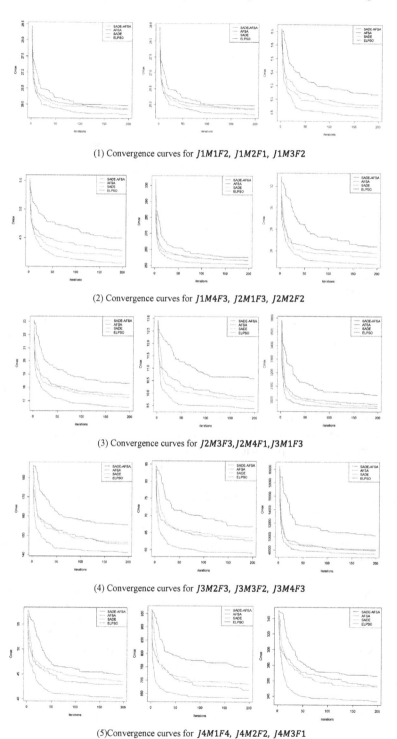

(1) Convergence curves for *J1M1F2*, *J1M2F1*, *J1M3F2*

(2) Convergence curves for *J1M4F3*, *J2M1F3*, *J2M2F2*

(3) Convergence curves for *J2M3F3*, *J2M4F1*, *J3M1F3*

(4) Convergence curves for *J3M2F3*, *J3M3F2*, *J3M4F3*

(5) Convergence curves for *J4M1F4*, *J4M2F2*, *J4M3F1*

Fig. 7. Average convergence curves of the main problem categories

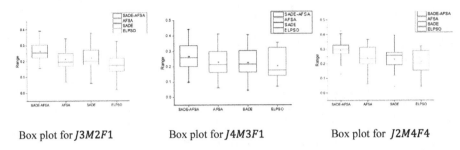

| Box plot for *J3M2F1* | Box plot for *J4M3F1* | Box plot for *J2M4F4* |

Fig. 8. Box plot of the modified proportional

analyses, it can be inferred that for the scheduling problem $pm|nr, ch, p_{ki}^A = p_{ki}(1 + at)|C_{max}$, regarding both the optimization capability and convergence rate, the hybrid SADE-AFSA is more effective and more robust.

6   Conclusion

In this paper, we investigate a parallel-machine scheduling problem with considering deteriorating jobs, incompatible job families, and module change simultaneously. For the objective of minimizing the makespan, we first explore the structural properties to obtain an optimal scheduling of jobs and families. The parallel-machine scheduling problem is proved to be NP-hard. Based on the structural properties and the proposed corollary, a hybrid algorithm combining Self-Adaptive Differential Evolution and Artificial fish swarm algorithm is proposed in which the individuals of a shoal are coded into job permutations. To evaluate the effectiveness and robustness, we develop a set of computational experiments. The experiments results reveal that the performance of the proposed algorithm lays over the other three algorithms in a reasonable time.

There are some research directions worthy to study in the future. First, we can consider other objectives such as total completion time and total tardiness. Second, updating changeover time as a family dependent parameter is of interest. In addition to that, we can develop some new heuristics and more efficient hybrid algorithms to solve other problems.

Acknowledgements. This work is supported by the National Natural Science Foundation of China (Nos. 71601065, 71231004, 71501058, 71690235, 71690230), and Innovative Research Groups of the National Natural Science Foundation of China (71521001), the Humanities and Social Sciences Foundation of the Chinese Ministry of Education (No. 15YJC630097), Anhui Province Natural Science Foundation (No. 1608085QG167). Panos M. Pardalos is partially supported by the project of "Distinguished International Professor by the Chinese Ministry of Education" (MS2014HFGY026).

References

1. Craig, I.K., Camisani-Calzolari, F.R., Pistorius, P.C.: A contemplative stance on the automation of continuous casting in steel processing. Control Eng. Pract. **9**(9), 1013–1020 (2001)
2. Browne, S., Yechiali, U.: Scheduling deteriorating jobs on a single processor. Oper. Res. **38**, 495–498 (1990)
3. Li, Shisheng, Yuan, Jinjiang: Parallel-machine scheduling with deteriorating jobs and rejection. Theor. Comput. Sci. **411**(40–42), 3642–3650 (2010)
4. Ji, M., Cheng, T.C.E.: Parallel-machine scheduling of simple linear deteriorating jobs. Eur. J. Oper. Res. **202**(1), 90–98 (2010)
5. Cheng, Wenming, Guo, Peng, Zhang, Zeqiang, Zeng, Ming, Liang, Jian: Variable neighborhood search for parallel machines scheduling problem with step deteriorating jobs. Math. Prob. Eng. **7**, 243–261 (2012)
6. Liu, M., Shijin W., Chengbin, C.: Scheduling deteriorating jobs with past-sequence-dependent delivery times. Int. J. Prod. Econ. **144**(2), 418−421 (2013).
7. Akturk, M.Selim, Ghosh, Jay B., Gunes, Evrim D.: Scheduling with tool changes to minimize total completion time: basic results and SPT performance. Eur. J. Oper. Res. **157**(3), 784–790 (2004)
8. Akturk, M.Selim, Ghosh, Jay B., Kayan, Rabia K.: Scheduling with tool changes to minimize total completion time under controllable machining conditions. Comput. Oper. Res. **34**(7), 2130–2146 (2007)
9. Chen, Jen-Shiang: Optimization models for the tool change scheduling problem. Omega **36**(5), 888–894 (2008)
10. Xu, D., Liu, M., Yin, Y., Hao, J.: Scheduling tool changes and special jobs on a single machine to minimize makespan. Omega **41**(2), 299–304 (2013)
11. Guo, Peng, Cheng, Wenming, Wang, Yi: Parallel machine scheduling with step-deteriorating jobs and setup times by a hybrid discrete cuckoo search algorithm. Eng. Optim. **47**(11), 1564–1585 (2015)
12. Wu, W.H., Xu, J., Wu, W.H., Yin, Y., Cheng, I.F., Wu, C.C.: A tabu method for a two-agent single-machine scheduling with deterioration jobs. Comput. Oper. Res. **40**(8), 2116–2127 (2013)
13. Wu, C.C., Wu, W.H., Wu, W.H., Hsu, P.H., Yin, Y., Xu, J.: A single-machine scheduling with a truncated linear deterioration and ready times. Inf. Sci. **256**, 109–125 (2014)
14. Mir, M., Salehi, S., Rezaeian, J.: A robust hybrid approach based on particle swarm optimization and genetic algorithm to minimize the total machine load on unrelated parallel machines. Appl. Soft Comput. **41**, 488−504 (2016)
15. Hu, W., Wang, H., Yan, L., Du, B.: A swarm intelligent method for traffic light scheduling: application to real urban traffic networks. Appl. Intell. **44**(1), 208–231 (2016)
16. Pei, Jun, Liu, Xinbao, Pardalos, Panos M., Fan, Wenjuan, Yang, Shanlin: Scheduling deteriorating jobs on a single serial-batching machine with multiple job types and sequence-dependent setup times. Ann. Oper. Res. **249**, 175–195 (2017)
17. Pei, Jun, Pardalos, Panos M., Liu, Xinbao, Fan, Wenjuan, Yang, Shanlin: Serial batching scheduling of deteriorating jobs in a two-stage supply chain to minimize the makespan. Eur. J. Oper. Res. **244**(1), 13–25 (2015)
18. Pei, J., Liu, X., Pardalos, P.M., Fan, W., Yang, S.: Single machine serial-batching scheduling with independent setup time and deteriorating job processing times. Optim. Lett. **9**(1), 91−104 (2015)
19. Pei, Jun, Liu, Xinbao, Pardalos, Panos M., Li, Kai, Fan, Wenjuan, Migdalas, Athanasios: Single-machine serial-batching scheduling with a machine availability constraint, position-

dependent processing time, and time-dependent set-up time. Optim. Lett. **11**(7), 1257–1271 (2017)

20. Pei, J., Liu, X., Fan, W., Pardalos, P.M., Lu, S.: A hybrid BA-VNS algorithm for coordinated serial-batching scheduling with deteriorating jobs, financial budget, and resource constraint in multiple manufacturers. Omega (2017). https://doi.org/10.1016/j.omega.2017.12.003

21. Liu, X., Lu, S., Pei, J., Pardalos, P.M.: A hybrid VNS-HS algorithm for a supply chain scheduling problem with deteriorating jobs. Int. J. Prod. Res. (2017). https://doi.org/10.1080/00207543.2017.1418986

22. Pei, Jun, Liu, Xinbao, Pardalos, Panos M., Migdalas, Athanasios, Yang, Shanlin: Serial-batching Scheduling with Time-dependent Setup Time and Effects of Deterioration and Learning on a Single-machine. J. Global Optim. **67**(1), 251–262 (2017)

23. Fan, W., Pei, J., Liu, X., Pardalos, P.M., Kong, M.: Serial-batching group scheduling with release times and the combined effects of deterioration and truncated job-dependent learning. J. Global Optim. (2017). https://doi.org/10.1007/s10898-017-0536-7

24. Yin, Y., Wang, Y., Cheng, T.C.E., Liu, W., Li, J.: Parallel-machine scheduling of deteriorating jobs with potential machine disruptions. Omega **69**, 17–28 (2016)

25. Graham, R.L., Lawler, E.L., Lenstra, J.K., Rinnooy Kan, A.H.G.: Optimization and approximation in deterministic sequencing and scheduling: a survey. Ann. Discret. Math. **5**(1), 287–326 (1979)

26. Jr, E.G.C., Garey, M.R., Johnson, D.S.: An application of bin packing to multi-processor scheduling. Siam J. Comput. **7**(1), 1–17 (1978)

27. Xiaolei, L.I., Shao, Z., Qian, J.: An optimizing method based on autonomous animats: fish-swarm algorithm. Syst. Eng.-theory Pract. **22**(11), 32–38 (2002)

28. Shen, W., Guo, X., Wu, C., Wu, D.: Forecasting stock indices using radial basis function neural networks optimized by artificial fish swarm algorithm. Knowl.-Based Syst. **24**(3), 378–385 (2011)

29. Li, Z., Zhang, H., Xu, J., Zhai, Q.: Recognition and localization of harmful acoustic signals in wireless sensor network based on artificial fish swarm algorithm. Dev. Neurosci. **9**(1), 53–60 (2013).

30. Dihua, Sun, Song, XiaoXiao, Zhao, Min: LinJiang Zheng.: Research on a JIT scheduling problem in parallel motorcycle assembly lines considering actual situations. Int. J. Prod. Res. **50**(18), 4923–4936 (2012)

31. Storn, R., Price, K.: Differential evolution: a simple and efficient adaptive scheme for global optimization over continuous spaces. J. Global Optim. **11**, 341–359 (1997)

32. Bhattacharya, S.S., Garlapati, V.K., Banerjee, R.: Optimization of laccase production using response surface methodology coupled with differential evolution. New Biotechnol. **28**(1), 31–39 (2011)

33. Cai, H.R., Chung, C.Y., Wong, K.P.: Application of Differential Evolution Algorithm for Transient Stability Constrained Optimal Power Flow. IEEE Trans. Power Syst. **23**(2), 719–728 (2008)

34. Santucci, Valentino, Baioletti, Marco, Milani, Alfredo: A differential evolution algorithm for the permutation flowshop scheduling problem with total flow time criterion. IEEE Trans. Evol. Comput. **20**(5), 682–694 (2016)

35. Brest, J., Greiner, S., Boskovic, B., Mernik, M., Zumer, V.: Self-adapting control parameters in differential evolution: A comparative study on numerical benchmark problems. IEEE Trans. Evol. Comput. **10**(6), 646–657 (2006)

36. Jordehi, A.Rezaee: Enhanced leader PSO (ELPSO): a new PSO variant for solving global optimization problems. Appl. Soft Comput. **26**, 401–417 (2015)

Author Index

Alba, Enrique 386
Ansótegui, Carlos 309
Archetti, Francesco 352
Arroyo, José E. C. 141

Baklanov, Artem 427
Barkalov, Konstantin 78
Bedenel, Anne-Lise 225
Beglou, Neema 154
Bhandari, Akshita 1
Biedenkapp, André 115
Biernacki, Christophe 225
Blot, Aymeric 241
Blum, Christian 199
Boduroğlu, İ. İlkay 257
Bossek, Jakob 184, 215

Camero, Andrés 386
Candelieri, Antonio 352
Caris, An 159
Chuin Lau, Hoong 98
Corstjens, Jeroen 159
Créput, Jean-Charles 16, 82

Dang, Nguyen 288
De Causmaecker, Patrick 241, 288
De Leone, Renato 180
Depaire, Benoît 159
Di Mauro, Anna 433
Di Nardo, Armado 433
Di Natale, Michele 433
Djukanovic, Marko 199
Doerner, Karl F. 64

Enaux, B. 175
Eremeev, Anton V. 337
Erzin, Adil 131

Fasano, Giovanni 180
Faury, Louis 271
Froehlich, Georg E. A. 64

Gaudrie, D. 175
Gergel, Victor 78
Gimadi, E. Kh. 402
Grimme, Christian 184
Gunawan, Aldy 98

Herbert, V. 175
Heymann, Britta 309
Huber, Sandra 370
Hutter, Frank 115

Jia, Linlin 36
Jourdan, Laetitia 225, 241

Kalyagin, Valery A. 304
Kampel, Ludwig 437
Karapetyan, Daniel 220
Kel'manov, Alexander 326
Kessaci, Marie-Éléonore 241
Khachay, Michael 427, 441
Khamidullin, Sergey 326
Khandeev, Vladimir 326
Kiechle, Guenter 64
Kirousis, Lefteris 49
Kobylkin, Konstantin 447
Koldanov, Alexander P. 304
Koldanov, Petr A. 304

Lalla-Ruiz, Eduardo 31
Lebedev, Ilya 78
Lindauer, Marius 115
Liu, Siwen 455
Liu, Xiyu 36
Livieratos, John 49
Locoratolo, Stefano 433

Marben, Joshua 115

Nesmachnow, Sergio 411
Neznakhina, Katherine 441

Özcan, Ender   154
Ozcan, Murat   437

Palomba, Andrea   433
Pardalos, Panos M.   304
Parkes, Andrew J.   154, 220
Pasynkov, Maxim   427
Perego, Riccardo   352
Picheny, V.   175
Pinheiro, Júlio C. S. N.   141
Plotnikov, Roman   131
Pon, Josep   309
Pyatkin, Artem   326

Qian, Xiaofei   455
Qiao, Wen-Bao   16, 82

Raidl, Günther R.   199
Riche, R. Le   175
Roma, Massimo   180
Rossit, Diego   411

Santonastaso, Giovanni Francesco   433
Segredo, Eduardo   31
Sellmann, Meinolf   309
Sergeyev, Yaroslav D.   180
Simos, Dimitris E.   437
Singh, Chandramani   1
Souravlias, Dimitris   370
Stolfi, Daniel H.   386
Stützle, Thomas   220
Sun, Yuwei   455

Tavares, Ricardo G.   141
Teng, Teck-Hou   98
Tierney, Kevin   309
Toutouh, Jamal   386, 411
Trautmann, Heike   215
Tsidulko, O. Yu.   402

Vasile, Flavian   271
Voß, Stefan   31

Xiang, Laisheng   36

Printed in the United States
By Bookmasters